Acoustics and the Performance of Music

Modern Acoustics and Signal Processing

For other titles published in this series, go to
www.springer.com/series/3754

Jürgen Meyer

Acoustics and the Performance of Music

Manual for Acousticians, Audio Engineers, Musicians, Architects and Musical Instruments Makers

Fifth Edition

Originally published in German by PPV Medien, Edition Bochinsky

Springer

Jürgen Meyer
Bergiusstrasse 2A
D-38116 Braunschweig
Germany
juergen.meyer-bs@t-online.de

Translated by

Uwe Hansen
64 Heritage Drive
Terre Haute IN 47803
USA
uhansen@isugw.indstate.edu

ISBN 978-0-387-09516-5 e-ISBN 978-0-387-09517-2

Library of Congress Control Number: 2008944095

Translation of the latest (fifth) edition, originally published in German by PPVMedien GmbH, Edition Bochinsky, Bergkirchen.

Printed on acid-free paper.

9 8 7 6 5 4 3 2 1

springer.com

Preface

Since the middle of the twentieth century, concert performance developments have created raised, and to some extent, new demands on musicians and architects. Reasons for changes in performance conditions can be found on the one hand in the tendency for ever larger concert halls and on the other in the fact that listeners, educated by quality recordings are used to a high degree of precision and subtle tonal nuances. These circumstances lead to acoustic and performance technical problems for the interpreter unknown in previous generations. These tasks must largely be mastered by the musicians themselves, yet in some sense Tonmeister (sound recording engineers) and builders of concert halls can have an essential influence on the tonal results of a performance. Thus it is important for all participants to be knowledgeable of those acoustic processes which shape the tonal development beginning with the tonal perception of the performer down to the aural impression of the listener.

With this background, the first German edition of the book *Acoustics and Musical Performance* appeared in 1972, in which those aspects of musical instrument acoustics, and room acoustics, relevant to music in general were considered. An important element of this book was the consideration of the degree to which approaches to performance practice could be derived from those principals. The great demand for these themes made several new editions necessary, which in each case were revised to include current knowledge. Thus, this English edition of the book is based on the 5th German Edition of 2004. In addition to new experimental results in the physical, technical realm, many personal experiences by the author, as presenter and conductor of demonstration concerts with large orchestras, not only in Europe, but also in the USA and Japan, relating to questions of orchestral arrangements have added significant insights.

In order to make this complex subject matter accessible even for readers without special knowledge in the physical sciences, the principal chapters are introduced by brief explanations of the most important fundamental concepts of acoustics as well as a selection of some of the more important hearing principles essential for understanding. The detailed representation of directional characteristics in the fourth chapter was originally intended especially for audio engineers. Since then, the new area of room acoustical simulation auralization has been developed which could not be anticipated in 1972. The sections concerned with room acoustics are deliberately limited to those aspects essential to musical performance. They are thus intended as an introduction for non-acousticians. On the other hand, it is likely more important for acousticians, when considering historic performance technical matters, to gain an insight relating to things which are routine for every musician.

Acoustic data which characterize tonal characteristics in sound radiation of musical instruments as well as room acoustics processes represent objective facts. In contrast, performance practical directions in many cases are only examples of subjective interpretations. They are intended merely to show the possibilities for utilizing acoustical facts for the realization of an artistic tonal perception. In this sense, the present volume occupies a position between the standard works of Fletcher/Rossing (*The Physics of Musical Instruments*) and Beranek (*Concert and Opera Halls – How They Sound*). They thus form a bridge between musical instrument acoustics and room acoustics, based on practical experience.

Scientific data obtained by the author are the result of continuing exchange between experimental investigations at the Physikalisch-Technische Bundesanstalt in Braunschweig and lectures in the framework of the audio engineering program (Tonmeister-Ausbildung) at the School of Music in Detmold (Music Academy). The author here again expresses gratitude to both institutions for decades of support. It is precisely the cooperation with generations of audio engineers that has significantly contributed to the fact that in recent years changes in attitude of conductors concerning the seating order of strings in the orchestra have been effected.

My personal thanks go to Professor Uwe Hansen of Indiana State University who has made enormous efforts to translate the voluminous text into English, while fine-tuning the formulation of numerous details in personal communications during several visits to Germany. Beyond that he has also taken it upon himself to interact with the publisher concerning production details. Special thanks also go to Springer-Verlag for the successful cooperation and the appealing final appearance of the volume.

Branschweig,
Germany

Jürgen Meyer

Translator's Preface

A number of comments are in order. It has been a great pleasure to be associated with this work, which so admirably bridges the gap between science and performance. As I was translating, at times it was as though I could hear Jürgen's voice speaking from the pages. The translation has progressed in three stages. The initial attempt was to preserve the integrity of the German original. Readers who feel that the final version retains too much of the German convoluted grammar have my sympathy and my apologies. The second step attempted to transform the literal translation into readable English. The third approach included a thorough review with the Author, to insure scientific accuracy and preservation of the Author's intent. As in the earlier translation of the second edition by John Bowsher and Sibylle Westphal we are using the American notation for octave assignment of notes, except we prefer the standard American usage of subscripts. Thus, what follows is a comparison of notations

American	C_1	C_2	C_3	C_4	C_5	C_6	C_7	C_8
German	C_1	C	c	c'	c''	c'''	c^4	c^5

Vowel association with formants in a spectral content occurs frequently throughout the book. The German edition uses the German letters a, o, u, e, i, and the modified "Umlaut" letters ä, ö, and ü for this purpose. Unfortunately English, and of course also American vowel usage frequently goes over into a double vowel or diphthong pronunciation. The use of international phonetic symbols was considered, but both the Author and I felt that easy access would be encumbered by that solution. We finally decided to retain the German vowel use and add the English pronunciation in parentheses such as a(ah) with the understanding that a detailed guide is included in this preface. This recognizes that English vowels, like a, o, and i are in fact pronounced as eh-ee, oh-oo, and ah-ee, where the second vowel generally is short. In this context the vowel relationship to the formant relates only to the first vowel of the otherwise colloquially used diphthong. Thus the sound represented by the o vowel should be sounded as only the "oh" sound with the usually short following "oo" sound entirely dropped. The following represents a pronunciation guide for the German vowels as used in connection with formant representation

u	"oo" as in tool
o	"oh" as in old
å	"aw" as in fall
a	"ah" as in father
ö	as in her with pursed lips
ü	as in French rue
ä	as in air
e	"eh" as in late without the final short ee sound
i	"ee" as in fleet

My thanks go to the Indiana State University department of Languages Literature and Linguistics for the use of language laboratory facilities and the help of numerous students.

Innumerable discussions with Dr. Ramon Meyer, formerly music director of the Terre Haute Symphony, Professor Emeritus of Music, and director of the Choral Program at Indiana State University, have been a great help in clarifying current use of musical terminology.

Terre Haute,
IN

Uwe Hansen

Contents

Chapter 1
Introduction to Acoustics

1.1 Fundamental Physical Principles

1.1.1 Sound Pressure

When we hear music, the perceived tonal impression is caused by sound carried to our ears by the air. Relevant in this context are the minute pressure variations which are superimposed on the stationary pressure of the air surrounding us. The pressure variations propagate as waves in space. These more or less periodic deviations from the stationary mean value, comprise the so called sound pressure variations, for which in practice the shorter term "sound pressure" is used.

Since our ear is capable of responding to a wide range of sound pressures, from a barely perceptible sound, to the intensity for which the hearing sensation becomes painful, generally a logarithmic scale is used to represent the range of sound pressure values of interest to the acoustician. This makes the scale accessible and inclusive. The relation of a certain sound pressure to a reference value is given in "decibels" (dB), and one speaks of the sound pressure level, where the concept "level" always refers to a logarithmic scale. To the unaccustomed reader this procedure may initially appear somewhat complicated, however, it has proven very advantageous in practice, particularly once a number of dB values are associated with their corresponding sounds as heard. Furthermore, the logarithmic dB-scale reflects hearing perception more closely than a linear representation.

So called "absolute" dB values for sound pressure levels are obtained when a reference value of 2×10^{-5} Pa (Pascal) is used. This value was chosen by international agreement. It corresponds approximately to the threshold of hearing in the frequency region where the ear is most sensitive. (Consideration of the reference value as well as the logarithmic calculations are carried out by the measuring instrument.) As an example, in a Bruckner Symphony, depending on concert hall size, and location in the hall, as well as the size of the orchestra, one can expect values between 90 and 100 dB for a *fortissimo*, on the other hand a *pianissimo* could result in 40–45 dB.

J. Meyer, *Acoustics and the Performance of Music*.
DOI: 10.1007/978-0-387-09517-2_1, © Springer Science + Business Media, LLC 2009

It is, however, possible to use some other, arbitrary sound pressure value as a reference. In that case one obtains a "relative" dB value, particularly suitable for characterizing the difference between two sound levels. A value of 0 dB, would indicate that the two processes being compared have the same sound pressure, not, however, that they are at the lower limit of hearing. If in the example given above a *fortissimo* is measured at 100 dB (absolute) and a *pianissimo* at 45 dB (absolute) the resulting dynamic difference would be 55 dB (relative). The addition "relative" is dropped in general usage, while in situations which are not clear from the context, the absolute measure is emphasized by an indication of the reference sound pressure.

1.1.2 Particle Velocity

The actual cause for generating and propagating pressure variations lies in the fact that individual air particles vibrate about their rest position, and thus collide with neighboring particles in the direction of their motion. The velocity with which the particles move, relative to their rest position, is called the particle velocity. As is the case with sound pressure, the particle velocity likewise is subject to fluctuations. However, as the particles vibrate back and forth, not only is the magnitude changed, but also the direction of motion.

Sound pressure and particle velocity together determine the so called sound field, which characterizes the all inclusive temporal and spatial properties of a sound process. Here one is not only concerned with the magnitudes of these two quantities, but also with their relative phase. This means that the maximum pressure will not necessarily always coincide in time with the highest particle velocity. A relative shift in time between variations in pressure and velocity is quite possible.

In the case of a propagating plane wave, however, as in the case for larger distances from the sound source, i.e., in the far field, pressure and velocity are in phase. Furthermore, there is a direct proportionality between those two quantities: when the sound pressure rises, the particle velocity increases in the same measure. The relationship between pressure and velocity is thus determined by the "resistance" presented by the air to the vibrations. This "characteristic field impedance" (earlier denoted as sound wave resistance) can be considered to be constant for practical purposes.

With that in mind, one can describe the sound field by sound pressure alone, when considering the far field, as is almost always the case for the listener in a musical performance. Furthermore, it should be noted that the ear responds exclusively to sound pressure. On the other hand, for recordings with microphones, it is entirely possible to come so close to the sound source that sound pressure and particle velocity are no longer proportional to each other and are no longer in phase, and thus both must be considered. The so called "near field effect" is well known. For certain microphone types it leads to an unnatural amplification of the low registers.

1.1.3 Sound Power

Describing a sound field by specifying sound pressure levels for a number of points in a room represents a view point oriented toward the listeners, or recording devices at those points. Naturally the measured sound level depends on the strength of the sound source. It is therefore also of interest to determine a characterization of the sound source, which describes its strength independently of spatial considerations and the distance from the listener. This relates exclusively to the sound source itself. Such a quantity represents the sound energy radiated by a source in all directions during a unit of time. This quantity is designated as the sound power of the source.

The physical unit for power is the Watt; however, since acoustic power values occurring in practice cover a large dynamic range, as is the case for sound pressure, acoustic power can be represented on the more accessible dB scale. At the same time this simplifies the connection between the power of a sound source and the resulting sound pressure, which is of particular interest in room acoustics.

A power of 10^{-12} Watt serves as reference value for the dB scale of sound power. This value is a result of the reference value for sound pressure and the characteristic field impedance for air. Numerically, the sound power level given in dB corresponds to the sound pressure level at the surface of a sphere with surface area of 1 m^2, which surrounds the center of the sound source, i.e., equal to the sound pressure level at a distance of approximately 28 cm from the center of the source.

Inasmuch as the dB scale is built on a logarithmic basis, the dB values for the power of individual sound sources can not simply be added to determine the combined power during simultaneous excitation. Rather, depending on the factor by which the total power exceeds that of the individual sources, a value must be added to the dB value of the individual sound sources.

Table 1.1 shows some value pairs needed for this calculation. For example, if the sound power is doubled – possibly by doubling the number of performers – the total radiated power is raised by 3 dB; however, on the other hand, if the number of players is raised from 4 to 5, this means an increase of only 1 dB in the radiated sound power.

Table 1.1

Power multiplier	Increase in dB
1.25	1
1.6	2
2.0	3
3.3	5
5.0	7
10.0	10

1.1.4 Frequency

The number of pressure fluctuations or vibrations occurring in a certain time period is designated as the frequency. In this context a single vibration is counted from an arbitrary starting point to the adjacent equivalent point where conditions prevail identical to the starting point. For example, in the case of a pendulum, the period of a vibration includes the time from the moment of highest excursion on one side until the subsequent instant of highest excursion on the same side.

The number of vibrations per second is given in Hz ("Hertz"). For very high frequencies 1,000 Hz can be replaced by 1 kHz ("Kilohertz"), to avoid large numbers. Vibrations perceived by our ear lie in the range of approximately 16 Hz–20 kHz. Frequencies above that are designated as ultrasound, frequencies below that as infrasound.

The musical reference note A_4, for example, has a frequency of 440 Hz; a summary of some additional frequency values associated with musical notes is given in Table 1.2, frequency values are rounded for the sake of clarity. The highest notes in this table occur only extremely rarely, nevertheless these frequencies, and the additional frequency range up to the threshold of hearing, is of interest in connection with tone color effects of overtones.

1.1.5 The Speed of Sound

While the particle velocity indicates the speed with which air particles move relative to their rest position, one designates the speed of propagation with which pressure fluctuations spread in the air space (or some other medium) as the speed of sound. It is thus this speed of sound which is relevant when considering the delay with which a sound process reaches the ear after being released from some distance away.

The speed of sound in air is independent of frequency, however, it depends in some small measure on the stationary air pressure, as well as on the carbon dioxide content, nevertheless, the latter effects are of little consequence for practical musical purposes.

Table 1.2

Frequency (Hz)	Note
16.5	C_0 C key in the 32′ register of the organ
33	C_1 C-string of a 5 string contrabass
66	C_2 C-string of a Cello
131	C_3 C-string of a Viola
262	C_4 lowest violin C
524	C_5 high tenor C
1,047	C_6 high soprano C
2,093	C_7 highest violin C
4,185	C_8 highest Piccolo C

Table 1.3

Air temperature (°C)	Speed of sound (m s^{-1})
−10	325.6
0	331.8
+10	337.8
+20	343.8
+30	349.6

On the other hand, the observation that the speed of sound increases with rising temperature, is more important. This influences the tuning of brass instruments, for example. This effect also plays a role in sound propagation through differentially heated layers of air, which can come about by less than optimal heating installations, solar influence, or over a cool water surface in open air performances. For this reason, values of the speed of sound in air at several temperatures are given in Table 1.3.

Accordingly, in the range of average room temperatures, a mean value for the speed of sound in air can be calculated as 340 m s^{-1}. This means that a listener at a distance of 34 m from the concert podium hears the sound with a delay of 1/10 s. This corresponds to a 1/16 note for a tempo of MM ♩ = 152. Such distances are quite common in large concert halls, however, the eye is no longer in a position to follow the motions of the performer exactly, so that the lack of temporal coincidence only becomes uncomfortably noticeable when using opera glasses.

1.1.6 Wavelength

Since sound waves spread with a certain propagation speed, naturally a spatial separation results between two successive pressure maxima (wave peaks). The faster these maxima follow each other (i.e., the higher the frequency), the shorter is the spacing. This spatial distance between two neighboring pressure maxima (or two neighboring pressure minima) is designated as the wavelength. It can be calculated from the formula

$$\text{Wavelength} = (\text{Speed of sound})/\text{frequency}.$$

For a speed of sound of 340 m s^{-1} a number of examples of wavelengths in air are given in Table 1.4.

Wavelengths ranging from several meters to a few centimeters occur in the region of audible frequencies. This means that at high frequencies the wavelengths are small compared to musical instrument dimensions, and the sizes of rooms, and surfaces on which the sound impinges. For mid frequencies, in contrast, the wavelengths are of the same order of magnitude as the dimensions mentioned. In the low frequency range the instrument and room dimensions must be considered

Table 1.4

Frequency (Hz)	Wavelength
20	17.0 m
100	3.4 m
1,000	34 cm
10,000	3.4 cm

as small in comparison to the wavelengths. These considerations are of great significance for sound radiation from sources, and for processes dealing with room acoustics.

1.2 Characteristics of the Auditory System

1.2.1 The Sensation of Loudness

A sensation of hearing arises when pressure fluctuations, which reach our ear, occur in a certain frequency region, and do not fall below a minimum sound level. The lowest frequency for which a vibration process is still perceived as a tone is approximately 16 Hz. This corresponds to a C_0 which is included in the 32′ register of some large organs. For yet lower frequencies the ear can already follow the temporal process of the vibrations, so that a unique tonal impression can no longer be formed. For younger persons the upper limit of hearing lies in the range of 20,000 Hz. This is, however, subject to individual variations, and decreases with increasing age.

The lower sound level limit for the ability to detect tones is also not the same for all people. However, from a large number of measurements, based on statistics, a mean value can be calculated, which characterizes typical behavior. In this context, the interesting result is noted that this so-called threshold of hearing depends in large measure on frequency. This characteristic is represented in the lowest curve of Fig. 1.1. In this diagram frequency increases from left to right, and the absolute sound level increases in the upward direction. One notes that the ear responds with most sensitivity to tones in the frequency range between 2,000 and 5,000 Hz. In this range the minimum required sound level is the lowest. For higher frequencies, but even more so for lower frequencies, the sensitivity of the ear is reduced, so that in these regions significantly higher sound pressure levels are required for a tone to become audible.

The same tendency is evident when at higher intensity, tones of different frequencies are compared in relation to their impression of loudness. The sound pressure level as an objective measure of existing physical excitation is by no means equivalent to the loudness as a subjective measure of sensation. In order to establish a connection between the two, the unit of a phon was introduced for loudness level. It is defined such, that the dB scale for sound pressure level and the

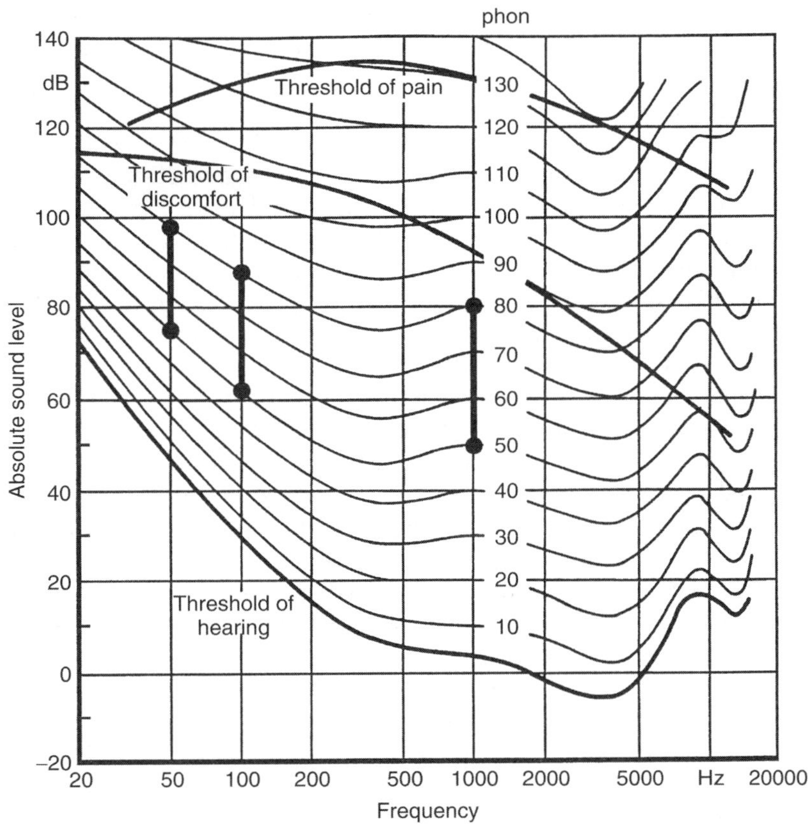

Fig. 1.1 Equal loudness curves for sound incident from the front, with Threshold of hearing and threshold of pain indicated, as well as threshold of discomfort (after Winkel, 1969). The heavy lines with end points show that depending on frequency, a different sound pressure level difference is required for the loudness level difference between 50 and 80 phons

phon scale for loudness level coincide at 1,000 Hz. When tones of different frequency are compared to a tone of 1,000 Hz, the so called equal loudness curves are obtained which represent the relationship between the objective sound pressure level and the loudness level as determined by the sensitivity of the ear. Several of these curves are recorded in Fig. 1.1.

For example, following the curve for 80 phons shows that this loudness level at 1,000 Hz (by dcfinition) was caused by a sound pressure level of 80 dB. On the other hand, at 500 Hz, approximately 75 dB are sufficient for the same loudness level impression, while at 100 Hz almost 90 dB are required. A level of 80 dB would accordingly be perceived as somewhat louder at 500 Hz than at 1,000 Hz. At 1,000 Hz it would be perceived as a loudness level of 84 phons, while the same sound pressure level would cause a loudness level of only 72 phons at 100 Hz.

In the region of lower loudness levels the curves rise more steeply with decreasing frequency. Here the sound pressure level differences are even larger between tones of equal loudness sensation: For a loudness level of 30 phons a sound pressure level of 30 dB at 1,000 Hz contrasts with a value of almost 50 dB at 100 Hz.

This differentially steep slope of the curves at low frequencies leads to the fact that the lines of equal loudness level come closer together, and thus a sound pressure level change here is responsible for a larger loudness level change than at higher frequencies. Thus a rise of the sound pressure level from 60 to 90 dB at 50 Hz corresponds to an increase in loudness level from about 25 to 70 phons, whereas in contrast, at 1,000 Hz, this would correspond to an increase from 60 to 90 phons; the perceived dynamic difference accordingly would be 45 phons at 50 Hz, at 1,000 Hz, however, only 30 phons.

The relationship of phon values to the corresponding sound pressure level in the equal loudness curves is strictly valid only for tones of long duration. When the duration of a tone impulse is less than approximately 250 ms (1 ms = 1/1,000 s) then the perceived loudness level is somewhat lower than expected from the corresponding sound pressure level. The difference in comparison to a long lasting tone is approximately 1 dB at 200 ms, and increases to a value of about 2.5 dB at 100 ms and to 7 dB at 20 ms (Zwicker, 1982). This indicates that the loudness impression for short tones or noises is not determined by the power (energy per time unit), but rather by the total acoustic energy (power times duration) (Roederer, 1977).

In contrast, a brief rise of the sound pressure level at the beginning of a tone of longer duration can increase the perceived loudness level above the level corresponding to the later sound pressure level of that tone. An increase of the sound pressure level by 3 dB during the first 50 ms raises the entire tone by approximately 1 dB, while a corresponding sound pressure level increase at a later point in time is practically not perceived (Kuwanao et al., 1991). From this electronic – and thus rather abstract – sound experiment one can conclude that instrumental tones played with firm attack have a loudness edge over tones played equally loudly, but with soft or uncertain attack.

In addition to the threshold of hearing and the equal loudness curves, the diagrams of Fig. 1.1 contain two additional curves which in a sense provide an upper boundary for hearing processes. The so-called threshold of pain indicates those levels above which the hearing sensation becomes painful. At this point a safety measure goes into effect in the middle ear, which protects the inner ear from a damaging overload, so that it no longer transmits the full vibration amplitude (Reichardt, 1968).

However, even below this threshold of pain there are loudness levels which in a musical context are no longer considered to be beautiful, but rather annoying. Such an esthetically determined boundary clearly is less exactly delineated than a transition to a sense of pain. This annoyance boundary, nevertheless, gives an indication for sound pressure levels which should not be exceeded. The decline of this curve for high frequencies is noteworthy, this feature takes on particular significance when tone color is under consideration.

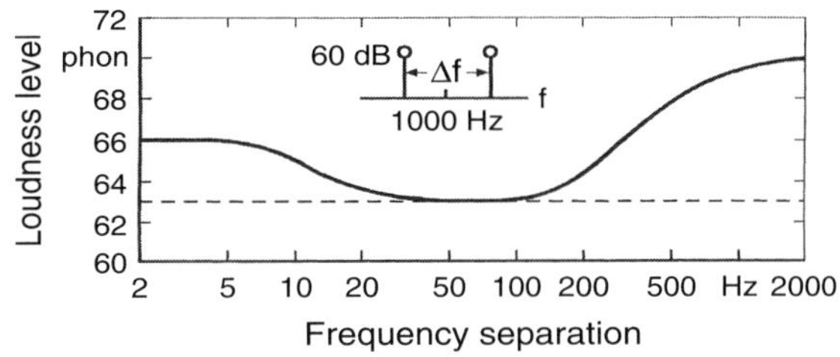

Fig. 1.2 Loudness level of two simultaneous sinusoidal tones in dependence on their frequency separation (after Zwicker, 1982)

As mentioned earlier, the sensitivity of the ear for high frequencies diminishes with age. This refers not only to the upper frequency boundary, but also to the shape of the threshold of hearing from 2,000 Hz upward (Zwicker, 1982). In comparison to the ear of a 20-year old (i.e., normal hearing), the sensitivity of the healthy ear of a 40-year old is approximately 8 dB less sensitive at 5,000 Hz, at age 60 the difference increases to 15 dB. For tones of 10,000 Hz the loss of sensitivity for a 60-year old is on the average about 25 dB, so that the threshold of hearing begins to come relatively close to the annoyance limit. This explains why older individuals often feel annoyed with strong high frequency tone components in loudspeaker reproductions, since these, if they can be heard at all, already overlap into the annoyance region.

When two sinusoidal tones are sounded simultaneously, one senses a difference in total loudness level depending on their frequency separation. This effect is represented in Fig. 1.2 for two sinusoidal tones, each of 60 dB sound pressure level, located symmetrically about 1,000 Hz, as a function of frequency separation. Three regions of different hearing reactions are noted (Zwicker, 1982). For small frequency separations below approximately 10 Hz an increase in loudness level of about 6 phons is noted when compared with a single tone; this corresponds to a rise in loudness level caused by doubling the sound pressure level of the single tone. When one considers that two tones with small frequency separation lead to beats, i.e., rhythmic variations of the sound pressure level, one can conclude that the hearing process is oriented toward the maximum values of sound pressure level rather than toward the time average value.

As the frequency difference between the two tones increases, the ear no longer senses the beats as direct fluctuations in time, it thus is oriented toward the sound energy of both tones. The energy doubling when compared to a single tone leads to an increase in loudness level corresponding to an increase in sound pressure level of 3 dB. The ear performs this energy addition as long as both tones fall within a so called "critical band" of the ear. The width of such critical bands is related to the structure of the inner ear (more precisely to the frequency distribution on the basilar membrane). Below 500 Hz the width of a critical band is about 100 Hz, above 500 Hz critical bands have a width corresponding to a major third, i.e., the lower and upper frequency limits of a critical band have a ratio of 4–5.

While the energy of all tones in a critical band is combined by the ear, a separate loudness level impression is formed for tones differing in frequency by more than a critical band, (one also speaks of partial loudness), and these individual partial loudness levels are combined to a total loudness level. For large frequency separation this leads to a doubling of loudness when two equally loud tones are sounded. The sensation that something is twice as loud corresponds to a loudness level increase of 10 phons, which in the region of 1,000 Hz is a sound pressure level rise of 10 dB.

Aside from the age dependent decrease in hearing sensitivity, illness or exposure to high level sounds can be responsible for influencing hearing ability to a greater or lesser degree. In this context it should be mentioned that the high loudness levels which occur in an orchestra can on occasion lead to a temporary or even permanent reduction of the threshold of hearing curve; high frequencies in the range of 4,000 Hz are of particular concern. Depending on instrument positioning, often only one ear is affected: for example, the left ear for violinists and trombone players, the right ear for piccolo players (Frei, 1979).

From a purely statistical standpoint, however, there appears to be no particularly increased risk based on the high sound levels in an orchestra. Apparently symphonic music has a different effect on the ear than noise, where the emotional attitude relative to the loudness level associated with music plays a role (Karlsson et al., 1983). Alone the fact, that musicians with a demonstrably diminished threshold of hearing are still valued members of an ensemble, shows that measurements of hearing thresholds do not form a unique qualification criterion for musicians. On the one hand musicians are able to compensate for age related limitations by experience, on the other hand it seems reasonable that hearing ability for medium and higher loudness levels remains normal, even if the hearing threshold has been lowered (Woolford & Carterette, 1989).

1.2.2 *Masking*

The threshold of hearing and the equal loudness curves of Fig. 1.1 are only valid for single sinusoidal tones which reach the head of the listener from the front under otherwise perfectly quiet conditions. If in contrast, two tones are sounded simultaneously, it can happen, that by reason of the loudness level of one of the tones, the other is no longer audible, in spite of the fact that it has a sound pressure level which exceeds the threshold of hearing for a single tone. In this situation the softer tone is masked by the louder one.

As an example for the masking effect of a disturbing tone of 1,000 Hz, Fig. 1.3 shows the so-called masking thresholds. These are the thresholds which must be exceeded by the softer test-tone to become audible. In the diagram, the frequency of the test-tone is plotted from left to right, and the test-tone sound pressure level increases to the top. Above the schematically simplified threshold of hearing curve, four lines are plotted. They indicate the required sound pressure level of the

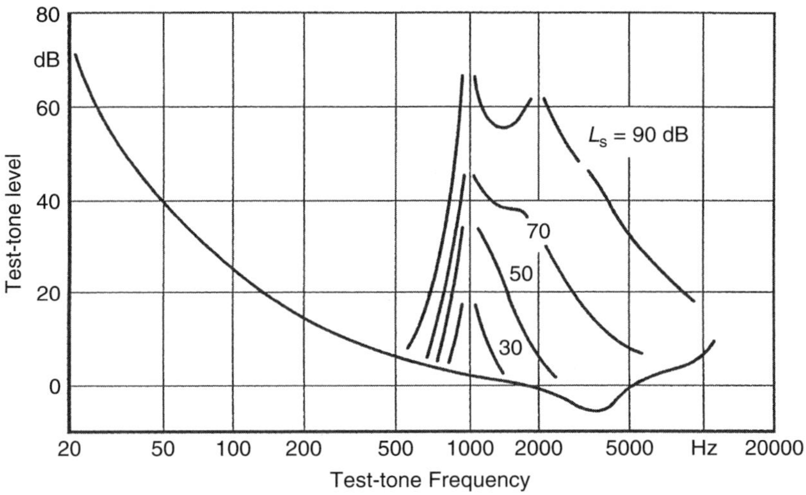

Fig. 1.3 Masking threshold of a test-tone which is masked by a masking-tone of 1,000 Hz with a level of L_S (after Zwicker, 1982)

test-tone for various levels of the masking tone. The fact that the curves are interrupted at 1,000 Hz, as well as partially at 2,000 and 3,000 Hz, results from the fact that beats, or difference tones between test-tone and masking tone, render an exact measurement impossible.

As noted, the masking effect is most strongly pronounced in the neighborhood of the frequency of the masking tone. Thus, for example, for a masking tone level of 90 dB, a test-tone of 1,200 Hz must exhibit a sound pressure level of at least 60 dB to become audible. It is particularly characteristic that the curves below the frequency of the masking tone drop off with a rather steep slope, while the slope above that frequency is significantly less, and for larger masking intensity a second maximum is noted at the octave of the masking frequency. This means that, by reason of the masking effect, predominantly high frequency contributions are diminished or even rendered inaudible by low tones in the context of hearing impressions. As is furthermore seen from the curves, the masking effect increases with increasing loudness, with strong dependence on the relation to the frequency region under consideration. This is one of the reasons why polyphonic music sounds more transparent when played softly, rather than at larger loudness levels (Lottermoser and Meyer, 1958).

The masking effect is not limited in time to the duration of the masking tone or sound. Inasmuch as the ear requires a certain recovery phase to regain its "undisturbed" sensitivity, a so-called post-masking effect is observed: after the cessation of the masking tone, the masking threshold remains initially unchanged for several milliseconds, and then goes over into an almost linear decline, reaching the original threshold after about 200 ms. These time indications are practically independent of the strength of the masking sound (Zwicker, 1982). For dynamic and temporally highly structured music, this effect can occasionally be significant. In most cases, however, it is insignificant because of the decay of the instrumental sound and the reverberations in the hall.

Since the ear processes softer sounds somewhat more slowly than louder sounds, a pre-masking is also possible, prior to the onset of the masking sound. This is, however, limited to a time frame of less than 20 ms. In a musical context such pre-masking can occasionally become significant when short attack noises, which precede the tone, are made inaudible or are at least weakened. It is of value to identify the point of attack, perceived by the ear, as associated with this pre-masking effect. This is the point which is relevant for determining a rhythmic structure in a tone sequence, in other words, the instant in which the ear senses the sound level to be already very close to the final value. This is illustrated by the fact that the perceived point of tone entrance lies about 10 dB below the final sound level, provided this lies by 40 dB above the threshold of hearing, or the masking threshold (in the presence of pre-existing noise), and this is relatively independent of the speed of the staccato attack. For very soft tones the point of attack can move as close as 7 dB to the final sound pressure level, i.e., it is sensed even later. For very loud tones, the tone entrance is already perceived at a sound pressure level of 15 dB below the final value (Vos and Rasch, 1981).

Naturally the masking effect influences not only shifts in hearing thresholds, but it is also noticeable when several audible tones influence each other in their relative loudness levels. In this context the apparent weakening of the higher tones by the lower ones is most significant, whereas masking in the other direction is relatively minor. This partial masking of loudness is particularly pronounced between sound components close in frequency. For example the loudness impression of a 1,000 Hz tone with sound pressure level of 60 dB in the presence of 30 dB rush noise, corresponds to a loudness level of a 50 dB 1,000 Hz tone without the rush noise. A rush noise level of 40 dB would suffice to render the 1,000 Hz 60 dB tone inaudible (Zwicker, 1982).

Nevertheless, partial masking is also experienced for widely separated frequencies. This phenomenon is represented in Fig. 1.4 for the interaction of two sinusoidal tones of 250 Hz (at a level of L_1) and 500 Hz (at level L_2). In the left portion of the diagram L_2 is constant (83 dB) and the level L_1 of the lower tone rises from 43 to 83 dB. In the right portion L_1 remains constant and L_2 drops to 63 dB as indicated by the lines. The shaded strips indicate the levels required for a tone sounded alone in order to be perceived of the same loudness as the corresponding tone when sounded concurrently with the other one. Inasmuch as the subjective impressions vary somewhat for different persons, the perceived levels L_{1E} and L_{2E} are represented as bands. The distance of these bands for L_{1E} and L_{2E} from the associated curves L_1 and L_2, which represent the objectively present levels, show by how much the two tones are perceived as softer when sounded simultaneously. The graph makes clear that the higher tone appears to be weakened by 5 dB when objectively it is 15 dB stronger than the lower tone, where it is still heard with almost its original loudness. When both tones are sounded at the same level, however, they mask each other equally: On the average they are sensed approxi-

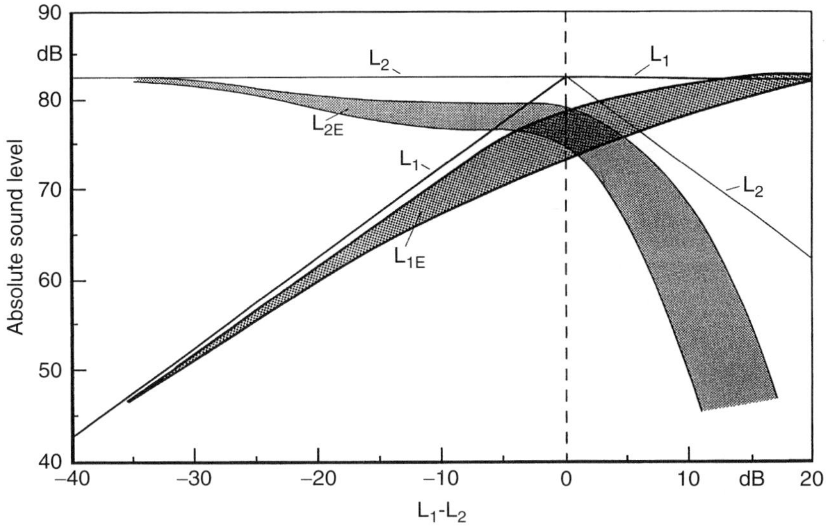

Fig. 1.4 Mutual masking of two sinusoidal tones of 250 Hz (level L_1) and 500 Hz (level L_2) sounded simultaneously. The shaded region indicates the perceived level

mately 6 dB softer, than when sounded alone. Finally if the level of the higher tone drops by more than 5 dB below the lower one, the upper tone is masked in preference.

1.2.3 Directional Characteristics

While the eye subtends only a limited angular region in its field of view, the ear senses sound events from all directions. There is, however, a certain dependence of the perceived loudness on the direction of sound incidence. This is mostly due to the fact that the particular ear, which is turned away from the sound source, is shaded by the head. In addition there is the contribution of the shape of the external ear and the ear canal which influence the sound pressure level formation directly in front of the ear drum. The sound pressure level at this location ultimately determines the impression of loudness.

The differing sensitivity of the ear for various directions of sound incidence is designated as the directional characteristic. For frequencies below 300 Hz there is no directional dependence of loudness impressions for the individual ear, however, for higher frequencies there is a clear preference for those directions from which the sound impinges on the ear without shadowing (Schirmer, 1963). In the normal case of binaural hearing a directional characteristic for loudness perception is present, which corresponds to the addition of the sensitivities of both ears (Jahn, 1963). Though a certain compensation due to the cooperation of the two ears is present, nevertheless, at higher frequencies a typical directional dependence of the loudness impression is evident.

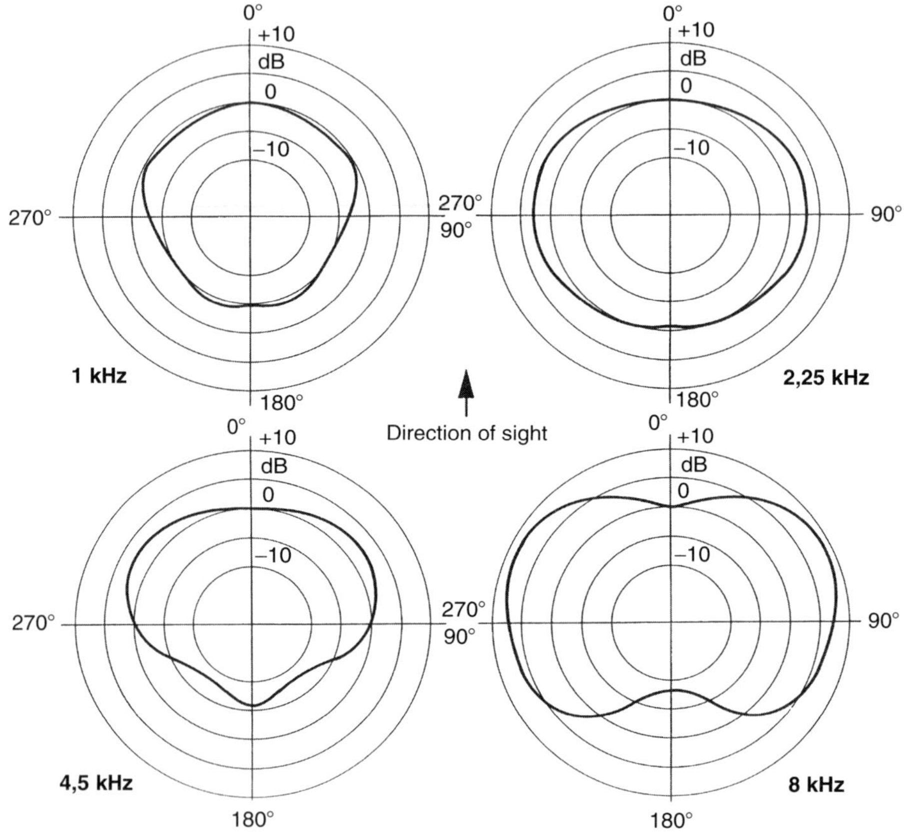

Fig. 1.5 Directional characteristics of the ear (binaural hearing) for various frequencies (after Jahn, 1963)

Several examples for the directional characteristic in a horizontal plane for binaural hearing are given in Fig. 1.5. It is noted from the diagrams that the 1 kHz sound is perceived as loudest when it arrives from the viewing direction. If it impinges on the listener from behind, it appears – in comparison to the same level in a free field – about 5 dB weaker. In contrast, the direction of greatest sensitivity at 4.5 kHz is from the side, the sound is perceived as approximately 3 dB stronger than if it comes from the direction of view or from behind. For frequencies around 4.5 kHz the angular region of greatest sensitivity is shifted somewhat more forward, while sound arriving from behind is noticed only very weakly. By 8 kHz the preferential hearing directions have again become oriented predominantly sideways.

The directional characteristics compare the sensitivity of the ears for sound incidence from a certain direction with that in the direction of view. For practical applications, however, another case is of interest: In enclosed rooms, generally a

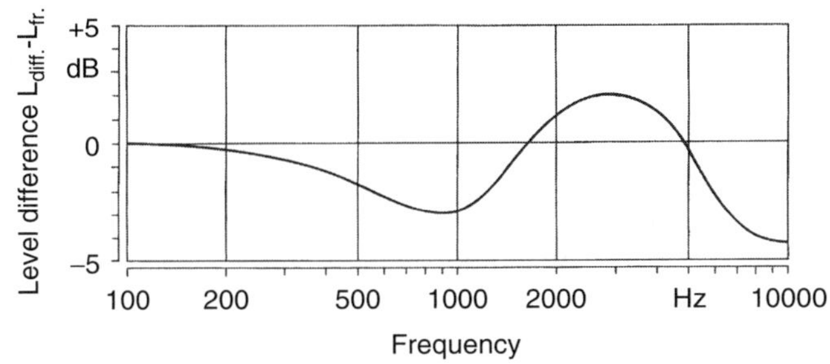

Fig. 1.6 Difference between sound pressure levels in a diffuse field, and for frontally incident plane waves for the case of equal loudness perception (after Zwicker, 1982)

diffuse sound field is formed, i.e., sound impinges on the listener from all directions. In such a diffuse sound field the sensitivity of the ear corresponds to an integral over all spatial directions, and departs therefore from the value for frontal incidence. This difference is shown in Fig. 1.6, where positive values indicate that a sound incident from the front is perceived as louder than the same level in a diffuse sound field. This is particularly the case between 2,000 and 4,000 Hz, while sound contributions between 300 and 1,500 Hz as well as above 5,000 Hz have the effect of appearing louder for uniform sound incidence from all sides.

1.2.4 Directional Hearing

When sound impinges on the head of the listener from a somewhat sideways direction, rather than directly from the front, a slight time differential ensues between the times of arrival at the two ears, since the path to the ear turned away from the source is slightly longer. This time difference is evaluated by the nervous system to determine the direction of incidence of the sound. In this context, the extremely short time span of 0.03 ms is sufficient to evoke a sensation of directional change; this corresponds to a sound path difference between the ears of about 1 cm. Thus, the ears are able to detect a departure of only 3° from the frontal sound incidence direction (Reichardt, 1968).

For incidence from the side, orientational resolution with this procedure would drop to 7.5°, furthermore, confusion with sound incidence from the rear would be possible if additional information based on directional characteristics of the ear were not available for these angular regions. For sound incidence from the side, both ears do not receive the same sound pressure level, but rather a difference results which is typical for a particular angle of incidence. This phenomenon plays a role especially at high frequencies, and thereby affects a certain relationship between tone color changes and directional changes. This cooperative relationship between sound running time difference, intensity variations and tone color changes, leads to a resolution on the part of the ears for sound incidence directions in the angular region of ±45° of the frontal direction which permits differentiation in steps of 3° and furthermore permits an orientation in steps of 4.5° in the region of 45–90°. It should also be mentioned that the ear reacts rather rapidly to a change between two sound sources coming from different directions. A jump from left to right (or right to left) is noticed in about 150 ms, a change from front to rear in less than 250 ms (Blauert, 1970). These time intervals correspond to the duration of very short notes.

When several equal sound signals arrive at the listener simultaneously, as can be the case with the use of loudspeakers for example, then only one sound source located between the original sources is perceived. For two speakers of equal intensity a median impression results. By changing the intensities, an apparent source between the speakers can be shifted. A comparable shift in location can be simulated by changing the running time difference, which however, must not exceed 3 ms (Hoeg and Steinke, 1972).

For running time differences of more than 3 ms, the sound source is located solely by the direction of the wave front arriving first, even if the following sound signal is stronger than the primary arriving signal. For running time differences between 5 and 30 ms, this level difference can amount to up to 10 dB without influencing the localization of the sound source in relation to the direction of incidence from the first source (Haas, 1951). This phenomenon is labeled as the precedence effect.

Finally, the spectral composition of the sound signal received from the source can influence the directional impression. This effect, among other things, determines the possibility to distinguish between incidence directions "front" and "back" as well as "up" in the median plane, i.e., the symmetry plane of the head. Frequency contributions below 600 Hz as well as those from about 3,000 to 6,000 Hz support the direction "front", components between 800 and 1,800 Hz as well as above 10,000 Hz, the direction "back", and components around 8,000 Hz the direction "up" (Blauert, 1974).

1.2.5 The Cocktail Party Effect

The high directional selectivity of the hearing mechanism rests on processing two distinct sound signals which are transmitted by the ears to the brain. This selectivity enables not only recognition of the direction of incidence of a single sound source, but also facilitates differentiation between multiple sound sources, located in different directions. In this, binaural masking plays an important role. When test sound and masking sound reach the listener from different directions, the masking is not as strong as when they come from the same direction. In the past, most investigations of masking were related to monaural masking, i.e., identical sound signals at both ears, or exposure of only one ear to test- and masking sound; the results quoted in Sect. 1.2.2 fall in this category.

Fig. 1.7 Hearing thresholds of a test-tone masked by pink noise (after Prante et al., 1990)
(**a**) sinusoidal tone and noise from the front
(**b**) sinusoidal tone 60° from the side, noise from the front
(**c**) sinusoidal tone 60° from the side, noise as diffuse sound field

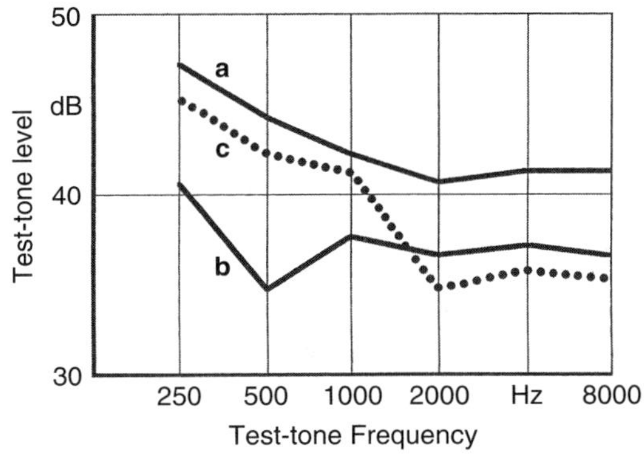

The effect of binaural hearing on masking is represented in Fig. 1.7. The upper curve (a) represents the usual monaural masking threshold for a (sinusoidal) test-tone which is masked by "pink noise", i.e., a noise with strong low frequency components. In this case, the sinusoidal tone and the noise arrive from the same frontal direction. Curve b applies to the case of the noise arriving at the listener as a plane wave from the front and the test-tone from the side with an angle of 60° relative to the direction of view. The directional difference of the two sound sources effects a lowering of the masking threshold by up to 10 dB for the mid frequencies, above 1,000 Hz the drop is still 6 dB, it thus raises the sensitivity of the ears for the test-tone. Curve c relates to the same position of the tone source, however, the noise reaches the listener as a diffuse sound field, it thus reaches the listener from all sides; in this case, which for example is of interest for locating a sound source in an expansive hall, the sensitivity is raised especially for high frequencies (Prante et al., 1990)

When multiple sound sources are distributed around a listener, the hearing mechanism (inclusive of further information processing in the brain) has the capability to concentrate selectively on one of these sources and emphasize it in comparison to the others. This phenomenon is referred to as the "Cocktail Party Effect", since Theile (1980) such a situation is particularly typical for a large number of distributed speaking voices. It is, however, required that the sound pressure level of the sound of interest lies about 10–15 dB above the masking level determined by the masking sound. Otherwise directional location is no longer possible. Through the Cocktail Party Effect, the intelligibility threshold for speech can be enhanced by up to 9 dB for several directionally distributed masking sources in comparison to having all sources come from the same direction (Blauert, 1974).

This concentration on one of many sound sources is particularly important for musicians playing in an ensemble, and the ability for such concentration is a matter of practice. In this context it is of value if the musician can visualize the sound without hearing it. It is this visualization which stimulates the relevant brain section, so that during further information processing in the brain, the already existing stimulation pattern needs only to be compared with the pattern arising from the arriving sound (Kern, 1972).

1.2.6 Masking for the Musician

Masking effects plays a particularly important role for those musician, for whom the instrument as a sound source is relatively close to the ear, who nevertheless need to hear the sound of other instruments and also the sound reflected by the room. Singers naturally have the same problem, however, they have the additional option of controlling the voice by sensing chest vibrations (Sundberg, 1979). The sound level, generated by the musician's own instrument near the ear can have substantial values. For a string or woodwind forte, this lies in the region of 85–95 dB, for brasses it can be an additional 10 dB. These values are in reference to the sound running directly from the instrument to the ear, without the amplification due to the surrounding room, they thus approximate a free field situation. This limitation is an advantage for subsequent room acoustics considerations. The level may not always be the same for both ears, since many instruments are held at the player's side. Figure 1.8 shows the level difference at the ears of several instrumentalists. As can be seen, the difference increases significantly with increasing frequency, where the differential path length of sound to both sides around the head introduces additional waviness to the curves.

The consequence of the directional characteristics of the ear, as previously described, and these additional level differences is, that the degree of mutual

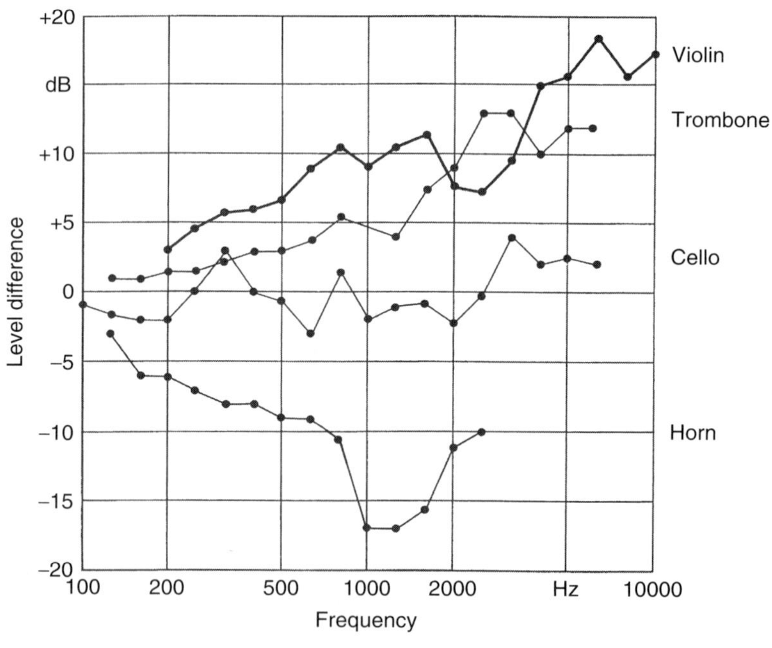

Fig. 1.8 Difference of sound pressure levels at the left and right ear of the player. Positive values mean that the level is higher at the left ear, negative that the value is higher at the right ear

Fig. 1.9 Directional Dependence of the Masking threshold for players with symmetrically held wind instruments (Oboe, Clarinet, Trumpet) in reference to the direction of lowest masking threshold (=0 dB)

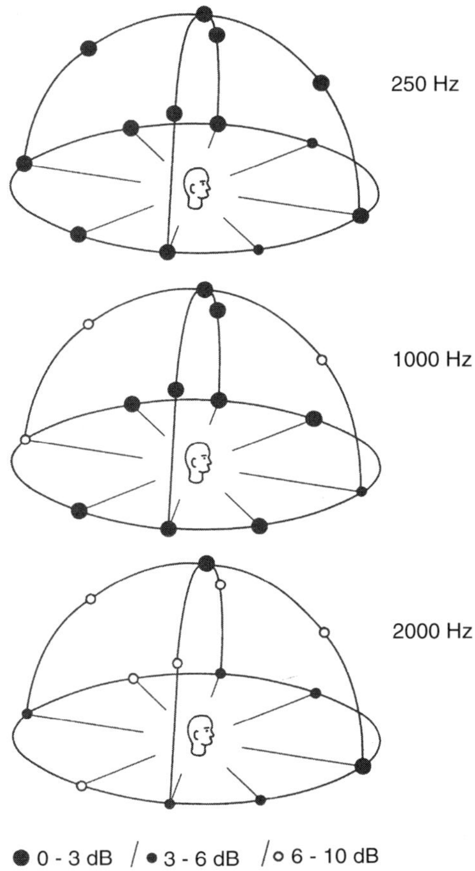

masking also depends on the direction from which extraneous sounds reach the player. The directionally dependent masking threshold varies by up to 9 dB around 250 or 500 Hz, and for frequencies above 1,000 Hz by values up to 20 dB. For the general description of this effect it is, however, necessary to average the relevant values over the entire tone scale. In all this, the question, for which direction of incidence the ear of the player is especially sensitive, and for which especially insensitive, is of particular interest.

For the case of symmetrically held wind instruments, such as the oboe, the clarinet or the trumpet, this effect is graphically represented in Fig. 1.9. Individual directions of incidence are indicated by marks which give the sensitivity in 3 dB steps in relation to the direction of greatest sensitivity. Measurement results are given for three frequencies of the extraneous sound and are valid for the entire tone range. Additionally it should be noted that for 500 Hz, all directions fall within the 3 dB range, there is here thus no practical directional dependence. Also at 250 Hz the influence of direction is small. In contrast, at 1,000 Hz, the median plane, except for vertically upwards, shows itself disadvantaged. At 2,000 Hz this direction and the direction of view are most advantageous, disadvantaged are all rising directions of less than 45° as well as the angled horizontal directions to the rear.

The cello already shows a slight asymmetry. Thus, for the cellist at 250 Hz, the direction left upwards is the only one outside the 3 dB region. For 1,000 Hz the direction vertical from above and angled behind are less sensitive (3–6 dB). On the other hand, at 2,000 Hz the direction left upwards is less sensitive than right upwards, the directions angled front and behind are likewise less sensitive than for the wind instruments of Fig. 1.9. The asymmetry is more strongly pronounced in the trombone, for which the 1,000 Hz sound contributions coming from the right, and in contrast the 2,000 Hz contributions from the left, are disadvantaged.

For the violin and the viola the uneven exposure of the players' ears to sound is naturally pronounced, yet even here, at 500 Hz, all directions remain within the 3 dB limits. For three other frequencies the relationships are represented in Fig. 1.10. Here, at 1,000 Hz, as at 2,000 Hz, several directions stand out, for which the masking threshold lies by more than 10 dB higher than for the optimal direction (Meyer and Biassoni de Serra, 1980). As a whole, a certain similarity with the pictures of Fig. 1.9 is recognizable, at least in the directions going up. It is noteworthy that the directional dependence of the masking threshold is practically unchanged when the player has an elevated threshold of hearing by about 10 dB, i.e., when the player has a slight deterioration of hearing ability, as is not uncommon for violinists.

For horn players as well, a distinct asymmetry exists when it comes to sensitivity to extraneous sound. True, at 250 Hz, the masking threshold is practically direction independent, however, at 500 Hz, the left side, namely the side away from the instrument, is less sensitive. For 1,000 Hz the directions left and right are the most favorable, while front and back are particularly insensitive. At 2,000 Hz there is a relatively low sensitivity for the side turned toward the instrument, particularly in the direction upwards and to the right, for which the masking threshold is raised by more than 10 dB in comparison to the direction of view as the most favorable direction (Meyer and Biassoni de Serra, 1980).

1.2.7 Sensitivity to Changes in Frequency and Sound Pressure Level

Periodic changes in frequency or amplitude of a tone are perceived differently by the ear, depending on how fast they occur. When the number of fluctuations per second remains below 5, the change in pitch or loudness, respectively, can be followed in its time sequence (Winckel, 1960). If a vibrato is carried out that slowly, it can therefore easily become a whine. From 6 Hz upwards, in contrast, one senses a uniform pitch or loudness, which is associated with an internal motion. The perceived pitch corresponds quite accurately to the central pitch about which vibrato fluctuates (Meyer, 1979); the loudness impression, on the other hand, is oriented toward the maximum level of the fluctuating sound (see Fig. 1.2). If the

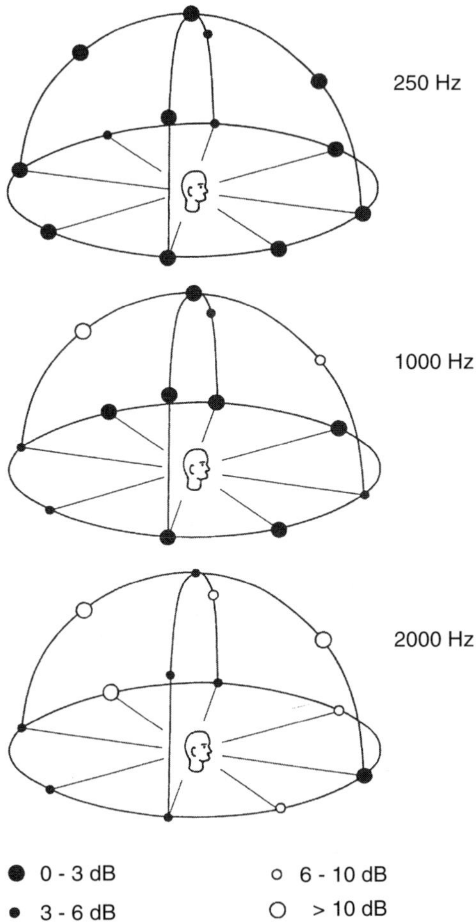

Fig. 1.10 Directional Dependence of the Masking threshold for a violin player, in reference to the direction of lowest masking threshold (=0 dB)

fluctuations occur more rapidly, the impression is given (approximately between 10 and 15 Hz) of a tonal roughness. The strength of this effect depends on the frequency position of the modulated tones, the fluctuation frequency, and also on the relative strength of the amplitude fluctuation; however, it is practically independent of the absolute sound pressure level (Terhardt, 1973, 1974).

The upper limit for fluctuation frequencies which produce roughness for low tones (below 400 Hz, i.e., below G_4) lies at about 100 Hz and rises to a value of 250 Hz for very high frequencies. The impression of roughness for tones around 100 Hz (G_2) is largest if the temporal fluctuations occur with a frequency of about 20 Hz. With rising tone frequency, the maximal roughness producing frequency increases, and for 3,000 Hz it reaches a frequency of 100 Hz. Considering that these temporal variations are caused not only by the superpositioning of several tones but also by the overtones of the very same tones, it follows that low tones can become rough if they are too rich in overtones. The phase position of the partials is relevant in this

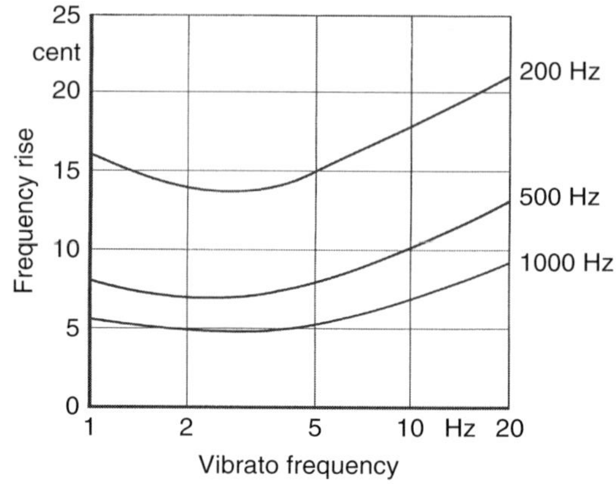

Fig. 1.11 Smallest noticeable frequency rise for tones of different frequencies in dependence on vibrato frequency (recalculated after Zwicker, 1982)

context: if the tone contains sharp impulses, as for example in brass instruments, the roughness is strong. In contrast, the roughness of a tone is weak, when individual partials stand out in relation to neighboring partials, particularly if in each critical band (see Sect. 1.2.1) one partial clearly dominates over all other partials, as for example in the plenum sound of the organ.

Naturally, all frequency- or amplitude-fluctuations must have a certain amplitude to become audible at all. For the case of frequency fluctuations of pure sinusoidal tones, three curves are plotted in Fig. 1.11, which show the smallest perceptible frequency rise in dependence on vibrato frequency. Two things are noted from these diagrams: The ear is most sensitive to pitch fluctuations between 2 and 5 Hz, all three curves exhibit a minimum at that point; if the frequency fluctuation is more rapid or slower, they are not as strongly noticeable. Furthermore, the sensitivity of the ear increases with increasing frequency of the vibrating tone; in the region between 1,000 and 2,000 Hz variations of ±5 cents are sufficient, whereas at 200 Hz variations of about three times that strength are required (100 cents correspond to one half step in a tempered scale). These sensitivity limits are of interest in connection with the strength of a vibrato, for example.

For periodic changes in sound pressure level, the largest sensitivity lies in the region around 4 Hz as well. The required measure of level fluctuation drops with increasing loudness. While in the neighborhood of the threshold of hearing, fluctuations by 4 dB first become audible, at sound pressure levels of 80 dB, fluctuations of approximately 0.3 dB are already sufficient (Reichardt, 1968). This sensitivity to amplitude fluctuations, however, depends somewhat on frequency. In the region of 1,000 Hz it is even larger than 0.3 dB yet, on the other hand, for low notes below 200 Hz it becomes lower. For the dynamic range available in musical practice, one can deduce from this sensitivity to sound pressure variations, that the ear can differentiate approximately 130–140 loudness steps (Winckel, 1960).

Chapter 2
Structure of Musical Sound

2.1 Introducing the Model

Every single tone which reaches our ear in the course of a musical work, contains a fullness of information. We perceive a pitch, loudness and tone color. We can also make statements about the steadiness of the pitch, or if the tone is enlivened with vibrato. Furthermore we notice changes and fluctuations in loudness as well as the nature of the tone entrance, be it attacked softly or sharply; similarly we can draw conclusions about the decay of the note. All these details give a characteristic tone picture from which we extract the musical content and also recognize what instrument generated the tone. In this, previously gathered listening experiences, play a not unessential role. Finally we can even draw conclusions about the nature and size of the room in which the music resounds.

This acoustic phenomenon can be described by a number of physical factors. In all this there is a certain difficulty posed by the circumstance that the individual characteristics perceived are not each determined by only a single physical quantity, but come about by the cooperation of several components. Nevertheless, the problem of finding an association between objective acoustical data and tonal impressions on the basis of practical experiences should not be overestimated.

A model, as schematically presented in Fig. 2.1, shall serve to clarify the complex sound processes which correspond to a single note. It proceeds from the concept that a tone is a vibrational process which changes in time, which is characterized by the number and strength of vibrations. Accordingly, the model contains a time axis (in the picture it runs from front to back), a frequency axis (from left to right) as well as an axis for sound pressure level (upwards), which is the quantity responsible for the loudness impression (Winckel, 1960).

Additionally the shape of the vibration is of great significance: Most often the sound does not consist of a simple sinusoidal oscillation, as known from the motion of a pendulum, but rather it exhibits a complicated time sequence. Thus the ear is placed in the position of needing to differentiate between tones of different timbre for the same pitch. However, such a complicated vibrational form can be considered as a superposition of a series of sinusoidal vibrations with different frequencies.

J. Meyer, *Acoustics and the Performance of Music*.
DOI: 10.1007/978-0-387-09517-2_2, © Springer Science + Business Media, LLC 2009

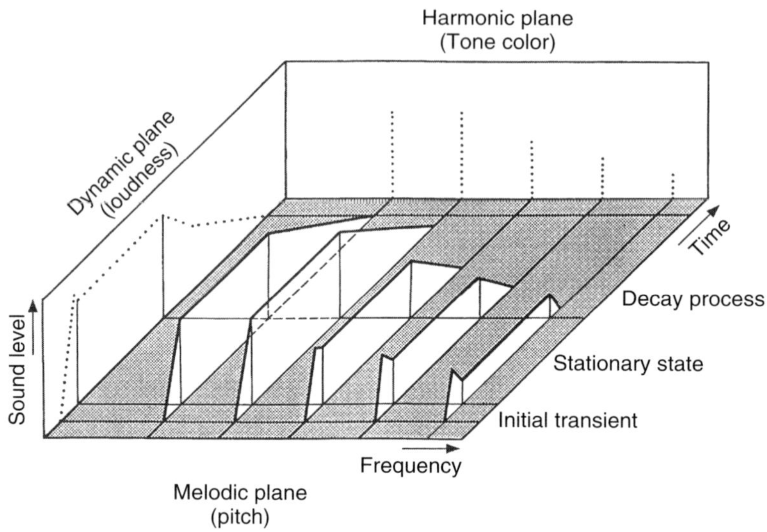

Fig. 2.1 Three dimensional representation of a single tone in schematic form

The entire vibration process, therefore, appears in our model as a sequence of several partial vibrations or overtones.

From a time standpoint, each tone can be structured into three sections:

1. The starting transient, i.e., that portion of time during which the tone is developed from complete rest to its final state.
2. The stationary condition, i.e., that portion of time during which the tone is practically not subjected to change.
3. The decay, i.e., that portion of time during which the tone, after completion of the excitation, dies out to complete silence.

The starting transient, in particular measure, contains characteristic features, which make it possible to distinguish between instruments. Often components are present which are no longer contained in the later tone picture.

A stationary condition is, strictly speaking, reached only for instruments with very uniform vibration excitation. These are above all those instruments which do not use the energy of the player but an external (constant in time) energy source, as for example the organ or many electronic instruments. Even the minimal air pressure variations of a wind instrument player, or the bow pressure changes of a string player lead to minute variations of the vibrational process. This is also the reason why one speaks of a quasistationary state in such cases. The most important features can, however, be describes as though it were a stationary state.

The decay process plays a special role for plucked- and percussion instruments, since in the absence of continuing excitation there is no stationary or quasistationary state. Yet, also for other instruments it has a certain significance for convincing connections between tones in flowing passages.

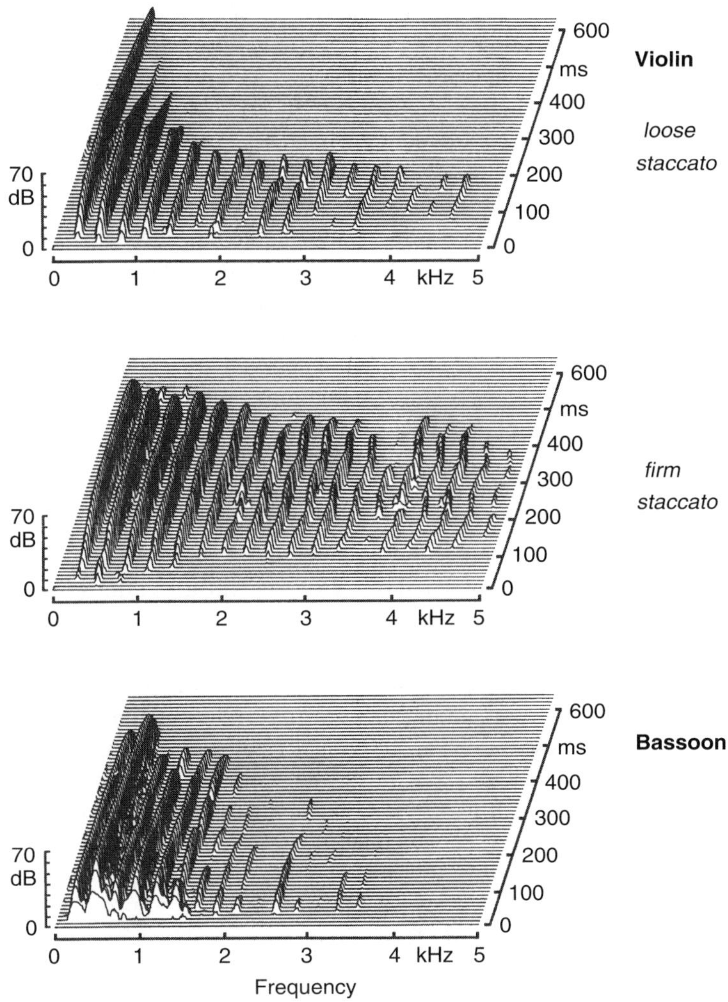

Fig. 2.2 Observed measurements of tone spectrum developments in time for three (C_4) *staccato* tones

Figure 2.2 shows the transition from abstract model presentation to real tones. It contains three dimensional sound spectra as measurement results of the analysis of three staccato tones with equal pitch, and approximately equal duration, however, of different tonal character. Differing numbers of partial vibrations are recognized. Furthermore, in the partial picture for the firm violin staccato, a vibrato becomes apparent in the form of a slight wave motion of the individual partials. The tone onset for example, occurs most rapidly in the bassoon, and is largely simultaneous for all partials, while the partials in the violin enter in part only with delay. Likewise a differential behavior is noted for the end of the tones, the long ringing of the loose violin staccato is particularly noticeable. Already these few suggestions make it appear rewarding to pursue a deeper consideration of such methods of analysis to gain detailed information about tonal characteristics of musical instruments. Aside from that, the examples shown in this picture show how difficult it is to compare the durations of tones of differing tonal characteristics.

For the sake of clarity it is often recommended to simplify the three dimensional representation of the tone by sacrificing one of the three quantities and projecting the model onto one of the planes represented in Fig. 2.1. Relinquishing the frequency dependence results in a time dependent sound pressure level graph, as indicated by the dotted line on the left sidewall of the model. Inasmuch as this concerns the time flow of the quantity most responsible for the loudness sensation, this plane is frequently referred to as the dynamic plane. If the tone model is projected onto the base plane with its frequency and time axis, one obtains the temporal pitch flow, including possible fluctuations by such things as vibrato; this plane is called the melodic plane.

A projection into the third plane with the frequency and sound level axes is equivalent to a cross section through the stationary part of the vibration process, more exactly a median value. For this viewpoint individual partials of the complex sound are represented with their respective frequency and strength, as the dotted lines on the back wall of the model show. Their height gives the sound pressure, or sound pressure level, their foot point on the horizontal axis indicates their frequency. Since this type of representation also makes it possible to visualize the frequency composition of cords, one speaks of the harmonic plane of the model. In analogy to the optical decomposition of light into individual portions of various primary colors, the representation of individual frequency components with their strengths is designated as a sound spectrum. Above all, this gives information about the tone color in the stationary part of the tone and in addition permits conclusions about the starting transient and decay behavior.

2.2 Frequency- and Level: Structures

2.2.1 The Harmonic Tone Structure of Sound Spectra

For periodic vibration processes, as they are present in the stationary and also quasistationary portion of sounds of almost all musical instruments, the individual vibration contributions form a so called harmonic series in reference to their frequencies, i.e., starting with the lowest frequency, partial tones are present whose frequencies are integral multiples of the fundamental frequency. As an example for such a partial tone sequence the note C_2, as sounded on a contrabassoon, is broken down in Fig. 2.3 into its first 16 harmonic partials. The note picture shows (without octave transposition) the location of the individual partials within the musical scale, below each note the relevant frequencies are indicated, the corresponding harmonic index is shown above the note picture showing that they form the 2-, 3-, 4-, etc fold frequency of the fundamental.

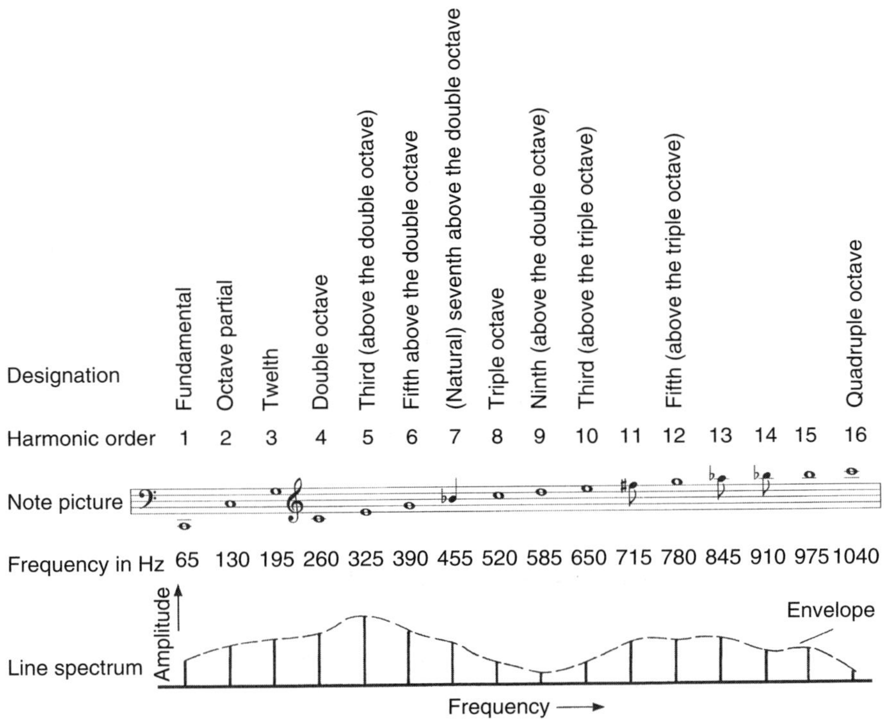

Fig. 2.3 Harmonic structure of the note C_2, represented schematically

This fundamental is principally responsible for the perceived pitch. In the example shown it is located at 65 Hz. A doubling of the number of vibrations corresponds to exactly an octave, so that the 2nd, 4th, 8th, and 16th partials again result in a C (or C_3, C_4, C_5, and C_6 respectively). Tripling the frequency leads to the fifth above the octave (G_3), correspondingly the 6th and 12th harmonics form fifths in higher positions. The 5th and the 10th partials are identified as notes E_3 and E_4 which form major thirds in relation to the corresponding C's, the 9th an 15th partials form the fifth and the major third respectively to the higher octaves of the 3rd partial, i.e., G_4–D_5 and G_5–B_5. All of these indicated intervals should be considered as "perfect" intervals.

Accommodating, however, the 7th, 11th, 13th, and 14th partials, which lie somewhere in between, proves more difficult, since their frequencies do not coincide with those of notes on a normal scale. Thus the 7th and 14th partials lie lower than B_4^b and B_5^b, they, along with the fundamental, form the basis for the interval of a so called natural seventh, which, however, is by no means sensed as dissonant in the context of an entire partial series since it is included in the full sound without beats. A similar situation exists for the 11th partial which lies below the $F_5^\#$, as well as for the 13th partial, which is just a little higher than A_5^b.

Clearly, sounds can consist of more than 16 partials, in the octave from the 16th to the 32nd partial, additional 15 intermediate values are found, which are increasingly crowded together within the scale. While a trained ear can still detect the lowest 6–8 partials individually from within a sound, since they fall within different critical bands (see Sect. 1.2.1), the higher harmonics melt together into a ton color impression, even for the experienced listener. In all this, the intensity relations between individual partials naturally play a significant role.

The amplitudes or sound pressure levels of the individual partials can be represented as a line spectrum in a schematic manner, as is already suggested in the harmonic plane of Fig. 2.1, and also in the lower part of Fig. 2.3. Every line by position and length denotes the frequency and strength of the relevant partial. Connecting the endpoints of the spectral lines results in the so-called spectral envelope. This curve gives a clear representation of the amplitude distribution as a function of frequency without considering the harmonic number of the individual partial. The spectral envelope is therefore especially suited for summary representations.

2.2.2 The Frequency Range of Sound Spectra

The lower limit of the spectrum of partials is always determined by the fundamental, that is, the frequency which corresponds to the written representation (possibly under consideration of transposed notation). Still lower, stationary, i.e., steadily vibrating tonal components, do not exist. Below the fundamental, only noise-like, or nonsteady tone components are found. Consequently, the lower limit of the spectrum of partials moves upward with increasing pitch.

In order to give a general view of the systematics of spectral composition for a musical instrument, an entire chromatic sequence for a horn is given in Fig. 2.4. In this type of analysis, each individual partial appears as a peak of certain width, which is connected with the sharpness of the instrumentation used. It is clearly recognizable how the first peaks of the spectra move toward the right with increasing pitch, i.e., toward higher frequencies. Similarly the spacing between the individual partials increases. For the low notes, the high frequency partials are so closely spaced that they assume a noise-like character ("metallic") for sufficiently high intensity, while partials of higher notes in the same frequency region are still sensed as "pure sounds" by reason of their larger spacing.

The upper boundary of the spectrum of partials is very different for individual instruments, it also depends in large measure on dynamics. Room acoustical conditions also play a more important role than for the low frequencies. For this reason it is most advantageous to be able to refer to sound power spectra when describing tonal characteristics typical of a particular instrument. Sound power spectra summarize the entire sound radiated by an instrument. They are independent of the surrounding room, as well as of the distance and direction of an observation- or measurement point. They are best recorded in a special room with high grade sound reflecting walls, a so called reverberation chamber.

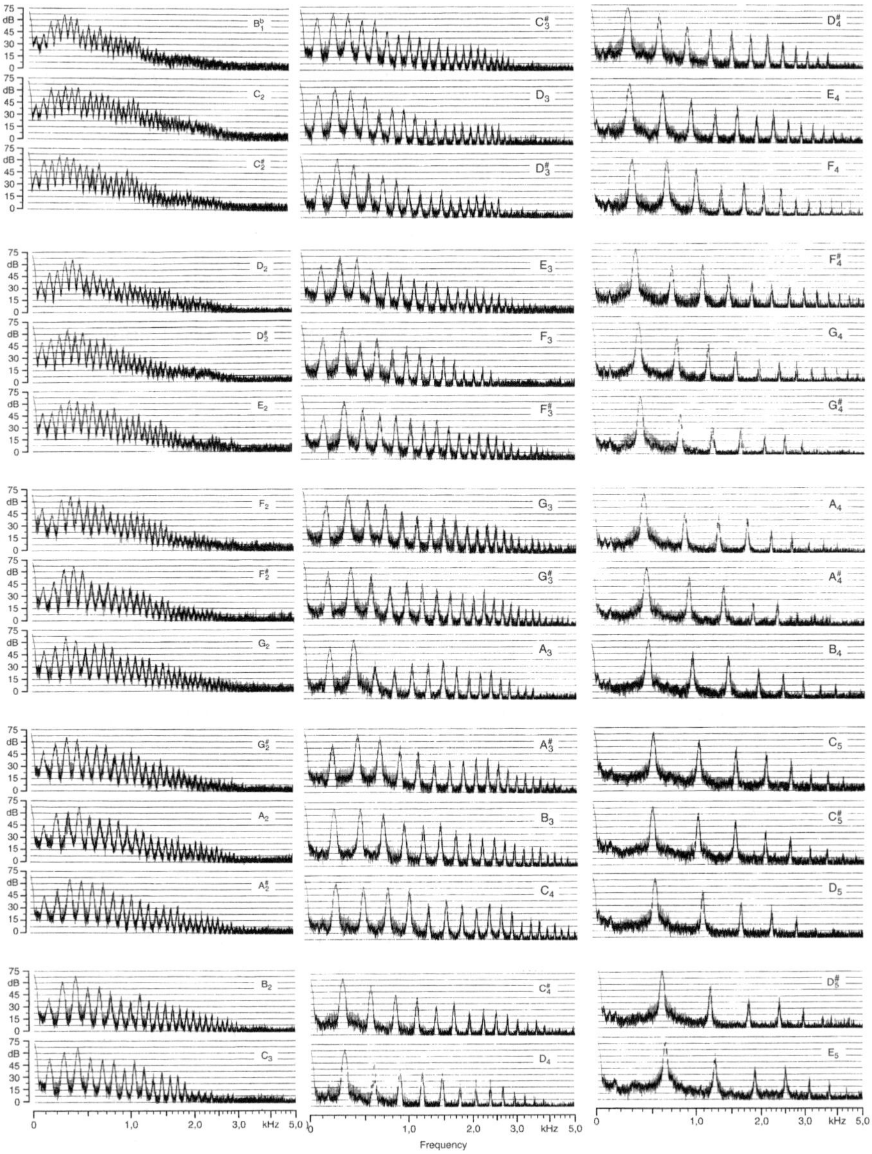

Fig. 2.4 Chromatic sequence of sound spectra for an F-Horn. Reproduction from experimental measurements

In order to be able to compare the behavior of different instruments at high and highest frequencies, it is desirable to establish a characteristic frequency as a reference point. For several reasons the frequency of 3,000 Hz appears particularly suited; one of the reasons is that sound power spectra – with only a very few exceptions – show a linear level decay above 3,000 Hz. Therefore, by specifying the slope of this decay in level above 3,000 Hz, on the one hand, and by specifying the level difference between the strongest partials and the 3,000 Hz component on the other, the course of sound power level spectra at high frequencies can be largely described.

Aside from the fact that tone color naturally changes over the tonal range of an instrument, as determined by the location of the fundamental, it can be said in general, that a tonal impression is brighter, and possibly sharper, as richness in overtones increases (in view of the frequency range and the intensity of the upper frequency components) (von Bismarck, 1974). For low tones, rich in overtones, the dense partial sequence in the upper frequency region leads to a rough character. This effect can occur for tones from the 3rd octave upward. While, for example, for G_3, overtones above 2,000 Hz effect a roughness, the corresponding limit for G_1 already lies at about 500 Hz (Terhardt, 1974). In contrast, overtone- poor sounds have a tendency for dark or soft timbre.

2.2.3 Formants

The fundamental certainly does not need to be the strongest partial in the sound spectrum. As the representation for the horn shows in Fig. 2.4, the fundamental dominates in this example in relation to other tone locations only in the upper register approximately from C_4 on upwards. For lower notes the intensity maximum is found at higher order partials. In this context it is remarkable that the location of the frequency maximum remains unchanged, and thus represents a particular characteristic for the sound of the instrument.

A similar phenomenon is also known from speech: When a singer sings a scale on a particular vowel, the region of strongest intensity remains fixed in its frequency location in spite of pitch changes. It is precisely for this region that all tones obtain the same (or at least similar) tone color. These amplitude maxima within a spectrum, which do not change their frequency with changing singing pitch, are called formants.

For the most important vowel colors the formant regions are assembled in Fig. 2.5 according to various authors. The individual maxima of the scheme mark the frequency region of strongest amplitude for the indicated sounds, between them transition colors are to be assumed, which can no longer be uniquely described by letters. They can be compared with the sound of the "same vowels" in various dialects. Generally for speech sounds, two or three more or less strong formants appear, for which, for the sake of clarity, only the most important ones are represented. Accordingly the dark vowels [u (oo), o (oh), a (ah), and å (aw). Translator's note: the German vowel is given first, the English equivalent sound is given in parentheses] are each characterized by one maximum, and the bright vowels each by two. To clarify the frequency values, the peaks are also entered as notes.

Through the connection between formant regions and tone color of vowels, a second possibility (in addition to associating frequencies with the scale) presents

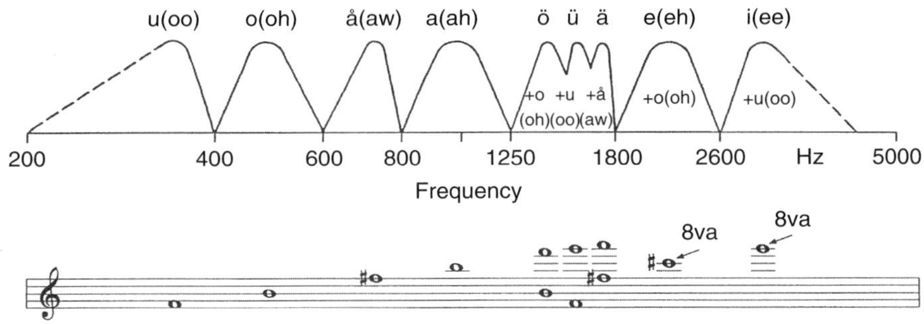

Fig. 2.5 Frequency location of formants for the vowels of the German language (after Thienhaus, 1934; Winckel, 1960; Trendelenburg, 1961)

itself to visualize the frequency regions of the sound of musical instruments. Strong components in the region of the u (oo)-formant (200–400 Hz) and above all the o (oh)-formant (400–600 Hz) are responsible for the fullness and sonority of the sound, while a pronounced a (ah)-formant (800–1,250 Hz) results in a forceful timbre. Especially the contributions between about 1,000 and 1,250 Hz prove very important for a striking tone impression. In contrast, excessively strong components in the region of modulated vowels (ö, ü, ä see translator's foreword) are sensed as uncomfortable, because they can lend a nasal character to the tone (Thienhaus, 1954), if the fundamental is too weak, and additionally insufficient intensity is present in the upper frequency region. Contributions in the region of the e (eh)-formant (1,800–2,600 Hz) and the i (ee)-formant (2,600–4,000 Hz) cause lightness and brilliance of tone. The typical regions for hissing sounds, which are perceived as voiced or voiceless, depending on whether they contain only noise components or also harmonic contributions, are located at frequencies from 4,000 Hz on upwards. Hissing sounds, however, possess pronounced formant character only in Slavic languages.

In order to achieve a vowel-like impression, a formant must be sufficiently clearly developed; i.e., the maximum of the frequency spectrum cannot be too broad. The so-called half-width can be used to characterize that. This is the difference between the two frequencies for which the sound intensity is just half the value at the maximum. Occasionally the width or clarity of a formant is described by the so-called logarithmic decrement which is calculated by f/f_m, where f is the half-width and f_m the mid-frequency. For example, the logarithmic decrement for the vowel "o (oh)" is 1.2, for the first and second formant of the vowel "e (eh)" it is 0.8 and 0.4. Raising the formants by 4–6 dB can intensify the character of a musical instrument, a further increase, however, is detrimental, making the sound rough and shrill (Mertens, 1975).

The esthetic effect of formants on the sound of musical instruments rests primarily on the similarity with singing, since here the human element itself, in a measure, becomes the standard. Furthermore, formants in the musical tone picture receive particular significance, since this characteristic of sound (in contrast to the upper frequency limit of the spectrum) is essentially independent of room acoustical influences. While it is true that the intensity of high frequency secondary formants can be weakened by room absorption, their frequency position, nevertheless, remains unchanged, so that the amplitude maximum continues to determine the tone picture.

2.2.4 The Effect of Individual Partials

The spectra of the horn in Fig. 2.4 show relatively smooth envelopes, which are largely characterized by the frequency range and the location of the formants. It is, however, entirely possible for individual partials to stand out within a spectrum, or on the other hand, have no intensity at all. In particular, sounds exist, for which odd numbered partials are more strongly developed than the even partials. A typical example of this is given by the gedackt organ pipes. For the clarinet in the low register, this type of spectrum also dominates. These lead to a covered and occasionally hollow tone, in this the absence of the octave components (2nd and 4th partials) also supports the dark timbre.

A similar hollow tone effect can also be achieved synthetically by appropriate instrumentation, as the score example 1 from the Bolero by M. Ravel shows. Naturally, because of the high pitched tone location of this passage, the tone color cannot be called dark, rather it resembles the Rohr-Flöte of an organ. The theme is played by the horn (in F) beginning from the C_5 in C-major. The two

Score example 1 M. Ravel, Bolero, measure 131 *ff*. Excerpts without strings and harp

(in octaves) piccolo-flutes supplement the spectrum of the horn by a strengthening of the 3rd and 5th partials by playing the theme starting from G_6 and E_7 in G-major or E-major respectively. In spite of the complicated way of writing the score, a fully harmonic tone is produced with the horn part as the key note, with the celeste pointing to the octave at each note entrance.

While the dominance of odd partials leads to a covered tone, strong even harmonics and, above all, octave components lead to an open and bright timbre. This phenomenon is often used to advantage in instrumentation of orchestral works by using parallel octaves or parallel motion in double octaves. Even triple octave combinations are found (e.g., bassoon – oboe – flute in the third movement of the 9th Symphony by L. v. Beethoven, measure 65 sqq.). A sound with very equally strong partials, in contrast, can sound very hard, especially if no formants are present. Also the typical snarling sound of the reed pipes of an organ results from a spectrum very rich in overtones, with no dominant individual partials in the upper register, in contrast to flue pipes.

2.2.5 Frequency Width of Partials

The representation of partials, using lines, is a schematic simplification resting on the assumption that the tones do not change in their frequency, so that an exact location on the frequency scale can be assigned. In most cases minute amplitude variations are present, which, however, are not analyzed by the ear in their fine structure. Since the spectra represent a median value for a (quasistationary) steady state, these frequency variations can lead to a broadening of the spectral lines, i.e., the partials each fill a narrow frequency band. This effect becomes particularly apparent in a vibrato, when the sound gains "fullness" for a steady sound pressure level; time averaging due to the room also plays a role in this.

A similar broadening of the spectral lines occurs in cases where several sound sources radiate with nearly the same frequencies. Since this process is especially pronounced when in a group the intonation of each singer or player is slightly different, it is also called the chorus effect. This effect, among others, is also responsible for the difference in tone between a string orchestra and a string quartet. In a good choir the half width of a tone, i.e., the frequency width within which the sound pressure level drops by no more than 3 dB, lies in magnitude from 1/5 to 1/3 of a half step. In contrast, one characteristic of the Don-Cossaks is that the half width of their low notes is a full half tone. For instrumental ensembles the intonation width is generally relatively small and does not exceed a value of 1/5 of a half step (Lottermoser and Meyer, 1960).

2.2.6 Noise Contributions

Because of the nature of tone production, most musical instruments exhibit some noise background in addition to the spectrum of partials. This represents an important part of the total sound picture (Winckel, 1969; Meyer, 1964a). Thus, for example with string instruments, the bowing noise will always be somewhat perceptible, and for wood wind instruments the blowing noise can never be totally suppressed. An attempt will always be made to minimize such extraneous noise, however, such a minimal contribution is actually essential for the sound to retain its natural character. As experiments with electronic tone synthesis have shown, instruments cannot be imitated satisfactorily with spectra of partials alone.

These background noises come about through the fact that by irregularities in excitation all resonances of the instrument are slightly stimulated each time. Consequently the admixture of noise has a unique character for each instrument type. A comparison between the violin and the flute in Fig. 2.6 is intended to visualize this: While for the violin below the fundamental (which lies at around 1,400 Hz) and between the partials, noise components of significant frequency width are present with only small dips, in the flute, at a very low noise level, only a few small noise peaks are present at the frequencies of a fundamental, fingered as F_6, and overblown at the twelfth.

The superposition of the noise components over the harmonic spectrum, as well as the broadening of the partials, naturally presents a departure from the mathematically exact vibration form. Esthetically, however, these phenomena are extremely important, since they not only enliven the tone picture, but also prevent the appearance of fatigue symptoms of the ear. An analogy of these somewhat less than sharp tonal contours to painting comes to mind, where a sharpness scale, or resolution grades can be followed from an extremely photographic-like precision of a Canaletto, to the pointillist manner of impressionism. Similar cross-connections to other art forms can also be found.

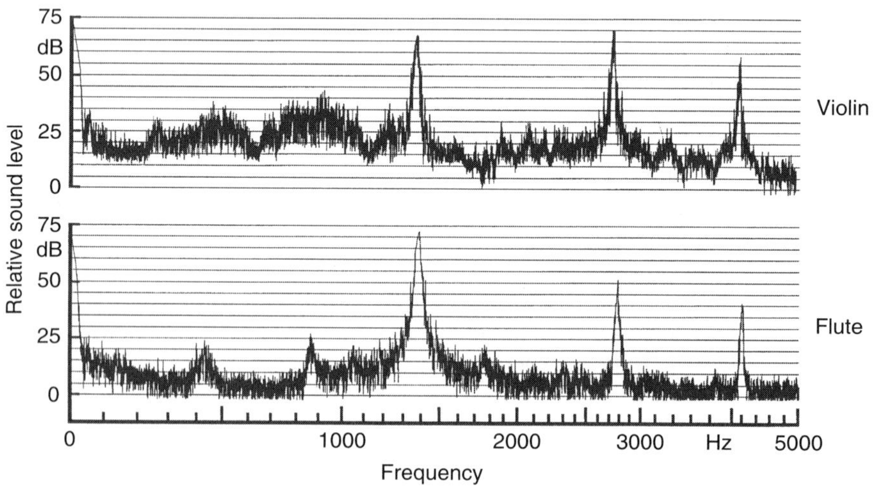

Fig. 2.6 Tone spectra with different frequency characteristics of the noise background. (Note played: F_6)

2.2.7 Dynamics and the Sound Spectrum

Playing volume is one of the most important factors influencing the sound picture. In this context the musical concept of dynamics refers principally to loudness, however, other factors such as relative motion or rest within a sound, or the nature of the attack can have a effect on the listener's impression of dynamics (Hadamowsky, 1958). The expressional value of dynamics thus by no means rests exclusively on differing sound pressure levels as they could be achieved, for example, by adjusting amplifier volume: Individual dynamic steps experience a particular characterization by the fact that for almost all musical instruments a change in playing volume varies not only the sound intensity in the stationary part of the sound, but also brings about clear changes in the spectral composition (Reinecke, 1953). By the way, it is this effect which has to be thanked for the fact that a *forte* is recognized as such for low volume sound reproduction by a loudspeaker even though it only has the intensity of a natural *piano*.

This phenomenon is illustrated in Fig. 2.7 for the upper register of a horn in the form of three spectra corresponding to the dynamic steps *fortissimo, mezzoforte* and *pianissimo* (Meyer, 1967b). The relative dB scale is set so that the strongest *ff* partial with 75 dB just reaches the upper limit of the registering level region. The very large overtone content of the upper spectrum is particularly obvious, here the partial contributions around 10,000 Hz still show amplitudes of about 45 dB, while in the *mf* spectrum only six partials and in the *pp* spectrum only four partials exceed the background noise. This already characterizes the tonal difference between the penetrating *ff* and the round and soft *mf* and *pp*. To that is added, that in the *ff*

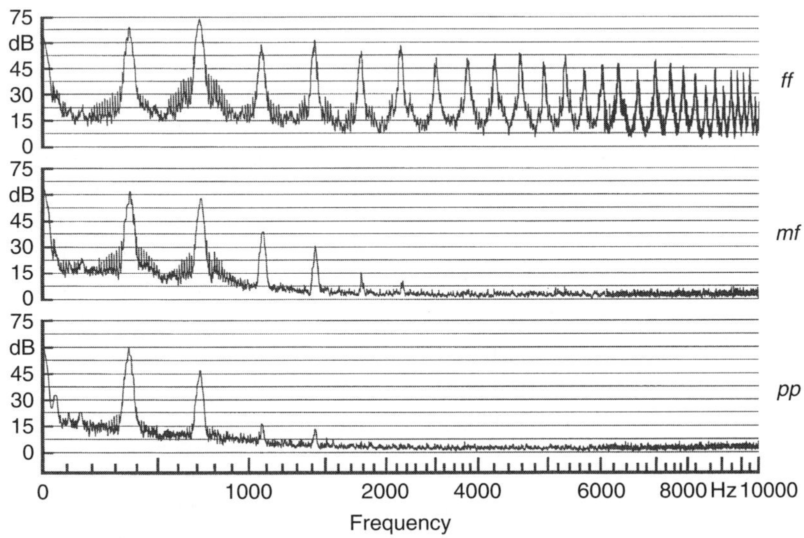

Fig. 2.7 Sound spectra of a French Horn played at different dynamic levels (F_4)

spectrum the maximal intensity has been shifted to the second harmonic (in contrast to the fundamental for *mf* and *pp*), so that the formant color becomes brighter. The importance of this spectral change for dynamics is recognized especially clearly through the fact that amplitude differences between *ff* and *pp* amount to 50 dB in the region of higher frequencies, while the difference between the corresponding strongest partials of the upper and lower spectrum lies at only around 15 dB. This results not only in a stepping of loudness (which can naturally still be subdivided), but also in a modification of the tone color.

When describing dynamic differences between instrument sounds, it is desirable, therefore, to take into consideration the level change of the 3,000 Hz component as well as possibly the change in slope of the level drop above 3,000 Hz, and also the change of the overall level, which is caused almost exclusively by the change in the strongest partials. All these data should be in relation to the sound power spectrum. A quantity, characteristic for each instrument group, is that level change of the 3,000 Hz component, which occurs when the strongest partials experience a change of just 1 dB. This quantity will be called the "dynamic tone color factor."

However, even when taking connections between loudness and tone color into account, individual dynamic steps such as *forte, mezzoforte, piano* etc. cannot be specified by a firm universally valid numerical value. The meaning of these performance specifications is as dependent on musical context as on the room environment. Finally the instruments themselves play an essential role, each instrument group has its own dynamic range between largest possible and minimum (still sounding) loudness. Furthermore, for many instruments this dynamic range is not the same over the entire tone range, where in addition even the quality of the individual instrument participates as a further factor.

2.2.8 Dynamic Range and Sound Power

Sound power level data are best suited to represent the dynamic performance range of an instrument in summary form, since they represent only the sound radiation from a particular instrument independently of the room, and can thus be recalculated for each room situation. Experience during measurements of the dynamic range accessible for performance have shown that there are noticeable differences in the accessible level limit, depending on whether the performers need to sound a quick tone sequence or can concentrate on each individual note. For a summary representation, both cases thus need to be considered.

In the sections on the dynamics of instruments (Chap. 3) the limiting values are given for the softest possible pp and the loudest possible ff, while playing fast scales covering two octaves, with the condition for the player that indeed each individual note is sounded. In addition, extreme values are given for individually played notes. Since the dynamic performance range of instruments often depends on the pitch – the low register of an instrument is often softer for pp, as well as for ff, than the high register – it is often the case that the softest playable tone is different than the loudest possible. These extreme values are especially of interest for microphone recordings. Under usual performance conditions it is, however, more important to be aware of a realizable dynamic range. Appropriate values should therefore be given as averages for fast and slow tone sequences.

Most of the sound power values presented in the following chapters are taken from sound power measurements performed using the reference sound source procedure in a reverberant chamber (Meyer and Angster, 1981; Meyer, 1990). They will be augmented by the earlier results of Burghauser and Spelda (1971), which were obtained in a medium size studio, as well as those of Clarke and Luce (1965), which were measured in a reflection poor room. All results by these authors were recalculated as sound power levels with consideration of all relevant boundary conditions.

For subsequent room acoustical considerations it is also necessary to know a characteristic value for the sound power of each instrument. For this purpose an "average *forte* sound pressure level" will be used, which is established on the basis scale, and single tone measurements in such a way that the playable dynamic range between *pp* and *ff* is divided by the steps of *p*, *mf* and *f* into equal segments. Starting from this *forte*- level the sound power of entire ensembles, or the sound pressure level to be expected in a room for arbitrary dynamics can be calculated (Meyer, 1990).

2.3 Time Structures

2.3.1 Deviations from a Steady Vibration Process

Decomposition of a sound into a sequence of harmonic partials is, strictly speaking, only possible for a stationary state. In contrast, a sudden change means an unsteadiness, which can no longer be described by a line spectrum. Such unique processes thus possess a spectrum which is not formed by discrete individual frequencies, but by a frequency continuum. This presence of arbitrarily closely spaced frequencies gives the ear a noise-like impression, which, during the onset, can assume the character of a crack.

Figure 2.8 shows the vibration processes and the spectra, next to each other, for a steady vibrating sinusoidal tone, and for the onset of a sinusoidal tone. The steady state can be represented by a line, while the onset of the tone exhibits a spectral broadening in the region of the tone, and beyond that, a decreasing amplitude with increasing frequency deviation. Also the end of a tone can be understood as such a process. The sound proceeds as though, in addition to the existing tone, a tone equal in amplitude and opposite in phase were suddenly turned on. The tones cancel each other, and the second crack remains. In practice, switching processes without some kind of a transition phase, occur only for some older types of electronic organs, in which the keys switch the already running tone generators to the amplifier and speaker. These instruments thus begin each tone with a noticeable crack, since the

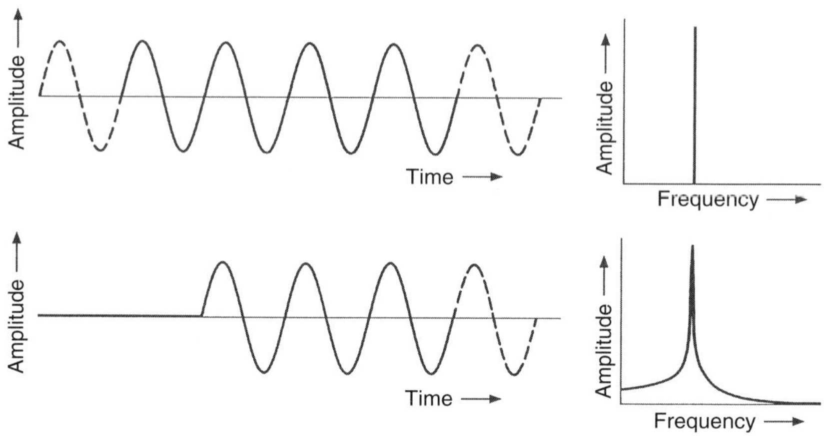

Fig. 2.8 Wave form and Spectrum for a steady tone and a tone entrance

eigenresonances of the speaker are excited. To some extent, similar cracks can also be heard at the end of the tone, which is perceived by the ear as an additional accent (Lottermoser and Meyer, 1962a).

Generally, such a switching process does not occur as suddenly as represented in Fig. 2.8, but is extended over a certain time period. Consequently the amplitudes of the noise components, which are present in addition to the actual vibration frequencies, become increasingly less as the duration of the switching process increases, and as the break in the vibration process at the onset is smoothed out. Naturally the final vibration amplitude in such a case is reached only after the full course of the onset process which is stretched in time. Since for such a soft transition from rest to the tone, the unsteadiness in the vibration process is reduced, the noise-like impression in the ear of the transient also decreases.

2.3.2 The Starting Transient

Musical instruments are complicated physical structures, for which in most cases, several coupled resonance systems influence the vibration process between excitation by the player and the radiated sound. A resonance system, however, cannot react suddenly to an excitation, rather, vibrations must slowly build up to their final strength. This is connected with the fact that a portion of the energy provided externally for the resonance system is in turn radiated, and a portion is absorbed by the instrument. As long as more energy is provided than used, the amplitude rises. Only when an equilibrium is reached between the input energy on the one hand and the absorbed and radiated energy on the other hand will the oscillation reach its ultimate strength. This starting transient is easily observed by placing a vibrating tuning fork on the corpus of a violin: the loudest volume is reached slowly after a gentle transition.

The more an instrument absorbs and the stronger it is damped by radiation, the sooner it will reach equilibrium, i.e., the shorter will be the duration of the starting transient. Thus damping plays an important role during the attack for an instrument. For most resonance systems, this damping is in large measure frequency dependent. This means on the one hand, that characteristic vibration processes proceed at different rates within the tonal range of the instrument, and on the other hand that individual partials grow at different rates in the process of developing the tone. The sequence of starting transients of the overtones belongs to those special characteristics which form the tone picture of the individual instrument groups. In all this, high tonal contributions with fast initial transients have the effect of suggesting a precise attack, similar to articulation in speech, while starting transients which are too slow, are perceived as a poor attack for the instrument.

Inasmuch as the amplitude during the starting transient rises very quickly initially, but then reaches its final value only with relatively small intensity increase, and since furthermore, this final value is difficult to determine precisely due to the appearance of fluctuations, the literature usually specifies the time by which the sound pressure level has reached a value 3 dB below the stationary state as the duration of the starting transient (Luce and Clark, 1965; Melka, 1970). This definition is generally applied to the total level without frequency weighting of individual sound portions, in practice this emphasizes the strongest partials.

Naturally, the duration of the starting transients can be influenced by the performer within certain limits. The harder a tone is attacked, the richer in overtones will be the unsteady contribution during the time process of the excitation; consequently the high tone portions will develop faster and more precisely. Nevertheless, natural limits are in place through the damping of the individual resonances. The more softly a tone is attacked, the slower the initial transient process of the instrument will be developed, and consequently the formation of the higher components will be weakened.

This difference is shown clearly in the juxtaposition of two starting transients for a G_4 on a flute in Fig. 2.9. As seen, the onset of the strongly attacked tone is marked by an attack noise of 40 ms duration. The fundamental and the third partial develop very quickly at the same time; the fundamental already reaches its full strength after 70 ms. The octave partial and the 3rd partial reach their final value after 90 and 100 ms respectively, the higher partials need 100–120 ms. In comparison to that, for the soft attack, the duration of the starting transient for the fundamental (120 ms) is not that much longer; the soft character primarily comes from the delay of the higher partials and the freedom from noise. Otherwise, the example also shows that the speed of transient development is a typical characteristic of performance style, which can be utilized as a means of expression. In contrast, possibly only the shortest possible duration of the starting transient of a staccato tone can be utilized to characterize an instrument.

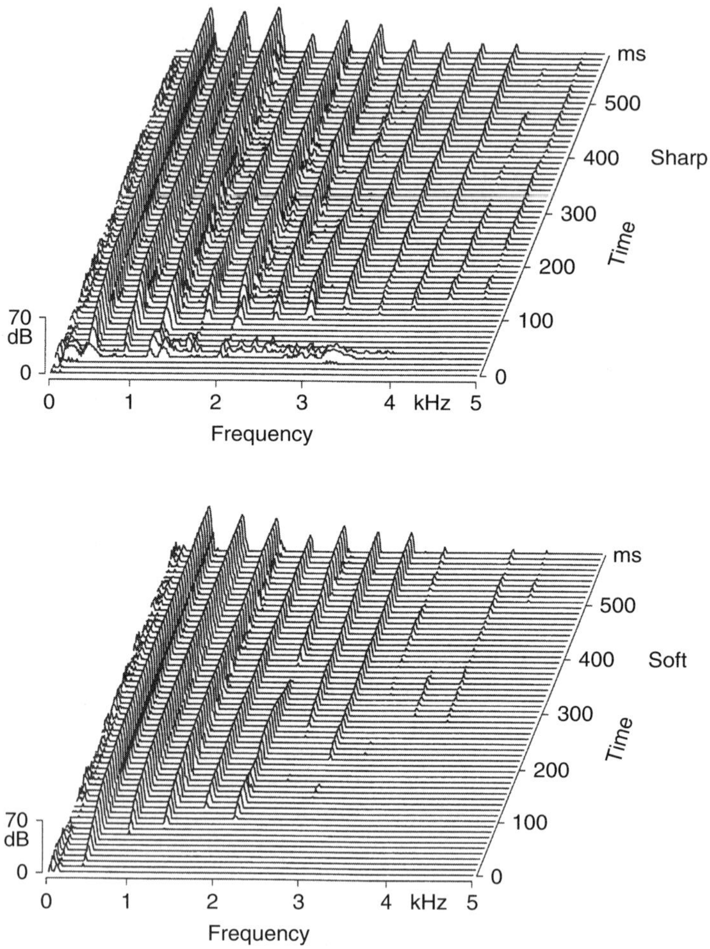

Fig. 2.9 Time development of a tone spectrum for different attacks (Flute, G_4)

2.3.3 Inharmonic Components

Since the sudden stimulation of an oscillating system represents an excitation of a frequency continuum because of unsteadiness in time, the starting transient includes not only the harmonic tone contributions, which will later be present in the stationary state, but the instrument will be caused to vibrate at all resonance frequencies. To the extent in which these frequencies do not coincide with later partials these oscillations form inharmonic tone contributions, which, however, essentially decay again during the initial transient (Lottermoser, 1958; Lottermoser and Meyer, 1966). In this way characteristic admixtures come about in the onset of the tone, which present properties typical of each instrument group by reason of the position and width of the resonances.

When resonances are broad and strongly damped – as for example in wide frequency ranges for string instruments – then corresponding broad band noise components arise during the initial transient, which cause an auditory impression of a tone color comparable to a whisper, however, without a definite sensation of pitch. Sharp resonances, which are not utilized for the formation of harmonics – as they are present, for example, in the overblown tones of woodwind instruments, are recognized by the ear at their pitch level, in spite of the noise character.

In the special case when the exciting frequency lies close to a resonance, beats are generated during the starting transient between the excitation frequency and the resonance frequency, which only disappear with the decay of the resonance frequency (Meyer, 1985). Depending on the separation of these two frequencies, the beats can enhance the tone picture or otherwise influence it through roughness. Such phenomena can occur particularly for the high partials of wind instruments when the upper resonances are not situated strictly harmonically; they are occasionally also found in string instruments when the corpus resonances are too sharp.

Though noise contributions in the initial transient should not be too obvious for tone esthetic reasons, they, nevertheless, contribute in an essential way to the precision of the tone entrance, as seen clearly in Fig. 2.9. In this respect they are comparable to the consonants in speech, which likewise form a noise-like introduction to the vowel vibrations, though in most cases with greater intensity. This articulation, emphasized by the nature of the tone entrance, can be further enhanced when tones in the frequency neighborhood are excited shortly prior to the actual tone. Some anticipatory grace notes serve in this way to accentuate the following main note, and are, so to speak, to be considered as "composed initial transients." In score example 2, there are even harmonically placed double grace notes which give the accent to the heavy beat.

While it is possible to use the ability to obtain an audibly precise, recognizable entrance, as a further criterion to evaluate the ability to address the instrument, the strength of the articulation as a means of artistic expression is subject to the playing technique of the musician. Yet it must not be overlooked, that pitch recognition by the ear is made more difficult through the noise components associated with very short notes, even if the rhythmic precision increases. The instrument groups with low noise level, but short initial transients are thus best suited for a sharp staccato.

Score example 2 Joh. Strauß, Polka "Leichtes Blut" 'Opus 319, 1st Violin, measure 5 sqq.

2.3.4 Decay of Resonating Systems

For no mechanical instrument does the termination of excitation occur as abruptly, as it begins for a staccato tone. Consequently, in contrast to initial transients, no new tone or noise components are excited. However, it is important for the decay process, that sound energy is still stored in the resonance systems of the instrument, which is radiated until it is used up. Of concern is a process opposite to the building of oscillations in the initial transient. Depending on their damping, each resonance decays more slowly or more rapidly. A tuning fork, for example, has a very sharp resonance, after the attack, its vibrations decay only slowly. In comparison, the resonances of wind instruments are strongly damped, so that they use the rest energy in a very short time, and practically no decay is audible.

Naturally, the decay can be influenced within certain limits by playing techniques. As was already recognized from Fig. 2.2, the nature of the decay for bowed string instruments depends on the bowing force, and is longest when the bow is lifted. It can also be shortened when the string is retuned by a change in fingering. Such a string frequency change can be compared to the opening or closing of keys or valves in wind instruments. These phenomena have particular significance for connecting two tones: by overlapping the decay of the first tone into the onset transient of the second, the melodic line gains continuity, while an all too sudden cut often leads to unwanted gaps.

Naturally, decay processes gain preferred interest in instruments for which, by reason of a very short excitation, no steady state can be developed. Percussion instruments fall into this category, as do struck or plucked string instruments. For these, the duration of the decay is particularly long, and it determines the tone picture more strongly by far than does the duration of decay processes in other instruments.

Since the amplitude decreases exponentially for damped resonances, a logarithmic representation of sound pressure level shows a nearly linear time dependence of decay, as is also shown in the model in Fig. 2.1. However, the slope of this drop is by no means equal for all partials, since the damping of the resonances is frequency dependent in most cases, where generally the high tone contributions drop faster than the low ones. Furthermore, tones, whose frequencies fall directly on a resonance, decay differently than tones which lie between two resonances.

2.3.5 Decay Time and Reverberation Time

The audible duration of the decay, on the one hand, naturally depends on the loudness of the tone, and on the other hand, on the ambient noise level in the room, since, as far as the ear is concerned, the end of the decay is determined by the point at which the tone drops below the ambient noise level. In analogy with room acoustics, the time span during which the decay is audible is referred to as the decay time. This measure is quite suitable for the description of the tonal

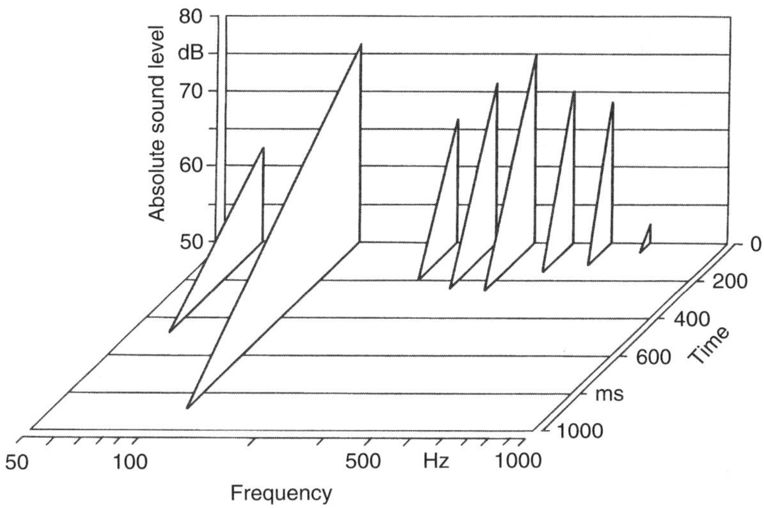

Fig. 2.10 Decay of a double bass tone (B_1, *pizz.* after Spelda, 1968)

impression, nevertheless, because of the dependencies mentioned above, it is not an objective measure of the slope of the level decrease.

The decay time for a pizzicato tone of a contrabass is reproduced in Fig. 2.10 as a typical example of a decay time. For this representation a noise level of 50 dB was assumed as a lower limit for the level drop, this level forms the base plane of the model. Time is plotted toward the front, the sound pressure level goes up, and the frequencies of the overtones increase toward the right. It is noted from this picture that the low tone contributions exhibit the shallowest decay curves, while the high components decay rather rapidly. The 2nd partial can be followed for the longest time, particularly since it also has the highest starting value. The overtone content accordingly changes in the decay in such a way that the tone color becomes increasingly darker and softer. As an alternative to this spectral representation, the decay time can also be given as a numerical value, which only gives the duration of audibility without frequency specification. The pizzicato tone of the pictured example would have a decay time of about 900 ms.

In order to obtain an objective measure for the slope of the level decrease, and thus for the tonal characteristic of the relevant instrument, often the time for which the level drops by 60 dB relative to its original value is determined, and this quantity is designated as the reverberation time -again in analogy with room acoustics- (Meyer and Lottermoser, 1961; Plenge and Schwarz, 1967). Since the level decrease is linear, this quantity can naturally also be calculated when the level record only covers a smaller region. For example, for the 2nd partial in Fig. 2.10, from a 900 ms time lapse and an associated level drop from 77 to 50 dB, i.e., of 27 dB, a reverberation time of 2 s is calculated (for a 60 dB drop).

Complicated resonance systems at times do not exhibit a linear level decay, but rather follow an initial steep portion with a later more shallow decay. This is most frequently explained by the fact that a strongly damped vibrating system radiates much energy, while at the same time a different section of the instrument, which absorbs less energy, radiates that energy only slowly. This causes a break in the level curve, and the decay process must be described by specifying two values for the reverberation time (for short and for long tones).

2.3.6 Fluctuations in the Quasistationary Part

Changes of the vibration excitation during a tone also need to be considered as unsteadiness in the vibration process, even if they are significantly less noticeable than during the starting transient and the decay. Persistent statistical fluctuations of the excitation are however, finally the reason for the noise admixture to the sound of the instrument, as discussed earlier. The measure of these variations transmits an impression of the stability or – in the negative case – the uncertainty of the tone. On occasion, however, such instabilities are deliberately utilized as tonal effects (*flautando*).

A particularly important structuring of the tone, introduced as an artistic means of expression is represented by the vibrato (Winckel, 1960; Gärtner, 1974). While it is used by singers, string-, and wind-players to very different degrees, yet, a good vibrato has something in common for all voices and instruments: The fluctuation frequency of the vibrato almost always lies in the region of 5–8 Hz, which is connected – as already mentioned – by the fact that the ear still senses a definite pitch for fluctuations at that frequency. In contrast, the width of the vibrato characteristic for singers, string- and wind-players is different. It can move by ±5 cents near the limit of the audible, or for singers it can certainly exceed the range of ±100 cents. The effect of the vibrato on the tone quality is also different: while the vibrato in all cases depends principally on a time modulation of the exciting frequency, still, depending on the resonance structure of the instrument, more or less pronounced modulations of the individual partial amplitudes result. If these occur in phase in a larger frequency region, a time dependent tone color modulation results. Occasionally this shapes the tonal impression more strongly than the original frequency modulation. The latter, for example, is the case for brass instruments and the flute (Meyer, 1991).

Chapter 3
Tonal Characteristics of Musical Instruments

3.1 Brass Instruments

3.1.1 The French Horn

3.1.1.1 Sound Spectra

It is typical for brass instruments to exhibit spectra which can be divided into two groups. In the upper register the fundamental is strongest, while for the lower positions a formant-like maximum is present. As noted in the chromatic representation in Fig. 2.4, the fundamental dominates from C_4 on upward in French Horns, while higher partials decrease in amplitude rather steadily. Below C_4 the maximum is initially relocated to the octave partial and then maintains its frequency position, so that in the lowest registers the 4th and 5th partials receive most of the energy. As a result, the main formant, typical for the French Horn, develops, which is located at approximately 340 Hz (Meyer, 1967b). It falls into the region of the vowel color "u (oo)" which is responsible for the round and sonorous sound of the Horn.

Below this maximum, the amplitude for the low registers drops rapidly with a slope of 12 dB/Octave. The lowest frequency possible corresponds to the note B_1 with approximately 62 Hz. The fundamental is roughly 25 dB weaker here than the strongest partial. This shows that the low frequencies only play a subordinate role in determining the tone quality of the Horn. Above the maximum, the amplitudes also decrease, however, there are several additional formants present which influence the tonal picture. Their frequency locations are represented schematically in Fig. 3.1. Below the Vowels are the results for the commonly used German double horn. The already mentioned main formant at 340 Hz is followed by the first ancillary formant at 750 Hz, also in the range of the vowel color "a (ah)," and further formants near 1,225 Hz, 2,000 Hz, and also (using the F-Horn) still near 3,500 Hz. This series of ancillary formants brightens the overall tone, so that the tonal character is not as dark as for a sung or spoken "u (oo)." In this, the higher frequency partials gain in importance with increasing overall volume, while only the lower formants contribute while playing softly.

J. Meyer, *Acoustics and the Performance of Music*.
DOI: 10.1007/978-0-387-09517-2_3,

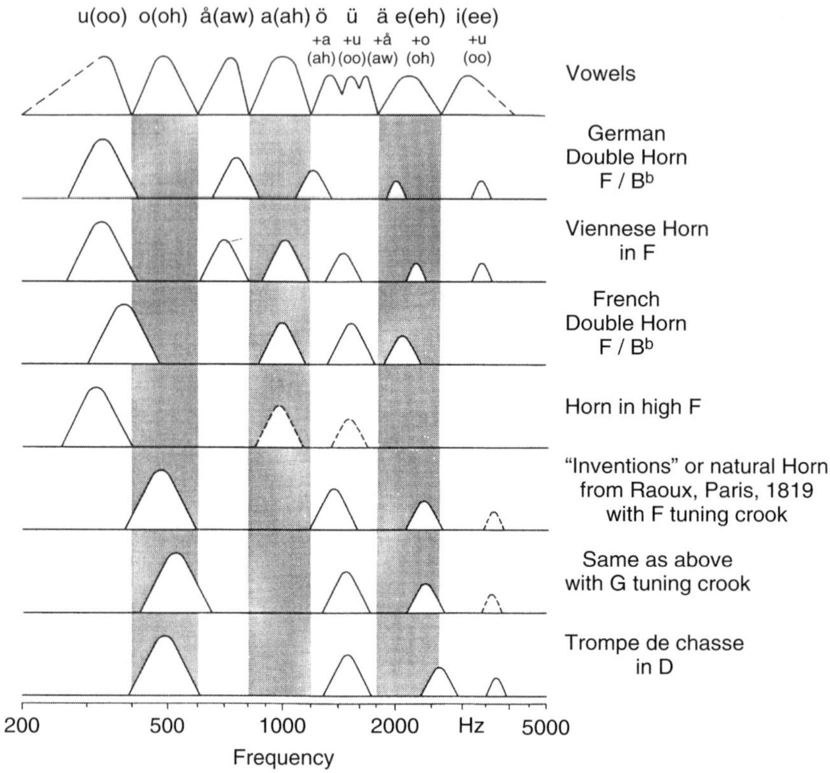

Fig. 3.1 Schematic representation of formant position for various horns

The intensity of the lower partials, and thus the fundamental tonal characteristics, however, also depend on performance techniques: When the mouthpiece is pressed relatively firmly against the lips, their vibrating portion is more clearly delineated, so that strong lip vibrations are possible without time dependent irregularities. Furthermore, instrument resonances are not damped as much as is the case for a low pressure contact. Consequently, for firm pressure, a fuller sound is produced, which carries better. In contrast the advantage of a low pressure attack lies in greater ease of playing in the upper register, which, however, is purchased with a thinner timbre lacking somewhat in substance. This also constitutes the typical difference between the European and American "schools" of brass performance technique.

The frequency range of the spectra extends upward to 1,500 Hz for the lower register at a dynamic level of mf. It increases to about 5,000 Hz for the highest notes. The overtone content of the F-Horn is somewhat larger than for the B^b-Horn, thus the F-Horn has a somewhat more pronounced tone color, particularly in the mid range. The valves also influence the higher frequency partials slightly: For fingerings with several valves the overtone content decreases so that the sound loses brilliance and becomes dull. As a result the player receives opportunities for tonal variations in addition to intonation corrections.

Noise contributions are especially weak for brass instruments, so that their influence on the sound picture of the horn is negligible. Only in the lowest octave of the tonal range could minor noise components, up to approximately 3,000 Hz, be noticed above the partials. This can be described as a hissing with an "i(ee)" like tone color in relation to the corresponding formant.

3.1.1.2 Dynamics

The strong dependence of overtone content on dynamics in the upper register of the horn has already been mentioned in the previous chapter (see Fig. 2.7) as an example for the note f_4. As a generalization, the power spectra can be described in terms of a level difference of the order of 10 dB for *ff* and 50 dB for *pp* when comparing the strongest partial with the 3,000 Hz component. For high frequencies a level drop of only 5 dB/octave for *ff*, and 15 dB/octave for *pp* is associated with this. Furthermore, the influence on the timbre connected with this effect is enhanced by the shift of the main formant toward higher frequencies with increasing loudness, thus brightening the main tone color from "u(oo)" for *pp* and *mf* to "o(oh)" or possibly "å(aw)."

In the low register as well, a similar influence of loudness on tone color is noted. Thus in the second octave, the power spectra show a level difference of 20 dB for *ff* and 50 dB for *pp* between the main formants and 3,000 Hz, which is associated with a level drop at high frequencies of 11 dB for *ff* and 15 dB for *pp*. However, by reason of the narrower partial positioning, there is still a group of approximately 8–10 harmonics present. The amplitude differences between the respectively strongest partial for *ff* and *pp* lie at roughly 20 dB. They are thus slightly larger than at higher registers. In addition, a shift of the dynamic dependent amplitude maximum is also noted in the low register. For increasing loudness the vowel color brightens, though this effect is not as pronounced as in higher registers. The dynamic and modulation possibility of the natural horn sound thus rests both on a change in tone color in the region of the lower partials and on the possibility of large amplitude changes for the higher frequency sound contributions.

For rapidly performed scales, the horn reaches power levels of 107 dB for *ff*, while for *pp* 86 dB are produced. For individual notes, the lowest values of 65 dB are reached. These, as well as the upper limits of playable dynamics are shifted over the tonal range of the instrument toward greater loudness, where, for a high range *ff*, 117 dB certainly is possible. On the whole, a practically realizable dynamic range of 35–40 dB can be considered as typical for the horn. However, for the highest notes a genuine *pp* can hardly be expected. (Meyer, 1990).

A sound power level of 102 dB can be given as a characteristic value for the *forte*-sound. The influence of dynamics on tone color expresses itself as a level change at 3,000 Hz. This increase ranges from 1.5 dB in the low register, to 3 dB in the higher regions – accompanied by a simultaneous change in the level of the strongest partials by 1 dB.

3.1.1.3 Time Structure

The initial transient for a tongued tone of a horn is characterized by a short pre-cursor impulse, well known from other brass instruments (see Fig. 3.5). The duration of this impulse, which contains primarily harmonic partials below 1,000 Hz, lies in the order of 20 ms. It begins, depending on sharpness of attack, between 10 and 30 ms after beginning to excite vibrations (Melka, 1970). Several such impulses can follow, which gives the onset of the tone a character of a rolled "r" which naturally is esthetically undesirable. As the attack is softened, the pre-cursor impulse diminishes in importance in the tonal picture. An excessively strong pre-cursor impulse with slow development of the fundamental, on the other hand, generates the notorious "blare."

The duration of the starting transient for tongued tones is shortest in the middle and high register, above F_3 it amounts to approximately 30–40 ms, in the low registers it rises to values between 40 and 80 ms. (Melka, 1970). These values are confirmed for tongued notes in the sonograms of Fig. 7.26, which will receive closer attention in the context of room acoustical effects. For soft attacks, the initial transient can last longer than 1/10 s, which does not include consideration of dynamic development for long notes.

Inasmuch as the air column in the instrument is capable of storing only small amounts of energy, the decay time of brass instrument notes is relatively short. Even the energy stored in the wall vibrations does not increase the decay time significantly. The horn typically has decay times around 150 ms.

When going from one note to another it is important to note whether the transition is accomplished by the lips or a valve. In the former case a continuous frequency transition is observed which serves to smooth out the connection, while the activation of the valve leads to a sudden frequency change in the resonator and a break in the vibrations. This makes the transition harder or more pronounced (see Fig. 3.3), so that either technique has tonal advantages depending on the musical context. In fact lip transitions are naturally only possible for certain tone sequences, furthermore, valve connections provide the player with an increased sense of security for the attack. A typical example for this is the triad motif in the large E major aria of Leonore in the first act of the Opera "Fidelio." This passage was originally written for natural horns, nowadays it is, however, mostly performed with valve transitions.

The impact of using a *vibrato* in horn playing is largely determined by stylistic considerations. When a vibrato is played, it appears mostly through amplitude modulation of higher partials of like phase, that is, through a pulsating tone color modulation, as shown in Fig. 3.6 for the lip *vibrato* of a trombone. The depth of modulation can reach approximately 10 dB for the higher frequencies. It thus contributes to the ability to notice the horn in an ensemble.

3.1.1.4 Special Playing Techniques

Normally the right hand of the player lies only loosely in the bell of the horn, causing a certain damping of the higher partials. The possibility of using the hand to close the bore almost entirely has been developed as a special effect to change the

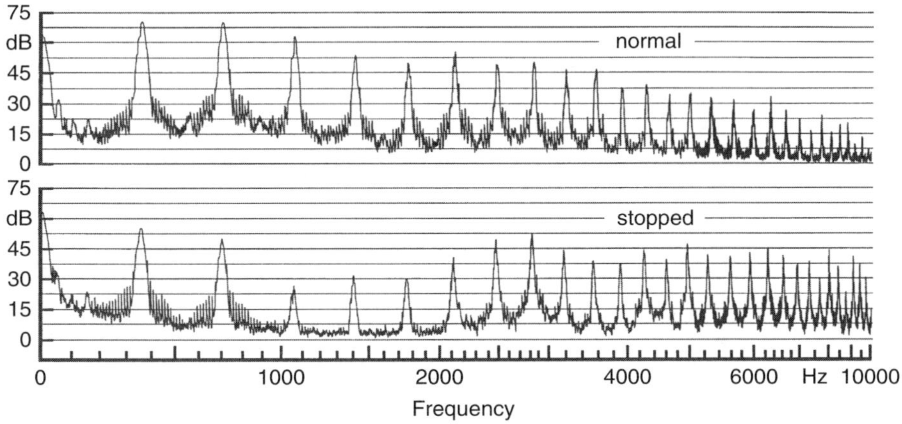

Fig. 3.2 Tonal Spectra of a Horn for the note F_4

tonal character in a fundamental way. This is particularly pronounced in playing *forte*; thus, a "stopped *sforzato*" is often indicated, the tonal effect of which is "metallic brittle and rough" (Kunitz, 1961). A typical example for this is the final chord in the Beckmesser-motif of the "Meistersinger" (score example 3).

As shown in the analyses of Fig. 3.2, the stopped sound lacks several essential components, while others are strongly developed. The gap from the 3rd to the 5th partials is noticeable, which covers the region of the a(ah)-formant, causing the pressed tone, lacking in strength. On the other hand, the metallic timbre is strongly emphasized by the maximum at 3,000 Hz and the strong partials above 10,000 Hz.

A comparison of stopped tones at different pitches shows that these frequency locations for the typical maxima and minima are always maintained. As a result, the *ff* – sound of a stopped horn becomes even less substantive in the upper registers. However, for a *mezzoforte,* stopped tones keep their strongest sound components in the usual location. Yet even here, a weakening in the region of the a-formant and an increase around 3,000 Hz leads to a tonal change in the direction of a metallic timbre.

A different tonal effect is caused by playing the horn with upward pointing bell and without a damping hand. The best known examples for this are found in the symphonies of G. Mahler. However, intonation suffers slightly with this playing technique, furthermore, the tone becomes hard and "coarse" as a result of the stronger contributions of the higher frequency components in the stationary and transient part of the sound. For these reason the technique is avoided by many players and conductors (Kunitz, 1961).

3.1.1.5 Horns of Special Design

In the Vienna Philharmonic, even today, pure F-Horns are played. They are distinguished from the German double horn by their narrow bore, the so-called

Score example 3 R. Wagner, Die Meistersinger von Nürnberg, motif of "Beckmesser" (3rd Act, 4th scene)

"Vienna valves" (Stechbüchsen-Ventile) and a particularly shallow cone in the mouth piece. This design does maintain the position of the main formant near 340 Hz, that is, in the range of the vowel color "u(oo)," as seen in Fig. 3.1. However, the number and positions of secondary formants is noteworthy. This is the only group of horns which shows five secondary partials, where the first two are located in the vowel range "a(ah)" and "å(aw)." This double a(ah)-formant give a particular strength to the sound and effects a song-like character; in its spectral distribution it is also reminiscent of the violins of Guarneri del Gesù (Lottermoser and Meyer, 1962b), famous for their rich and powerful sound. The two following secondary formants are positioned slightly higher than in the German Horn, however, the high components in the Vienna school are damped slightly more using the hand, so that the sound in spite of its fullness displays a certain softness.

For valved note transitions, noticeable differences between the Vienna Horn and the German double horn also occur, as shown in Fig. 3.3. The rotary valve of the double horn provides several paths for the air stream, so that short duration turbulences occur. These are noted as crack-like transition noise, as recognized in the three-dimensional spectrum (Windholm and Sonneck, 1988). In contrast, the note transition with a pump valve progresses more smoothly; i.e., the Vienna Horn has the advantage of a softer transition and a better legato, where the advantage of the German Horn can be seen in the better articulated transition and the better staccato.

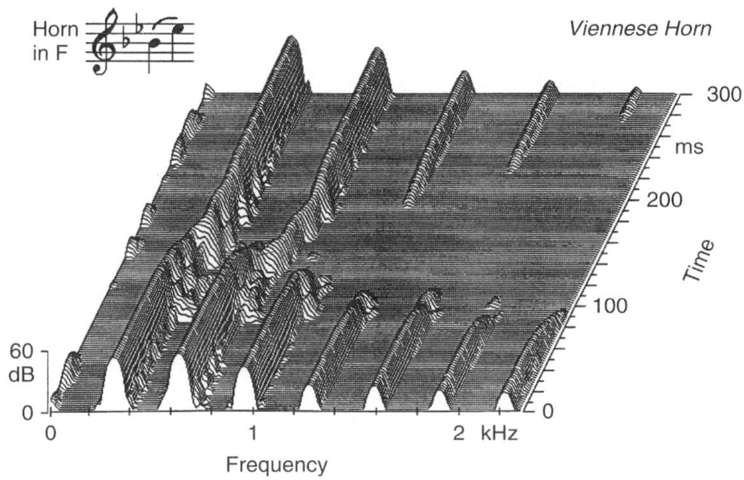

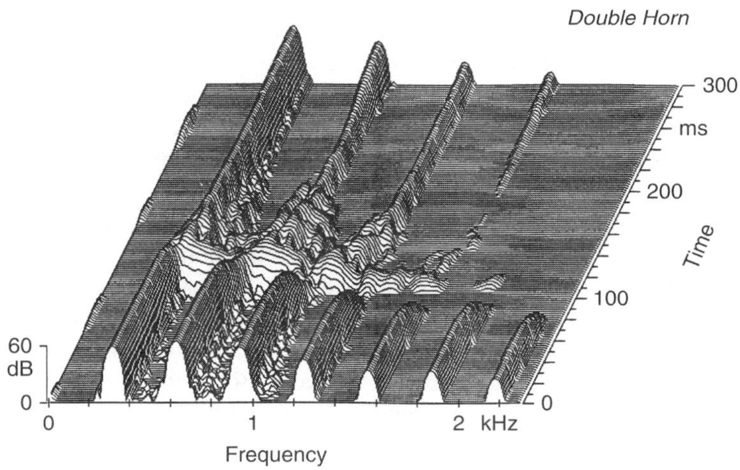

Fig. 3.3 Time development of a tone spectrum for a connected tone transition, represented for two horns of different construction (after Widholm and Sonneck, 1988)

Out of the tradition of horn virtuosity common in France, a double horn was developed, in use there today, which is distinguished by its slender tone and ease of attack. High intensities for upper frequencies create a relatively bright tonal character, somewhat reminiscent of a hunting horn. A change in the frequency location of the formants in comparison to other horns has a similar effect. The principal maximum is moved approximately to the boundary between the "u(oo)" and "o(oh)" regions, causing the basic vowel-like characteristic to become brighter. Two strong secondary formants are located in the region of the vowel "a(ah)," and also in the region of nasal components. It is precisely these strong tone contributions around 1,500 Hz which form a particular characteristic of the typically French timbre. They also appear in bassoons in similar fashion. A further secondary formant in the frequency region of the vowel "e(eh)" brightens the tone of the French horn additionally.

Horns with high tuning are occasionally used, because they allow certainty of attack for high passages with greater reliability. However, this advantage is purchased with a reduction in tonal quality. For example, in a horn in high F, partials in the mid-register only reach up to 2,500 Hz, while in a normal F – horn they are present up to 4,000 Hz for the same notes at the same loudness. Connected with this is the lack of secondary formants, as clearly represented in Fig. 3.1. It is especially this circumstance which contributes significantly to the lack of color in the character of high F-horns, which can be described as dull and blunt. Only from C_5 on up can the high F-horn be considered to be equivalent in tone to the lower instruments, so that it appears, its use can only be justified for high passages.

3.1.1.6 Historic Horns

Double horns of today's standard design have been available since approximately 1900, when the development of valves around 1835 made it possible to construct chromatic instruments. The typical horn of the period from about 1755 to 1845, on the other hand, was the so called "inventions" or natural horn, for which the tuning of the natural tone series was adjusted by inserting different lengths of tuning crooks. This instrument was blown with the right hand in the mouth of the bell, as are today's horns, however it had a brighter tonal character.

The formant locations for an instrument from the beginning of the nineteenth century are shown in Fig. 3.1 as an example of a natural horn. The relatively high location of the main formant is particularly noticeable: For an F-tuning it is located at approximately 480 Hz, and moves to about 525 Hz for tuning in G. Additionally it should be noted that it is lowered to about 425 Hz for tuning in E^b. This means that these horns correspond to a brighter or darker "o(oh)" in their basic tone color, in contrast to an "u(oo)" for today's horns. The secondary formants of the natural horn also lie relatively high, and the nasal components are more prominent. As a whole, the overtone content is greater than in current horns, at *mf* the harmonic content of the vibrations of the lower natural tones reaches up to 3,000 Hz, and go above 5,000 Hz for the upper registers, where the notes are richer in overtone content for the longer horn, i.e., for the lower tuning. However, the differences in tonal characteristics for different tunings are not as pronounced as the contrast with today's horn. With respect to the tonal brilliance and the richness of the tonal picture, the natural horns in view of their timbre are not as far removed from the trumpet as today's instruments, nevertheless, the typical horn character is basically preserved.

The horn playing technique as employed in Baroque times with the corni de caccia can best be compared with today's French trompe de chasse, which is also blown without the damping hand in the bell mouth. Because of this technique, a tone rich in overtones is produced; the spectrum extends almost up to 10,000 Hz for intermediate intensities, and even for *piano* it contains partials up to approximately 4,000 Hz. The main formant lies in the region of the vowel sound "o(oh)" – much like in the natural horn, the secondary formants also show frequency locations which compare to the natural horn. Since noise components are relatively strong for higher frequency tone contributions, the open hunting horns give a rough and distinct metallic impression. As a result of the richness in overtones, the trompe de chasse is brighter in timbre than the natural horn, so that it comes close to the trumpet, and particularly the bass trumpet.

3.1.2 The Trumpet

3.1.2.1 Sound Spectra

Among the instruments of the orchestra, the trumpet is one of the richest in harmonics. Already for a *mezzoforte* the harmonic tone contributions of the low and middle regions of the tonal range extend above 5,000 Hz. In the upper regions, the boundary of the spectrum is pushed to approximately 8,000 Hz (Mühle, 1965). This results in a radiant and brilliant tone, with the further characteristic, that the region of strongest partials lies in a relatively high frequency range. However, below this maximum, the spectrum drops at the relatively flat rate of 6 dB/octave.

For today's standard B^b trumpet the playing range begins with an E_3 (165 Hz) neglecting rarely used pedal tones. The main formant of the sound spectrum is located at about 1,200 Hz and is pushed up to about 1,500 Hz in the fifth octave. As noted in Fig. 3.4, this means an emphasis on the vowel "a(ah)" for the largest part of the range, responsible for the strong tone of the instrument. For the higher regions the nasal components become more apparent, without, however, removing the sparkle from the tone. In this, especially the secondary formants in the vowel regions "e(eh)" and "i(ee)" play an important role in the brightening of the sound. The prominence of these two groups of partials prevents an extreme sharpness which could arise from a tone so rich in overtones in the absence of formants.

The light and brilliant tonal effect is roughly uniform over the entire tonal range since the fundamental does not dominate the trumpet sound even while playing *mezzoforte,* except for the highest notes. The brilliance of the timbre is furthermore supported by the fact that the noise contributions are very weak, so that hearing impressions are hardly influenced at all.

3.1.2.2 Dynamics

With increasing loudness, the overtone content increases dramatically, so that for *ff,* tone contributions to the threshold of hearing are present. Under these circumstances the trumpet becomes the orchestral instrument richest in overtones. On the other hand, in the lower loudness regions the spectrum of high notes, similarly to the case of the horn, is reduced to a few partials, so that the timbre can become softer for example than that of the oboe.

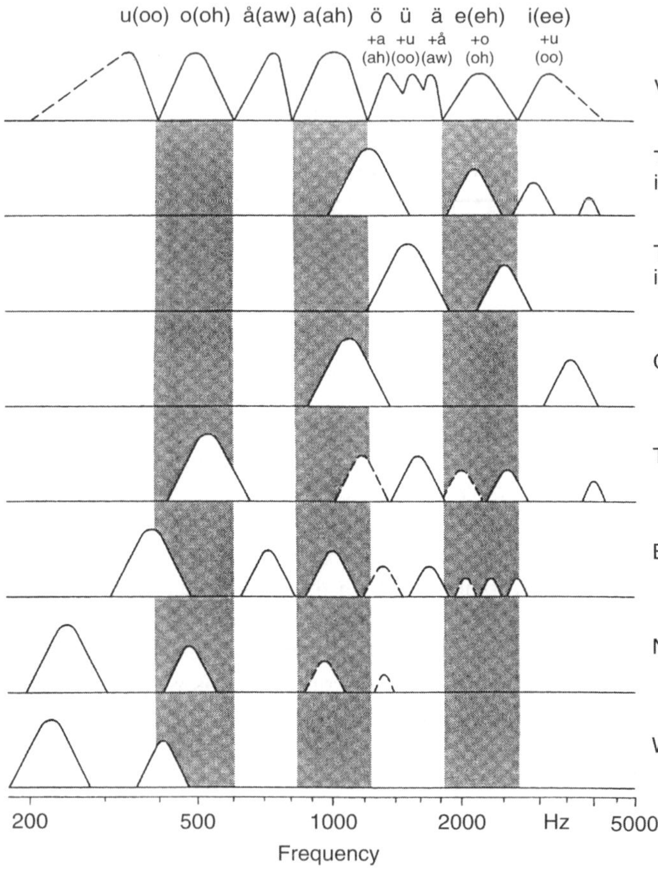

Fig. 3.4 Schematic representation of formant location for various brass instruments

In the power spectra, the 3,000 Hz level at *ff* lies by only 12 dB below the level of the strongest component, for *pp*, however, by around 40–50 dB. While the spectra for *ff* in the low register drop at 11–16 dB/octave by about 2,500 Hz, for *pp* this drop for the low notes already occurs around 1,600 Hz and for the higher notes around 2,000 Hz, and can reach a slope of 25–30 dB/octave, and for the highest notes even 50 dB/octave. The directional dependence of the sound radiation, however, is also very important for the tonal effectiveness of the high frequency contributions. This can lead to a significant intensity increase in the axial direction, when compared to the sound power levels averaged over all directions (see Sect. 4.2.1).

The dynamic range for rapidly played note sequences is characterized by a sound pressure level of 89 dB for *pp* and 104 dB for *ff*. For single notes the *pp* level may be dropped to as low as 78 dB in the low range and raised to 111 dB for high notes at *ff*. From this, a practical dynamic range of from 25 to 30 dB is obtained. The highest notes of the trumpet, however, lend themselves to less dynamic variation, which results from the relatively loud *pp* with a sound pressure level of 100 dB.

A level of 101 dB should be considered as a typical value for the sound pressure level of a trumpet at *forte*. The influence of dynamics on tone color is particularly extreme for the trumpet: For a variation of the strongest partials by 1 dB, the 3,000 Hz components of the low register change by 2.2 dB, while for the high notes, this results in a change by more than 4.5 dB (Meyer, 1990).

3.1.2.3 Time Structure

The initial sound of tongued notes is marked by an extraordinary incisiveness in a trumpet. Relevant for this is not only the comparatively short initial transient before the steady state is achieved, but also a very strong preliminary impulse. The initial transient for tongued notes in the fourth octave is accepted as from 25 to 30 ms. For higher frequencies this is shortened to values of below 20 ms (Luce and Clark, 1965; Melka, 1970). A first sharp peak in the amplitude representation appears, however, after 10–15 ms, depending on the sharpness of the attack. This preliminary impulse has a duration of only 5 ms in the trumpet, it is thus shorter than in the horn. It is already suggested in the amplitude development of the fundamental and the octave partial. The maximum amplitude occurs between 2,000 and 3,000 Hz, yet even for still higher components the pre-cursor impulse shows a higher intensity than the subsequent steady state. Accordingly, the precision of a trumpet staccato, is particularly achieved by the high frequency contributions. Even though the attack contains relatively few noise components, the attack can cause a crack-like impression by reason of the extremely rapid amplitude development in the pre-cursor impulse.

With a soft attack the trumpet tone develops very slowly. In the low register the initial transient can last for nearly 180 ms, in the middle register this is shortened only insignificantly to 150 ms (Melka, 1970). Only for very high tones is the difference between the duration of the initial transient for sharp and soft attacks no longer as large, and even a softly attacked tone develops within 40 ms, however the preliminary impulse is not as strong as for a staccato. Therefore, a tonal character which, on the whole, is soft, can be achieved in spite of the high overtone content.

3.1.2.4 Mutes

Use of a mute with the trumpet leads to a modification of the spectrum, this refers to the overtone content in general, as well as to the relocation of formant regions (Meyer, 1966c). The so-called normal (conical) mute strongly diminishes the intensity from the fundamental up to above 1,500 Hz. As a result, the "a(ah)" formant, which is so important for the open sound and for the tonality associated with the fundamental, is missing from the tonal picture. Accordingly, the timbre loses substance and it gives the impression of lacking strength. At the same time, the nasal contributions gain in importance, so that the muted trumpet sounds somewhat squeaky. In addition, an increase in intensity above 4,000 Hz adds to the metallic character of the tone. A typical example for the use of this tonal effect is given by the excerpt from “Pictures at an Exhibition” orchestrated by M. Ravel, as shown in the score example 4. The color of the muted trumpet represents the verbosity of the imploring Jew Schmyle in conversation with the self-confident Goldenberg.

In contrast, the high frequency contributions are strongly reduced by the use of a cup mute, so that above 2,500 Hz, practically no partials appear (with the exception of notes in the highest playing register). Inasmuch as the intensity reduction is already effective from 1,000 Hz on, the formant is moved, depending on tone height, into the region of the vowel color “o(oh)” to “å(aw)” so that the tone is relatively round and without sparkle. The fast and precise initial transients, typical for a trumpet, however, remain essentially intact.

The so-called Wah-Wah mute finally makes it possible to influence the frequency location of its Helmholtz resonance by changing its position in the bell mouth. This causes time-dependent tuning shifts which are sensed as tone color transitions or flowing tone colors.

3.1.2.5 Trumpets in Other Keys

Inasmuch as composers in the classical era could only write their trumpet parts for natural instruments without valves, their works call for many different key trumpets. In today’s orchestra trumpets with tuning other than B^b (with valves) are used only in special situations, and then particularly for very high parts, as they frequently occur, above all, in Baroque music.

The D trumpet, which in its tuning is located a major third above the normal B^b trumpet, shows in its tone picture a corresponding shift in its formant region to higher frequencies (Mühle, 1965). As recognized from Fig. 3.4, already for the largest portion of the tonal range the main formant moves to a position around 1,500 Hz, i.e., into the region of the nasal tone colors. As a result, the timbre loses strength, and shows an otherwise thinner effect. This tonal impression is further emphasized by the fact that the higher tone contributions, roughly above 2,500 Hz, are stronger than in the B^b trumpet. True, the difference in the fourth octave, on the average amounts to only to 2 dB, however, in the mid and upper registers it rises to more than 5 dB.

The high B^b trumpet, which is also used occasionally for high Baroque parts, no longer has a pronounced main formant, its highest intensity lies at around 900 and 2,000 Hz. Between these, we find a dip in the nasal region. All of this gives the instrument an open clear tone, and also, undoubtedly because of its high voice location, it guarantees a certain security in the attack. Furthermore, inasmuch as the intensity between 3,000 and 5,000 Hz falls by about 20 dB below that of the normal B^b trumpet (for the same total loudness), and also a secondary formant appears around 5,700 Hz, a radiant, bright tone picture is the result, without the hardness, which for the normal B^b trumpet is often unavoidable at that high register.

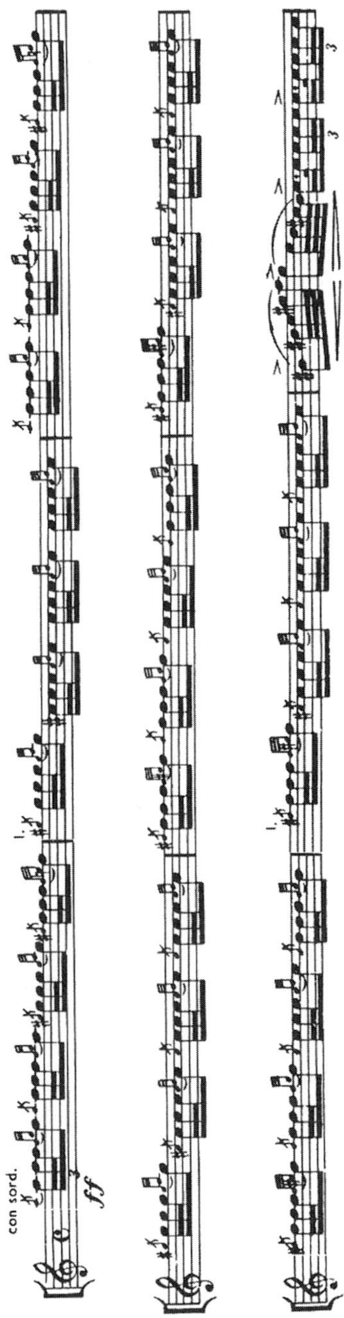

Score example 4 M. P. Moussorgsky, Pictures at an Exhibition, "Samuel Goldenberg and Schmyle" trumpet part measure 9 ff. (in the orchestration by M. Ravel)

3.1.2.6 The Clarino

The Clarino is a valveless brass instrument, built as a reconstruction of a Baroque instrument, which in its outward appearance is reminiscent of a natural horn. Its basic key of D is located a sixth below the normal B^b trumpet, so that the actual playing region is situated in the region of high order natural tones. The clarino has two overblowing holes, which are intended to increase intonation certainty for high notes.

In the low register the tone is characterized by a formant between 1,000 and 1,200 Hz. This is in the region of the vowel color “a(ah),” and also somewhat below the B^b trumpet (Mühle, 1965). From the fifth octave upward the spectral structure is dominated by the fundamental.

When comparing tones of the same total level, the overtone content of the Clarino (which also has a secondary formant at around 3,500 Hz) is stronger than in the Trumpet because of its greater instrument length (Müller, 1971). However, the individual dynamic steps of the clarino, looking at the intensity, are significantly lower than for the B^b trumpet. The difference for those tones played with both overblowing holes closed amounts to approximately 10 dB; upon opening the overblowing holes the sound level again drops by about 5 dB. In consequence, the absolute *forte* of the trumpet is richer in overtones, when compared to the Clarino, after all. The overall result is a bright, yet soft tone color for the Clarino, which is not as brilliant as that for the trumpet. The initial transient of the Clarino is somewhat longer than for the trumpet, which further supports the soft tonal characteristic.

3.1.3 The Trombone

3.1.3.1 Sound Spectra

The tonal range of the tenor trombone extends to E_1 on the low end, so that the spectral range of this instrument begins at 41 Hz when pedal tones are included. Similarly to the trumpet, for the *mezzoforte* tones of the trombone, the fundamental in the spectrum dominates only in a few cases, rather, the spectra drop below the maximum at a rate of 10 dB/octave for the pedaltones, and 5 dB/octave for the others. The frequency region of highest partial intensity is found in the area of 520 Hz, as shown schematically in Fig. 3.4. This formant location corresponds to the vowel “o(oh),” however, it is more pitch dependent than in the trumpet. While the maximum for low registers is found at 480 Hz, suggesting a clear “o(oh)” sound, in the upper registers the main formant is shifted to 600 Hz, i.e., to a transitional tone color between “o(oh)” and “å(aw).” The sonorous fullness of tone in the low register is transformed into an open, forceful timbre at the high end. However, it needs to be pointed out that the formants of the trombone (with a logarithmic decrement of 2.1) are not as sharply defined as those of the bassoon for example, which for that reason comes significantly closer to the character of a sung vowel.

Above the main formant, the overtone intensity decreases only relatively slowly. At the same time several secondary formants develop. The first one of these contributes particularly to the strength of the striking timbre by emphasizing the components in the "a(ah)" region. Additional weak maxima in the nasal region, as well as in the brightening regions of the vowels "e(eh)" and "i(ee)" complete the tone picture. Furthermore, individual timbre differences can be caused by the instrument itself (bore and bell mouth width), by the player, and by the mouth piece, and this apparently in the order mentioned with decreasing importance (Pratt and Bowsher, 1978).

3.1.3.2 Dynamics

At *ff* the trombone develops a sound extraordinarily rich in overtones, where the 3,000 Hz components are only 5–10 dB below the strongest partials. For an extremely strong attack, this can certainly fall into the region of the second formant. For high frequencies the sound power spectrum drops at a rate of only 3–6 dB/octave. Since the partials in the spectrum are very closely spaced by reason of the instrument's low pitch, a noise-like impression can be caused by the high overtone density at high frequencies. This is perceived as a metallic tone. At low loudness levels only a limited number of partials appear. When compared with the low registers of the horn, this can still be considered as an ample spectrum. Thus the sound power level at around 3,000 Hz lies at about 50 dB below the main formant and furthermore drops by 20–30 dB/octave. The *pp*-tone of the trombone is therefore not very soft, nor is it possible to reduce the overall loudness as much as for a horn.

An excessive weakening of the overtones causes the trombone sound to become dull and lacking in outline. In the low loudness range, therefore, those instruments will sound best, which by reason of a narrow bore are relatively rich in overtones. However, these instruments will then sound hard in *ff* and will not develop a desirable carrying ability. A strong and sonorous *f*-sound is better obtained with larger bore instruments, which, on the other hand, sound less convincing at *p* (Wogram, 1979). This means that the tonal esthetics require less pronounced spectral differences between *f* and *p* than are usually generated by a trombone.

For rapid note sequences one can count on a sound power level of 89 dB at *pp* and 105 dB at *ff*. In the low register individual notes can be reduced down to 73 dB, *ff* notes in the upper register can reach 113 dB. The dynamic range accessible to the performer includes 30–35 dB. The median sound power level for a *forte* lies at 101 dB. The 3,000 Hz components vary by 2.1–2.9 dB for a 1 dB shift of the strongest partials (Meyer, 1990).

3.1.3.3 Time Structure

In contrast to a trumpet, the onset of the tone in a trombone is primarily characterized by the short time span required to reach the final amplitude, rather than by the sharpness of the preliminary pulse, even for tongued notes. In the low registers the starting transient for tongued notes is about 40 ms, in the upper registers this is reduced to as little as 20 ms (Luce and Clark, 1965; Melka, 1970). Interestingly, trombones reach the steady state in the region of the fourth octave somewhat faster than trumpets. Naturally, for a soft attack the duration of the starting transient is increased; however, with times around 70 ms, the trombone reaches the steady state significantly faster than the trumpet or the horn. For these two instruments the performer obviously has more flexibility to shape the attack than for the trombone.

Figure 3.5 represents two typical pictures for the initial transient behavior of the trombone. In these pictures the amplitude development in the upper representation is to be considered as the normal case, thus as the esthetically more satisfying attack, while the lower picture represents a poorly attacked tone with an excessively strong preliminary impulse. It is clearly noticeable that the amplitude again decreases significantly before the tone finally is developed.

While these pre-cursor tones are frequently pointed out in the literature as characteristic of all brass instruments, it must be noted in contrast, that at least for the lower instruments good players always strive to achieve good tone development. An excessively strong pre-cursor gives a hardness to the trombone sound

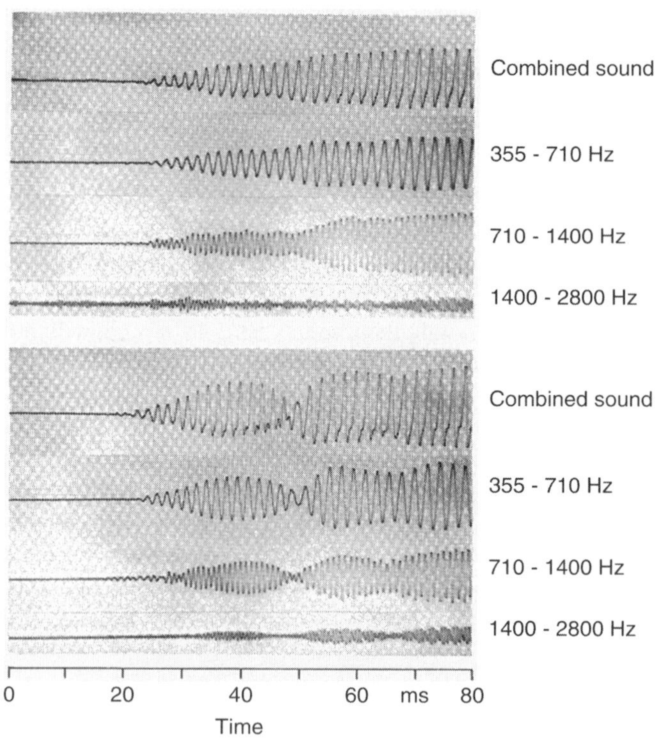

Fig. 3.5 Octave filter oscillogram for initial transient processes in a trombone (played note: B^{b}_{4}) *top*: good attack; *bottom*: poor attack

which is mostly disturbing. This can even assume the character of crackling. Nevertheless it is interesting that residuals of a pre-cursor can also be found in a smooth attack. In the frequency region between 710 and 1,400 Hz the vibration amplitude develops in two steps. The first phase displays about the same duration as the pre-cursor in the lower picture. This shows an instrument-dependent tendency for such a pre-cursor effect of roughly 25 ms duration. This corresponds to a round-trip of the sound within the instrument.

In the trombone, in contrast to the other brass instruments, the *vibrato* is not only executed through the lips, but also through movement of the slide. The tonal effect of these two performance techniques is juxtaposed in Fig. 3.6 for the note F_3. The strong time variation of the upper frequency limit is particularly noticeable. Evidently the tone color modulation dominates the frequency modulation.

Typical for a pronounced lip *vibrato* is a *vibrato* width of about ±10 cents; it causes variations of about 3 dB in the lower partials, this increases to 9 dB at

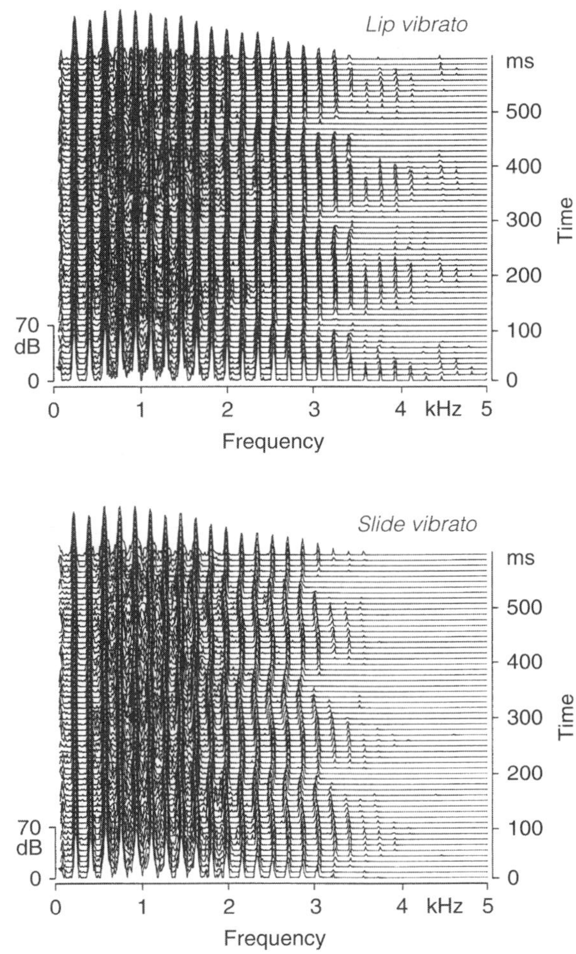

Fig. 3.6 Time variations of a sound spectrum for a trombone *vibrato* (note F_3) for different playing techniques

1,000 Hz and to roughly 15 dB at 3,000 Hz. All partials experience this amplitude deviation in phase. For a slide *vibrato* a width of ±20 dB is not unusual, however, the associated amplitude modulation is less, it only reaches 10 dB at 2,500 Hz. Thus for the slide *vibrato*, the tone color modulation is not as strong as for the lip *vibrato*, furthermore the pitch modulation is frequently implemented through a relatively slow *vibrato* (under 5 Hz.).

3.1.3.4 The Bass Trombone

For orchestral parts specifying bass trombone voices, usually instruments with F valves are used. These, in contrast to normal tenor trombones, have slightly larger bore and sometimes also a somewhat larger bell. This causes the tone to change in the direction of a darker tone color. As example for a very marked bass trombone sound, Fig. 3.4 includes the formant locations for a slide trombone in F (with two valves), which was especially developed for the low passages in Verdi operas, and is therefore designated as "Cimbasso."

As a comparison with the tenor trombone shows, all maxima are shifted slightly toward lower frequencies, however, the characteristic dense formant sequence of the trombone is maintained. The fundamental formant in the region of the vowel color "u(oo)" provides the substance to the tone so necessary for a bass voice, the actual strength, however, comes to the instrument from the double å(aw)-a(ah) formant which is found in similar form in the Vienna F-horn. The repositioning of the first maximum to about 370 Hz in contrast to 520 Hz for the tenor trombone signifies a lower tuning by nearly a fourth, so that the pitch location and tone color roughly correspond.

The trombone character of the Cimbasso-sound is evidenced also in the greater frequency range of the spectrum, which was already expressed by the long series of formants. Already for the lowest notes the partials reach up to 3,000 Hz, then, from the second octave on, an overtone series is formed, which also in *mezzoforte* goes beyond 4,000 Hz. The fact that the amplitudes are less than for the tenor trombone, corresponds to the low pitch of the instrument. As the averaged envelopes for the upper region of the second octave show in Fig. 3.7, the difference in partial intensity for equal loudness of tenor trombone and Cimbasso between 1,200 and 3,000 Hz is approximately 10 dB, and above this frequency rises to about 15 dB. The tenor trombone thus has a more brilliant effect.

3.1.4 The Tuba

3.1.4.1 Sound Spectra

The bass tuba and the contrabass tuba are the lowest instruments of the orchestra. The lower limit of their range are at B_0^b (29 Hz) or possibly at A_0 (27.5 Hz), however, these low notes present extraordinary demands for the performer. The

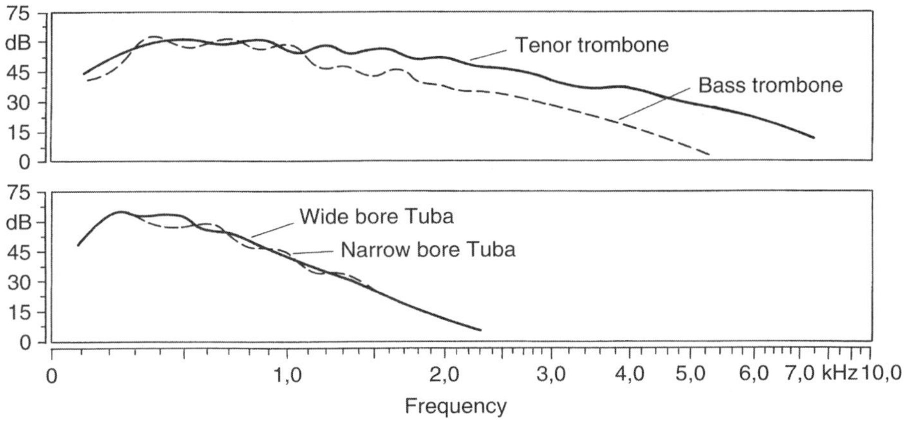

Fig. 3.7 Averaged spectral envelopes of several brass instruments for the tonal region of F_2 to B_2^b

bass tuba in F is naturally located a fifth above the contra bass in B^b, however, it can practically be played equally low, provided its bore is sufficiently wide (Kunitz, 1959). Thus the essential difference between a bass tuba and a contrabass tuba rests in the sound, ignoring the seldom needed upper region of the playing range.

The spectra of the tuba differ from those of the trombone in the low register mostly in the significant decrease of overtone content. For low notes at *mezzoforte,* the upper frequency limit for harmonic contributions lies between 1,000 and 1,500 Hz, depending on the structure of the instrument. In the middle and upper registers the partial series is broadened to frequencies between 1,500 and 2,000 Hz. This steep amplitude drop for higher tone contributions is clearly expressed in the spectral envelope of Fig. 3.7, showing a characteristic totally different from trombones. At the same time they also show that the lower partials are significantly weaker than the strongest components. Below the 1st formant, the power spectrum of a tuba drops by 10–15 dB/octave depending on the bore.

The main formant, which is thus generated between 210 and 250 Hz depends somewhat on the design, for the wide instrument it is lower than for the narrow one. As Fig. 3.4 shows, the tone color for both cases is somewhat darker than a normal "u(oo)". For the wide bore instrument a secondary formant is located on the boundary between "u(oo)" and "o(oh)," while for a narrow bore the corresponding secondary maximum is related to an open "o(oh)." Inasmuch as, particularly in the upper registers, there are no additional secondary formants for the wide bore instrument, the narrow tuba produces a somewhat slimmer timbre, while the wide bore emphasizes the dark, soft and occasionally muffled tonal impression.

3.1.4.2 Dynamics

The *pp* tone of a tuba is determined by a spectral decrease with a slope of 20 dB/octave above 250 Hz, it is thus very dark and soft. In contrast, above the actual overtones, disturbing noise components become noticeable at greater loudness levels. As a result the tone should not be forced, lest it become raw. Esthetically determined upper limits should therefore take precedence over technically attainable upper limits for *ff*. With that in mind, Richard Strauss (1905), in his text on instrumentation, specifies that the tuba not play beyond *mf*, and it is therefore very understandable that Berlioz, in the last movement of his Symphonie fantastique, has two bass tubas play the "Dies Irae" in unison.

The sound power level of the tuba at *forte* lies at 104 dB; rapid note sequences can vary between 93 and 108 dB. Low register individual notes can be considered as relatively subdued at a sound power level of 77 dB, particularly when considering the reduced sensitivity of the ear at low frequencies. Generally, a playable dynamic range of 25–30 dB is obtained, where the 3,000 Hz components change less than for other brass instruments, i.e., only by a factor of 1.5–2 of the amplitude change for the strongest tone contributions. The technically possible *ff* of 112 dB in the high register is in practice of no significance (Meyer, 1990).

3.1.4.3 Time Structure

In spite of the low pitch, a very fast initial transient is achieved in the tuba. The duration of the starting transient for tongued notes in the region of C_2 is about 40 ms. For the upper registers this is reduced to as short as 25 ms. For the lowest registers, however, this is increased to over 60 ms (Luce and Clark, 1959; Melka, 1970). If a *staccato*, even for a good player, does not sound as crisp as for other brass instruments in spite of the short initial transient, then the cause is found primarily in the fact that the tone is not rich in overtones. Furthermore, the preliminary impulse is not as steep in its rising slope as for other brass instruments, so that it does not characterize the attack as precisely.

Surprisingly, softly attacked tuba tones reach their steady state more rapidly than is the case for horns and trombones. While the initial transient for C_1 can last for over 130 ms, corresponding values for higher registers lie around 60 ms, and again the limited frequency range of the spectrum supports the soft tone onset. This also explains why the overtone-rich trombone requires a slower initial transient process. The horn, on the other hand, already occupies a special position with the particularly wide modulation region for its tone development.

3.2 Woodwind Instruments

3.2.1 The Flute

3.2.1.1 Sound Spectra

The tone of transverse flutes is characterized by the very uniform overtone structure of the spectra. With few exceptions, particularly for the notes between C_4 to or E_4^b or E_4, the fundamental is the most strongly developed of all partials for the entire range of the instrument. For no other orchestral instrument is this characteristic as clearly marked. Above the fundamental the intensity of the overtones drops quite linearly with increasing frequency. Secondary formants appear only occasionally, and then very weakly. They are thus not typical for this instrument group, but rather characterize tonal idiosyncrasies of individual instruments.

The intensity relationship between the lower partials, and thus the tone color, can be varied by the performer within relatively wide limits. The performance technical parameters available are the blowing pressure (i.e., the air stream velocity through the lips), the degree of coverage of the mouth hole (and thus the distance between lip opening and blowing edge) and the blowing direction. The size of the lip opening affects only the dynamics and not the tone color. Raising the blowing pressure leads to a better coincidence of the first overtones with their associated resonances, increasing their strength relative to the fundamental. The tone thus brightens in its color, which supports the impression of the changed timbre caused by the dynamic change resulting from the increased pressure. Reduced coverage (increasing the distance s in Fig. 3.8), in addition to influencing the intonation, leads to a softening of the air stream at its edge. This causes a weakening of the overtones for unchanged fundamental strength; the shorter the distance the brighter will be the tone color. Furthermore, by varying the coverage, the intonation of the higher resonances can be adjusted, within limits, to minimize noise contributions. Finally, the blowing direction determines the strength relation between even and odd numbered partials. As Fig. 3.8 shows, an air stream directed symmetrically toward the blowing edge ($y = 0$) enhances the fifth, while for a slightly more outward or inward directed air stream the octave or double octave is strengthened and the fifth is diminished. The flute sound is perceived as having particular tonal beauty when the fundamental and octave are approximately equal in strength, and the twelfth is weaker by about 10–15 dB (Bork and Meyer, 1988).

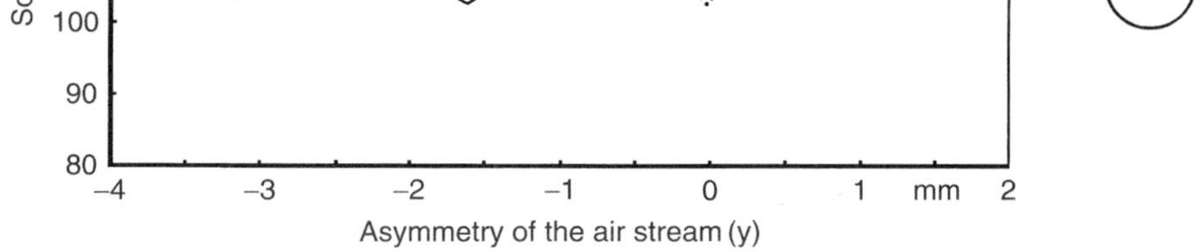

Fig. 3.8 Strength of the first four partials in the sound spectrum of a flute (played note: C_4) in their dependence on blowing direction

As already suggested in Fig. 2.6, the noise in the tone of a flute consists not only of flow noise with intensity independent of frequency, but it also contains components which influence the tone. They appear in overblown notes and result from continuous statistical excitation of "unused" resonances. In the region of E_5 to D_6, therefore, such flow noise peaks appear at one half of the fundamental frequency and odd multiples thereof; also from E_6^b on up, at 1/3 and 2/3 of the frequency of the fundamental and at the corresponding multiples not divisible by three; in particular, for cross fingering, the flow noise peaks can also be inharmonic. The height of these peaks depends largely on the quality of the instrument and the playing technique. Contrary to a widespread opinion, for metal flutes the material appears to play a subordinate role in relation to tonal impressions, especially for the listener (Coltman, 1971). Special "preferences" for gold or even platinum flutes rest predominantly on a psychological effect, which of course, in individual cases may certainly influence the performance technique of the flautist.

3.2.1.2 Dynamics

As already mentioned, the player can influence the dynamics within relatively narrow limits (about 6 dB) without changing tone color, by variation of the lip opening. The predominant dynamics contributions, however, are accompanied or supported by changes in tone color. In contrast to most other instruments this is above all relevant to the lower partials, and less so in the upper frequency region. Thus at *ff* the 3,000 Hz components lie about 20–30 dB below the fundamental, and for *pp* only by 30–40 dB. Above that, all spectra drop with a slope of 15–22 dB/octave. The dynamic range of the flute is relatively small. For rapidly performed scales through two octaves it is only 12 dB with a sound power level of 82 dB at *pp* and 94 dB at *ff*. This is also related to the fact that for a flute the playable dynamic range depends especially strongly on pitch. The low register with a *pp* around 67 dB and a *ff* in the neighborhood of 86 dB is particularly weak; in the highest octave the sound power level can reach 100 dB for *ff*, while for *pp* it can hardly be lowered below 83 dB. When one further takes into account the low sensitivity of the ear at low frequencies, then these values suggest that a low frequency *ff* is perceived as equally loud as a *pp* of the highest notes. Consequently, for a flute, the tonal time structure receives particular significance by its influence on tonal emphasis.

As a whole one can count on a practically realizable dynamic range of from 15 to 20 dB in a flute. The average *forte* power level lies near 91 dB. A change of the fundamental by 1 dB is connected – depending on performance technique – to a change of the 3,000 Hz components of up to 1.5 dB in the lower and middle register and 2 dB in the upper register (Meyer, 1990).

3.2.1.3 Time Structure

Among wind instruments, the flute requires the longest time for tonal development. In addition, the attack contains particular characteristics which appear in response to initial blowing techniques. Dominant among these are so-called preliminary tones, which are formed by higher resonances (Rakowski, 1966).

For the lowest notes, these preliminary tones have a duration of about 50 ms. They are located in a frequency region around 2,000 Hz, that is approximately three octaves above the fundamental. Their intensity is about 10 dB greater than the subsequent stationary state. For the lower partials it is typical for the tone picture to have the octave partial exhibit a relatively fast initial transient and the actual fundamental to develop substantially more slowly thereafter.

In the mid-range the initial transient is as fast for the fundamental as for the next higher tone contributions. In addition to the preliminary tone, which raises its frequency to 4,000–6,000 Hz in correspondence to the three octave rise above the fundamental, noise-like components also appear below the fundamental, which disappear after about 100 ms. The duration of the high preliminary notes is of the order of 50 ms. Finally, the initial transients for high overblown notes consist initially of sub-resonances before the energy passes to the actual partials after about 50 ms.

The strength of the preliminary tones naturally depends on the sharpness of the attack, or on the articulation syllables used; however, it should by no means be considered as a negative characteristic, to be suppressed by performance technique. It enhances the sharpness perception of the attack, which is especially important in view of the relatively long initial transients for flute staccatos (low register about 100 ms, mid-register around 30 ms.). This was already shown very clearly in Fig. 2.9, the juxtaposition of a sharply articulated and a soft tone attack. Finally, preliminary tones, as well as noise components generated by the attack, participate substantially in the tone effect known as the "flutter tongue," as used, for example, by R. Strauss in characterizing the wind in tone painting fashion in Variation VII of Don Quixote (see score example 5). Here the individual tongue beats of the "flutter tongue" follow each other with a frequency of 25 Hz, and for higher frequencies lead to an amplitude modulation of 15 dB.

While the end of a tone for brass instruments and reed instruments is characterized by a very short decay time, because of the sudden cessation of lip or reed vibration, the decay process for a flute can be influenced somewhat by the performer. This is unique among wind instruments. For a "normal" flute tone ending in the mid register, the decay time (above 60 dB) is at about 125 ms for the fundamental and 80–100 ms for the next three overtones. A soft termination of the tone prolongs the decay time to about 200 ms for the fundamental and the octave, and to 120 ms for the following overtones. Naturally, even these values are short in comparison with string instruments.

Of particular importance for the flute sound is the effect of the *vibrato* on the spectrum. Two characteristic examples are shown in Fig. 3.9. The width of the *vibrato* for a flute is relatively small. Frequency fluctuations of from ± 10 to ± 15 cents can already be considered as a strong *vibrato*. For this amount there are practically no amplitude variations in the lower partials; the effect therefore is one of a stable tone. Only the higher overtones move to the resonance wings by reason of the *vibrato* and therefore experience strong amplitude modulation. Above 3,000 Hz this level

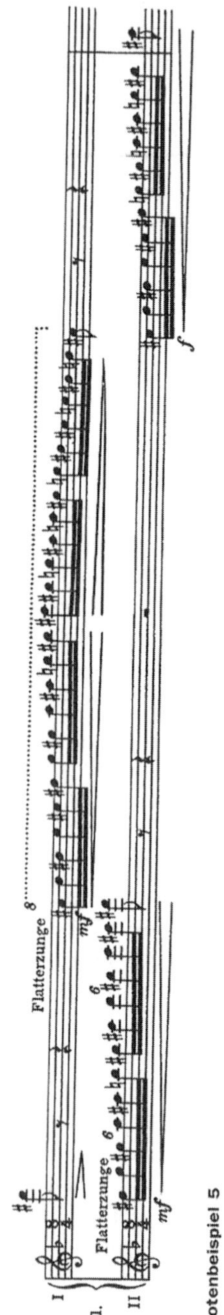

Score example 5 R. Strauss, Don Quixote, Variation VII, flute passages

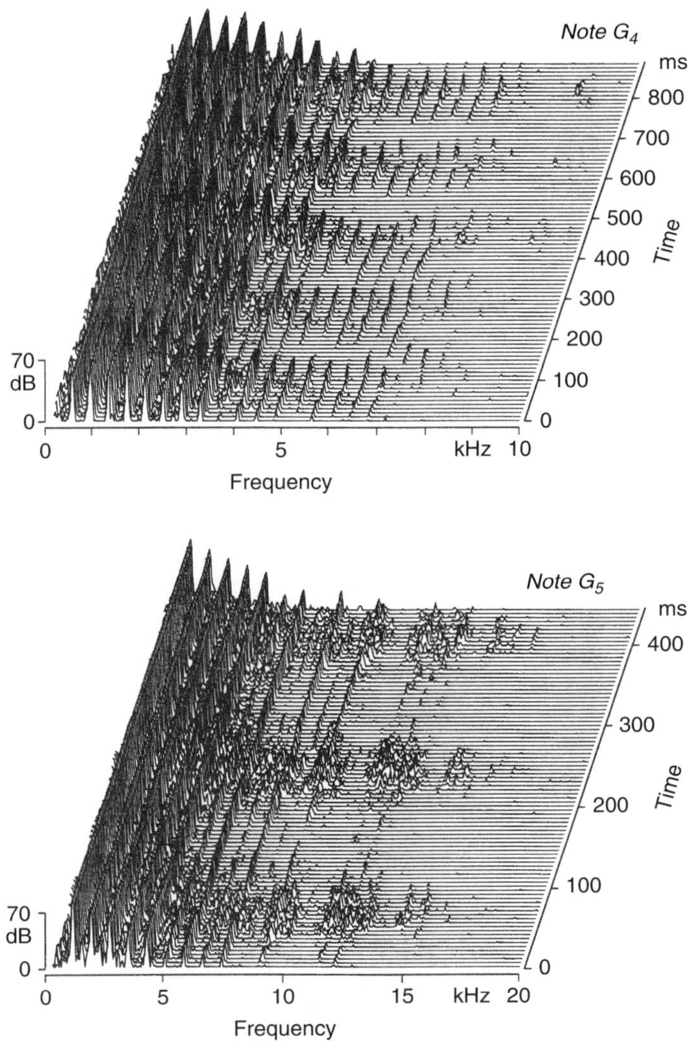

Fig. 3.9 Time variations in the tonal spectrum of a flute vibrato for different pitch regions

fluctuation reaches 15 dB and effects a pulsing tone color variation. For example, this effect is clearly pronounced for Note G_4 and also recognizable for Note G_5. Added to this is a very pronounced modulation of the blowing noise which reaches well above 10,000 Hz with a level fluctuation of 15 dB. The tone color modulation, as well as the noise modulation, lend the fluctuating flute tone a particular prominence within an ensemble, even when the blowing noise is not as strongly pronounced as in the pictured example (Meyer, 1991).

3.2.1.4 The Piccolo

As is the case for the flute, the piccolo in its spectrum also shows a steady amplitude drop beginning at the fundamental, which is the strongest partial, to the higher tone contributions. In this context, a secondary formant can appear in individual cases in the region around 3,000 Hz, which, corresponding to its location in the vowel "i(ee)" region, supports the very bright tone of this instrument. In the middle and upper regions of the tonal range, harmonic components are formed up to about 10,000 Hz which nevertheless do not give metallic sharpness to the tone since the frequency separation of the partials for the piccolo is very large, due to the high pitch. The timbre is better described as bright and penetrating, which is particularly connected to the fact that the strongest tone contributions of this instrument fall precisely into the frequency range of highest sensitivity for the ear.

In addition, it should be noted that the dynamic performance range of the piccolo is particularly narrow. On the average it barely covers 15 dB, and it is just that wide over the entire tonal range (Burghauser and Spelda, 1971). At *pp* the piccolo is therefore very loud in comparison to other instruments: the lower dynamic limit rises from a sound power level of 78 dB for the low notes, to a value of 88 dB in the highest register. Correspondingly an *ff* produces values between 93 and 103 dB, so that in high passages the piccolo can certainly be made to stand out in a full instrumentation orchestra sound: Particularly typical examples of this are found in the symphonies of D. Shostakovich.

3.2.2 *The Oboe*

3.2.2.1 Sound Spectra

As determined by the different process of generating vibrations and the conical nature of the bore, the oboe has a totally different tonal character than the flute. Acoustically this is borne out in the spectrum which is rather rich in overtones. The frequency distribution of this spectrum is determined by a series of formants. Even though the spectral region reaches down to about 233 Hz (the fundamental for B^b), the strongest tonal contributions do not appear until about 1,100 Hz. This principal formant, as shown in Fig. 3.10, leads to a basic tone color similar to the vowel "a(ah)." In addition, for the lowest notes a certain sonority is obtained by a subformant between 550 and 600 Hz which can shift the tone color in the direction of a bright "o(oh)."

The fundamental of the oboe sound is relatively weak especially in the low registers, since the sound power level below the formant maximum drops by 4–6 dB/octave. The fundamental thus lies up to 15 dB below the main formant. Its vowel character is therefore enhanced. Two secondary maxima at around 2,700 Hz, i.e., in the boundary region between the colors "e(eh)" and "i(ee)," and around 4,500 Hz lend a pronounced brightness to the oboe, which is also enhanced by further partials up to 9,000 Hz (at *mf*). Finally, in the low register, the frequency location of the sub- and main formant results in the effect that the even partials

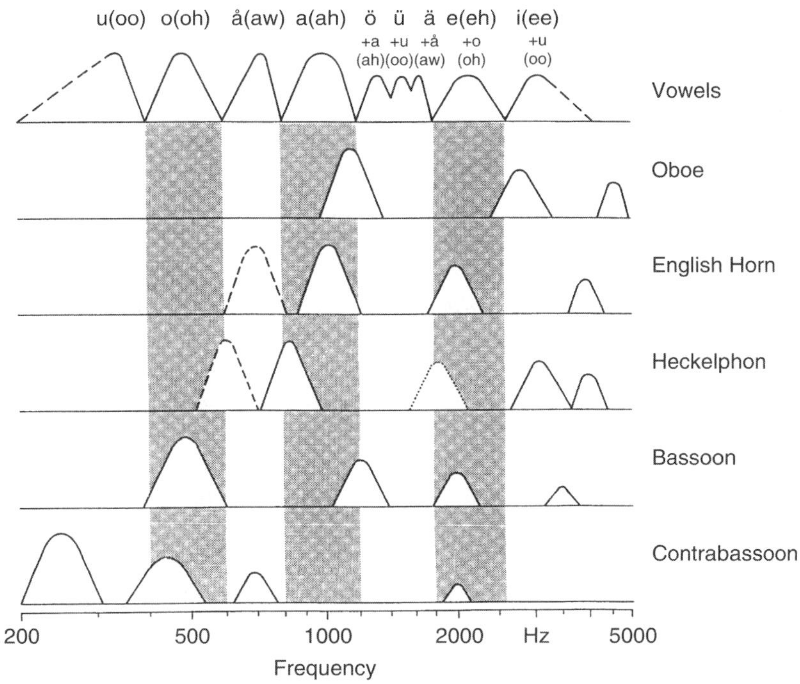

Fig. 3.10 Schematic representation of formant locations for double reed instruments

dominate in intensity over the odd partials. This differentiation (of the order of 8 dB) leads to a particularly open sound.

At the boundary for overblown tones, the formant region becomes discontinuously broader, above D_5 the formant thus loses in character (Smith and Mercer, 1974). Up to about B_5^b the fundamental and octave are nearly equally strong, so that the maximum no longer exhibits a clear peak, or the octave partial may even dominate somewhat. From B_5 on up the fundamental dominates; for the higher registers, therefore, the vowel-like tone picture of the oboe increasingly gives way to a rather hard and less expressive timbre. How critical these high notes of the oboe are in view of tonal considerations is noted, for example, when in the slow movement of the G major violin concerto of W. A. Mozart, the voice carried up to the D_6 by the flute is taken over by the oboe, as happens occasionally, because in the outer movements only oboes are present.

The typical tonal difference of Vienna oboes in contrast to generally played instruments of the French style, rests in the fact that because of the somewhat narrower cone, the main formant lies by 50–100 Hz lower, and the higher overtones are more pronounced. Thus the tonal character becomes somewhat more pointed and less nasal. For Baroque oboes the main formant, depending on instrument construction, lies more or less far below 1,000 Hz, so that the tone region determined by this formant is stretched by several half steps in comparison to today's instruments (Benade, 1976).

3.2.2.2 Dynamics

When considering dynamics, varying the tone color does play an important role for the oboe, however, the nature of reed vibrations imposes limits. While the *ff* for brass instruments is determined by strong tone components even above 10,000 Hz, the center for change in the power level spectrum of the oboe lies in the region around 3,000 Hz: for the oboe *ff*, these components are located only 10 dB below the strongest tone contributions; above 3,000 Hz the envelope drops relatively steeply with 23 dB/octave. For *pp* the linear amplitude drop already begins at 1,500 Hz and has a slope of 20 dB/octave or less, so that the 3,000 Hz component lies up to 30 dB below the strongest tone contributions. As a result, for *pp*, tones are produced with only very few strong partials: In the low register the spectrum is reduced to four to five harmonics, where the amplitude maximum is shifted to the second partial (in contrast to the 4th or 5th in *mf*). In the upper registers, actually only two partials appear, and the fundamental dominates in intensity by far. For *pp* the main formant is thus shifted in the direction of a somewhat darker basic color, in addition, the tone becomes more tender due to the lack of higher frequency contributions.

In comparison to the influence of the tone color on the dynamic range, the difference between the power level at *ff* and *pp* is relatively narrow. For rapidly performed scales the level at *ff* rises to 95 dB in contrast to 83 dB at *pp*. For isolated notes the power range expands to values from 70 dB for *pp* in the mid range to 103 dB for *ff* of very high notes. One can thus count on a practically usable dynamic range of up to 30 dB, with only 20 dB in the extreme registers. For the characteristic value of the oboe *forte* sound, a sound pressure level of 93 dB can be given. The "dynamic tone color factor" which is responsible for the relative shift of the 3,000 Hz components moves between 1.7 and 1.9, and is thus a little smaller than for flutes and clarinets.

3.2.2.3 Time Structure

The attack for double reed wind instruments is distinguished by particular precision and clarity. The reason for this lies in the facts that on the one hand initial transients are extraordinarily short, and on the other, that there are no noise-like or inharmonic contributions. In fact, in the initial transient, the individual partials experience a nearly exponential amplitude development so that the auditory impression due to the smooth envelope is clear and pure. Oboes are therefore especially suited for a short, and yet tonally precise *staccato*, which can lead to difficulties in performance alongside other instruments which are not in a position to give their *staccato* passages the same pearlescent clarity.

The initial transient for tongued notes, even for the lowest notes of the oboe, is shorter than 40 ms and is lowered to less than 20 ms with increasing pitch, where the high frequency contributions will already have reached their final strength in about 10 ms. Yet, with "*cantabile*" playing, tones can also be developed softly.

In this, the initial transient can be stretched to 100 ms and can last in the upper registers for about 40 ms (Melka, 1970). Thus the initial transients for the oboe with soft attack can reach similar values as the violin for sharp attack. This comparison makes the influence of the initial transient on the tone picture of the instrument especially clear. Its complement is found in the short decay time of the oboe which lies in the order of 0.1 ms.

3.2.2.4 English Horn and Heckelphone

As lower instruments in the oboe group, the English Horn, and for particular compositions also the Heckelphone, are used in the orchestra. Corresponding to its tuning in F, the English Horn is located a fifth below the oboe, while the Heckelphone lies an octave lower than the oboe. Both instruments possess a pear-shaped bell, which is especially important for the characteristics of the lowest notes.

From a tonal standpoint, the English Horn and the Heckelphone represent a transposition of the oboe timbre into a lower range, as can be seen from the survey of formant regions in Fig. 3.10. In the oboe the main formant can be described as a bright "a(ah)," in the English Horn the vowel color of a dark "a(ah)" dominates and in the heckelphone a transition color between "å(aw)" and "a(ah)."

This formant shift relative to the oboe, however, is not as strong as the shift of the tonal range. In contrast to the pitch shift by a fifth, the shift in color only involves about a whole step, while the heckelphone main formant lies a third below that of the oboe.

In the lower registers of the tonal range an additional formant comes in, which in the English horn corresponds to a somewhat brighter "o(oh)" and in the Heckelphone to a darker "o(oh)." This generated the basic color for the "rather wailing" tone of the English horn and the relatively dark timbre of the heckelphone.

As the location of the higher formants in Fig. 3.10 proves, the English horn and the heckelphone form a decided expansion of the oboe group in the direction of lower pitch. The continuity of this tonal line is illustrated particularly impressively by the score excerpt from "Salome." On the other hand a comparison of formant locations also permits recognition of the difference between the heckelphone and the bassoon from a tonal standpoint. In addition, it is noted that these two lower oboe instruments do not require more initial transient than the oboe itself (Meyer, 1966a). However, the dynamic range in comparison to the oboe is narrower; for the English horn the limits for the sound power level lies near 79 dB at *pp*, and 94 dB at *ff*.

Score example 6 R. Strauss, Salome, Part excerpt: Oboe group

3.2.3 The Clarinet

3.2.3.1 Sound Spectra

The clarinet presents a typical example for sounds, for which the odd partials outweigh the even ones (Backus, 1961 and 1963; Meyer, 1966c; Strong and Clark, 1967). However, they do not preserve these characteristics over the entire pitch range, they change their spectral composition in the region of the upper register in favor of the even tone contributions. Fundamentally the playing region of the B^b clarinet can be divided into three segments, which show different characteristics in relation to their spectral construction and thus their timbre. The boundaries between them are somewhat fluid and depend to some extent on construction details, reed strength and performance technique. Fig. 3.11 contains a juxtaposition of three typical sound spectra for the individual regions, which, however, show the dominance of the fundamental over all other partials, as an essential common characteristic.

In the low octave of the range from D_3 (147 Hz) to D_4 the odd partials are significantly stronger than the even ones. The difference can be followed for the lowest notes up to about the 15th partial; the second and fourth partials are particularly weak in their amplitude, the difference between them and the neighboring odd partials is generally greater than 25 dB, it can even take on values up to 40 dB. As a result the tone becomes dark and hollow; in this register the clarinet is therefore especially suited for "achieving of dark and sinister tonal effects," as for example at the beginning of the first movement of the 5th symphony by P. I. Tchaikovsky (Kunitz, 1957).

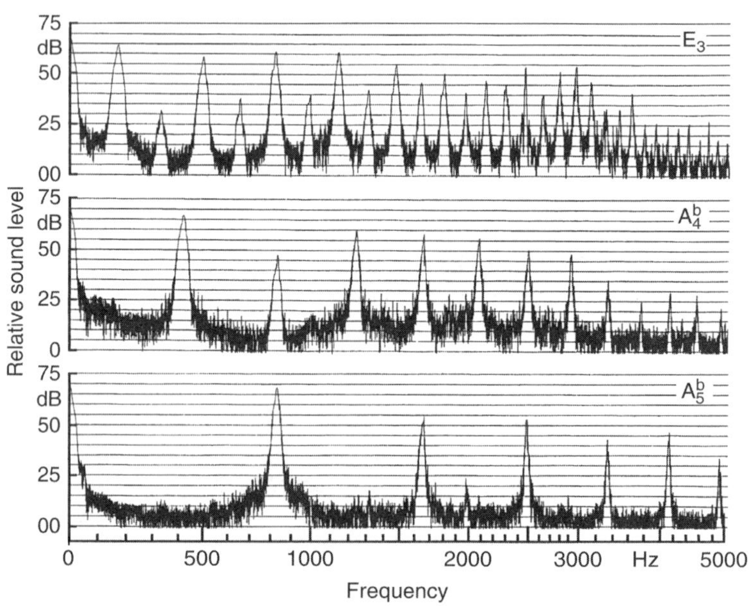

Fig. 3.11 Sound spectra of a clarinet

In the region from approximately E_4^b to G_5 the 1st and 3rd partials are decidedly more strongly pronounced than the octave partial, from the 4th harmonic on upwards the odd and even contributions are, however, equally developed. For a good player, the overblow boundary, which is located near A_4, cannot be recognized from its spectral construction. Nevertheless, the so called short notes up to $G_4^\#$ can occasionally sound somewhat dull, when the amplitudes above 3,000 Hz are too weak. In this pitch range the indifferent location of the octave partial also plays a role: with increasing pitch its difference from the 1st and 3rd partial diminishes, so that cases can arise where it is too strong for a decidedly hollow timbre, and too weak for a forceful sound.

Above $G_5^\#$ the fundamental dominates in a particularly strong measure. It is associated with a steadily decreasing overtone series, which gives a "full round substance" to the high register (Kunitz, 1957). Luster and brilliance are additionally achieved by a formant between 3,000 and 4,000 Hz, i.e., in the region of the vowel color "i(ee)." The large intensity difference between the harmonic partials and the noise background is responsible for the clarity of the sound. It gives the tone an especially pure sound. This becomes particularly clear in comparison to the flute, which in this pitch range exhibits a similar overtone structure, but produces considerably higher noise contributions, and furthermore presents stronger fluctuations in the temporal micro-structure of the sound level, so that the tone of the clarinet gives the effect of being more steady, and firmer and thus also stronger.

3.2.3.2 Dynamics

Of all wind instruments, the clarinet can produce the softest *pp*. At that point the power level drops to about 65 dB and in the region of the fifth octave can even be lowered to 57 dB. The result of this is a sound pressure level, which in a larger hall approaches a value with an order of magnitude around the ambient background noise. For fast scales the *pp* power level does rise to 77 dB. In the highest registers, like above D_6, such a pronounced *pp* can no longer be blown.

For *ff* the power levels for rapid note sequences reach a value of 97 dB. Individual notes, especially in the fifth octave, can rise to 106 dB. This results in a dynamic range, which in its breadth is hardly found in any other instrument. In the lower registers it is measured at around 30 dB, in the mid-registers at a scant 50 dB, and in the upper registers at about 25 dB. The characteristic *forte* for the clarinet lies near 93 dB.

The extraordinarily wide-spanned dynamic expression possibilities of the clarinet are further enhanced by the strong tone-color variations existing between dynamic steps. (Meyer, 1966c). For *ff* the intensity maximum of the low notes is shifted to frequencies above 1,000 Hz, i.e., into the region of the bright a(ah)-formant. For lower frequencies the spectrum drops with 3 dB/octave, so that in this exceptional case the fundamental does not dominate. While in the low register the power level spectrum drops with a slope of 12 dB/octave above 2,500 Hz, in the mid- register the very strong partials persist up to about 5,000 Hz, followed by a spectral drop of 23 dB/octave. This richness in overtones effects a pronounced brilliance of the *ff*; however, this can also present a certain hardness in the low registers, which is associated with the fact that the amplitude difference between even and odd partials becomes larger with increasing performance strength.

In contrast, for *p* the clarinet tone becomes softer in character through the intensity equalization between the two partial groups. While the spectra in the low register, even at *pp*, are still relatively rich in overtones, since above 600 Hz they drop only by 15 dB, in the fourth octave decidedly tender tones can be produced with only three or four partials, i.e., a spectrum of less than 1,500 Hz. The dynamics are therefore predominantly determined by the overtone content. The partials in the frequency region around 3,000 Hz lie from around 40 dB (low register) to 50 dB (high register) below the fundamental for a *pp*, while the level difference at *ff* only amounts to 10–12 dB. At the same time the "dynamic tone color factor" rises from 2 for the low register to above 2.5 for the upper registers.

3.2.3.3 Time Structure

As is the case for the double reed instruments, the attacks with the clarinet can be very clear and precise. Here, staccato notes practically do not contain higher frequency preliminary sounds – as was the case, for example, with the flute, but rather exhibit a quite uniform amplitude growth in all overtone regions. As Fig. 3.12 shows for the lowest note of the clarinet, even the fundamental reaches its final strength within a few vibrational periods. Thus the attack does not give the pointed effect of the oboe. As a whole, the initial transient is completed 15–20 ms after the attack, yet it can be stretched to more than 50 ms for a soft attack. These values are valid for the entire pitch range of the clarinet (Melka, 1970). The decay time, even in the low register, is not longer than 0.2 s and drops to about 0.1 s in the upper register.

In addition to the slower initial transient of the lower partials, the features of the soft attack are above all formed by the fact that higher overtones increasingly delay the initial transient with increasing frequency. Thus, special importance goes to the transfer of the high tonal contributions for differentiation in articulation, even though, as in the flute, they do not dominate the transient process.

Occasionally noise-like components appear in the attack, when closure of the flaps triggers resonances which have not yet been excited by reed vibrations. This phenomenon plays an important role, especially for connected note transitions.

In the clarinet, a *vibrato* can be produced either with the diaphragm or through the lips. In the rare event that a clarinet *vibrato* is used in classical music, the associated frequency variations are only slight. For a diaphragm *vibrato* a width of about ± 7 cent would be typical, this is thus at the limit of audibility for frequency modulation. A lip *vibrato* can reach a width of ± 15 cents in the upper registers. These frequency fluctuations are connected to amplitude changes of higher partials, particularly at frequencies around 2,000–3,000 Hz, which can amount to up to 20 dB, however these are, at least partially out of phase. Thus their effect in the ear

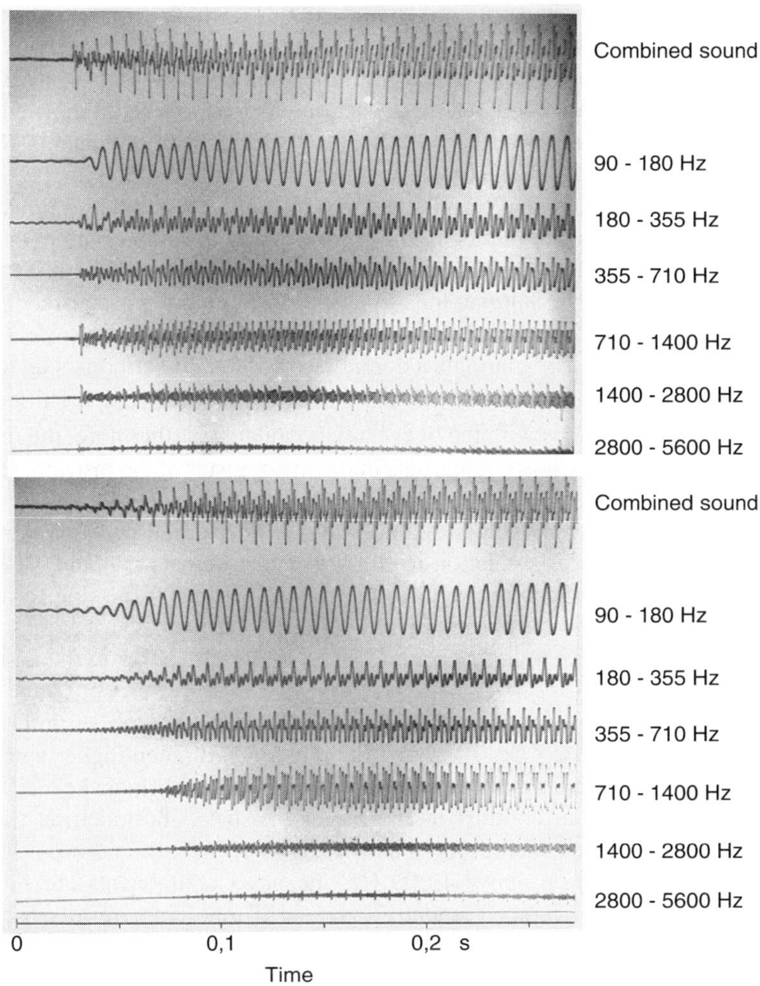

Fig. 3.12 Octave filter oscillogram of initial transient processes in a clarinet (played note: D_3, sharp and soft attack)

is compensated to a lesser extent: within individual frequency bands, the energy of the partials varies only by 5–6 dB for a diaphragm *vibrato* and by 7–8 dB for a lip *vibrato*. A lively tone color effect is thus created in the tonal impression of the clarinet *vibrato*, which overshadows the effect of frequency modulation.

3.2.3.4 Clarinets of Different Pitch

The influence of the instrument's pitch on the tonal character is particularly pronounced for the clarinet (Meyer, 1966c). Above all the difference between an A and a B^b clarinet rests on the fact that – especially in the low register – the intensity of the A-clarinet in the overtone region around 1,000 Hz, which is in the region of the vowel color "a(ah)," is about 5 dB lower, also the contributions above about 3,000 Hz are weaker. A further nuance arises from the fact that in the A-clarinet the even partials are even more reduced. All these phenomena lead to a very "dark and song-like" timbre, which at times is also specified as "holding back and tender" (Kunitz, 1957). In contrast, the B^b-clarinet gives a more brilliant and forceful impression as a result of its somewhat richer partial spectrum. It is noteworthy, how much even small intensity differences matter in differentiating in the tonal picture, they are nevertheless sufficiently significant for Richard Strauss to employ both instrument types at the same time in Salome, the A-clarinet for melodic passages and the B^b-clarinet for brilliant figures and ornaments.

While for the A- and B^b-clarinets a dominance of the odd harmonics up to about the notes $F_5^\#$ or G_5 is observed, for the clarinet in C this boundary is shifted up to about B_4^b; thus it coincides with the over blow boundary. Furthermore, the intensity difference between even and odd partials in the low register of the instrument is less than for the larger clarinets. Thus its tone color is not as hollow and covered. The C-clarinet exhibits significantly higher amplitudes in the region between 1,500 and 4,000 Hz, the difference in relation to the lower instruments is about 10 dB and effects not only a significantly brighter sound, but because of missing formants a "cooler and harder" timbre. These characteristics render the C-clarinet particularly suitable for folklore related tasks such as, for example, the polka in the first act of the "Bartered Bride."

Still more strongly pronounced is the difference between the clarinets in D and E^b, and the normal clarinets. Already from 1,000 Hz upward much higher amplitudes appear in the high clarinets, which in the region around 2,500 Hz make a difference of more than 25 dB, in contrast to the larger clarinets. Inasmuch as the level difference between even and odd partials shrinks to 10–15 dB, and because of the diminished intensity drop above 1,000 Hz, the nasal components are relatively strongly developed, a bright, frequently even shrill tone color results, which on occasion can have hard and pointed effects. The high clarinets are thus predestined, above all, for special tonal effects, as they are demanded in "Feuerzauber" (Walküre), or "Eulenspiegel." The dynamic range of the small clarinets is relatively narrow, with median power levels of 80 dB at *pp* and 96 dB at *ff*, a range of only 16 dB is noted. The possibilities for expression are thus significantly less – especially for *piano* – than for normal clarinets.

The bass clarinet is characterized by properties already considered in the A-clarinet. In particular, the difference between even and odd partials grows to more than 30 dB, where the fundamental is especially emphasized. Through this the timbre becomes particularly hollow and dark, it obtains, as it were, something "mysterious and melancholy." The *pp* can be played extremely discreetly, in the lower register it drops to a power level of 59 dB, where for subjective tone impressions one must still consider that the loudness impression is further reduced by the decreasing aural sensitivity at low frequencies. At *ff* the instrument is also not very strong, a power level of only about 97 dB can be expected.

3.2.4 *The Bassoon*

3.2.4.1 Sound Spectra

At the low end, the range of the bassoon extends to B_1^b (58 Hz), with an extended bell joint A_1 can be reached, which for example is required for "Tristan und Isolde." The lowest frequency contributions in the spectrum of the lowest notes, however, are developed only relatively weakly. The intensity maximum is not formed until the 8th or 9th overtone. This frequency location corresponds to the region of the main formant, which lies at about 500 Hz, and which particularly for bassoons is very pronounced. (Lehman, 1962; Meyer, 1966c; Strong and Clark, 1967).

The tone color which is very similar to the vowel "o(oh)" is not only caused by the central location of the amplitude maximum in the region of this vowel, but also rests on the fact that the width of the bassoon formant nearly corresponds to that of the vowel "o(oh)." For all other instruments the formant width is wider than the width of the vowel with corresponding frequency location. The logarithmic decrement has a value of 1.4 (Meyer, 1968) for the prevalent construction of a German bassoon, the corresponding value for a sung "o(oh)" is 1.2 (Tarnóczy, 1943). A fitting example for the dramatic exploitation of this similarity of the bassoon with the human voice is given by the duet of Aida and Amonasro from the Nile scene in the opera Aida (see score example 7), where also during the singing of Aida the presence of her father is given musical expression.

This formant location around 500 Hz characterizes the tones of the bassoon in the region up to C_4. Below this formant, the power spectrum drops with a slope of approximately 8 dB/octave, so that the fundamentals of the lowest registers are developed correspondingly weakly. In the fourth octave the amplitude maximum is shifted along with the 2nd partial to somewhat higher positions, so that the tone color approaches an "a(ah)." From about B_4 on up, the fundamental dominates, so that the vowel color is no longer as pronounced.

Score example 7 G. Verdi, Aida, 3rd Act, Duet Aida – Amonasro, score excerpt

Additional tonal characteristics of the bassoon include a very high number of overtones, which form a series of secondary formants in several strong groups. As already recognizable from Fig. 3.10, and represented again by Fig. 3.14 in an other context, these secondary formants lie in the regions of 1,150, 2,000, and 3,500 Hz. Of these, the first one falls into the region of the vowel color of a bright "a(ah)" which contributes to a strong timbre. In the higher regions of the tonal range it is shifted more into the region of the nasal components, so that the tone picture no longer gives such an open impression. A typical example for this somewhat nasal color of the bassoon in the upper registers is given by the Beckmesser-motif in the "Meistersinger" (see score example 3, p. 48 Sect. 3.1.1.4) The two higher frequency formants account for a tonal brightening effect corresponding to their position near the vowels "e(eh)" and "i(ee)," and prevent the tones' becoming too dull or blunt.

The frequency location of the main formant, as also the intensity of the secondary formants, depend within certain limits on the performance technique. If a higher intonation is required it can be achieved by raising lip pressure. The drop of sound level associated with that must be compensated by a simultaneous increase in air pressure. While the sound level remains the same, the formant rises slightly along with the upward pressed tone, and the overtones gain in intensity (Smith and Mercer, 1973). The tonal character as a whole, thus becomes brighter. This effect must be considered when choosing the reed: the lower the reed is tuned, the brighter will be the tone color, assuming the same intonation.

3.2.4.2 Dynamics

The brightening effect of the secondary formants is most noticeable for mid-loudness levels. For *ff* the overtone series reaches to above 12,000 Hz, so that these very high tone contributions largely determine the tone, and the maxima around 2,000 and 3,500 decrease in importance. Here, however, directional effects also are relevant (see Sect. 4.3.4), since the power level spectrum decreases above 1,000 Hz with a slope of 20 dB/octave. The high frequency overtones of the lower registers are distributed so densely that they assume a noise-like character, which can lead to a certain hardness when the tone is too forced. Additionally, the secondary maximum, which for *mf* lies near 1,150 Hz, and is shifted into the region of the nasal components for *ff*, exhibits a greater intensity in the upper registers than the principal maximum. Through this the tone loses sonority.

In contrast, at lower loudness levels, the principal formant is shifted toward a darker tone color, furthermore, for *pp* the power spectrum above approximately 600 Hz already drops by 25 dB/octave in the low register, and by 35 dB/octave in the middle and higher registers, so that the tone becomes rounder and more damped. However, for *pp*, noise contributions in the frequency region around 3,000 Hz can be formed in the middle register, which come about because of Eigenresonances of the reed. They depend in their intensity within wide limits on reed properties (Meyer, 1966c). The consequence of this is that the dynamic tone factor of the bassoon – at least in the lowest octaves of the range develops tendencies which are contrary to usual expectations: When the strongest tone contributions in the region of the main formant are changed by 1 dB, the components in the region around 3,000 Hz are strengthened or respectively weakened by 0.6–0.9 dB, only from the third octave upward does the "dynamic tone color factor" follow the usual tendency with values of 1.2–1.5.

In the bassoon, the dynamic range depends in particular measure on the speed of performed notes. For rapid scales the power level can be varied between 81 and 96 dB; in the mid register long notes can be weakened to 72 dB and raised to 102 dB, which corresponds to a dynamic range of 30 dB. For the low and high registers this is narrowed to a dynamic range of about 25 dB. The power level for a mid- range *forte* lies near 93 dB, as for the oboe and the clarinet.

3.2.4.3 Time Structure

In spite of the low pitch of the instrument, the attack of the bassoon is very precise. The reason for this lies in the fact that the overtones in the middle and high frequency regions have a very short initial transient; they already reach their final strength within about 20 ms as already shown in the example in Fig. 2.2. Through this, the beginning of the tone is clearly defined. As is the case for other reed instruments, no additional noise accents appear, so that the attack gives a very pure impression. Nevertheless, as seen in Fig. 3.13, the low frequency contributions (below about 200 Hz) for such sharply attacked notes require a longer time for the initial transient – a value of 20 ms would only involve about two vibrational periods for the fundamentals of the lowest notes.

Since the lower tone contributions are relatively weak from an intensity standpoint, their initial transient time of 50–80 ms does not reduce the precision of the attack, however, a short staccato in the lowest registers can influence the sonority. For the bassoon, most references in the literature give an overall value, without frequency weighting, for the initial transient time in the order of 30–40 ms, which characterizes the precision of the attack in comparison to other instruments (Luce

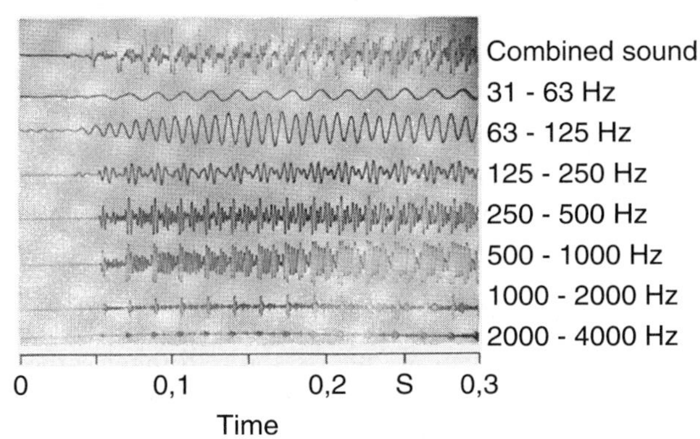

Fig. 3.13 Octave filter oscillogram of the initial transient for a bassoon (played note B_1^b)

Score example 8 C. M. von Weber, Der Freischütz, Beginning of the overture

and Clark, 1965; Melka, 1970). Even for a soft attack, the initial transient is not lengthened significantly above 60 ms, however, the tonal impression in that case is primarily determined by the fact that the higher frequency components enter later than the lower components and that the initial transient is slower than for sharply attacked notes. This difficulty, to achieve a soft entrance, is noticeable, for example, at the beginning of the Freischütz overture, when the bassoon is expected to merge unnoticeably into the "from Nothing" swelling string sound, without lending an accent to the joint beginning. (Score example 8).

The decay processes are short, as is the case for all wind instruments. The decay time lies at around 0.1 s for the higher notes (as was recognized from Fig. 2.2) and can be lengthened to 0.4 s for the lowest registers (Rakowski, 1967). The influence of the player is minor.

When bassoon players play a pronounced *vibrato*, the width is about ± 15 cent. This results in level fluctuations of only 4 dB for the strongest partials, i.e., in the region around 500 Hz. The strongest level fluctuations occur between around 1,400 and 1,800 Hz and can reach up to 15–20 dB. Since, by reason of their time structure, they draw the attention of the listener, they emphasize the nasal quality of the sound. While all overtones fluctuate in phase, one could speak of a tone color modulation, still, the tone receives an inner stability through a number of weakly modulated partials with fluctuations of only 5–6 dB, which do not permit that effect to become as noticeable as is the case with the brass instruments and the flute.

3.2.4.4 The French Bassoon

While the bassoon of so called German construction is preferred in almost all countries, in France and also in a few east European countries instruments are played with different fingering and different tonal characteristics. These so called French Bassoons were historically developed alongside the German Bassoon and represent a certain parallel to the Horns in France. In the tone picture, the nasal components are also strongly pronounced, as the juxtaposition of the formants shows in Fig. 3.14. While the basic formant in comparison to the German instruments is shifted in the direction of a darker "o(oh)," all secondary formants, however, are located noticeably higher. As a whole, the tone of the French Bassoon is richer in overtones, and the vowel color of the basic formant (with a logarithmic decrement of 1.6) is somewhat less firmly determined, since it is broader than in the German model.

By reason of all these characteristics, French bassoons sound less forceful, and have a more slender tone, which in connection with the good attack at the pitch of these instruments is particularly suited for virtuoso passages. In their tonal character, therefore, they serve the compositions of the French impressionists particularly well, as also the older, preferably virtuoso wind literature, while the German bassoons are based more in the tonal conception of the romantic epoch.

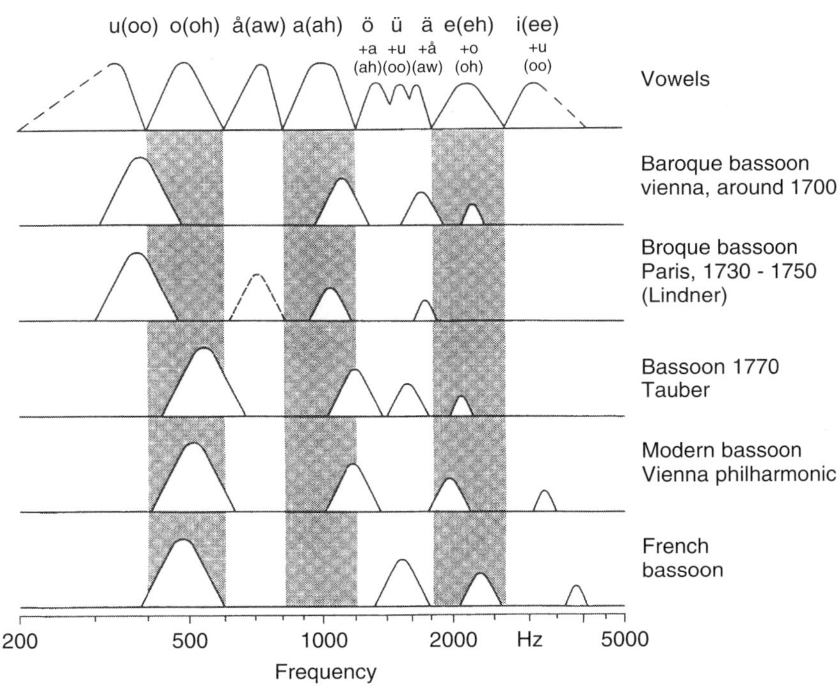

Fig. 3.14 Schematic representation of the formant location of several bassoons

3.2.4.5 Historic Bassoons

While data about instruments in use today are based on a large number of instruments investigated, and thus are generally applicable, for historic instruments, naturally only individual examples can be cited. For the bassoon, results are available for two baroque instruments, and one example from the classic period, all three instruments are well preserved. They are still being played in ensembles for old music. While they are not as balanced as today's instruments, some typical characteristics can be drawn from their formant locations (see Fig. 3.14) (Meyer, 1968).

Particularly noticeable is the low frequency position of the main maximum for Baroque bassoons, which results roughly in a bright "u(oo)" as the basic color. This, combined with the lower location of the secondary formants and the already relatively small overtone content, brings about a tonal effect darker than for today's instruments; it can on occasion even become somewhat muffled. To this is added, that the logarithmic decrement of the main formant in the lower octaves of the tonal range lies at around 1.2, while in contrast, it is above 2.0 for the upper registers. This signifies, that the tonally best regions for the Baroque bassoon are located in the lower regions; thus, these instruments seem especially suited for supporting the continuo foundations or for characterizing sinister moods in operas or oratorio scenes, as for example the appearance of Samuel's shadow in Händel's "Saul."

In contrast, the tonal picture of the bassoon looks entirely different in Mozart's time. The main formant is situated in the region of the vowel "o(oh)" as with today's instruments, however, for the upper locations of the tonal range, it moves right up to the boundary with "å(aw)." The secondary formants are concentrated near a very bright "a(ah)" nasal component, and finally a middle "e(eh)," so that they add a character which is not too weak but slender, to the already relatively bright fundamental color. The logarithmic decrement of the main formant with a value of 2.1 in the lowest register, is less favorable than for the higher notes with values of 1.4–1.6. The best location from a tonal standpoint, i.e., the region with the most pronounced formants, is therefore in the middle and the upper registers, while the low register is somewhat duller. With that type of timbre, the instrument is particularly suited for supporting precision in the bass groups as also in higher song-like passages.

3.2.4.6 The Contrabassoon

With a lower limit of the tonal range at B_0^b or even A_0 (e.g., in "Salome") the contrabassoon, next to the contrabass tuba, is the lowest instrument in the orchestra. Its spectral region thus begins at 27.5 Hz. However, the radiated energy at that pitch is very low. Below the first maximum in the power spectrum the level drops – as also for the bassoon – with 8 dB/octave. This first maximum in the overtone structure is located at about 250 Hz. Secondary formants are found near 400–500 Hz as well as in the region around 800 Hz for greater loudness (Meyer, 1966c). The basic color can, therefore, be described as a dark "u(oo)" with an addition in the region of a dark "o(oh)."

Because of the double bend in the body, the tonal development over the range of the instrument is not as uniform as for the normal bassoon. Particularly in the lowest register the strongest tone contributions switch between partials of different order. Since the ear is significantly more sensitive in the region around 400 Hz than in the region of the lowest partials, the strongest partials in the sensitivity spectrum become more prominent and make the pitch orientation more difficult for the lowest notes. For example, in the descending tone sequence B_0– B_0^b the rise between the strongest partials, namely from about 395 Hz (13th partial of B_0) to about 405 Hz (14th partial of B_0^b) can be heard very clearly. This results in a perceived uncertainty about whether the note sequence rises of falls.

Noteworthy, however, is the relatively short initial transient for the contrabassoon. For tongued notes in all registers this amounts to only from 30 to 35 ms, so that the instrument gives an impression of rather high agility in spite of its low pitch (Melka, 1970). On the other hand, its dynamic range in comparison to a bass clarinet, is relatively narrow: for individual notes it does go up to 15 dB, yet in a melodic context it drops to 10 dB. This is caused by a relatively high lower limit of the dynamic range with a power level of 86 dB at a *pp,* while for *ff* the 96 dB value comes close to the upper limit of a normal bassoon.

3.3 String Instruments

3.3.1 The Violin

3.3.1.1 Sound Spectra

For wind instruments the tonal picture is primarily determined by resonance effects of the enclosed air column, the vibrational characteristics of which are determined by the dimensions of the bore diameter, the bell, the mouthpiece etc; the material, in contrast only plays a subordinate role. For string instruments, on the other hand, the tone is primarily formed by the resonance characteristics of the corpus. Since, however, the acoustical properties of the wood differ from instrument to instrument, the result is a multitude of variants.

The basic structure of the power spectrum of string instruments is determined by the spectrum of the string vibrations excited by the bow. In this spectrum the fundamental dominates, and the subsequent overtones decrease with increasing frequency at a rate of 6 dB/octave. Above the bridge resonance, i.e., above approximately 3,000 Hz for violins, the power level decreases by 15 dB/octave. Superimposed on this is the shaping of the tone by the corpus, which possesses a large number of resonance regions, fixed in frequency, they thus are not shifted with performed pitch – as is the case for wind instruments. For a chromatic scale, therefore, the excitation spectrum moves over the resonance chain, which remains fixed in frequency, so that the tonal character can change from note to note. The tonal spectra of string instruments are thus not as uniform and systematically structured as those of wind instruments, but rather exhibit a greater range of color (Leipp, 1965; Lottermoser and Meyer, 1968).

By reason of this cooperation between string vibration and resonance body, the fundamental is almost without exception the strongest partial of the spectrum for the violin in the upper registers. This is valid for the entire range of the E-string, for the A-string from about E_5, and for the D-string from about C_5 on up. In addition, a dominating fundamental appears in two especially well defined resonance regions of the violin, namely around G_4 and between C_4 and D_4. This lowest resonance is associated with the normal mode of the air volume enclosed in the body and is designated as the air resonance. Below this resonance, the fundamental drops in the power spectrum by about 4 dB per half step, so that it lies around 20–25 dB below the strongest partial on the G-string.

As an example of sound spectra of the violin, Fig. 3.15 gives measurement results for the four lowest notes of an instrument with very good tonal qualities. In comparison with the chromatic sequence for a horn in Fig. 2.4, the greater multiplicity in the distribution of partial intensities in the violin is clearly recognized. Thus, neither a steady course of the envelope nor the domination of odd partials can

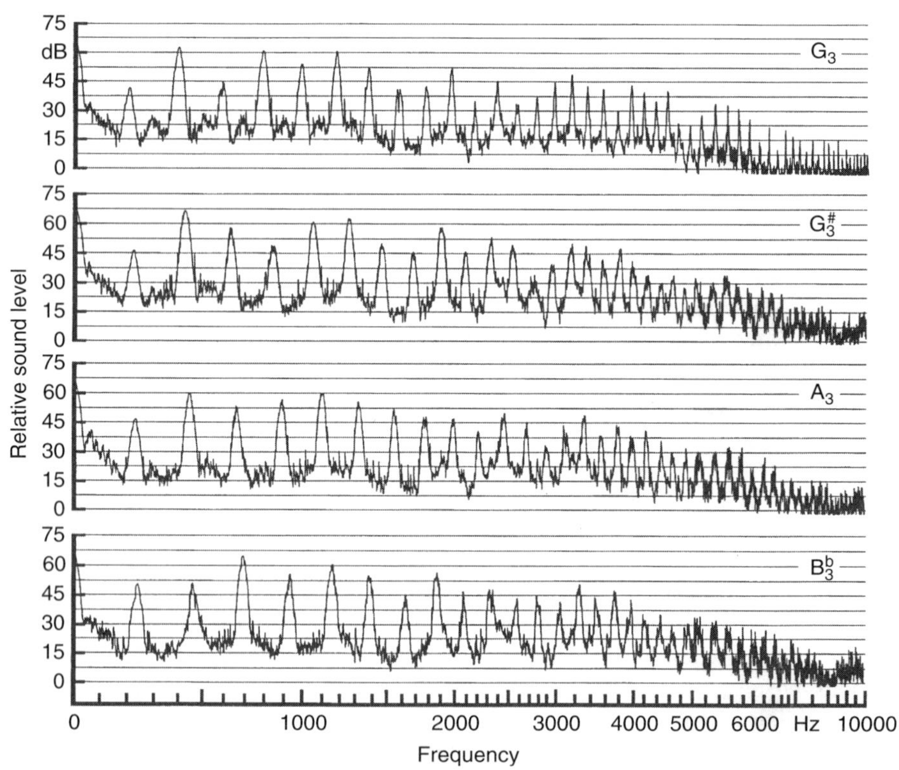

Fig. 3.15 Sound Spectra of a violin by Guarnerius del Gesù (1739)

be read from the curve, rather the spectra are characterized by individual prominent partials which happen to coincide with a resonance of the corpus. For G, $G^{\#}$, and A, pronounced octaves follow weak fundamentals, while for B^{b} the third partial is strongest. Beyond that, up to nearly 10,000 Hz the spectra then show individual partial groups of larger intensity which show a formant-like character. For example, this occurs between 1,000 and 1,200 Hz as well as between also between 3,000 and 4,000 Hz.

As evident, the formant location for the individual notes, varies somewhat, it is entirely possible that for certain notes, in relation to the location of body resonances, no such predominant partial groups are formed. The number of notes within the full chromatic range, for which formants appear, is therefore a typical quality criterion for the instrument, and the vowel-like tone color supports the song-like tonal character. For example, in the well known Stradivarius violin "Prince Klevenhüller" for 40 of a total of 52 well studied notes, formants could be demonstrated (Lottermoser and Meyer, 1968). In this context it needs to be emphasized that the production of these strong partial groups, in contrast to other tonal characteristics, depends only in very small measure on the quality of the performance.

Typical violin formant locations for the low notes are around 400 Hz; as a dark "o(oh)" these sounds provide the sonority for the lower G-string. In the region from 300 to 350 Hz, old Italian violins of the Stradivarius type already show a higher intensity than French violins or most newer instruments (Meyer, 1982a; Dünnwald, 1988). Through this, the tonal difference between the notes in the region of the air-resonance (about B_3 to D_4) and the lower notes of the D-string is especially diminished. A second formant region, covered by the vowel color "a(ah)," i.e., between 800 and 1,200 Hz, is very essential. This partial group must be considered as particularly characteristic for the violin sound, it lends strength and substance, and prevents the nasal timbre. The exact frequency location of this formant is very significant for individual traits within the tonal picture: in violins of Guarnerius del Gesù tonal contributions between 1,000 and 1,250 Hz are almost always especially emphasized, as also noted in Fig. 3.15. These instruments are therefore even stronger than the Stradivarius violins, for which the maximum clearly lies below 1,000 Hz. Instruments with a pronounced dark timbre have the corresponding formant even below 1,000 Hz, that is at a tone color tending more toward an "å(aw)"; This characteristic is also found in several Stradivarius violins. In this context it should be noted that a spectrum with a del Gesù-like sound, with maxima near 400 and 1,200 Hz, was perceived as particularly "well-sounding" in subjective hearing tests among a multitude of tones. (Terhardt and Stoll, 1978).

The next higher frequency region around 1,600 Hz, which is responsible for the covered tone coloring and also for a nasal timbre, is radiated more strongly by a decided majority of old Italian violins in contrast to other instruments. This removes a certain hardness or directness from the sound. A series of further formant regions are responsible for the brilliance and brightness of especially the upper tones, the most important of these are between 2,000 and 2,600 Hz, i.e., in the vowel region of "e(eh)," and between 3,000 and 4,000 Hz in the region of the vowel "i(ee)."

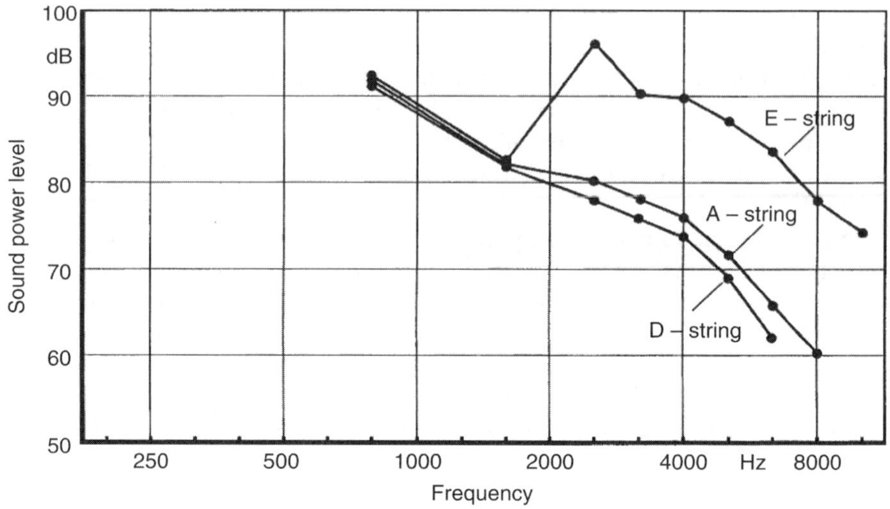

Fig. 3.16 Power level spectra of a violin, the note A_5 is played *ff* on three different strings (after Meyer and Angster, 1981)

The frequency range of violin tones depends on various components of performance technique. The largest number of overtones appear with an open string, since the string termination is more sharply defined at the nut than is possible with a pressing finger. However, the more firmly the finger is pressed, the richer in overtones, and the more precise other notes become. Naturally the relative position of the note also is of importance. Thus, the change from the D-string to the G-string shows two characteristics typical for the transition to a darker tone color: the formant maxima shift to somewhat lower frequencies, where especially the region around 1,000 Hz is weakened, furthermore, the overtone content above about 2,000 Hz decreases strongly. While the difference between D and A strings is least pronounced, the overtone content increases significantly in the transition from the A to the E string. Fig. 3.16 shows a comparison of power spectra for A_5 played on the D –, A – and E – strings with equal strength. Above about 2,500 Hz, i.e., above the bridge resonance, the overtone level of the E – string lies about 15 dB above the corresponding tone contributions of the A – string, in contrast the D – string has an overtone radiation of only about 3 dB less than the A – string.

The sound spectrum depends strongly on bowing technique. The three influential quantities to be modified by the performer are: the speed of the bow, the pressure exerted by the bow hair on the string, as well as the point of contact at which the bow touches the string. The bowing speed influences fundamental and overtones equally, it is, therefore, the most important means of influencing dynamics. In contrast, bow pressure has no influence on the fundamental, assuming equal bowing speed, an increase in bow pressure primarily raises the intensity of higher overtones (Bradley, 1976; Cremer, 1981).

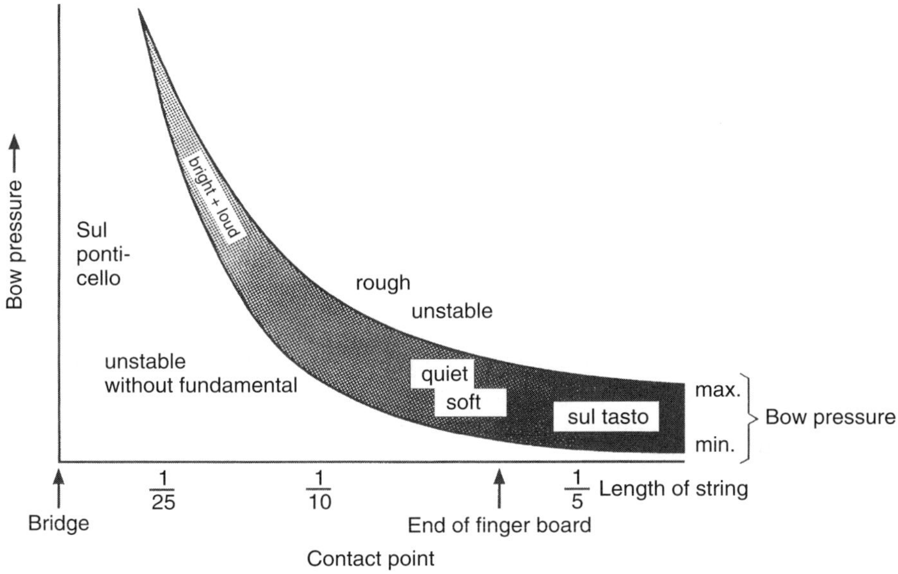

Fig. 3.17 Dependence of bow pressure on contact point for constant bow speed, The shaded area represents the actual performance region with its tonal and dynamic expression possibilities

The point of contact, on the other hand, influences the entire spectrum. Thus, the closer the contact point is to the bridge, the more bow pressure is needed. In Fig. 3.17 this relationship is represented. Below the line of minimal bow pressure, the available energy is insufficient to maintain a stable vibration. Above the line of maximal bow pressure, the restoring force of the string is insufficient to force strictly periodic vibrations. Increased fluctuations and added noise contributions, therefore, cause the tonal picture to be rough, which can create the impression of scratching. The closer the contact point lies to the bridge, the brighter and louder the sound becomes. However, the variational range of the permissible bow pressure also becomes narrower. Soft sounds with covered coloration arise from a contact point near the finger board (Schelleng, 1973; Meyer, 1978b).

The attack noise, generated by the bow, in addition to the harmonic partial spectrum, belongs to the typical tone characteristics of string instruments (Lottermoser and Meyer, 1961). The fact that this hissing noise is a specific component of the tonal picture has become particularly evident in experiments with electronic imitation, where the harmonic spectrum alone could not create the impression of a string instrument. As already suggested in Fig. 2.6, the bow noise receives a typical coloring through the body resonances. This is the same for all notes of an instrument, and it stands out, especially in hearing impressions in the upper registers, since here the partial spacing is greater. Irregularly excited vibrations of the air volume (around C_4) of unique pitch are audible as a color noise, and so is a broader resonance region above 400 Hz. In comparison to wind instruments, these noise contributions – particularly for lower dynamic levels – are relatively strong for strings. For equal loudness of the harmonic tone contributions, the noise components of string instruments are by about 20–30 dB stronger than for wind instruments, with the exception of the flute.

The shaded area represents the actual playing region with its tonal and dynamic expression possibilities.

3.3.1.2 Dynamics

Dynamics for string instruments are essentially responsive to bow speed and the location of the contact point. Assuming that the bow speed, in practical terms, can vary between 10 and 125 cm s^{-1}, this corresponds to a dynamic range of about 22 dB. The distance of the contact point from the bridge, between the closest and most distant point increases by a factor of six, which corresponds to an additional 16 dB. In principle, therefore, one can count on a dynamic range of just under 40 dB. This range, however, cannot be fully exploited. For some notes, strong resonances make a *forte* easier, they also make a *pp* very difficult. Thus, near the air resonance it becomes difficult to make an extreme *pp* speak.

The softest *pp* which can be performed on a violin depends on the strength of the basic noise. This, in turn, maintains a fixed intensity from *pp* to beyond *mf* and only increases for very large playing volumes (Lottermoser and Meyer, 1961). The lower limit of the dynamic range of the violin is therefore given by the condition, that the partials must extend sufficiently clearly beyond the noise. As a result, some favored isolated notes have a *pp* power level of only 58 dB, however, this rises to 74 dB for rapid tone sequences. For *ff* a violin reaches power levels of 94 dB for fast sequences and 99 dB for slow ones, so that one can count on a practical dynamic range of about 30–35 dB. The power level of a median *forte* lies near 89 dB. (Meyer, 1990).

3.3.1.3 Time Structure

The excitation of string vibrations by the bow offers far greater flexibility for shaping the initial transient than is the case for most wind instruments. Nevertheless, even for sharp attacks on a violin, initial transient times are longer than for example on an oboe. On a G string, a fast initial transient process lasts almost 60 ms (see also the upper example in Fig. 2.2), this, however, is shortened in the upper registers. In the mid-range values between 40 and 50 ms can be expected, on the E string the initial transient time is shortened down to almost 30 ms (Melka, 1970). For gentle attacks the duration of the initial transient process can be extended to 200–300 ms without having the tone disrupted in its slow development. Thus a very rounded tone picture emerges.

Inasmuch as the strength of the higher overtones depends essentially on bow pressure, their development is somewhat delayed, which does not aid a *staccato* attack. Deliberately high bow pressure in an attack (e.g., in a *detaché*-stroke) creates a broad-band articulation noise, with a duration of about 50–100 ms.

Since this noise contains preferentially the frequencies of the strongest resonances, the vibrations of these resonances can be heard within the total sound, partly with an almost tonal effect, which leads to a basic color of the tone, common to all sounds of the instrument. These initial transient effects of the resonances are most noticeable while playing *col legno*, because in that case the string vibrations do not come to full development.

Finally, it also needs to be mentioned, that during an attack or string change, the frequency of the note played is lowered by 10–20 cents when compared to the final value. This also influences the aural impression of the total "initial transient" phenomenon, even if a pitch variation does not become directly noticeable. What is perceived is merely the less precise attack when compared to an oboe or clarinet.

The decay time for string instruments depends in large measure on whether the bow remains on the string or is lifted. As the middle example of Fig. 2.2 has shown, the decay time in the first case is only about 0.1 s. If the bow is lifted, the tone will continue to sound longer, depending on the length and mass of the string. For a violin, decay times of 1 s in the lower and 0.5 s in the upper registers are typical. For open strings, decay times lie between 2 and 3 s. All these values are given in relation to the relevant fundamental; the first overtones continue to vibrate at most half that long, and the higher overtones are practically damped immediately (decay times of less than 50 ms).

A pronounced *vibrato* for violinists has a width of the order of ±30–35 cent. These frequency fluctuations of the played note lead to the circumstance that the individual overtones in their increasing and decreasing frequencies move up and down on the flank of corpus resonances and are modulated in intensity. Depending on whether they fall on the rising or falling flank of the resonance their intensity pulsations occur in equal or opposite phase. This effect is clearly recognizable in

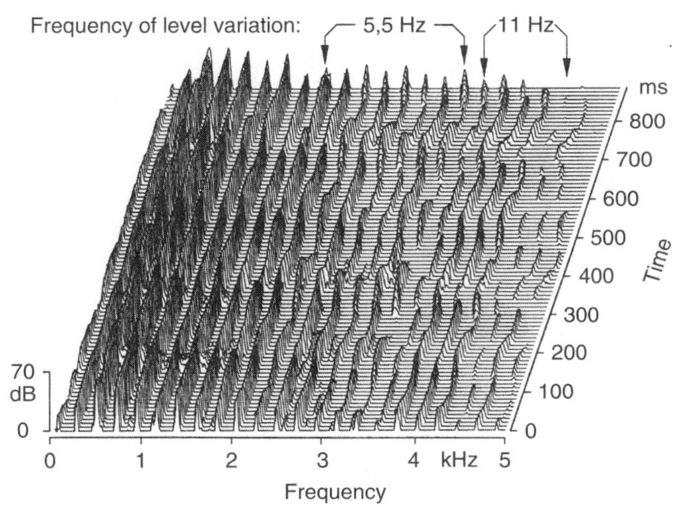

Fig. 3.18 Changes of the tonal spectrum with time for a *vibrato* in bowed strings (Violin, note A_3)

Fig. 3.18: The 8th partial (at about 1,800 Hz) has its maximum at the frequency maximum, while, for example, the 15th partial (at about 3,300 Hz) has its maximum at the frequency minimum. It can also happen that the frequency of a partial straddles a resonance peak, and that thus during a period of the frequency motion two amplitude maxima and minima appear. An example of this amplitude modulation at twice the frequency is the 20th partial (at about 4,400 Hz) while the 16th partial (at about 3,500 Hz) runs through a low point, and thus correspondingly pulsates with a phase opposite to that of the 20th partial.

When a pronounced higher partial is modulated in this fashion at twice the frequency, it can stand out in a penetrating way and lead to a kind of roughness of the sound. While in exceptional cases individual partials can exhibit fluctuations of up to 25 dB, in most cases for frequency groups relevant to the ear, a partial compensation between components of opposite phase occurs, so that the ear only senses fluctuations of the order of 10 dB. The lower partials, still perceived by the ear as separate, fluctuate typically by 3–6 dB below 1,000 Hz and by 6–15 dB between 1,000 and 2,500 Hz at a vibrato of ± 35 cents (Meyer, 1992).

The time structure plays an important role in polyphonic chords, since they can either be attacked sharply or be broadened by arpeggiation. The left picture of Fig. 3.19 shows the temporal spectrum development for the notes G_3 (green), E_4 (yellow) and C_5 (red). Disregarding the attack noise, one can recognize that G_3 and E_4 enter simultaneously and that C_5 follows with a delay of about 40 ms; the overtones are developed slightly later. After 170 ms the G_3 is no longer excited, as is evident by the detaching overtones. For the deliberately broadly attacked *arpeggio*-triad in the right picture the notes E_4 and C_5 only follow the low G_3 by about 250 ms, this note in turn decays quickly, and only continues in the fundamental and the octave partial. These two examples provide the limits of the width of possibilities for a temporal structuring of such chords.

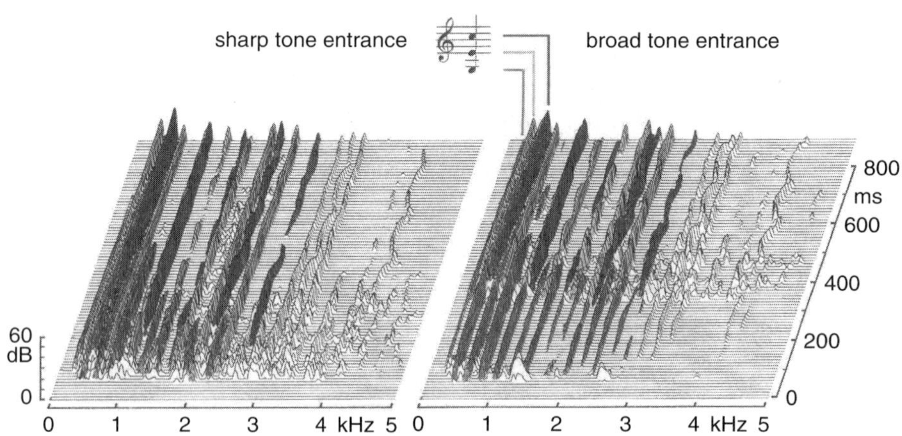

Fig. 3.19 Time evolution of tone spectrum for a three note violin chord with two different playing approaches (See Color Plate 1 following p. 178)

3.3.1.4 Special Performance Techniques

For a *pizzicato* the string is displaced slightly from its rest position and can vibrate freely after release. This results in an extremely short initial transient, which for the violin involves less than 10 ms in all registers (Melka, 1970). Inasmuch as the noise components, which are also excited, lie far below the actual partials in strength, the lower limit of the dynamic region can be reduced to 51 dB. For *ff*, in contrast, sound pressure levels up to 90 dB can be achieved, albeit only for short durations. Accordingly the dynamic range of 39 dB is of the same order of magnitude as for *arco* performance.

The spectrum for pizzicato tones depends largely on the plucking location. For plucks near the bridge, the sound becomes rich in overtones and hard. A softer timbre results above the fingerboard. It assumes a somewhat covered character when one plucks in the middle between the bridge and the fingering touch, since at that position the even partials are only excited weekly and the odd partials dominate.

The audible decay, and thus also the duration of *pizzicato* tones, naturally depends on the loudness of a performance. For *pp* it varies between 40 and 150 ms and for *ff* between 350 and 800 ms. Open strings have the longest decay times, and notes of equal pitch decay more rapidly when fingered in high positions in comparison to low positions (Spelda, 1968). Furthermore the upper overtones decay more rapidly than lower contributions, as is noted from Fig. 3.20. Thus, the decay time, as noted in the upper picture, is approximately 3 s for the fundamental and 1.5 s for the first overtones. The influence of the *vibrato* on the tone of the string *pizzicato* is noteworthy. As illustrated by the lower picture, the tone initially sounds more lively. It becomes notably dryer, during the *vibrato* since the motion of the fingers provides additional dampening. The decay time is thus cut in half when compared to the note without *vibrato*. In the example pictured, to 1.5 s for the fundamental and from 0.7 to 0.8 s for the subsequent overtones. This effect is unnoticed by many string players (Meyer, 1992).

Placing a mute on the bridge dampens primarily the high-frequency contributions. However, there are some additional effects, the details of which depend on the nature and the weight of the mute. With increasing mute weight, the resonance of the body, normally found at 400 Hz, is shifted to lower frequencies. It thus approaches the air resonance. This leads to an increased intensity of the fundamental for the lower notes of the D string. A determining factor in this is also the strength and the weight of the bridge, so that the effect varies from instrument to instrument. Resonance shifts can also occur in the region of higher frequencies, without, however, causing higher intensities than for normal play.

The amount of pitch-damping is important in its effect on tone color. Most mutes dampen components down to those frequencies which lie above nasal contributions. As a result, the nasal tone is created which frequently is considered typical for violins with mutes. Particularly light mutes naturally cause the least change, creating a somewhat lighter and barely nasal tone. They are especially suited for

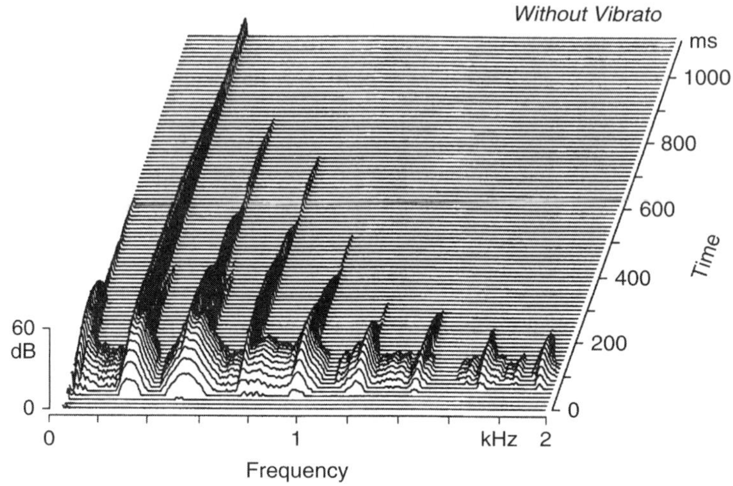

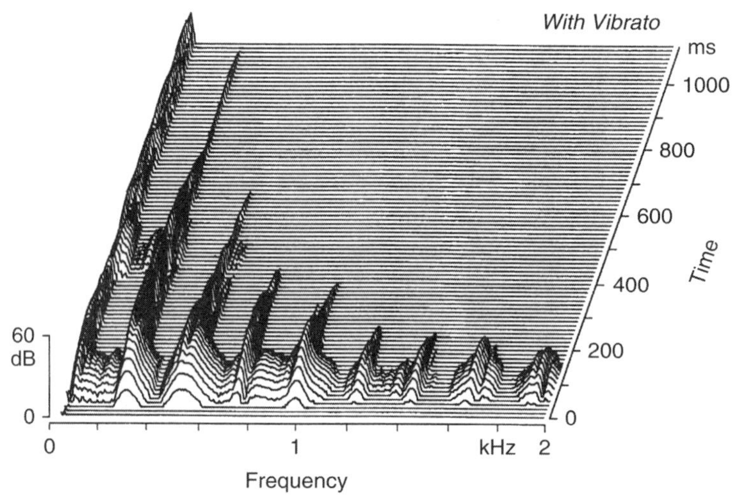

Fig. 3.20 Influence of *vibrato* on time evolution of the tonal spectrum of a string *pizzicato* (Violin, B_3)

low passages, for which the resonance shifts, mentioned earlier, toward the lower notes of the D string caused by heavy mutes would be disturbing. On the other hand this type results in a very soft and covered sound for the A and E strings, since they also reduce the nasal contributions. However, this does require a weight of about 6 g which is available in a five prong steel mute, for example. Through the choice of suitable mutes, a number of different tonal variations can be achieved. For most mutes a reduction of radiated sound power of the order of 6 dB results. For very heavy mutes this can amount to 10 dB.

"Harmonics" (also called *flageolet* tones) are played by gently touching the string. The fundamental of the open string is also excited very weakly by the bowing noise. The noise background thus receives a character similar to the flute, resulting in the unique timbre of such tones. Since the excitation for these tones is rather critical, the initial transient is somewhat longer than for normally fingered notes, however, because of the relatively large energy of the entire string vibration, the decay time is quite large. The dynamic range is around 20 dB, significantly narrower than for normal tones. It is possible to play a very gentle *pp* with a power level of 64 dB, however, the upper limit of an attainable power level of 84 dB is very low (recalculated from Burghauser and Spelda, 1971). For so called "artificial harmonics," which are created above a fingered note, the dynamic range is yet more stringently reduced.

3.3.2 The Viola

3.3.2.1 Sound Spectra

Tonal characteristics of the viola naturally resemble those of violins in essential features (Fletcher et al., 1965). Lower tuning (by a fifth) of the viola extends the range down to C_3 (130 Hz). This means that for the lowest radiated frequency the wavelength is 1½ times larger than for the violin. The dimensions of the viola, depending on body size, however, are only from 1.15 to 1.2 times larger than for violins. Thus, the typical resonances are lowered by significantly less than a fifth when compared to a violin, and the tone color is not that much darker, which would correspond to a tuning lowered by a fifth. Therefore, many attempts have been made to build larger violas in an effort to increase the fullness of sound in the low register.

The air resonance of violas generally lies in the neighborhood of A_3, around 220 Hz. Correspondingly, the amplitudes of the fundamental of the C string are very weak. For C_3 they can lie by more than 25 dB below the strongest partial. The low notes of the viola have their largest intensity in the two resonance regions around 220 and 350 Hz. These two fall into the vowel region of "u(oo)." They provide substance for the tone. Further formant regions, which primarily affect the mid-register and high notes are located around 600 Hz, which is near the transition between the vowel colors of "o(oh)" and "å(aw)," as well as around 1,600 Hz. The last maximum brings out the frequently noticed nasal timbre of this instrument group. As a whole, the intensity of the higher overtones is less than for violins. The resulting tone is therefore not as brilliant. In good violas, however, the timbre is additionally brightened by a secondary formant in the region of the vowel color "i(ee)" (between 3,000 and 3,500 Hz). This also weakens the nasal effect.

3.3.2.2 Dynamics

The lower limit of the usable loudness region for the viola lies slightly above that of the violin: In the low registers slow notes can go down to a power level of about

67 dB at *pp*, and in the regions of the D and A strings to about 63 dB. For a *ff*, in contrast, the radiated power level is somewhat lower than for the violin. It lies at about 95 dB, and only the rarely used pitch range above C_6 drops to values below 88 dB. For rapid note sequences the usable dynamic range drops to a region of 73–91 dB. Accordingly, the dynamic range of the viola is slightly less than that of the violin, which is also related to the somewhat more sluggish initial transients in the extreme registers. As with other bowed string instruments, the influence of performance technique on the higher overtones is small. The "dynamic tone color factor" lies near 1.1. With a value of 87 dB the power level for a median *forte* is by 2 dB lower than for violins.

Only when playing harmonics does the viola exceed violin radiation by about 3 dB. The reduction of radiation by using a mute generally does not reach the same values as for violins, since mutes in relation to bridge mass are not as heavy. A level reduction of about 4 dB can be expected.

3.3.2.3 Time Structure

In spite of its larger dimensions, for sharply attacked notes the viola does not require a longer time for full tone development when compared to the violin, in fact, frequently even shorter initial transients were noted. Accordingly, initial transients for *staccato* notes occupy about 30 ms. These values are valid even for the lowest registers, where the open C string can exhibit an initial transient of 20 ms (Melka, 1970). This result not only requires optimal conditions for string and bow, but it is also deceptive in as much as it covers the fact that the already weaker lower tonal contributions do not reach their final intensity in such a short time. This is hardly annoying for long notes. However, for very short notes the tonal character of the instrument is shifted toward a brighter, and often nasal tone color.

The possibilities for expressive shaping of an attack are not equal for all pitch ranges of the instrument. Thus, in the range of the C string the initial transient can be broadened only to 100 ms by an appropriately soft attack. With rising pitch, the variation possibilities increase. On the A string, tone development times of more than 200 ms can be achieved in the upper registers. In this context it is noteworthy that these soft attack values for the viola are also less than for the violin, and that they exhibit a tendency opposite to the violin: in a violin the lower notes are more strongly influenced than the high ones.

For the viola, the decay time is negligibly longer than for the violin. For detached bowing, values around 1 s are observed for fingered notes in the mid-register, and 3–4 s for open strings. As is the case for violins, overtones decay more rapidly than the fundamental.

3.3.2.4 Pizzicato

The initial transients of plucked strings on a viola are nearly as short as for the violin. Only on the C string is this transient slightly longer than 10 ms, for higher strings this time is not exceeded (Melka, 1970). However, violin intensities are not achieved. For *ff* the power level is about 90 dB, and for chords over all four strings this value can be raised by 3 dB. The softest *pp* can be assumed to be about 58 dB, so that even in regards to the lower limit a narrowing of the dynamic range results, when compared to the violin.

The decay of *pp* tones lasts 50–150 ms, for *ff*, *pizzicato* tones reach a duration of 280–600 ms (Spelda, 1968). These values are similar to those for violins. A certain difference comes from the fact that with the viola the open strings decay noticeably longer (500 ms on the average) than fingered notes (about 400 ms), while this difference is not as pronounced for violins.

3.3.2.5 Viola d'amore

Even though the Viola d'amore blossomed already in the time of the Baroque and the early classic period, this instrument, with its peculiar tone color is found occasionally in modern operas such as "Katja Kabanova" or "The Affair Makropoulos" by L. Janacek, or "Louise" by G. Charpentier. These examples follow H. Pfitzner, who used the instrument in his "Palestrina," and G. Puccini, who used its peculiar sound effect as an accompaniment of a choir behind the stage in "Madame Butterfly."

The large tonal range from A (110 Hz) up to the height of the 6th octave, as well as the drone strings constitute the essential difference of the seven-string viola d'amore, in contrast to other bowed string instruments. When the instrument is played, the drone strings respond with their fundamental frequency or overtones and thereby emphasize various partials in the tonal picture. When – as customary – these strings are tuned to a D-major or d-minor chord, the tonic and dominant of this key receive particular brilliance, however, sympathetic resonances are found also for other keys. If, on the other hand, the drone strings are tuned to a group of neighboring half tone steps – e.g., from E_4^b to B_4^b without A_4 (Stumpf, 1970), a formant-like tonal effect is achieved in the region of the vowel color "o(oh)," however, the purity of the tonal picture is easily compromised by beats of decaying drone string vibrations. Good instruments already have a strong body resonance near F_4, which effects a tone coloring in the direction of the vowel color "o." In the upper register two additional formants are added near 650 Hz and 1,000 Hz, which shift the basic tone color toward "a(ah)," while in the low register an air resonance between 210 and 249 Hz leads to a timbre similar to a viola.

3.3.3 The Cello

3.3.3.1 Spectra

With a lowest note of C_2 (65 Hz) in the tonal range of the Cello, it is located an octave below the Viola. This difference in tone location also corresponds quite closely to the frequency distribution of the main resonances, and thus to the basic tone color of the cello sound. The air resonance for most instruments is located near

110 Hz (A_2). Below this resonance the radiated power level drops at a rate of 6 dB/octave. The lower contributions, particularly the fundamentals of the C-string, therefore have lower intensities. They can be up to 12 dB weaker than the strongest partials, when they fall on pronounced resonances as is the case, for example, for $F_2^{\#}$. The limited radiation of very low frequencies is especially apparent when the lowest string is tuned to B_1^b, as required, for example, in the slow movement of the R. Schumann piano quartet.

The richness and sonority of good instruments is achieved by two formant regions, which are located around 250 Hz and between 300 and 500 Hz. This causes a vowel color between "u(oo)" and "o(oh)" of the lower two strings. In the upper registers, an additional partial group gains in significance. Their maxima, depending on the character of individual cellos, lie between 600 and 900 Hz, and thus fall into the color region between "å(aw)" and a dark "a(ah)." In the frequency region of the bright a(ah)-formant, that is between about 1,000 and 1,200 Hz, the sound spectra of cellos have a pronounced depression, which is followed by the bridge resonance around 2,000 Hz.

This resonance structure of the cello body leads to the circumstance that in the region between about 200 and 2,000 Hz, a pronounced wavy nature is superimposed on the approximately 6 dB/octave drop of the power spectrum, as expected from the excitation. This causes the power level to fluctuate by ±5 dB from the steady 6 dB drop/octave (see also Fig. 7.16). This wave nature is more strongly pronounced than in the violin and therefore leads to greater tone color differences between notes separated by about a fourth or a fifth. Above about 2,000 Hz, the spectra of the lower or middle registers drop with a slope of about 16 dB/octave. In the upper register this drop is more shallow with a value of 10 dB/octave, so that the tones in the whole can become very rich in overtones.

3.3.3.2 Dynamics

By reason of its size, a cello certainly can radiate more sound energy than a violin, however, at higher frequencies it becomes clearly noticeable that the relatively large front and back plates vibrate in highly subdivided patterns, where neighboring sections vibrate out of phase, causing an acoustic short circuit. The overall sound radiation is thus of similar order of magnitude as for a violin. For rapid scales over a wide pitch range, the power level can be varied between 74 and 96 dB, for slow individual notes, between 63 and 98 dB. Above G_4 the *ff* levels, however, move around 90 dB. Thus the practically useable dynamic range covers from 25 to 30 dB. The median *forte,* with a power level of 90 dB lies by 1 dB above the violin. The "dynamic tone color factor" again lies for most notes around 1.1. When the strongest partials fall on pronounced resonances this is reduced to 1.03. This means that a connection between tone color and dynamics is almost non existent for these tones.

It is worthy of mention that played harmonics are especially easily addressed, and consequently can be played particularly softly: down to below a power level of 62 dB, which is softer than for violins and violas. The influence of mutes is also stronger on the lower limit of the loudness region. A *pp* can be dampened down to 55 dB with a mute, without concurrently reducing the dynamic range: for *ff*, 89 dB are nevertheless possible, resulting in a variation possibility of 34 dB.

3.3.3.3 Time Structure

As is the case for higher pitched string instruments, the initial transients of higher frequency partials reach their final values more rapidly than the lower components. Furthermore, noise components arising from the attack of all resonances are especially pronounced at the beginning. Since the lower partials, however, exhibit significantly longer initial transients than in a violin, i.e., between 60 and 100 ms (Melka, 1970), it can happen for notes of short duration, that, in spite of a sharp staccato attack, only the high frequency partials reach their full amplitude, while the low partials radiate only weakly. As a result, as already suggested for violas, the tonal picture of fast passages lacks the desired sonority, and emphasizes the nasal and noise-like contributions excessively. In interaction with other instruments, this characteristic of the cello must be considered during fast staccato runs, so that wind players should not perform their staccato too pointedly. This should help to achieve a uniform expression for the passage. A typical example is given by the solo quartet of Flute, Oboe, Violin and Cello in the "Martyr-Aria" in Mozart's "The Abduction from the Seraglio," where both strings frequently have difficulty in adapting their tone production to the staccato of the flute and the oboe.

In contrast, for slow passages the cello tone can be attacked and developed very softly. In the low register, initial transients of over 300 ms occur in that setting. For the high A string the tone development takes about 200 ms.

The decay time for a cello is significantly longer than for a violin and a viola. This is caused primarily by the longer and heavier strings. When the bow does not remain in contact with the string at the end of the note, values of 2 s are obtained in the middle and lower registers and 1 s in the upper register for fingered notes. For open strings the decay time lies near 10 s for the C-string (since no resonance can extract energy from the fundamental) and near 5–8 s for the higher strings.

As is the case for the violin, for a *vibrato*, the frequency modulation caused by hand motion results in level fluctuations of individual partials. However, except for Wolf – tones, these fluctuations are not very pronounced in the low register, i.e., below 300 Hz. Furthermore, there are hardly any individual notable high frequency partials which are perceived as penetrating. The Performer's attention is therefore demanded primarily for partials in the range of 400–1,200 Hz when they fall on the edge of a very pronounced resonance.

3.3.3.4 Pizzicato

The large string length of the cello provides the performer with the possibility of varying the *pizzicato* over a particularly broad range of expression. It is thus possible to create a *pp* with a short duration sound power level of 51 dB. Such a low level is rarely found for other orchestral instruments, particularly when realizing that in this context the strongest components occur at frequencies for which the ear is relatively insensitive. In contrast, for *ff* the same levels can be performed as *con arco*, for individual partials then lie in the region of 90 dB. A four string chord can even reach nearly 100 dB (recalculated after Burghauser and Spelda, 1971).

This large dynamic range naturally also involves great variation in decay time. In a *pp* individual partials can be heard for 50 to 200 ms, for high loudness a tone persists from 400 to 1,400 ms. For open strings these values move somewhere above 1 s, while for fingered strings they are shorter, depending on the length of the vibrating string portion (Spelda, 1968). If a very dry, i.e., short *pizzicato* is desired, this can be achieved by damping the string by means of harmonic fingering.

3.3.4 Double Bass

3.3.4.1 Sound Spectra

The double bass is among the lowest instruments of the orchestra. Four-string instruments reach down to E_1 (41 Hz), the lower limit of the tonal range for a five-string instrument lies near C_1 (33 Hz) or B_0 (31 Hz) (Planyavsky, 1984) depending on tuning. The air volume resonance, the lowest resonance of the instrument, lies, depending on instrument construction, between approximately 57 and 70 Hz, which is nearly an octave below the lowest note. Consequently the fundamental of C_1 has a level which lies about 15–20 dB below the strongest tonal contributions. This difference increases with increased air resonance frequency.

For the low double bass tones, the most important tonal contributions are located between 70 and 350 Hz. They provide the dark color and richness for the sound. However, they do not create a vowel-like character, since the formant region of the "u(oo)" begins above this range. A secondary formant near 500 Hz rounds out the tonal picture in the low registers with a tendency toward a dark "o(oh)." Even for *ff*, the spectrum above the bridge resonance, which lies near 1,250 Hz, drops with a slope of 15 dB/octave. In the upper pitch ranges the spectrum widens, where frequently an additional formant appears near 800 Hz, which in color lies on the transition between "å(aw)" and a dark "a(ah)." However, the frequency location of this additional maximum varies, depending on size and construction of the bass. As a whole, in the double bass, the tendency of a drop in power spectrum between the fundamental and the bridge resonance by 6 dB/octave, is evident as well, and furthermore, a resonance dependent waviness of about ± 3 dB is superimposed on this uniform decrease.

Noise contributions to the tone of the bass frequently reach upper frequencies beyond harmonic partials. Thus a specific mixture appears in the tonal picture, often described as a "buzzing." This becomes especially prominent when the basses play alone, since this effect is mostly masked by simultaneous sounding of higher instruments.

3.3.4.2 Dynamics

When compared to a cello, the sound power level of the double bass as a whole is about 2 dB higher: The average *forte* lies near 92 dB, for *ff*, the double bass reaches a level of 96 dB for rapid tone sequences, and for individual notes it even reaches 100 dB, where especially the region around A_1 and A_2, as well as some notes in the third octave are supported by pronounced body resonances. At *pp* the levels move around 79 dB for fast note sequences, and 66 dB for individual notes, where the notes of the second octave can be played particularly sensitively. Thus, in practice, a dynamic range of 25–30 dB can frequently be realized. Especially critical are fast soft passages, for which the level of the double bass can hardly be lowered to *pp*, when all notes are to be addressed (Meyer, 1990).

The relationship between dynamics and tone color in the double bass is noteworthy. Only in the upper registers does the overtone content rise as the dynamic level increases: As is the case for other string instruments, the "dynamic tone color factor" has a value of 1.1. However, in the first octave it drops to 0.8, and in the second octave to as low as 0.6. This means that the strongest tone contributions change more than the higher components. The reason for this lies in the fact that for low playing volume the bow pressure is reduced so much, that the fundamental vibration of the string is no longer fully developed. The tone thus loses substance and becomes more tender and gentle, and can also obtain a nasal timbre.

The greater loudness of the double bass in comparison to other string instruments also appears for harmonic fingering, for which the power level can rise to 91 dB at *ff*, and can be weakened to 74 dB for *pp*. However, the overall level of the double bass can be reduced especially by employing a mute. The reason for this appears to be in the fact that the tone contributions which are effectively damped by the mute, lie in the frequency region of large ear sensitivity, while the low tone components in the region of relatively closely spaced equal loudness curves (see Fig. 1.1) decrease noticeably in their perceptive impression even for a smaller decrease in level. The dynamics of a muted double bass cover a region of about 68–88 dB (recalculated after Burghauser and Spelda, 1971).

3.3.4.3 Time Structure

As a result of its size, the double bass, in its low registers, requires significantly more time for the initial transients than the higher string instruments. Even for sharp attacks, the initial transients in the region of C_2 require over 120 ms, only from C_3 on up are values of below 100 ms achieved (Melka, 1970). However, the higher tone contributions have a shorter initial transient, so that in an even stronger measure than in the cello, a tone color change results for notes of very short duration: The low components, which otherwise determine the fullness of the tone, remain too weak, and the contributions in the region of nasal formants determine the tonal impression. Furthermore, for fast staccato passages, the relatively strongly pronounced noise development during the attack becomes especially noticeable, yet in a positive fashion, this brings about a rhythmic articulation of such phrases.

For a broad tone development, the initial transient in the upper register of the double bass requires about the same time as in other string instruments (150–250 ms), in the lower registers the spectral development requires more time. Thus the notes around C_3 can reach initial transient times of about 350 ms, and near C_2 even more than 400 ms. Open strings, however, do not permit such a soft attack, even the C_2 string does not have initial transients longer than 180 ms (Luce and Clark, 1965; Melka, 1970).

The decay times of the double bass are slightly longer than for a cello. For fingered notes they are of the order of 3 s, and for open strings near 10 s. As noted in Fig. 2.10, these data refer to the lower partials. The decay time of higher partials lies in the region of 0.5 s.

3.3.4.4 Pizzicato

In spite of the low frequencies, the pizzicato tones of the double bass have extraordinarily short initial transients. The open C – string requires only about 35 ms. For higher positions the initial transient time drops steadily: for C_2 it is still about 25 ms and drops to less than 15 ms above C_3. Thus in all regions, a very pronounced tone placement is possible (Melka, 1970).

Dynamic limits lie about 3 dB higher than for the cello. A sound power level of 93 dB can be achieved for short durations at *ff*, for *pp* this can be reduced to 60 dB, which corresponds to a range of 33 dB; accordingly, the dynamic possibilities for a *pizzicato* are greater than for *con arco* (Spelda, 1968).

With a maximal perceived tone duration of 1.6 s at *ff*, the double bass exceeds all other string instruments. However, this value applies only for the three middle open strings, yet a median value of at least 1 s can be assumed for all registers. For notes, which are fingered in the higher registers, the decay time is reduced by about 2/3 of the value for the same note on the next higher string. Even when playing very softly the notes continue to sound for 400–500 ms, which is significantly longer than for a cello.

As is the case for the other string instruments, the low tonal contributions have a significantly longer decay time than the higher ones. This results in a very homogeneous tonal effect in *pizzicato* for the entire string section, since this decrease in decay time in relation to the spectrum finds a parallel in the decay time decrease in the higher instruments. For *pizzicato* – chords which include all string groups, a uniform tendency of frequency dependence in decay time results, which can lead to a bell-like effect. A pronounced example for such a tonal effect is found in the 5th Symphony of P. Tchaikovsky, an excerpt of which is represented in score example 9, where, in spite of the dynamic indication of *mf*, the skilled compositorial use of open strings enables a wide vibrational amplitude for each attack. Yet it is all the more astonishing to experience orchestras which deliberately avoid open strings by appropriate fingerings and *divisi*-performance, in order to create as soft a tone as possible, which likely does not correspond to the intent of the composer.

Score example 9 P. I. Tchaikovsky, Symphony No. 5, 2nd Movement, measure 108 ff

3.4 The Piano

3.4.1 Sound Spectra

No stationary state is created for the piano sound, since there is no uniform continuous excitation. Nevertheless, quasi-stationary conditions can be assumed as an approximation at least for short durations. As a result, spectra of partials can certainly be used for the tonal description of the sound during its initial phases, however the time structure, and above all the decay behavior play a much more important role than in string and wind instruments.

The sound spectra of the piano are in large measure determined by the sound radiation characteristics of the sound board, which has its strongest resonances in the region of about 200–1,000 Hz. For larger instruments, resonances can even appear between 100 and 200 Hz, which lend additional fullness to the lower register of the tonal range. Furthermore, depending on instrumental construction details, resonance effects between 100 and 200 Hz can be intensified, which aids the brilliance of the tone (Wogram, 1984).

Accordingly, for the largest portion of the piano sounds, the fundamentals of the partial spectra dominate. Only in the two lowest octaves of the tonal range, the intensity maximum is shifted to overtones in the frequency range of about 100–250 Hz. Below this amplitude maximum, the strength of the partials decreases with a slope of from 12 to 15 dB/Octave, so that the fundamental of the lowest note A (27.5 Hz), with a level of 25 dB, lies below the strongest component.

Above this amplitude maximum, i.e., in the middle and upper registers, the envelope for most pianos decreases quite steadily. The average level decrease of the envelope below 1,500 Hz is of the order of 10 dB/Octave, individual partials in contrast, depending on their location near resonance peaks, or between resonances, can be enhanced or attenuated by several dB. Above 1,500 Hz, the spectra drop by 15–20 dB/Octave. For tones near C_8 (4,100 Hz) this means that the spectra only contain three partials, where the octave component is so weak that the sound looks almost like a pure sine wave.

Tone color determining formants are rarely found in grand pianos or uprights. They only appear in very few models between 500 and 2,000 Hz. In contrast to grand pianos, upright pianos have a tendency to emphasize certain frequency ranges between 100 and 350 Hz. This is based on resonance effects of air enclosed in the piano housing and, depending on the location of the piano in the room, similar resonance effects of air between the wall and the piano sound board. These lend coloration to the tone which is independent of the fundamental tone. It is particularly noticeable in multi voice playing. A grand piano on the other hand is more suitably adapted to variations in register (Bork, 1992).

A peculiarity of the spectra for some models, or at least for some limited tone range, consists in the fact that certain partials are suppressed, or are reduced in level by locating the hammer impact point at a whole number fraction of the string length. For example it is often attempted to weaken the seventh partial and its multiples by locating the hammer impact point at 1/7 of the string length. The object is to avoid the roughness of the seventh. This concept, however, is only fully effective for short tones, since the amplitudes of these partials adjust themselves to neighboring partials by mutual energy exchange for vibrations of longer duration. (Meyer and Lottermoser, 1961).

Two further characteristics, which participate in the development of the piano sound timbre, are the inharmonicity of the overtones, and the noise contributions. The latter are significant primarily during the attack, however, they are also observable in the spectra. Their frequencies reflect the resonance distribution of the instrument. Thus, the strongest noise contributions are to be expected in the range from about 300 to 750 Hz. In the highest registers of the tonal range, their level can reach up to 6 dB of the fundamental of the partial spectrum (Wogram, 1984). Inasmuch as the frequency of the noise lies far below the fundamental, it is no longer masked by the actual tone, and thus becomes particularly noticeable.

In contrast to instruments with stationary excitation, the overtones of the piano sound do not have a strict harmonic frequency location, but are stretched, i.e., they are located slightly higher than the whole number multiples of the fundamental frequency. This stretch is especially evident in the upper registers, however, it can also be noticed in the mid and lower registers. The amount of the stretch does not differ significantly between grand and upright pianos, so that inharmonicity, contrary to earlier opinions, does not make a contribution worthy of mention, to the general tonal difference between them (Bork, 1992). Rather, it represents a tonal characteristic for both instrument types.

Finally, longitudinal vibrations need to be mentioned. They can lend a unique coloration to the lowest registers. Usually the longitudinal partial, which is most frequently inharmonic, lies between the 12th and 20th partial of the harmonic spectrum. Its frequency location cannot be influenced by the usual tuning procedures. It is thus located in a region of good sound board amplification as well as high ear sensitivity. As a result it is perceived as disturbing when playing scales, since the longitudinal partials for different strings occur at different frequencies and are thus perceived as nonsystematic frequency jumps (Bork, 1989; Conklin, 1990).

Relief, i.e., shifting longitudinal partials to harmonic frequencies, is only possible through appropriate choices of string materials.

3.4.2 Dynamics

While the dynamic range of the grand piano is primarily determined by the key attack, it is also influenced by the use of the pedals and the position of the lid. When playing scales in two voices the sound power level at *ff* can reach around 104 dB. In the low registers it can be 1–2 dB higher than in the upper registers. The values apply with open lid, without the use of the right pedal. Use of the pedal raises the power level in the low register by 4 dB, in the upper register by 3 dB. Closing the lid, on the other hand, lowers the level by only 1–2 dB. In *pp* dual scales lead to a power level of the order of 88 dB without essential influence by register. Use of the left pedal lowers the level only by 1 dB, closing the lid effects a further decrease by 2 dB. For individual notes the *pp* can be further reduced, so that it is possible to drop to a sound power level below 65 dB. When one further considers the level increase due to a full two handed performance, the dynamic range of a grand piano rises to roughly 45 dB.

The felt surface of the hammer is increasingly hardened by contact with increasing strength of attack. Consequently, the exciting impulse becomes richer in overtone content with rising dynamics (Askenfelt and Jansson, 1990). This means, that for the spectrum, an increase in 1 dB of the strongest partials, a corresponding rise of about 2 dB (for some grand pianos as much as 2.5 dB) is observed for overtones in the region of 3,000 Hz, so that the tone of the open grand piano gains both in brilliance and brightness. When closing the lid, however, the higher frequency contributions are damped about twice as much as the strongest partials.

3.4.3 Time Structure

Inasmuch as the excitation of the strings is caused by the impact of the hammer, the speed of impact, on the one hand, and the contact duration, on the other, determine the tonal development, where the hardness of the hammer felt also plays a role. For upright, and grand pianos, therefore, an initial transient results, the duration of which has the same order of magnitude as is found for a pizzicato in bowed string instruments. In the lower registers the initial transient lasts around 20–30 ms; for individual notes, favored by resonance conditions, this is reduced to 15 ms. In the upper register the initial transient is reduced to values between 10 and 15 ms (Melka, 1970).

The attack noise is a characteristic peculiar to the piano. The example of the initial transient for the C_6 in Fig. 3.21 shows various noise components in addition to the partials. It is noted that the low resonances of the instrument are excited, they vibrate over a time period of 100 ms, or even more, and give a certain color to the tone. Furthermore, between 600 and 2,500 Hz, a clicking noise of short duration is

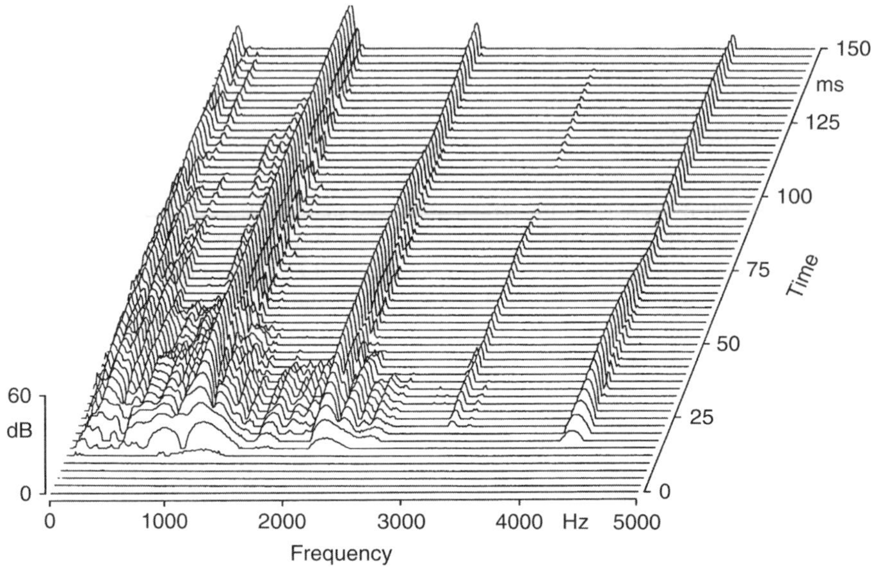

Fig. 3.21 Time evolution of a tonal spectrum for a grand piano (played note C_6)

recognized. It only lasts for 25–40 ms, and provides the attack with a certain kind of articulation. Finally, the actual partials are accompanied by side bands, which are caused by a very brief excitation of neighboring strings. These, however, are masked in the perception processes in the ear.

Noise components in the attack provide important contributions to the tonal variation possibilities obtainable by the nature of the attack. They result from the various motion processes in the keyboard mechanism. While, during a legato attack, the key is accelerated uniformly, the staccato motion of the key is characterized by layered fluctuations, which are transmitted on to the hammer motion (Asklenfelt and Jansson, 1990). However, the high speed of the key motion is also transmitted to the frame and sound board in the form of a force impulse through the key support, so that even before the hammer impact on the string, an audible noise can be created (Askenfelt, 1993). Figure 3.22 shows the initial transient for a staccato and a legato attack (Koornhof and van der Walt, 1993), the time of hammer contact with the string is given by "0." This representation, where a coarse frequency resolution was chosen in favor of better temporal resolution, clearly permits observation of the preliminary knocking noise, the duration of which is barely 40 ms. Listening tests in a concert hall have shown that the presence or absence of such articulation noise is clearly discernable. The assumption for this is of course, that this noise is not masked by an earlier tone. In this sense, legato performance, rests in large measure on the temporal overlap of tones by the nature of the connection, including pedal technique.

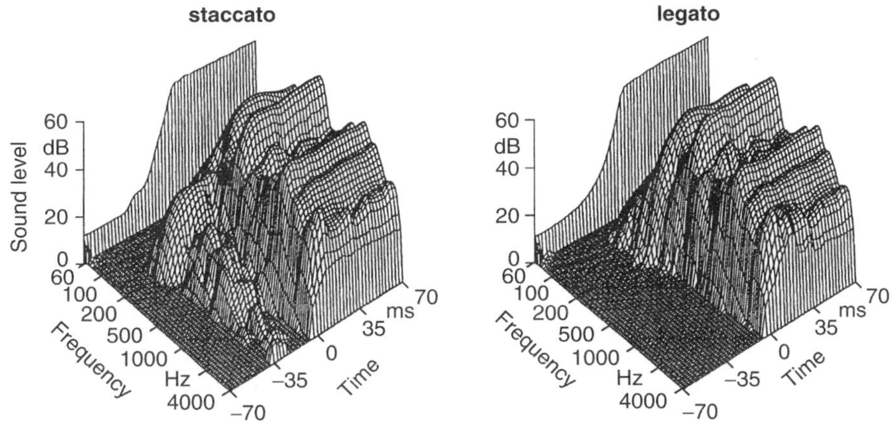

Fig. 3.22 Influence of attack technique on the noise components of initial transients for a grand piano (played tone: C_3; after Koornhof and van der Walt, 1993)

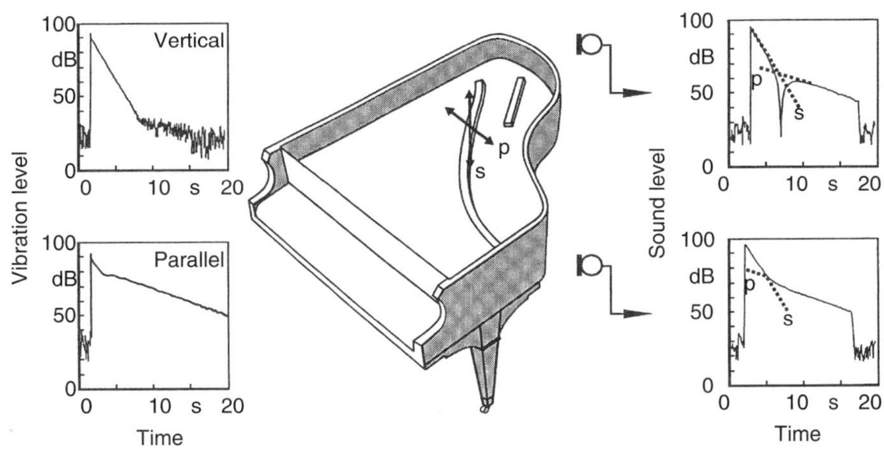

Fig. 3.23 Decay behavior of a grand piano (after Weinreich, 1977). *Left*: Sound board vibrations in two planes, *Right*: radiated sound at two microphone positions

The temporal fine structure during the decay belongs to the most important tonal characteristics of the piano sound. When the dampers are lifted from the strings, i.e., by depressing the right pedal, the sound can be followed for 10 s or more. Initially the intensity decreases more rapidly, and subsequently decays over a longer period of time in the region of decreased loudness. The reason for this can be explained in the context of Fig. 3.23: the initial impact on the string occurs in a direction perpendicular to the sound board; in this direction the sound board is in a position to extract energy from the string in relatively strong measure, as shown in the upper left partial picture indicating the amplitude of the sound board vibrations. In addition, string vibrations parallel to the sound board are formed, though much

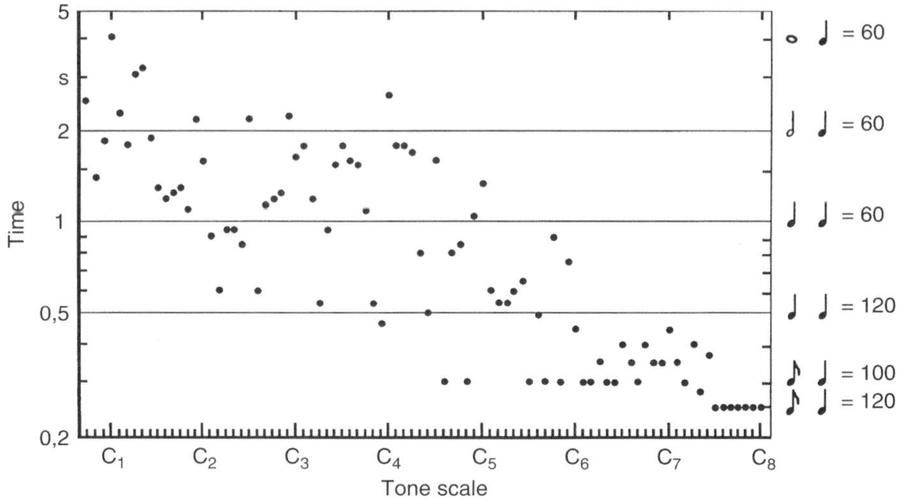

Fig. 3.24 Duration of the first, linear decay phase for individual notes of a grand piano (after Meyer and Melka, 1983)

weaker. Since the sound board presents a much higher impedance for transmitting such vibrations, this energy transmission process is much slower (left lower partial image). The radiated sound field includes a superposition of these two different forms of vibration. Depending on the relative phase of the two components the decay process of the sound field surrounding the instrument can produce a different time structure (Weinreich, 1977). In principle the time evolution of the decay process can be represented by two straight lines. The first of these drops with a relatively steep slope from the maximum amplitude, while the second continues the decay with a noticeably shallower slope. This time structure is independent of the dynamic level of performance.

The decay of short notes is determined exclusively by the slope of the first decay phase. The duration of this first linear phase is represented in Fig. 3.24 for all notes of a grand piano, where the time scale is indicated by note values and metronome indications at the right side of the graph. The strong scatter of individual points is noteworthy. An explanation for this is found in the variations of the initial slope, and also in the variations in the level for the onset of the subsequent slower decay. This contributes significantly to the tonal animation of the instrument. In general it can be said that in the low register the first decay phase corresponds to a half note or at least a quarter note at a slow tempo, while in the mid register it is sufficient for fast quarter notes, and is limited in large measure to allegro eighths notes in the upper register.

The determination of the decay time for this first decay phase is best based on the slope for a level decrease of the first 10 dB (subsequently recalculated for 60 dB). This is in analogy to the "Early-Decay-Time" in room acoustics (see Sect. 5.3). Since this initial decay time can differ by up to a factor of 2 for neighboring notes,

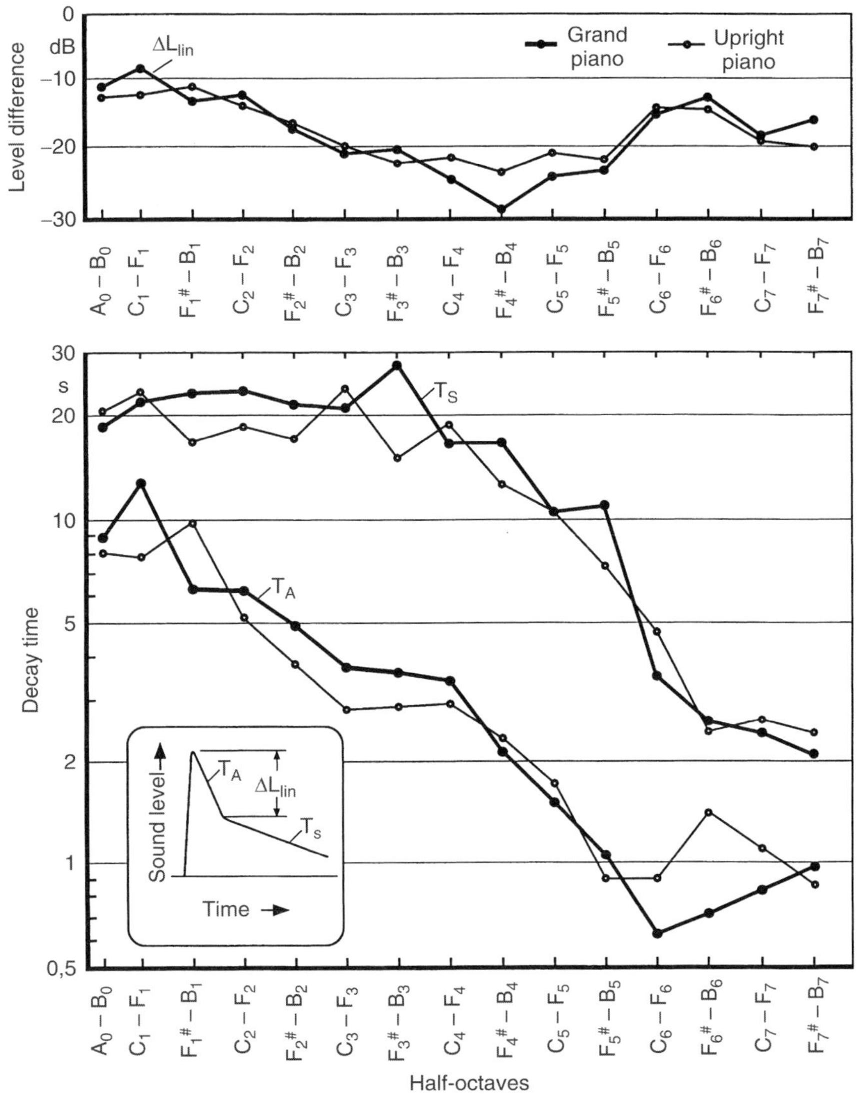

Fig. 3.25 Decay times for grand pianos and upright pianos (Median values for half octave regions). *Top*: Level difference between peak level and onset of the later decay phase. *Bottom*: Initial and later decay time

the typical sequence for a tone scale is represented in Fig. 3.25 in such a manner, that in each case a median value was formed for all notes within half an octave: Starting with values of 10 s, the initial decay time drops with rising pitch by a factor of 1.7 per octave. This results in values around 3 s in the middle register, and between 0.6 and 1.4 s in the highest registers. Grand pianos and uprights exhibit the same tendency (Meyer and Melka, 1983).

The duration of the initial decay time is primarily responsible for the song-like characteristic of the melody line and the connection between the notes of flowing music. When individual notes of an instrument exhibit particularly short decay times, they give a dull and dry impression and they drop out of the overall tonal picture. The threshold for detection of differing initial decay times by the ear depends on their duration. From a simplified approach it can be said that for decay times in excess of 4 s, changes of about 25% are necessary to detect a difference. For decay times of between 1.5 and 3 s, changes of 15% are sufficient, for decay times of less than 1 s, changes of 10% already suffice.

Notes of long duration are needed for the subsequent temporal fine structure of the decay process to gain significance. The essential characteristics of the later decay time (again referred to a level decrease of 60 dB) and the level difference between the peak value and the break value are represented in Fig. 3.25. On the average, the values of the later decay time move around 20 s for the lower half of the tonal range, where individual notes can have a decay time of as long as 30 s. In the upper register the later decay time is reduced by a factor of about 1.9 per octave, and thus reaches a value of between 2 and 3 s for the highest notes. Individual differences between grand pianos of different manufacture result from the fact that the longest values of decay time occur in different registers. Thus for instruments which emphasize the low notes the longest decay times fall into the 2nd octave, while others, by preference of the mid register, have a more rounded, but in the whole, less-full sound.

The level difference between the peak value and the breaking point separating early and late decay is shown in the upper portion of Fig. 3.25 as additional decay information. Again, mean values of half octaves are considered. The dB values of adjacent isolated notes can vary by up to 30%. The threshold of hearing distinction lies at around 3 dB. It is a characteristic of almost all instruments that the slower decay in the low registers begins at a relatively high level. The breaking point, however, drops steadily up to the mid register, so that the first decay phase clearly shows a linear drop of over 30 dB within the middle (C_4) octave. The tone thus gains clarity, without causing the flowing character to suffer, assuming a sufficiently long decay time. In the upper registers the level of the slower decay rises noticeably. Because of the shortness of the decay time, this does not detract from the clarity of the tone picture, and furthermore, it supports the tonal content of the higher notes. In addition, particular emphasis needs to be given to the fact that the level difference between the maximum and the transition from the more rapid to the slower decay does not depend on the strength of the attack. In contrast, the level of the slower decay can be raised by several dB by activating the left pedal. This enhances the effect of the decay.

The representation of the decay process by use of a broken line (as in the sketch of Fig. 3.25) does not yet give a complete description of the time dependence of fine structure. It is true that level trends of that kind are the norm in low registers, yet in the upper registers, level curves frequently show a superposition of beat-like variations onto the regular level decay curve. Already in the mid-registers from 30 to 50% of the notes are affected by this. From the C_5 octave upward this includes from 70 to

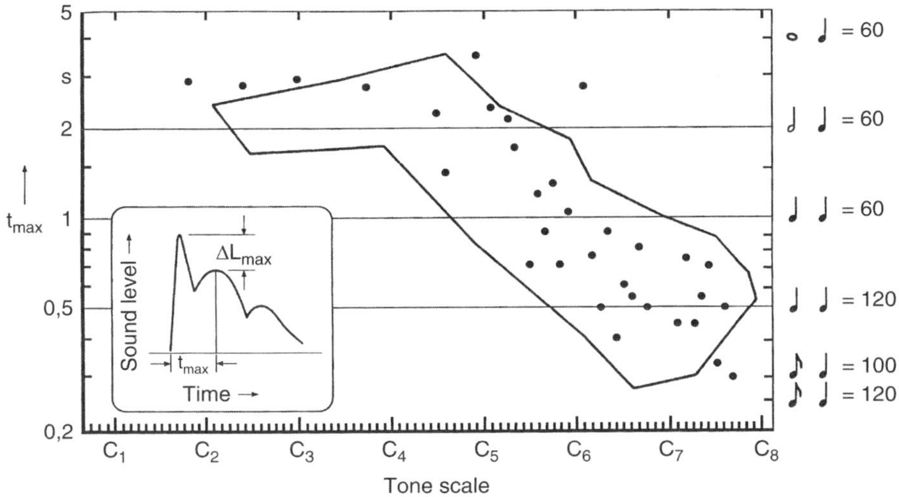

Fig. 3.26 Time location of the first beat maximum in the decay processes for a grand piano (measurement points for individual partials) and an upright piano (enclosed region of spread)

over 90%. The first maximum of these "beats" follows the tone onset after about 2–3 s in the lower half of the keyboard, for higher registers this delay is reduced by a factor of about 2 per octave. As shown in Fig. 3.26, this breathing of the tone is only noticeable in the lower registers for relatively long notes, while in the upper register it already becomes significant for rather rapid tone sequences. In general, this first "beat" maximum lies about 10–20 dB below the level of the initial peak, however, for individual notes this level difference can assume values between 8 and 35 dB (Meyer and Melka, 1983). It should also be noted that the temporal level fluctuations occur even for well tuned pianos. This is caused by changes in the direction of string vibration, furthermore, differences in inharmonicity between the three strings associated with the same note can play a role.

3.5 The Harpsichord

3.5.1 *Sound Spectra*

In the harpsichord, as in the piano, the frequency range of the most strongly radiated sound intensities are determined by the main resonances of the sound board. They lie in the range of 200–800 Hz. Depending on construction details, the maximum of the radiated sound is found between 300 and 600 Hz. As determined by the process of string excitation through plucking, the string excitation is much richer in overtones than for a piano. In addition, the inharmonicity of the strings is significantly less than for piano strings. In the low registers of the harpsichord, at the 50th partial, a deviation from the harmonic frequency of about 15 cents is usual, while a deviation of 30 cents is already considered to be relatively high. For the same pitch on the piano already the 6th partial lies 30 cents above the harmonic frequency.

Below 200 Hz, the radiated sound energy drops with a slope of more than 40 dB/octave, so that only very weak partials can be expected. Consequently in the lower registers the strongest partial can be found between 200 and 500 Hz, while the fundamental always dominates in the mid- and upper registers. The intensity distribution at higher frequencies is in large measure time dependent. For a numerical description it is therefore recommended to base energy content on the first second of the harpsichord sound (Elfrath, 1992). Above 800 Hz the spectra drop initially with a slope of about 7 dB/octave, and subsequently pass through a dip between 2,000 and 2,500 Hz, which lies barely 15 dB below the strongest partials. This is followed by a secondary maximum, which lends its particular presence to the harpsichord. Above about 5,000 Hz the spectrum drops at a rate of about 15 dB/octave.

In comparison to the otherwise minor spectral differences, this formant, which lends a certain presence to the instrument, is rather individually pronounced for harpsichords of different construction. The rise, in contrast to the previous dip can vary between 2 and 6 dB, so that finally the secondary formant lies about 7–12 dB below the strongest spectral components. This frequency varies also. It is most often located near 4,000 Hz; for some instruments, however, around 5,000 Hz. However, it is always situated clearly above the so-called singer's formant (see Sect. 3.8.1.), which, in its significance for the presence, and the ability to carry the sound of the voice, has a similar function. As an example, this effect is clearly observable for a harpsichord continuo within an orchestra, when these high frequency contributions, in their rhythmic structure, stand out from the overall sound, while the harmonic chord foundations are only of subordinate importance or are even totally inaudible – at least to the listener in the hall.

3.5.2 Dynamics

The dynamic range of the harpsichord is very limited, since the nature of the key attack has no essential influence on the process of plucking the string. Different dynamic steps are therefore only accessible through registration, i.e., by playing one or more strings for each key. The combination of two 8′ registers results in a power level increase of 2–3 dB in comparison to the single register, combination with a 4′ register effects a change in tone color in the direction of a brighter timbre, which also gives the impression of a dynamic increase. In the mean, one can count on a power level (calculated after Burghauser and Spelda, 1971) of between 71 and 87 dB, depending on registration and performance technique. These are values which cannot only be reached by a single violin, but can certainly be exceeded. In comparison it should be mentioned that the clavichord is even softer by about 10 dB.

The low power level of the harpsichord frequently leads to the desire for electroacoustic amplification, to improve the dynamic balance with the remaining ensemble. It is important for such a reproduction of the harpsichord sound with speakers, to counteract a nasal sound caused by an additional lowering in the frequency region around 1,500 Hz, when the harpsichord is amplified above its original loudness (Thienhaus, 1954).

3.5.3 Time Structure

The duration of the initial transient of individual harpsichord tones is very short. In the middle and upper registers it is only 10–25 ms. In the low registers this can be stretched to the range of 45–75 ms, depending on structural characteristics of the instrument (Neupert, 1971; Weyer, 1976). In addition, an articulation peak of short duration, approximately 20–30 ms, occurs with a principal intensity mostly above 2,000 Hz. This lends a degree of precision to individual harpsichord tones which can lead to an uncomfortable hardness in the performance of several simultaneous notes. This is at least one of the reasons why chords on the harpsichord are usually performed somewhat arpeggiated.

The superposition of vertical string vibrations and vibrations parallel to the sound board determines a decay process similar to that in the piano. However, the initially steeper amplitude drop is less dominant for the overall tone than it is in the piano, since on the one hand the parallel vibration is more strongly excited by the plucking process, and on the other hand the vibrational energy is transferred more rapidly from the vertical to the parallel vibrations. The temporal division of the decay process is thus not necessary (unlike the piano) and the specification of a single value for the decay time is sufficient.

In Fig. 3.27, the decay times of four historic harpsichords of different styles are represented in relation to tone location (Elfrath, 1992). On the whole, values are found, which in order of magnitude can be compared to the late decay times of

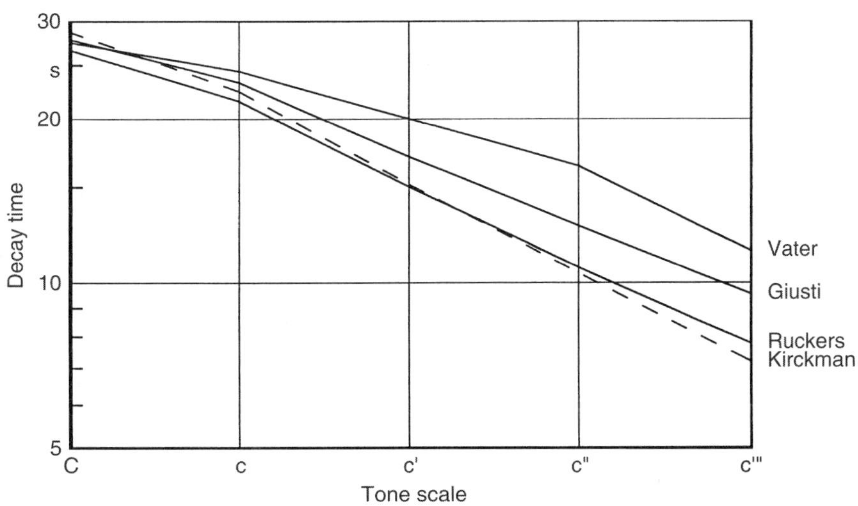

Fig. 3.27 Decay times for four harpsichords of different origin (after Elfrath, 1992). A. Ruckers, Flanders 1628; G.B. Giusti, Italy 1681; Ch. Vater, Germany 1738; J. Kirkman, England 1783

pianos; not, however, to piano initial transients. In the low octaves, the curves for the individual instruments are located very close to each other, but drop subsequently with differing slopes (by a factor of 1.25–1.5 per octave), so that the decay times differ by up to 70% in the highest register. Even though these curves represent mean values, where, for each instrument, decay times can be clearly differentiated from tone to tone, they show the individual characteristic of the instruments more clearly than the time averaged spectral composition. In addition, the different decay times of the higher overtones lead to a change in tone color within less than 100 ms. This also contributes to the characterization of the instrument.

Newer harpsichords have decay times closer to the Kirckman instrument in Fig. 3.27, where the possible spread is relatively large. To generalize, one can stipulate a decay time of about 20 s for C_3 with a drop by a factor of 1.4 per octave (see also Neupert, 1971; Fletcher, 1977). Assuming an initial sound power level of 70 dB and a noise level in the room of 30 dB, the tone can be followed by the ear for about 2/3 of the decay time. The real tone duration, however, is almost always significantly shorter than the decay time, so that the tonal character is essentially determined by the level drop of the first 10 dB or at most 20 dB.

3.6 The Harp

3.6.1 Sound Spectra

Sound radiation by the harp is essentially determined by a few distinct body resonances. The three most important resonances lie between 200 and 450 Hz. These are followed by two additional strong resonances up to 850 Hz (Firth, 1977). Above 1,000 Hz, the resonance sequence is more closely spaced, however, the individual resonances steadily decrease in strength. Inasmuch as the tonal range of the harp extends down to a C_1^b (31 Hz) the fundamentals of the spectra are relatively weak in the low register, the overtones which fall on the resonances between 200 and 450 Hz dominate. In the range of from G_3 or C_4 upward the fundamental becomes the strongest component of the spectrum.

As is the case for *pizzicato* playing on string instruments, the overtone content of the harp sound depends on the plucking location. For a pluck near the center, a complete spectrum is formed with strongly decreasing overtones, thus in the mid and upper registers, already the octave partial is from 10 to 15 dB below the fundamental, while in the low register, the decrease of the spectrum begins above 450 Hz. For an attack exactly in the middle of the string, the odd partials clearly dominate in contrast to the even ones, and the sound becomes full and soft. An attack at 1/3 of the string length suppresses the third partial and lends brightness and brilliance to the tone through the relatively strong octave partials (2nd and 4th order). Attack near the end of the string *(presso la tavola)* leads to a spectrum which drops only by 20 dB for the first eight partials, and thus has a guitar-like or possibly even metallic sound.

3.6.2 Dynamics

The harp and the piano are similar, in that the sound power radiated by the harp can only be indicated for the initial time period of the tone, i.e., in the time region of the strongest sound development. At the lower limit of the dynamic range the sound power level lies at around 60 dB (as calculated after Burghauser and Spelda, 1971), only when reaching the second octave does it rise to a level of about 70 dB. The upper limit rises from about 88 dB for the lowest notes to 100 dB in the region of C_4 and then again drops to about 80 dB in the C_6 octave. This results in a maximum dynamic range of 40 dB around C_4.

3.6.3 Time Structure

The attack for the harp is characterized in large measure by a sharp precision. This is partly determined by very short initial transients. For the lowest notes they are about 20 ms. In the upper register they drop to less than 10 ms (Melka, 1970). Tones, for which the fundamental falls on a resonance, have an initial transient of longer duration than neighboring tones. The coincidence of the neighborhood of a fundamental with one of the five main resonances can lead to short term beats during the initial transient.

The second reason for the precision in the initial transient can be found in the decay behavior of the harp. The higher partials, caused by the precise attack decay much more rapidly than the lower partials resulting from the full tone. In the lower register these have a decay time of the order of 4–6 s, in the mid-register about 2 s. Tones, whose fundamental falls directly on one of the main resonances have a noticeably shorter decay time than their neighbors, which causes the relevant tones to become dull and blunt.

Because of usual string lengths and strengths, a string in its fundamental tuning is longer and thinner than its neighboring, next lower string, raised to the same note by a pedal shift. This enables a tonal differentiation. Strings in fundamental tuning decay more gently, strings with raised tuning result in a harder sound (*"secco"*). The particular performance technique of increasing the decay time rests on the ability to tune two strings to the same pitch by a pedal shift, to couple them acoustically. For example Puccini specified this effect in "Turandot."

Inasmuch as "unused" strings are not damped, in contrast to the piano, they experience mutual coupling to the vibration processes through the resonant body. In this function they contribute significantly to the increase in decay time. In a given setting they must be damped by hand to interrupt the decay. It is also possible to excite string vibrations by a strong external sound field through the sound board. Orchestral chords can initiate such a decay without harp participation. This is clearly audible, at least in the neighborhood of the harp, and possibly needs to be damped. This is particularly important if there is a microphone located in the vicinity of the harp.

3.7 Percussion Instruments

3.7.1 *Timpani*

For percussion instruments, even more so than for the piano, the tonal character is largely determined by the time structure of the spectral composition. On the one hand it could be of interest to consider the tonal spectra at the instant of strongest sound radiation, on the other hand, the different decay of individual spectral components plays an important role. This is particularly clear for tympani, where after initial impact noise, harmonic components dominate, evoking a clear pitch impression.

The membranes of timpani can vibrate in a multiplicity of different vibrational shapes (so called vibrational modes). For the simplest mode, the rim forms a nodal line, and the membrane vibrates with its entire surface area in phase. Added to this lowest "ring mode" are higher modes, for which additional nodal lines are formed as concentric circles. A further group of modes is formed by nodal lines running radially across the membrane which cross in the middle ("radial modes"). This group has the important property that the frequencies of the first three to five modes are in reasonably good harmonic relation to each other: their frequencies ratios approach the numerical ratios of 2:3:4:5:6; these modes thus are essentially responsible for the pitch impression (Rossing, 1982b; Fleischer, 1991).

The perceived pitch lies an octave below the main tone, that is for the large concert kettle with a tonal range of F_1 to D_2 between about 44 and 73 Hz, and for the small concert kettle (A_1 to G_2) between 55 and 98 Hz. The low D kettle reaches down to D_1 (37 Hz), the high A kettle up to C_3 (130 Hz). However, pitch perception is not as precise as it is for string and wind instruments, because partials are not as accurately harmonic, and the frequency location is so low. Consequently, Verdi for example, maintains the original tuning for complicated modulations, when time does not permit retuning and uses the notation of G for the tonic in G^b major. This problem has been eliminated in the meantime by the invention of the pedal tympani.

Timpani tuning is accomplished by changing tension in the membrane. For a good instrument, this changes the frequencies of the radial modes in proportion to each other, i.e., harmonic relationships are retained. This property of timpani is related to the kettle size. Its volume influences the vibrational frequencies of the membrane within certain limits. The lowest ring mode experiences a smaller frequency change than the radial modes, thus it does not move harmonically with the other modes when tuning.

The strength of the partials radiated by these vibrational modes depends on the location of the impact. Impact near a nodal line is almost totally ineffective in exciting the corresponding mode. Impact at the center of the membrane would therefore excite the (inharmonic) ringmodes strongly, and the (harmonic) radial modes only weakly. In contrast, the usual impact location, located more near the rim favors the harmonic components and reduces the inharmonic ones. The player can selectively excite the lowest radial mode (the fundamental) or the next higher radial mode (the fifth) most strongly.

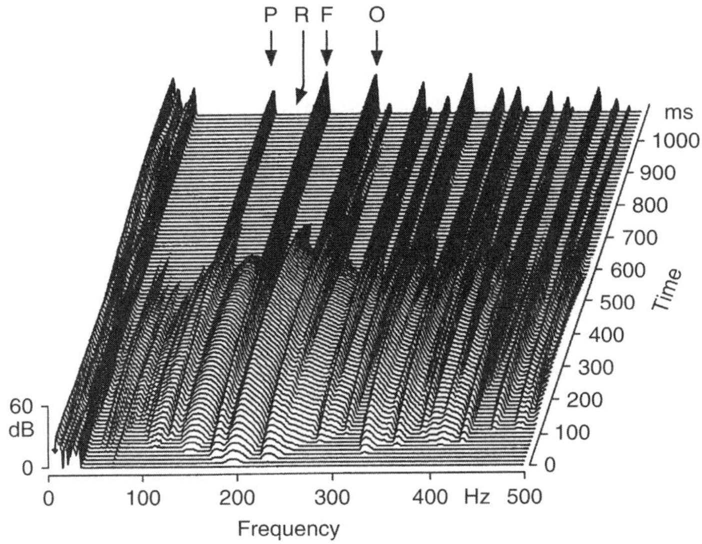

Fig. 3.28 Time evolution of a timpani spectrum of pitch A (after Fleischer, 1991). P Principal mode, R 1st Ring-mode, F Fifth, O octave

Figure 3.28 shows the change of the spectral composition of a timpani tone with time. Approximately the first second of the sound is represented. For the duration of approximately the first half second, many quickly decaying components are recognized. These constitute the impact noise. In this example the lowest ring-mode lies at around 140 Hz. After that, a series of slowly decaying partials remains. The principal mode at 110 Hz, as well as the fifth and the octave stand out prominently. Usually the decay time of the principal mode moves between 7 s for the low and 1.5 s for the highest register. The decay times of 10 s and 3 s for the fifth and the octave respectively are significantly longer, so that these partials, in time, dominate over the principal mode. For natural skin membranes this difference is not as pronounced as for man made materials. For natural membranes the pitch dependence of decay times is also less pronounced, consequently, as a whole, it gives a more even impression (Fleischer, 1991). With hand damping the decay time is reduced down to 0.7 s for low, and to 0.2 s for high pitches.

The strong harmonic partials mask the inharmonic components relatively early in the auditory impression, contributing to the purity of the timpani sound. This is the case, provided that, – aside from the correct impact location, the frequency of the lowest ring mode does not lie below the fundamental. This condition is no longer satisfied for high tuning of the drum head, therefore the high timpani tones sacrifice clarity (Fleischer, 1991). The timpani sound can also lose clarity, when the vibrational modes become less precise due to uneven membrane thickness. In that case the tonal character depends especially strongly on the point of impact, a circumstance which the player can utilize for particular nuances.

The initial transient is characterized by the relatively slow development of the low pitch contributions, which move in the neighborhood of 100 ms. At the same time, higher frequency contributions can prefer the entrance point at the time of the timpani impact to some extent, if they are sufficiently pronounced. For this, Melka (1970) indicates initial transients of less than 20 ms.

The playable dynamic range of about 45 dB is very large. For ff a sound power level of 115 dB can occur. For *pp* this reduces to 67 dB in the low registers and down to 70 dB in the upper registers (recalculated after Burghauser and Spelda, 1971). The strongest tonal contributions develop between about 100 and 250 Hz, depending on tone location and impact characteristics.

3.7.2 The Bass Drum

In contrast to timpani, the bass drum belongs to the unpitched percussion instruments. The muffled tone, with its strongest contributions in the frequency range near 100 Hz, is characterized by a multitude of inharmonic partials, some of which are closely paired. The usual soft mallets prevent excitation of higher frequency components. The heavier the mallet, the more energy can be transferred to the membrane, however, this also increases the contact duration, which in turn suppresses the higher frequency components. In addition, the more closely centered impact location leads to a preference of the ring-modes of the membrane vibrations (if possible, these modes are to be avoided in the timpani), which likewise supports the dull non-distinct tone character. For an impact point closer to the edge, higher, and mostly asymmetric modes are formed, causing the sound to become harsher. The dynamic range encompasses a sound power level from 79 dB at a *pp* to 108 dB for *ff* (calculated after Burghauser and Spelda, 1971). In relation to the loudness impression for *pp*, one needs to take into consideration that at the low frequencies, relevant for the bass drum, the ear is not very sensitive at low levels, so that the bass drum can certainly reach the lower limit of an audible *pp*.

The time structure of the tone is characterized, on the one hand, by the drop in frequency of up to 140 cents, i.e., more than one half step, during the first second after the impact. On the other hand, beats of the mode pairs, mentioned earlier, in the frequency range above the fundamental, create amplitude variations, which result in the breathing character of the bass drum tone (Fletcher and Bassett, 1978). The median decay time lies around 8 s for the strongest tone contributions, in the region of 200–400 Hz it is around 4 s, at higher frequencies it drops by a factor of 2 per octave. Components below 50 Hz can ring for 15 s or longer (Plenge and Schwarz, 1967).

In addition to the felt mallet there is a second mallet, the so called "brush" used in Turkish music, it is formed from a split reed stick. It produces shorter, harder impacts of higher precision. It serves to subdivide the measures struck by the felt mallet. The symphony 100, the "Military Symphony," of Jos. Haydn is among the best known examples for this. In this symphony, for the most part, the dull felt impact marks the first beat of the measure, and the brush impacts are used for the

Score example 10 *Top:* J. Haydn, Symphony Nr. 100, 2nd movement, measure 174 *ff.* (without winds). *Bottom:* W. A. Mozart, The Abduction from the Seraglio, 3rd Act, Chorus of the Janissars (without winds and low strings)

additional beats. There are, however, places where *sforzando* impacts of the brush are found on the first beat of the measure. Mozart, in contrast, in his "Abduction from the Seraglio" lets the hard brush stroke run through as a uniform rhythm, and combines it on the first and occasionally on the third beat of the measure with the impact of the felt mallet (see score example 10).

3.7.3 Snare Drum

Since Rossini made the snare drum acceptable in his opera "La gazza ladra" in 1817, it has also found entrance into the symphony as a rhythm instrument. The rhythmic precision is accomplished by the short transient of about 7 ms (Melka, 1970), as well as by the lack of very low tonal contributions. The maximum of the radiated sound lies between 300 and 1,000 Hz, depending on the nature of the impact. For a *forte*, an impact near the middle of the membrane favors the low components, and for a *piano* an impact near the edge favors the higher partials. Both, the inharmonic location and the frequency width of the partials, prevent a unique pitch impression.

The high frequency components of the spectrum are further strengthened by the so-called snare strings, which are stretched below the low membrane. By tuning these strings below the membrane, they collide in a pulsating manner with the membrane, and thereby excite additional high frequency vibrational modes. This increases the noise impression of the snare drum significantly (Rossing et al., 1992). The decay time of the strongest partials is of the order of magnitude of 1 s (Plenge and Schwarz, 1967), so that even a very rapid impact sequence (drum roll) is recognized as such without going over into a uniform noise. For a *pp* the sound power level lies at around 74 dB, which, because of the short duration of the individual impacts, can be perceived as very soft. For *ff* the sound power level reaches about 100 dB, which suggests a rather wide dynamic range (as calculated after Burghauser and Spelda, 1971).

3.7.4 Gong

In a number of operas with large orchestras, tuned gongs are used to create exotic tone colors. The best known example for this, – next to Saint-Saens' "La Princesse jaune" and Strauss' "Frau ohne Schatten" – could be Puccini's "Turandot," where nine gongs are required with pitches between A_2 and A_3, as well as a "Gong Grave" in A_2. Their pitch is determined by a precise fundamental and an octave partial, and occasionally supported by a double octave (4th partial). Often the pitch of this 4th partial corresponds to a seventh. The third partial is always inharmonic. It lies by a whole step to a fourth above the octave, a major third is perceived as tonally optimal. Additional higher inharmonic partials complete the spectrum, and with greater loudness a strong rushing noise is an essential part of the tone color.

While for higher pitched gongs the fundamental always dominates, in the low gongs, the 3rd and 4th partials can become by up to 8 dB stronger than the fundamental; the octave partial always lies by about 10 dB below the fundamental. The lower the gong is pitched, the louder the rush noise becomes, with an intensity maximum around 1,000 Hz, which is in the region of the vowel color of "a(ah)." Around 3,000 Hz the rush noise level for a low gong at ff is about 10 dB below the fundamental and above 3,000 Hz it drops off with 10 dB/octave, while for the higher gongs at 3,000 Hz it already lies by 20 dB below the fundamental.

The strength of the impact exerts a strong influence on the tonal picture of gongs. While the intensity of individual partials rises steadily from *pp* to *mf*, non-linear effects appear for yet stronger excitation. These lead to energy transfer between modes, whereby especially rush noise contributions become more pronounced. Thus the accessible dynamic range for a clear tone without rush noise from *pp* to *mf* or *f* only covers 17 dB with a sound power level of 91 dB at *pp* and 108 dB at *f*. A further increase by 8 dB is possible, however, that will not raise the strength of the fundamental, the additional energy appears in the higher frequencies, particularly in the 3rd and 4th partials, as well as in the rush noise. Furthermore, a consequence of these non-linear effects is that the fundamental and partial frequencies, and thus their pitch, start at up to 80 cents higher for very strong impact, and only in the course of about 4 s reach their final value. For *mf* the tone starts about 20 cents high and needs approximately 2 s to reach its final value.

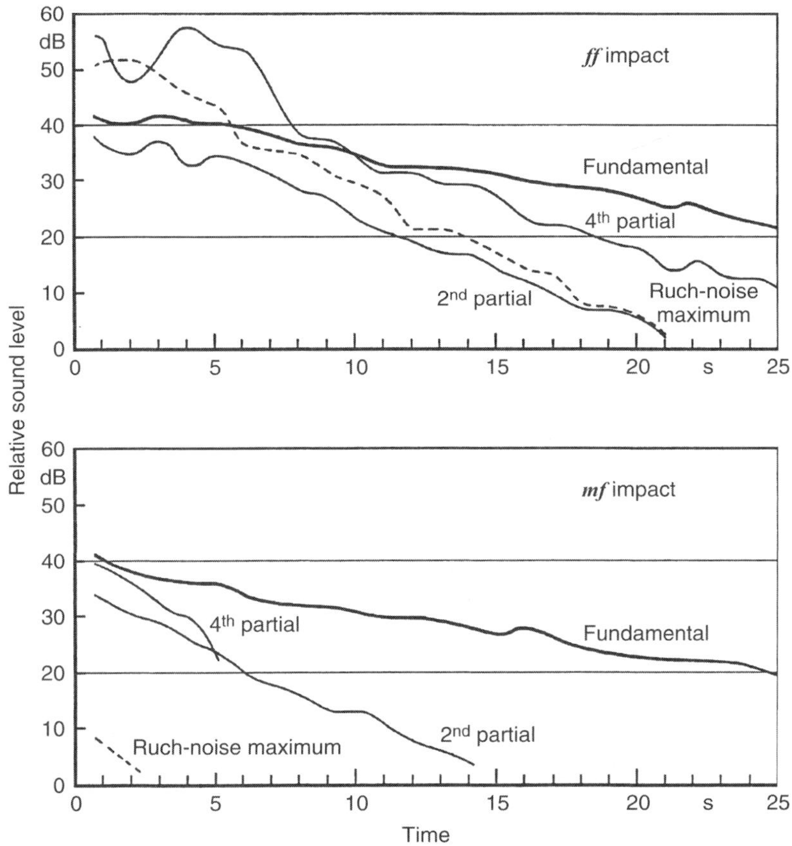

Fig. 3.29 Time dependence of the level of the most important tone contributions of a Gong Grave in A_2 for two different dynamic levels

The time dependence of the level structure for individual tonal components during the first seconds after an ff impact is particularly interesting. As Fig. 3.29 shows, the discrete partials initially become lower, while the rush noise reaches its maximal strength after 2 s, and thus exceeds all other tone contributions. After that, the rush noise decays steadily, and after 6 s becomes weaker than the fundamental, which only then becomes the strongest component. This slow spectral development endows the gong with its magnificent sound. At mf and especially at pp the individual partials exhibit a smooth level drop from the beginning. The same occurs for *ff* after a later point in time. The decay time of the fundamental is shortened from about 75 s for a Gong in A_2 to 30 s for a Gong in A_3. The decay time of the following partials, up to about five times the fundamental frequency, is also about half as long. Above that, the decay time drops uniformly down to the value of 4 s at 8,000 Hz. For "*secco*- impacts" this extremely long decay is reduced to a few seconds by damping with the free hand.

3.7.5 Cymbals

As is the case for the gong, the tonal picture of cymbals is determined in large measure by the time development of the different tonal contributions. The large number of inharmonic partials, which to some extent are densely spaced, do not permit the emergence of a pitch. At first, during the initial transient time of 10–20 ms, strong vibrations of a few radial modes are formed at frequencies around 400 Hz and also in the region between 700 and 1,000 Hz. After 50–100 ms they pass their dominant role to high frequency rush noise contributions between 3,000 and 5,000 Hz, which at times can expand to 10,000 Hz. Again, the energy exchange between different vibrational modes, or also between longitudinal and bending waves becomes a determining factor. This preferred sound radiation of high frequencies forms the tonal impression in the time frame of about 1–4 s after the impact, and results in the bright shrill sound of the cymbal. Thereafter the maximum of the sound intensity reverts back to the frequency contributions around 400 Hz. This is primarily determined by the damping of the vibrational modes, i.e., the decay behavior (Fletcher and Rossing, 1991).

The decay time of vibrational modes around 400 Hz is about 30–40 s; for components around 3,000 Hz it amounts to approximately 10 s, and for components around 6,000 Hz it is still 5 s (Plenge and Schwarz, 1967). The lowest partials in the region of 50–100 Hz, which are relatively unimportant for sound impressions in a room, can even have decay times of the order of 100 s, however, this becomes noticeable only for close microphone positions. It should be noted, that for frequencies below about 700 Hz the decrease in level initially occurs relatively rapidly, because of energy transfer to other modes, and the decay times mentioned earlier only take effect after about 200 ms, thus a time-level plot shows a break (as is the case for the piano) (Müller, 1982; Fletcher and Rossing, 1991).

Within certain limits the dynamic range of cymbals depends on the nature of the excitation. For an impact with a felt mallet, a calculation after Burghauser and Spelda (1971), yields sound power levels between 73 dB for *pp*, and 101 dB for ff. For a wooden mallet these values rise to 82 dB for *pp* and 111 dB for *ff*. Different mallets have more influence on the strength of the high frequency contributions than differences in impact. When two cymbals are crashed against each other, sound power levels between 74 dB for *pp* and 108 dB for *ff* can be expected.

3.7.6 The Triangle

Closely spaced inharmonic partials at very high frequencies play the most important role in determining the tone of a triangle. For the normal impact direction (perpendicular to the triangle plane) the spectrum reaches to over 20,000 Hz without significant level drop. The maximum of the spectral envelope is formed around 6,000 Hz. Only one partial is found below 1,500 Hz, it is located near 400 Hz. When the triangle is hit in a direction parallel to the plane, the number of excited vibrational modes is reduced, and the partial sequence is not so dense. When, in addition to a fundamental near 400 Hz, there are two nearly harmonic partials, for example near 1,600 and 2,000 Hz, it is entirely possible that a certain pitch is noticeable in the sound of a triangle (Rossing, 1982a).

The initial transient of about 4 ms is extremely short particularly because of predominantly very high frequency contributions (Melka, 1970). For the fundamental, the decay time is about 30 s. For higher frequencies it drops rather uniformly by a factor of 2 per octave (Plenge and Schwarz, 1967). This long decay affects that for a rapid impact sequence (usually hit as a roll between two different sides of the triangle) the sound goes over into a fluctuating steady tone and assumes the character of a silvery shimmer for the overall sound of the orchestra.

The dynamic range is characterized by a sound power level of 66 dB for *pp* and 91 dB at *ff* (as calculated after Burghauser and Spelda, 1971). Because of the predominantly very high frequency contributions, the triangle is not only heard easily above the orchestra, but it is also easily located by the listener in the hall, since these high frequencies are only reflected relatively weakly by the hall.

3.8 The Singing Voice

3.8.1 Sound Spectra

By nature, the spectral envelope of the singing voice is mainly characterized by formants. In each case, the sung vowel determines the location of the strongest partial. A male voice can vary the lowest formant in the region of 150–900 Hz and the second formant between 500 and 3,000 Hz. For female voices the lower limit lies higher corresponding to the tonal range. The constant change between vowels within the text consequently leads to constantly changing envelopes. Above about A_5 (880 Hz) the fundamental begins to exceed the first formant of the vowels, the tuning of the oral cavity is therefore undertaken by female singers for optimal enhancement of the fundamental and the octave (by the 2nd formant). This technique can also be used for lower voices. It serves to raise volume, however, it reduces vowel recognition, and to some extent causes tone quality to suffer. It is also used extensively with peak tones of tenor and alto voices, to lend sufficient strength (Sundberg, 1977; 1991). For tenor voices, however, generally the 2nd and 3rd partials are predominantly amplified (the fundamental less so), otherwise the voice assumes a more female character, as is the case for countertenors (Titze & Story, 1993).

The spectrum can drop strongly below the first formant. In the low register of male voices, the fundamental can lie by 15–20 dB below the strongest partials. High frequency contributions in the range between about 2,300 and slightly above 3,000 Hz have special significance. Not only is this the location of those secondary formants which make it possible to differentiate between different voices (for the same vowel), but it is also the location where for trained voices the so-called singer's formant develops. This singer's formant can provide a quality criterion

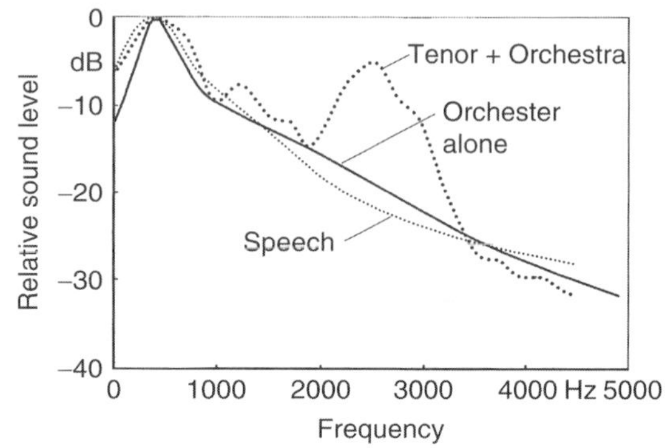

Fig. 3.30 Typical envelope for Speech, Orchestra, and singing voice with pronounced singing formant (after Sundberg, 1977)

for the singing voice (Winckel, 1971; Sundberg, 1977). It can reach a level of within 5 dB of the strongest partials. Since the orchestral instruments radiate much weaker overtones in this frequency region, the singer's formant lends to the voice an ability to carry the tone and transcend the orchestra (see Fig. 3.30). As determined by the varied lengths of vocal tracts, a typical frequency location of the singer's formant occurs around 2,300–2,500 Hz for a bass, 2,500–2,700 Hz for a baritone, and 2,700–2,900 Hz for a tenor. The typical frequency location of the singer's formant for a female voice lies somewhat higher, like 2,900 Hz for a mezzosoprano. Above 3,500 Hz the spectrum drops steeply at a rate of approximately 25 dB/octave. Exceptions lead to a timbre which is too metallic. In "belting," the vocal technique frequently used in musicals, in which, by elevating the larynx, the singer's formant is raised to above 3,500 Hz, the high frequency components are deliberately strengthened (Estill et al., 1993). It is of interest that this frequency placement finds a parallel in formants of the harpsichord (see Sect. 3.5.1). Very high frequencies are nevertheless important for recognition of consonants. Voiced sibilant sounds reach up to about 8,000 Hz, unvoiced sibilants even up to 12,000 Hz.

3.8.2 Dynamics

The dynamic range of all singing voices is characterized by a clear rise from the low register to the higher positions. As shown by the measurement results of Burghauser and Spelda (1971) – converted to sound power level – in Fig. 3.31, for solo voices the lower dynamic limit of low male voices in the lower register lies around 70 dB, for a dramatic tenor and for female voices around 60 dB, for high registers it rises to values between 85 and 110 dB. The upper dynamic limit lies between 85 and 95 dB for the low register, and it rises up to 110–125 dB in the upper region. This results in a median dynamic range of 25–30 dB, which, for favorable tone locations can be widened to more than 40 dB.

Since, once during each vibration cycle, the vocal cords close completely at higher dynamic levels, while for soft singing there is always a residual opening, the overtone content increases with increasing dynamics. This effect is very

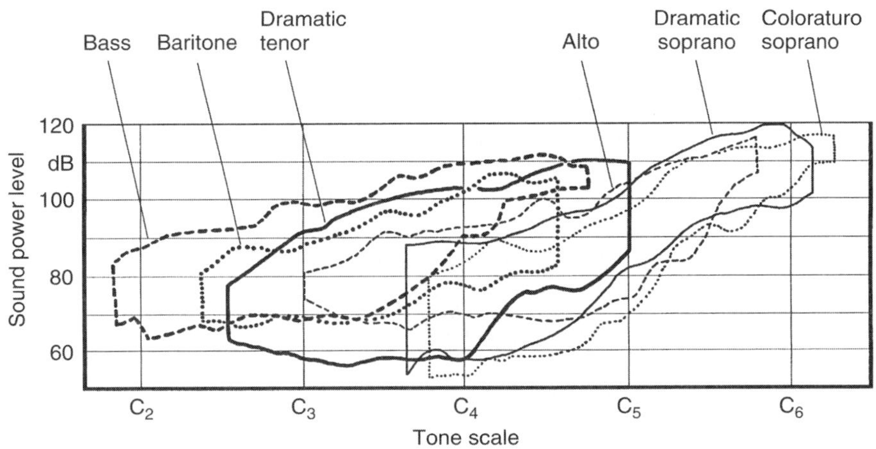

Fig. 3.31 Dynamic range of solo singing voices (after Burghauser and Spelda, 1971)

pronounced for a singer's formant which increases by about 1.5 dB for every 1 dB increase of the strongest partials. For individual voice control in a noisy environment, it is of interest to note that the sound level at the ear of the singer lies 10 dB below the radiated sound pressure level (Ternström and Sundberg, 1983).

With choral singers, the sound power level generally is not as high as for soloists, though the difference should not be that large for professional opera choruses. For certain lay choirs, Ternström (1989) found a dynamic range which led to an average sound power level for individual singers from 71 dB at *pp* to 97 dB at *ff*. For boys choirs the dynamic range of individual singers is narrower and reaches from about 80 dB at *pp* to 91 dB at *ff*. This suggests for an average *forte* an order of magnitude of 88 dB for boys and 91 dB for adult choir voices.

3.8.3 Time Structure

For a singer, the initial transient is determined by the nature of the beginning consonant. Explosive sounds lead to a very short noise impulse of from 20 to 30 ms duration; already after 40–60 ms the full harmonic sound can be developed. In contrast, sibilants are characterized by a duration of about 200 ms. For an initial "m," a 40 to 50 ms noise is immediately followed by a humming phase – for a closed mouth – lasting up to 150 ms, before the full tone is developed. It is interesting to note that the singer's formant already comes in during this humming phase. An initial "r" is characterized by a noise impulse sequence, whose individual impulses follow each other with a 35–45 ms separation.

The singers' vibrato usually moves in the frequency region of 5–7 Hz (Winckel, 1960), whereby the vibrato frequency typically increases slightly toward the end of the tone (Prame, 1993). For a *vibrato* width from about ±40 to ±80 cents, a pure frequency modulation results without change in the envelope. At most, individual

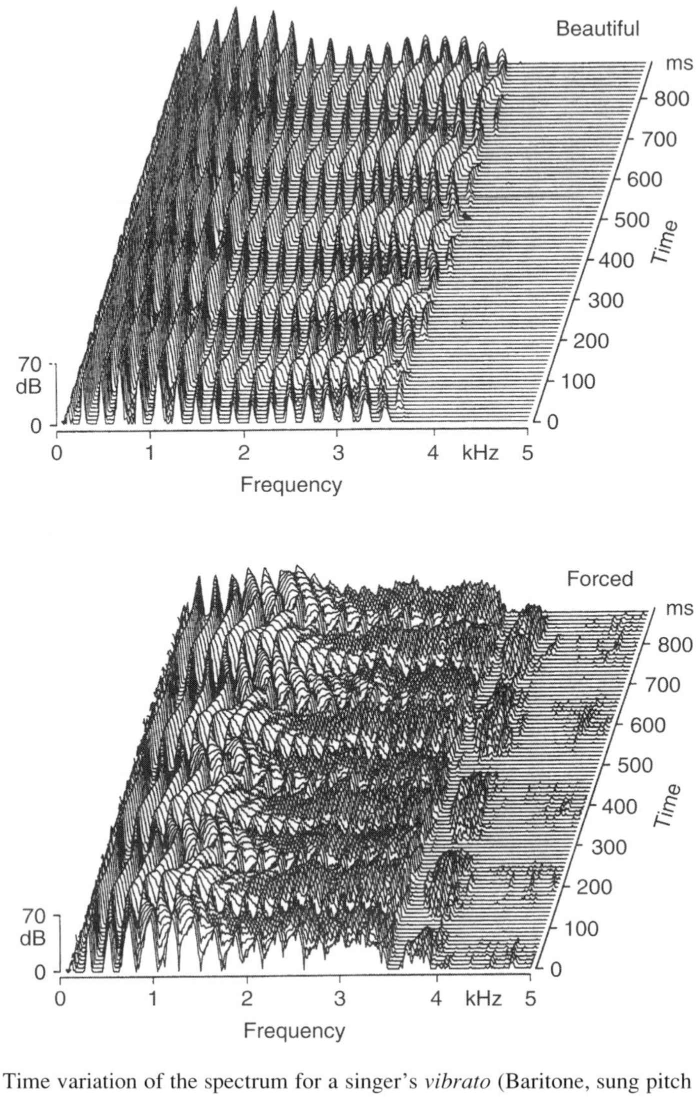

Fig. 3.32 Time variation of the spectrum for a singer's *vibrato* (Baritone, sung pitch G_2)

partials will slide up or down on the flank of the formant resonance curve, which will actually contribute to the clarification of the formant (Benade, 1976). Nevertheless, under those circumstances the tone color remains constant in time. For a very forced vibrato, as shown in Fig. 3.32, it is a different matter. This vibrato, which – not only by reason of its width of more than ±200 cents – is sensed as esthetically unsatisfying, is additionally characterized by a constant phase amplitude modulation with high tonal and noise contributions. This makes the tone especially noticeable, if not even penetrating. Depending on musical context, a vibrato tending in that direction can occasionally be very appropriate, to increase the voices power to stand out.

3.8.4 Choral Singing

Fusing numerous voices into a homogeneous choral sound demands not only mutual adaptation on the part of the singers, but also the avoidance of all effects which make an individual voice stand out. Therefore, the singer's formant, so essential for the soloist, becomes an annoyance for choral singers, unless it is present in comparable strength in all individual voices. In professional choirs it is approximately 5–15 dB weaker than in soloists (Sundberg, 1990), while it is almost not found at all for lay choirs. As a result, individual voices in a lay choir, which emphasize the singer's formant, stand out in the overall sound, furthermore, an additional vibrato is particularly dangerous. In lay choirs, generally the 3,000 Hz (at *mf*) components are from 20 to 25 dB weaker than the strongest tonal contributions. This can reach 30 dB for boys' choirs (Ternström, 1991b).

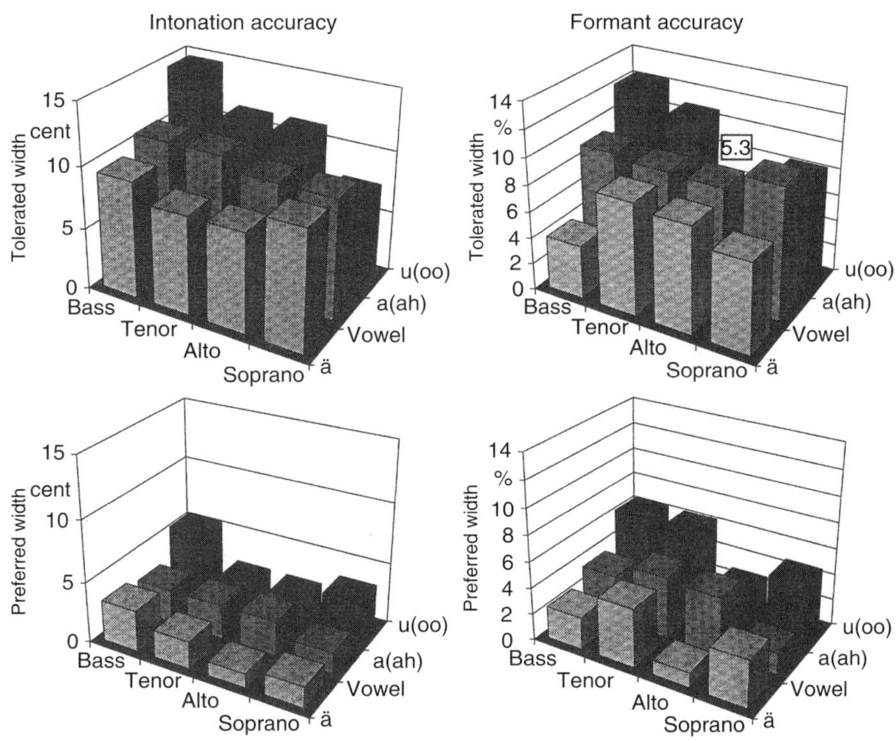

Fig. 3.33 Subjective evaluation of intonation accuracy and formant determination (after Ternström, 1991a)

Basic for an effective all-inclusive choral sound is the accurate intonation of all singers. In hearing tests, Ternström (1991a), using synthetic choral sounds, determined what accuracy is preferred, or alternately is tolerated. The results are represented in Fig. 3.33 for each of four voices and three vowel colors. Accordingly, intonation is considered good, if the standard deviation (width of intonation spread) clearly lies below ±5 cents (i.e., two thirds of the voices should fall within theses limits). An intonation accuracy of ±10 cents is still tolerable. Bass voices are relatively insensitive to these limits, particularly for dark vowels. In contrast Lottermoser and Meyer (1960) found significantly larger deviations. An extreme case is given by the Don Kossaks with up to ±60 cents. Tuning the formant frequencies, i.e., the matching of vowel sounds between individual singers of a choir, is also important. As indicated in the pictures on the right of Fig. 3.33 for the lower two formants, efforts should be made to keep the standard deviations significantly below ±6%, while ±9% can still be tolerated. For the higher formants ±12%, should be maintained, a standard which can be reached only by very systematic choral training.

Chapter 4
Directional Characteristics

4.1 Foundations of Directional Sound Radiation

4.1.1 Directional Effects and Polar Diagrams

In the previous chapter, individual tonal characteristics of musical instruments were considered without concern for possible influences by the room. Furthermore, the fact that most instruments do not radiate sound in all directions with equal intensity, but rather exhibit more or less pronounced directional effects, was not taken into consideration. This dependence of the radiated sound pressure on direction is referred to as the directional characteristic.

In the simplest case, like for example in the setting of a spherical sound source which expands and contracts uniformly in all directions, the sound radiation is also equal in all directions. Therefore, for such a "breathing sphere" one speaks of sphere-like directional characteristics, or omnidirectional characteristics. This case also arises when the sound source is small in comparison to the wavelength of the radiated sound, i.e., mostly for low frequencies. In Fig. 4.1, the frequency regions for which the respective orchestral instruments radiate the sound spherically are represented. As noted, mostly the fundamentals in the lowest octave of the relevant range are of interest. No spherical sound radiation is found above 500 Hz.

The presence of two such spherical radiators with some separation already leads to very complicated sound field relationships which depend on the distance between the two sources, as well as the frequencies, phase relations, and strengths of the radiated vibrations. Preferred directions are found, for which the contributions of the two sources reinforce each other, and there are other directions for which they weaken each other, or possibly cancel entirely. For simple arrangements, these directional characteristics can be calculated (Lessig, 1965; Franz et al., 1969, 1970), in less obvious cases, such as vibrating plates in string instruments, a mathematical treatment becomes inordinately complex, and it also presupposes accurate measurements of vibration shapes. Consequently, the following assembled representations of directional characteristics for musical instruments were all determined experimentally, except for the tympani. The necessary measurements

J. Meyer, *Acoustics and the Performance of Music*.
DOI: 10.1007/978-0-387-09517-2_4, © Springer Science + Business Media, LLC 2009

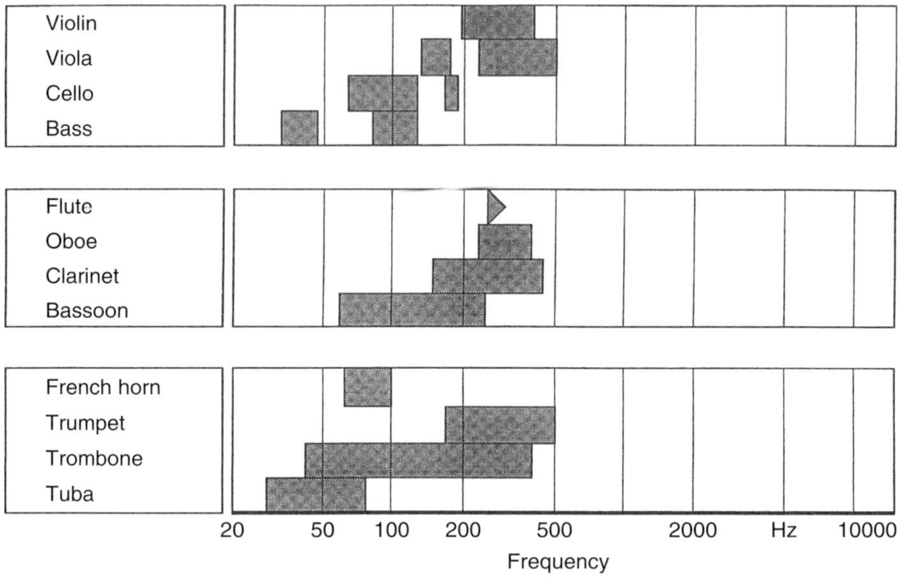

Fig. 4.1 Frequency region for omnidirectional sound radiation by orchestral instruments

were carried out in an anechoic chamber with a microphone distance of 3.5 m (also see Meyer, individual publications from 1964 to 1970).

As an example of such measurement results, the polar diagrams for an oboe at a number of frequencies are given in Fig. 4.2. In the individual pictures, the relative sound level in dB is represented in angular dependence. The 0°-direction corresponds to the axis of the sound column. The indicated frequencies are not those of only the fundamentals of the played tone, but rather, they represent the relevant partial of the spectrum falling in the indicated region. In particular, the curves for higher frequencies naturally relate only to overtones. These few diagrams already show that not only the overall sound strength changes with direction, but also the spectrum and thus the tone color.

4.1.2 Evaluation and Representation

The relatively complicated curves of the many frequencies that must be included in order to characterize the directional effect of instruments, lead to the necessity of extracting the fundamental characteristics from the multitude of individual results, in order to represent them in a clear form. This opens the possibilities of determining the level difference between the largest and smallest value in each diagram. This "Dynamic" of the directional characteristics, however, it is of relatively minor importance in relation to the tonal effect in the room, because the deep structures are mostly very narrow (see for example in the picture for 1,750 Hz at about 75 and 170°). For microphone recordings, on the other hand, such steep flanks of dips become uncomfortably noticeable, particularly when the performers are moving.

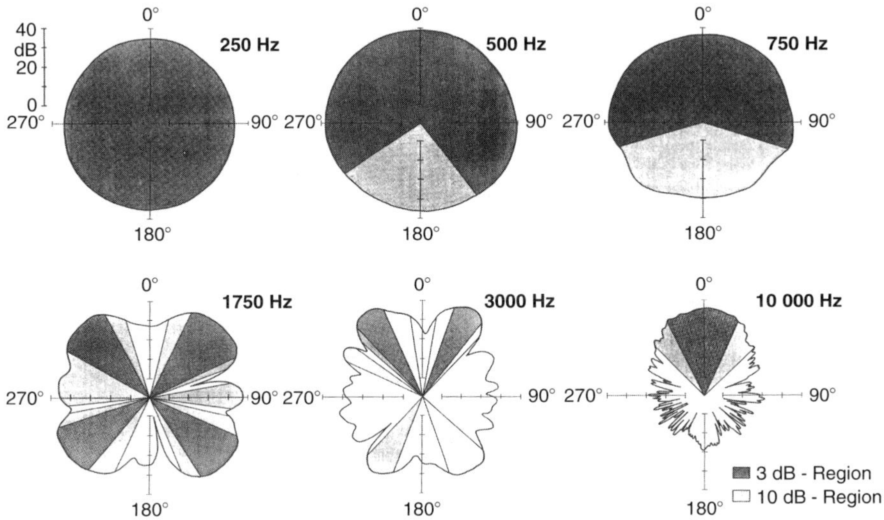

Fig. 4.2 Polar diagrams of an oboe at different frequencies. The 0° direction corresponds to the axis of the instrument

For the discussion of spatial effects, it has been an advantage to determine those angular regions for which the sound level does not sink by more than 3 dB or more than 10 dB respectively, below the maximum of the curve. At the same time, it has to be remembered that a level difference of 3 dB represents cutting the energy in half – which corresponds to a sound-strength difference of one-half the numbers of performers – and a level difference of 10 dB is perceived as approximately one-half the loudness. The width of the 3 dB regions is designated as the half width.

In the polar diagrams of Fig. 4.2, the boundaries of the regions are represented, and the regions of differential intensity are shaded. In the example of 250 Hz, it is clear that there is no angle for which the amplitude drops by more than 3 dB below the maximum. In practice, therefore, one can still speak of a directionally uniform or round characteristic. In contrast, pictures for higher frequencies exhibit pronounced preference regions of various width and angular locations.

Another interesting quantity is the difference between the sound level radiated into the forward and backward directions. It is designated as the front/back ratio given in dB. This ratio indicates the effectiveness of the sound reflecting wall behind the performer. In order to obtain numbers of practical use from a highly structured diagram, forming the front – back ratio from an average of $\pm 10°$ for the two directions under consideration is recommended in each case. Occasionally, it is also an advantage to specify a certain front – side ratio in the appropriate format. For example, this can give useful information about different seating arrangements of wind players in opera or concert orchestras.

Finally, a quantity called the statistical directivity factor is important for room acoustical considerations. It represents a relationship between sound pressures actually present, to those which would be caused by a sound source of equal total power with omnidirectional characteristics at the same distance. The statistical directivity factor can be given in dependence on direction: Values larger than 1 indicate directions with, on the average, stronger radiation; values less than 1 indicate directions of below average radiation. For example, an ideal dipole reaches a value of approximately 1.7 in the direction of strongest radiation. On the boundary of the 3 dB region, the statistical directivity factor drops to 0.7; on the boundary of the 10 dB region, to 0.3 of the maximum value. A survey of characteristic values of statistical directivity factors for orchestral instruments is found in Figure 7.3. For sound level considerations it is advantageous to convert the statistical directivity factor to a dB value. The quantity is designated as "directivity index." It specifies by how much the sound level is higher in the direction considered than it would be for an omnidirectionally radiating sound source of equal power.

In the case of the oboe, one can assume with some degree of accuracy, that the directional characteristics are rotationally symmetric about the long axis of the instrument. This means that by rotating the curve, represented in Fig. 4.2, about an axis along the 0–180° line, full characteristics over all spatial angles are obtained. In order to obtain the tonal effect in the room, only the shadowing of the instrument by the performer, particularly in the rearward direction, needs to be considered additionally. For instruments, which do not present a symmetry axis of that nature, such as for example string instruments, one is limited to the specification of several interesting planes. Fig. 4.3 shows several models of directional characteristics of the horn (including shading by the performer). The results are represented here in four planes: Horizontal plane, vertical plane passing through the direction of the bell, a vertical plane perpendicular to the previous one, and finally, a further vertical plane which include the axis of the sound column. Thus, largely all spatial radiation relationships are included.

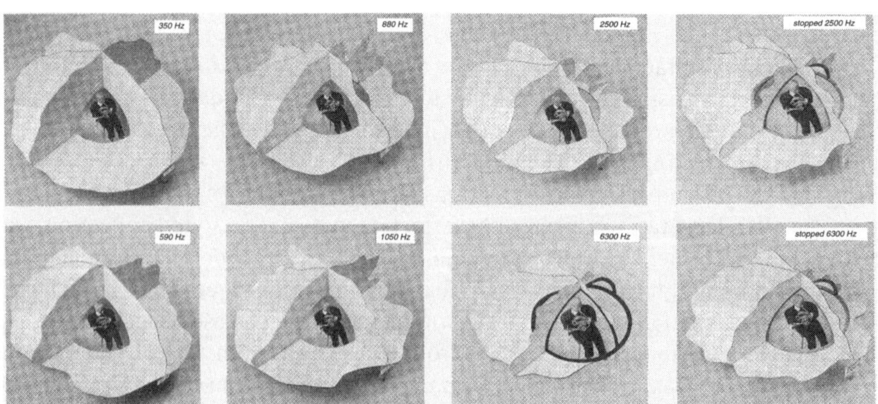

Fig. 4.3 Spatially represented directional characteristics of the horn with player. The amplitude region from the maximal value to the inner boundary of the model is 30 dB

4.2. Brass Instruments

4.2.1. *The Trumpet*

For open brass instruments (for all groups except the horn), the directional effects of sound radiation are only determined by the shape and size of the bell and the connected conical part of the bore. There is practically no influence of material wall vibration. The directional characteristic is therefore essentially rotationally symmetrical about the bell axis. In as much as the dimension of the bell in relationship to the wavelength of the radiated sound is important, the directional characteristics are frequency dependent. In that context, it is of no consequence if the vibration of a particular frequency are those of the fundamental or an overtone of arbitrary order. Thus, the directional characteristics can be considered a function of frequency without regard to valve position or overtone order (Martin, 1942; Meyer and Wogram, 1970).

As already shown in Fig. 4.1, the sound radiation for trumpets at frequencies below 500 Hz is directionally uniform. In the direction of higher frequencies, side constrictions are formed while the largest intensity is radiated in the direction of the bell axis.

From about 2,000 Hz upward, a well defined bundle of energy is noticeable in the direction of the bell, while in a sideways and rearward directions, a multiplicity of secondary maxima are observed. These are separated from one another by deep cuts. Their number increases with increasing frequencies. However, their intensity drops (in relation to the value of the main radiation direction).

Near 2,000 Hz, the amplitude of the side maxima is about 16 dB less than in the axial direction. At 5,000 Hz, this is about 25 dB, and for 10,000 Hz, this level difference rises to somewhat more than 25 dB. In the rearward direction, the intensity also decreases in similar manner: At 2,000 Hz, it is by about 10 dB weaker than coming from the bell, at 5,000 Hz, by about 17 dB, at near 10,000 Hz, around 22 dB. These values for the front/back ratio assume free radiation to the front without shadowing by a music stand.

Inasmuch as the greatest intensity is radiated in the direction of the bell for all frequencies, a general view of the principal radiation regions can be obtained from the many polar diagrams by representing the width of the angular regions for which the amplitudes do not drop more than 3 dB and not more than 10 dB in relationship to their maximum value, as a function of frequency. These curves are given in Fig. 4.4.

One recognizes that the trumpet has a limiting frequency of 500 Hz, the actual directional effect only begins above that frequency. The 3 dB curve initially drops rapidly and after several intermediate maxima, reaches an approximate steady value of 30° for frequencies from 4,000 Hz upwards. These intermediate maxima, come about by the fact that the sound within the bell is no longer propagated as a plane wave, and that therefore, a linear phase propagation is no longer present across the bell area. The first maximum near 800 Hz corresponds to a width of the main radiation region of nearly 270°, the second maximum near 1,200 Hz, possesses a value of 135°.

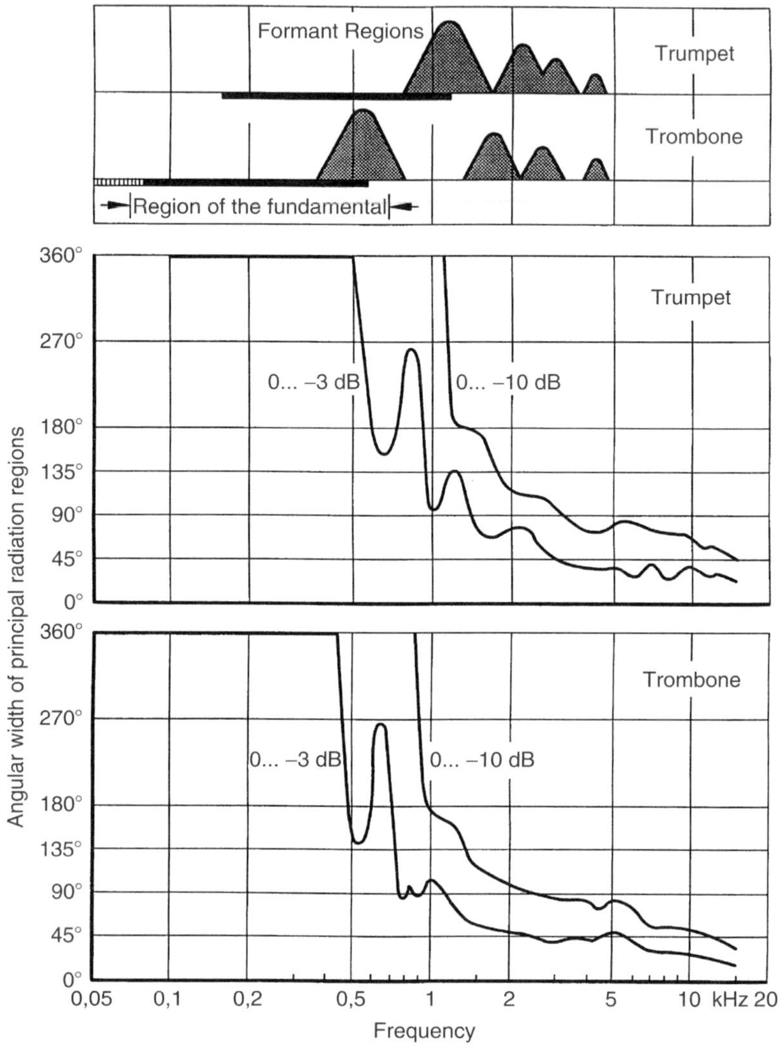

Fig. 4.4 Width of the principal radiation regions of trumpet and trombone, in dependence on frequency with indication of the location of the fundamental and formants

Assigning this directional characteristic to the region of the fundamental and formant of the trumpet sound, enables the upper picture of Fig. 4.4. The fundamentals of the higher registers exceed the region of omnidirectional radiation by more than an octave; the strongest tone contributions in the region of the main formant already form a clear lobe: The 3 dB region has a width of approximately $\pm 60°$.

The 10 dB curve up to 1,100 Hz runs near 360°, i.e., only above these frequencies appear amplitudes which lie more than 10 dB below the maximum value. This curve falls even more steeply and exhibits only relatively minor waviness in the direction of higher frequencies. In the region near 1,300 Hz, it has already dropped to 180°. From this it is recognized that the amplitude to the side and above the performer is just 10 dB less than in the direction of the axis. For very high frequencies, the 10 dB width of the main radiation region narrows to values between 45° and 75°. All these data refer to a concert trumpet of standard bore. For narrower jazz instruments, these curves are shifted to higher frequencies: The first maximum moves from 800 to 890 Hz, and the second from 1,200 to 1,260 Hz. However, the half widths are not changed for the higher frequencies.

The sharper concentration which occurs with rising frequencies also becomes noticeable in the magnitude of the statistical directivity factor: While at 2,000 Hz it only has a value of 2.3 in the direction of strongest sound radiation, it rises to over 4.4 near 6,300 Hz and up to 6.6 at 15,000 Hz. This corresponds to a directivity index of 7.3 dB at 2,000 Hz, 12.8 dB at 6,000 Hz, and 16.4 dB at 15,000 Hz. Theses values characterize the extremely pronounced directional effect of the trumpet. A tabular summary of the angular dependence of the statistical directivity factor, as needed for the calculation of curves for the so-called, diffuse field distance in rooms (see Sect. 6.1.3) is given in the appendix (page 323 of original German Text).

4.2.2 The Trombone

Polar diagrams for trombones exhibit fundamentally similar configurations as those for trumpets, all typical characteristics are merely shifted toward lower frequencies by reason of the larger dimensions of the flare (Meyer and Wogram, 1970). In comparison to the main radiation direction, the intensity decrease toward the side amounts to 18 dB already at 1,500 Hz. It then rises to above 25 dB at 4,000 Hz, and to values slightly more than 25 dB in the region near 8,000 Hz. Correspondingly, already near 1,500 Hz, the radiation toward the back is also more than 10 dB weaker than from the bell. Near 4,000 Hz, this level difference rises to about 18 dB, and at 8,000 Hz it reaches a value of 25 dB. Again, these values from the front/back ratio assume that the sound radiation is not essentially hindered by the music stand.

The diagram for the 3 and 10 dB regions in Fig. 4.4, shows a structure similar to that for a trumpet, the curves are simply shifted toward lower frequencies. Thus, the radiation is omnidirectional up to about 400 Hz. The first secondary maximum of the 3 dB curve (near about 650 Hz) also reaches a width of nearly 270°, the second however, rises less strongly above the course of the curve. In the region between 2,000 and 5,000 Hz, the half width amounts to approximately 45°. Accordingly, the concentrations for frequency contributions near 2,000 Hz, are sharper for the trombone than for the trumpet. These relations are reversed in the region near 5,000 Hz. In the context of considering the sound radiation of the trombone, it is noteworthy that the fundamentals fall largely into the omnidirectional radiation region, and that the main formant is radiated in a less concentrated pattern, i.e., broader, than for a trumpet: For the strongest tone contribution of the trombone, the

3 dB region has a width of about ±90°. A corresponding tendency is also evident for the 20 dB curve. Reference should be made to the fact that for a trombone, already at a 1,000 Hz, the 10 dB line runs through 180°, and that below 900 Hz it no longer deviates from 360°. It is furthermore noticed, that for trombones of varying bores and bells, no noticeable differences in directional characteristics are observed.

This statistical directivity factor reaches values which lie only slightly below those of trumpets of comparable frequencies (i.e., for the trombone, one octave below the trumpet): For the direction of the strongest radiation at 1,000 Hz it is 2.1, for 3,000 Hz 4.5, for 10,000 Hz it is 6.1. This corresponds to a directivity index of 6.3 dB at 1,000 Hz, 13 dB at 3,000 Hz, and 15.6 dB at 10,000 Hz. Again, the angular dependence of the statistical directivity factors is given in tabular form in the appendix (page 323 of original German Text).

4.2.3 The Tuba

In the tuba the bore is wider and conical, consequently the sound radiation relationships are somewhat different than for the trumpet and the trombone (Meyer and Wogram, 1970). Even though directional characteristics can be considered as spherical only below 75 Hz, discrete secondary maxima and deep cuts only occur for relatively high frequencies (roughly above 1,000 Hz). In the region near 500 Hz, the radiated amplitudes toward the side, the front and back reach an attenuation value of about 10 dB in comparison with the principal direction. At 800 Hz the level drop in a plane perpendicular to the bell axis amounts to 20 dB, at 2,000 Hz even 28 dB. In contrast, in the direction pointing away from the bell, the intensity drops less strongly with increasing frequency: at 500 Hz the drop is 10 dB as well, however, at 800 Hz it is only about 15 dB, and at 2,000 Hz around 22 dB.

By reason of the large cone shape, and the relatively weak loading of the bell, a large separation between the 3 dB and the 10 dB curves results for the tuba in the region of low frequencies, as shown in Fig. 4.5. Thus the half width above 75 Hz drops sharply, and already at 100 Hz reaches a value of 180°. The first maximum is rather weak, and the subsequent ones are almost unnoticeable. In the region between 300 and 400 Hz, the principal radiation region covers approximately a right angle and narrows to about 30° from 1,100 Hz on upward. Up to about 450 Hz the 10 dB curve runs at about 360°, and then drops even more rapidly than for other instruments. Below 450 Hz it intersects the 180° line and reaches a value of 90° near 1,000 Hz.

The statistical directivity factor has a value of 1.45 already at 125 Hz for the direction of strongest sound radiation, however, it rises to 2.0 at 400 Hz and 4.5 at 1,000 Hz, and at 2,000 Hz with a value of 6.6, it reaches the same magnitude as a trumpet at its highest frequencies. Values of 3.2 dB at 125 Hz, 6 dB at 400 Hz, 13 dB at 1,000 Hz, and 16.4 dB at 2,000 Hz correspond to this for the directivity index.

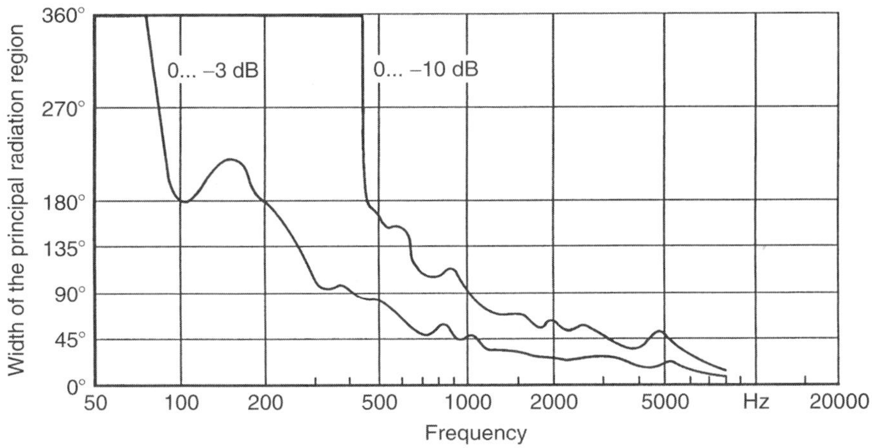

Fig. 4.5 Width of the principal radiation regions of a tuba

4.2.4 The French Horn

The directional characteristics of a French horn are not quite as transparent as those for other brass instruments, since they are not solely determined by the bell. Since the horn player inserts the right hand into the bell opening while playing, the soundfield is already influenced at that point. The bending of the sound around the body of the performer also plays a role, since the instrument is held very close to the body. Finally, the angled position of the horn further complicates relationships. For all these reasons considerations of directional characteristics demand the combined treatment of instrument and player as a unit (Meyer and Wogram, 1969).

Because of the absence of rotational symmetry, it is necessary to restrict consideration of directional effects to a few planes, as already suggested in Fig. 4.3. In order to be able to describe the location and width of the principal radiation regions, the angular coordinates used are represented in Fig. 4.6. In the horizontal plane the 0° line coincides with the direction of sight, from there, angles are measured in the clockwise direction while looking at the player from the top. In the vertical planes the 0° direction lies in the horizontal plane, so that the angular measures indicate elevation above the horizontal.

As can be recognized from the models in Fig. 4.3, the directional characteristics at 350 Hz as a whole give a relatively round impression, however, the amplitude in the upward direction is by about 7 dB less than in the principal direction. At 1,500 Hz this upward attenuation increases to about 15 dB, and at 6,000 dB to above 25 dB. The front/back ratio, relative to the direction of sight of the player, shows the peculiar tendency to possess the largest value of almost 20 dB in the region of 1,700 Hz, and to decrease both in the direction of increasing and decreasing frequencies (with a minimum at 1,000 Hz and a secondary maximum at 800 Hz). In this context the energy radiated toward the back is always larger than in the direction of sight, as

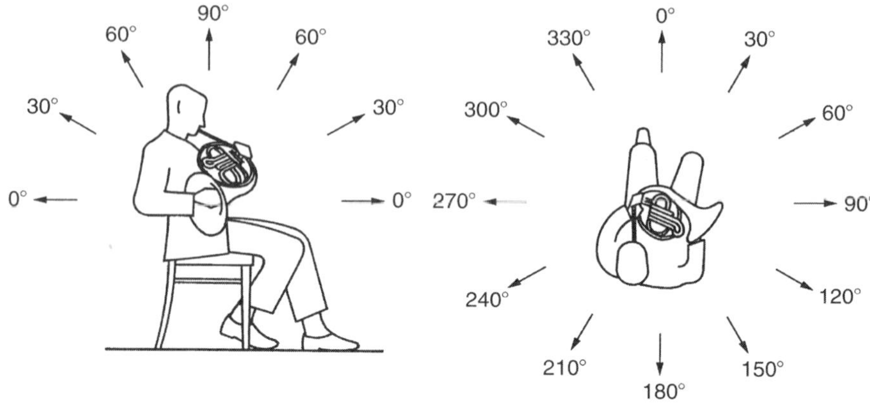

Fig. 4.6 Polar Coordinates for directional characteristics of the French horn

will be illustrated in Fig. 7.27 in a later context. The region of smallest sound radiation is located toward the left side of the performer. At most frequencies it is oriented at an upward angle, only in the region around 1,000 Hz is it oriented located nearly horizontally. Around 6,300 Hz, amplitudes smaller than −30 dB occur for a wide angular region. In these directions, the model no longer shows values, but rather consists only of black bows; the minimum of −40 dB lies at 250° in the horizontal plane.

In Fig. 4.7, those angular regions are graphically represented for which the amplitudes do not drop below the maximum values of the corresponding planes by more than 3 dB or 10 dB respectfully. In the partial pictures for the individual planes, the frequencies of the fundamental or overtones respectively, run from left to right, the direction of sound radiation corresponding to the coordinates given in Fig. 4.6 is given on the upward axis. In the horizontal plane, the directional characteristics below 100 Hz can be given as uniform for all sides. Initially, a weakening on the left side of the player is noticeable with increasing frequency. Already at 200 Hz, this leads to a narrowing of the principal radiation region to less than a semicircle. After a subsequent widening near 400 Hz, the region of strongest radiation between 600 and 900 Hz is concentrated in an angular region somewhere between 100 and 200° Above 1,000 Hz, the maximum is relocated at angles between about 80 and 130°, and at the same time, it becomes significantly smaller. Finally, at high frequencies, only a narrow zone near 140° is formed.

In the horizontal plane, the amplitude drops below the 10 dB boundary only above 500 Hz. The location of the white areas then clearly shows, how, with increasing frequency, the sound radiation is narrowly concentrated near the maximal direction. In that context, the weak secondary maximum in the region around 330°, i.e., toward the left front, needs to be pointed out. It still contains frequencies up to approximately 1,700 Hz.

In the vertical planes, the 3 dB regions below 100 Hz are also closed; thus, at these low frequencies the horn, including the influence of the player, can be

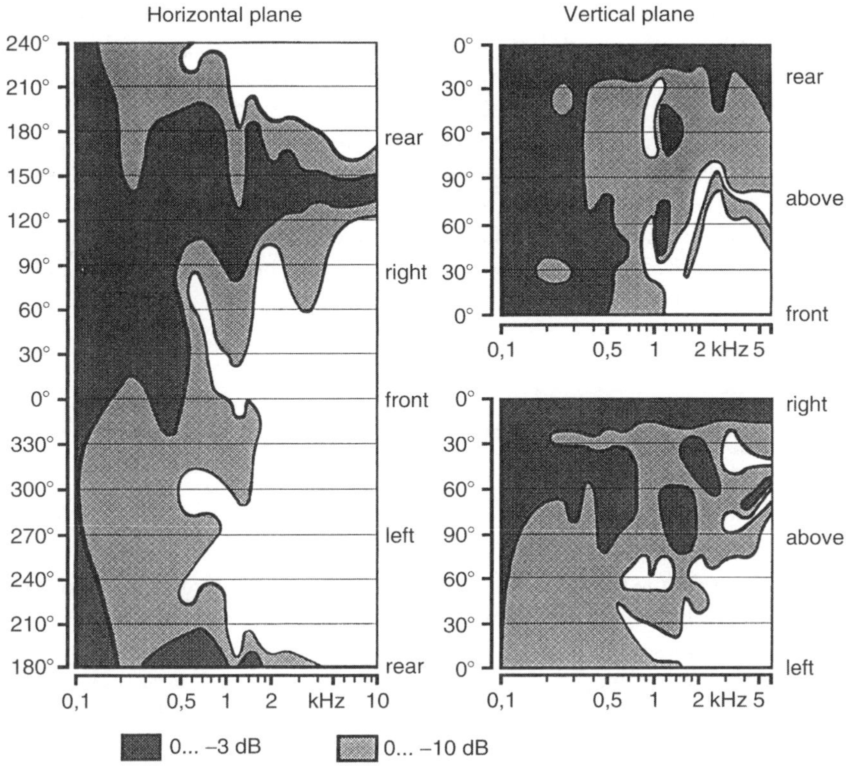

Fig. 4.7 Angular range of the principal radiation regions for the French Horn (with player)

considered as a spherical radiator. The plane running from left to right indicates a preference toward the right side as a whole. In a shallow angle of 0–15°, the strong amplitudes are maintained over the entire frequency regions. Furthermore, several maxima with various frequencies and angular locations are directed upwards toward the right. Toward the left of the player, already starting with 500 Hz, several dips occur which fall below the 10 dB boundary (similar to the horizontal plane).

The vertical plane, oriented from front-to-back, shows a larger area for the half-value region; aside from two dips, which are angled upward at about 30° at frequencies around 200 Hz, the characteristic is round up to 300 Hz. Up to 600 Hz and again at around 1,000 Hz, strong sound radiation is found toward the front upwards. Above that however, the main intensity is concentrated in a rather shallow angle toward the back. In the region of higher frequencies, the amplitudes in the frontal region from the horizontal up to nearly the vertical, lie by more than 10 dB below the maximum value.

At 500 Hz, the statistical directivity factor takes on a value of 1.7 for the relevant direction of strongest sound radiation, at 1,000 Hz, it goes up to 2.4, and at 3,000 Hz, it reaches a value of 4.8. The corresponding values for the directivity index are 4.5 dB at 500 Hz, 7.5 dB at 1,000 Hz, and 13.6 dB at 3,000 Hz.

For stopped tones, directional characteristics are changed in such a way that sound contributions around 1,500 Hz, as well as 6,000 Hz, are radiated in an increased measure upward and toward the front, while components between 700 and 1,000 Hz, as well as between 2,000 and 4,000 Hz are more sharply clustered than for open blowing, where the preferential direction is essentially determined by the bell in spite of the damping hand (see Fig. 4.13). When the horn is blown in an open manner (i.e., without hand in the bell), a definite concentration in the direction of the bell axis occurs. In the region around 1,000 Hz, the half value width is about 60° which is narrowed to 22° for the highest tone contributions. A uniform radiation in all directions appears below about 175 Hz.

4.3 Woodwind Instruments

4.3.1 The Flute

Sound radiation of the flute is characterized by the fact that energy is given off by the blow hole as well as by the first open tone hole (the open end for the lowest note). Thus, the flute as a sound radiator, functions as a dipole. For the fundamental tones, which are not overblown, these two sound sources have a separation of approximately half a wavelength. Consequently, a directional effect similar to that of open labial pipes of an organ are observed (Franz et al., 1969, 1970), however, the dipole characteristics are even more typically pronounced, since in contrast to organ pipes, the two partial sources are practically equally strong for the cylindrical transverse flute and furthermore, the two sound sources approach the half wavelength value of the fundamental frequency more closely because of the narrower bore.

For the not overblown tones, i.e., in the region from C_4 to D_5, the standing waves in the instrument (including the end corrections on both ends), include half a wavelength whereby both ends vibrate in phase. Consequently, the vibrations originating from the two partial sources nearly cancel in the direction of the instrument by reason of the time delay which corresponds to half a vibrational period. In the direction perpendicular to the instrumental axis, they add however, so that this direction is expected to be the direction of strongest radiation.

Similarly, the directional characteristics for overtones of the flute sound can be explained from the dipole behavior. Thus, the relative separation of the two partial sources increases with the ordinal number of the partial tones: For the 2nd partial, it amounts to two half wavelengths for the 3rd partial, three half wavelengths, etc. In addition, the phase relationship between the two sound sources needs to be considered; whenever there is an odd number of half wavelengths in the instrument, the ends vibrate in phase, for an even number of half wavelengths however, out of phase. Thus, for example, for the 2nd partial, a maximum attenuation occurs both in the direction of the instrument axis and perpendicular to it, because the vibrations of the two out of phase vibrating partial sources show a time delay of a full period in the first case, while in the second case, it occurs without time difference (Bork, 1991b). The dipole characteristics explained in this manner, can be clearly observed up to the 6th partial.

For the tones, overblown as an octave, from Eb_5 to D_6, the same sound radiation relationships occur, as for the even partials of the tones played an octave lower. The two partial sources always find themselves in opposite phase, for the overtone series, their separation always is an even multiple of the half wavelength. A typical dipole character is formed, especially for the first three partials. At higher frequencies, i.e., approximately from 3,000 Hz upwards, other effects are super-imposed, which, by reason of a traveling wave, within the instrument shift the angular regions of maximum sound radiation more into the direction of the open end (Bork, 1991a).

A general view of the principal radiation regions is given in Fig. 4.8. Since, as explained, the directional effect not only depends on frequency, but also on the order of the overtones, the 3 dB region and the 10 dB region are given in separate partial pictures (in contrast to corresponding representations for other instruments). In each diagram, frequency runs from left to right, directions are recorded from bottom to top, where 90° indicates the instrumental axis in the direction of the open end, 0°, the direction perpendicular to that, toward the front, which is approximately in the direction of view for the player (not exactly because the flute is mostly held with a slight turn against the axis of the head.

In contrast to all other wind instruments, the directional characteristics of the flute depends almost exclusively on the order of the resonances and not on their frequency location. Thus, for example, for all fundamentals of not overblown tones, this results in a 3 dB region from about 327 to 33° and from 147 to 213°. This region is entered and shaded in the left partial picture between 260 and 590 Hz, corresponding to the frequency location of the fundamentals. For the octave partials, four preferential regions arise which follow from 520 to 1,180 Hz. The number of maxima increases with rising order, however, as individual fields, they become ever narrower. Furthermore, shadowing by the head becomes noticeable toward the back and the left. In order to be able to assign a specific sound field to a particular overtone, the frequency regions of the partials of 1st to 6th order are shown above the 3 dB diagram for tones which are not overblown, as well as for the 1st through 3rd order octave partials. Furthermore, the even partials of the not overblown tones (valid also for the partial series of overblown tones) are represented with bold boundaries to make them stand out in contrast to the odd harmonics.

The diagram for the 10 dB region contains only fields for the first four partials, since there are no lowered dips in the subsequent frequency region. Only above 5,000 Hz are the amplitudes lowered below the 10 dB boundary in the rearward direction. However, the regions of weaker sound radiation are important for the lower partials. Thus, in the 0° direction a drop of 10 dB is present for the even partials, while the odd partials possess a maximum in that region as shown in the partial picture on the left. Accordingly, the odd partials can dominate the spectrum in this direction, so that the tone has a hollow effect. In contrast, the largest intensity for the entire series of partials (below about 5,000 Hz) can be expected toward the

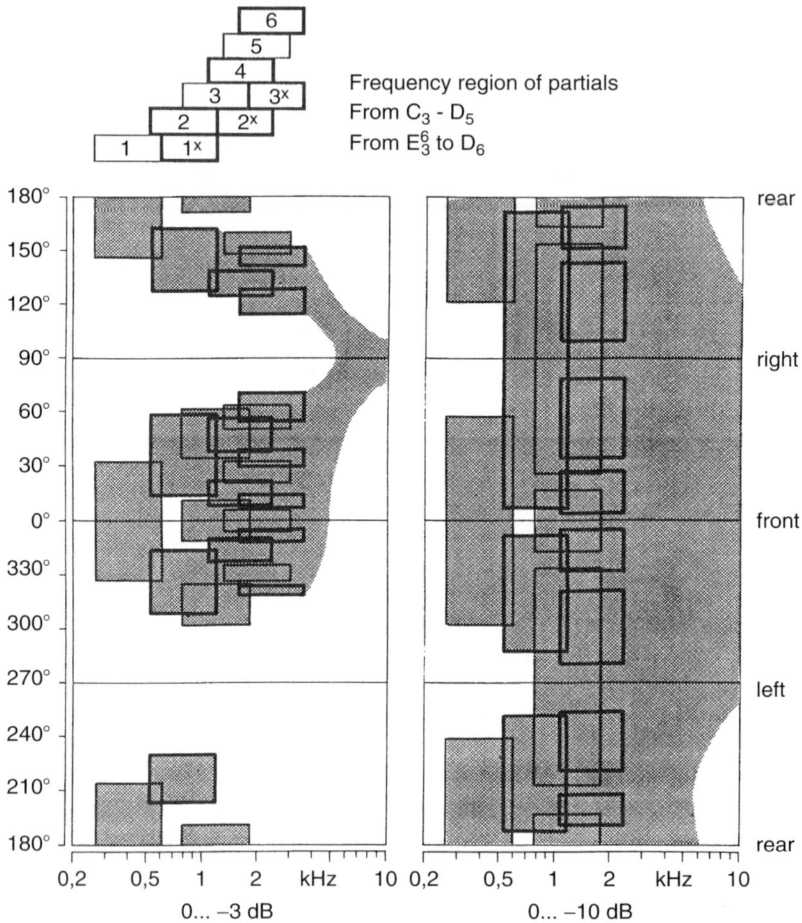

Fig. 4.8 Angular range of the principal radiation regions of the flute (with player)

side in a region between 20 and 30°, as well as the directions located symmetrically toward these. As recognized from a comparison below 0° and below 180° the front-back ratio for the flute is very small. For the first partial it is approximately 0 dB and remains below 10 dB even for the highest frequencies.

For the flute, the statistic directivity factor does not reach values as high as for brass instruments. For the relevant principal directions a value of 1.45 is found for the not overblown fundamentals (in the direction of sight only) as well as for the octave partials, i.e., overblown fundamentals (below approximately 35° on both sides of the direction of sight), this corresponds to a value of 3.2 dB for the directivity index. For those frontal directions in which both fundamental and octave partials are radiated strongly, the statistical directivity factor is approximately 1.3 (corresponding to 2.3 dB). Even for the very high frequencies at about 6,000 Hz, the

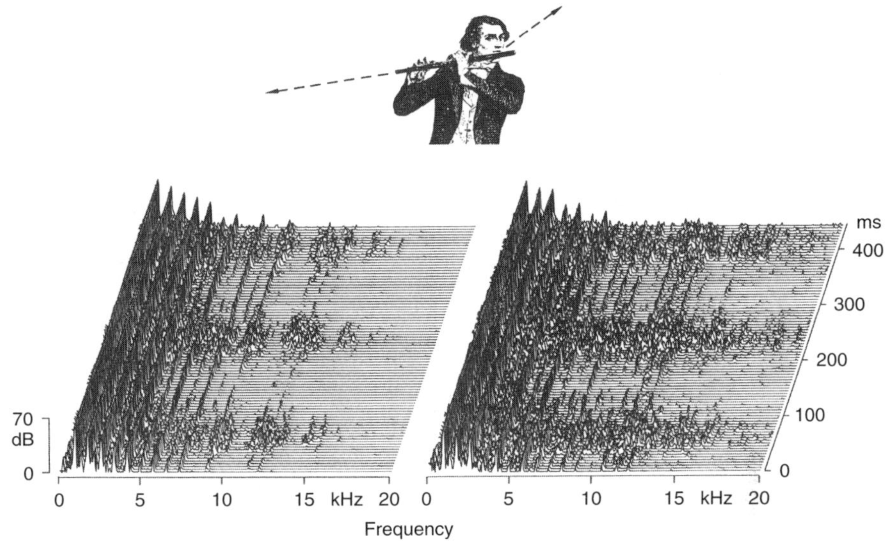

Fig. 4.9 Time changes of the sound spectrum for a flute *vibrato* (note G_5). *Left*: toward the side in the direction of the instrument axis. *Right*: upward in front of the player

statistical directivity factor only rises to a value of about 2.2 which corresponds to a directivity index of 6.8 dB.

All data given so far concerning the directional characteristics of the flute refer only to the sound radiation of harmonic partials. For the attack noise, which, in larger or smaller measure, can give a particular character to the flute sound, no dipole characteristic is formed. The attack noise is radiated almost exclusively from the mouth, because in this case the mouth hole functions almost like a spherical radiator. As shown in Fig. 4.9, the noise level is therefore higher toward the front (and in a wide angle region of unimpeded radiation) than toward the side where significant shadowing is observed. Nevertheless, the strength of the noise contributions radiated toward the side is noticeable for frequencies of more than 10,000 Hz.

In addition it should be noted for historic flutes with a strong conical bore, that the sound radiation of the harmonic components comes predominantly from the mouth hole. As a result, the directional characteristic is relatively round, above all, the dips below 90° are no longer pronounced. Furthermore, the odd partials no longer dominate in the 0° direction above the even ones. This difference from normal flutes is above all important for microphone recordings at relatively small distances.

4.3.2 The Oboe

Inasmuch as in reed instruments the upper end of the exciting reed does not function as a sound radiator – in contrast to the mouth hole of the flute – no dipole nature is observed in the directional characteristics, they are determined by other phenomena (Meyer, 1966b). Since the actual sound source is small in comparison to the wavelength, the low frequency components are radiated spherically, as recognized from the polar diagrams of Fig. 4.2. In the region of the mid frequencies an effect comes to play which is known from the group radiators in loud speaker technology. It is not only the first open side hole but all open flaps which participate in the radiation. Because of their separations and the phases of the vibrations of these individual point sources, preferred directions result, the angles of which depend on frequency. The partial picture for 1,750 and 3,000 Hz in Fig. 4.2 includes directional characteristics of this nature, where a clear drop in the direction of the instrument axis (0°) is observed. For very high frequencies, the sound radiation occurs predominantly from the bell, i.e., from the open end of the instrument, where again, with increasing frequency, traveling waves in the instrument dominate over standing waves. This results in a relatively sharp concentration in the direction of the instrument axis, as the polar diagram shows for 10,000 Hz in Fig. 4.2.

The group radiation effect in the region of mid frequencies is naturally influenced by the number of open tone holes. For the lowest note of the instrument, for which all flaps are closed, it consequently does not enter at all. In this special case all tonal contributions have their greatest intensity in the direction of the instrument axis. For all other notes the preferred regions of sound radiation show certain similarities, both with regards to their width as also to their direction. Three examples for a low, a mid, and a high, not overblown note are shown in Fig. 4.10 to illustrate this phenomenon. As in the representation for the horn (Fig. 4.7), the individual pictures show the frequency running from left to right; the radiation directions are entered from bottom to top. The 0° line corresponds to the direction of the axis from the direction of the bell (see Fig. 4.12). The area with dark shading represents the region in which the sound level does not drop by more than 3 dB below its maximum value for the relevant frequency. The 10 dB region is represented in the partial images for the individual notes by thin lines corresponding to the frequency location of the overtones.

From these summary diagrams it can be recognized that in the lower end of the tonal range, already at 500 Hz, the radiation is no longer spherical. While in the mid range up to about 600 Hz uniform radiation in all directions is present. In contrast, at $C_5^\#$ the preferred region for the fundamental (approximately 550 Hz) is already relatively narrow. Disregarding a few variants in the nature of depressions around 0°, the possibility is opened, by virtue of the similarity of these three tones, chosen as extremes, to determine the average value for all notes as a typical characteristic for the oboe. This result is represented in the right partial image, where the 10 dB region is represented by the light shading.

From this representation it can be determined that the typical sound radiation of the oboe is spherical up to 500 Hz. The principal sound radiation region becomes narrower with increasing frequency, where, beginning at approximately 800 Hz a decrease occurs in the 0° direction. Therefore, in the region around 1,000 Hz the preference directions lie sideways in the region of ±60°. For further increasing

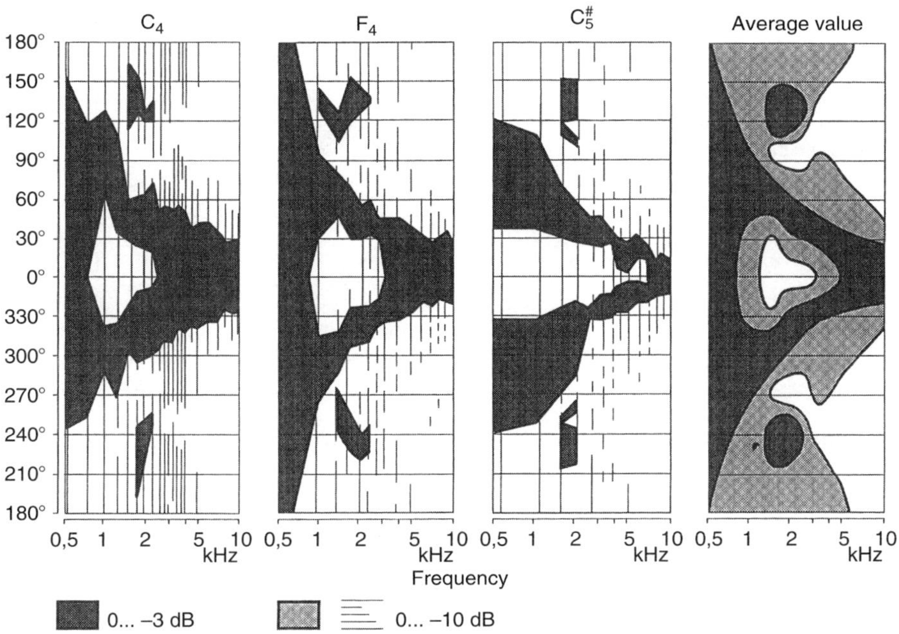

Fig. 4.10 Angular region of the principal radiation directions the oboe for different notes as well as an average value for the entire tonal range

frequencies they again move close to the central axis and they come together at around 5,000 Hz. Between about 1,000 and 2,000 Hz the 0° drop is deeper than 10 dB.

Since the oboe is small in comparison to the size and body of the player, the sound is blocked by the musician toward the rear and also toward the side. This effect is in addition to the directional characteristic of the instrument. This results in a front to back ratio of approximately 15–18 dB for the tone contributions from 1,000 Hz on upwards. To what degree this shadowing effect is also noticeable toward the side can be observed from the diagrams in Fig. 4.11, where the level difference for these directions is represented in relationship to the level in the direction of sight of the player. It is notable that the oboe is approximately 5 dB weaker toward the side than in the direction of sight.

The statistical directivity factor is not very large in the region of the group radiation effect since, for a rotationally symmetric directional characteristic about the 0–180° axis, the preferred region lies near the equator and thus for a width of 30° already encompasses a significant portion of the entire sphere surface. In contrast, for very high frequencies a relatively narrow region around the pole results with concentrations around 0°. Consequently, the statistical directivity factor and the principal radiation direction amounts to only 1.2 at 1,000 Hz which corresponds

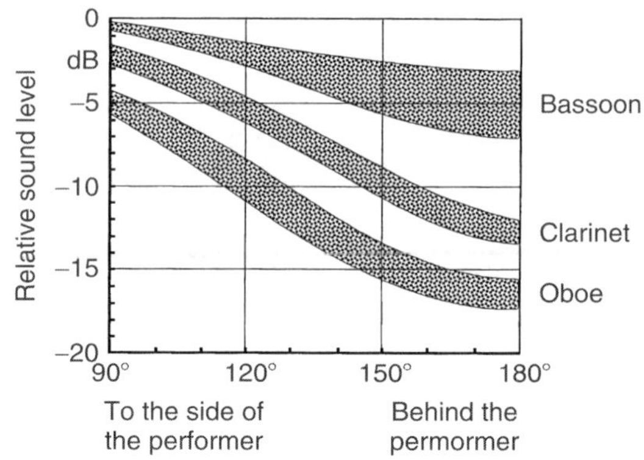

Fig. 4.11 Weakening of the sound level in comparison to the direction of sight of the player. The spread applies for frequencies above 1,000 Hz (for oboe and clarinet) and 500 Hz (for the bassoon)

to a directivity index of 1.5 dB. These numbers, however, do not take into account the shadowing by the player. In practice, therefore, one can assume a statistical directivity factor of around 1.5 with a directivity index of 3.5 dB for the tonal contributions around 1,000 Hz. Finally, for 10,000 Hz a statistical directivity factor of 3.6 is found (corresponding to 11 dB), which is a value smaller than for brass instruments.

4.3.3 The Clarinet

In spite of the different nature of the mouthpiece with a single reed and in spite of the nearly cylindrical bore, the clarinet has directional characteristics similar to the oboe. As a comparison of the principal radiation region shows in Fig. 4.12, the radiation of the clarinet alone can be called spherical up to approximately 700 Hz. In practice, however, because of the shadowing by the player, one should assume a frequency limit below approximately 500 Hz. For increasingly higher frequencies, the preferred regions of the clarinet are somewhat narrower. Here the directional properties of the group radiator effect are determined by the open tone holes. However, the central dip near the 0° line is extended over a broader angle. In particular the area for which the sound level lies by more than 10 dB below the maximum is also larger. For the lower notes of the tonal range this drop possesses a depth of nearly 20 dB, in the upper registers this drops to values which lie only slightly above 10 dB (Meyer, 1965b).

A noticeable departure from the oboe, and also the bassoon, consists in the fact that for the clarinet the island-like maxima are missing, which for the other instruments in certain frequency regions lead to a clover leaf-like directional characteristic (compare the polar diagram of the oboe at 1,750 Hz in Fig. 4.2). The reason for this behavior can be found in the fact that in the clarinet the bell is more pronounced and consequently the sound radiation in this direction is stronger than in the direction located symmetrically to that in the rear. These secondary maxima are therefore only observed in the subdivision of the 10 dB areas. For very

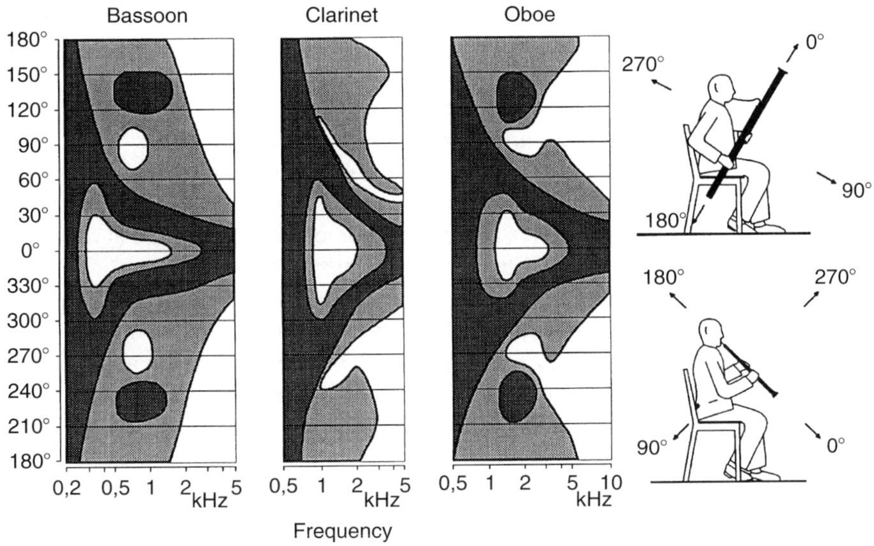

Fig. 4.12 Angular region of principal radiation for oboe, clarinet, and bassoon (average value for entire tonal range)

high frequencies, the characteristics of oboe and clarinet are practically no longer distinguishable after the central depression becomes closed at around 3,500 Hz.

A shadowing of the sound by the player is not as strong for the clarinet as it is for the oboe since its size is slightly larger. Furthermore, the fact that the clarinet normally is held slightly steeper than the oboe during performance, becomes an advantage for the front-back ratio. As can be seen from Fig. 4.11, a level for tonal contributions above 1,000 Hz is approximately 13 dB below the level in the direction of sight. Toward the side a difference of only 1.5–3 dB is noticed in comparison to the direction of sight.

The statistical directivity factor is slightly larger than for the oboe, as already expected by the slightly narrower principal radiation region of the clarinet. It has a value of 2.0 corresponding to a directivity index of 6 dB at 1,000 Hz in the direction of the maximum level (approximately near ±75°) with consideration of the shadowing by the player. At 5,000 Hz a statistical directivity factor 4.5 (corresponding to 13 dB) results for the 0° direction. Here the effect of the bell is clearly noticeable.

4.3.4 The Bassoon

Corresponding to the larger dimensions of the bassoon and its lower pitch, the typical marks of the directional characteristics in comparison to the higher reed instruments have been shifted to lower frequencies (Meyer, 1966b). Thus the spherical radiation is present only below 250 Hz. Above that, preferential regions are formed toward the side which rather quickly approach the 0° axis with increasing frequency, which, however, at about 500 Hz change the location only slowly. Consequently, the central decrease is relatively broad in the region around 300 Hz and extends in narrow form to above 2,000 Hz. Here the area contribution for which the level lies by more then 10 dB below the maximum value is relatively large, however there are no values with more than 15 dB.

In the frequency region between 550 and 1,300 Hz, similarly as for the oboe, four angular regions of preferred sound radiation appear which are distributed in clover leaf fashion. Above 2,000 Hz the lobes become very narrow and at 5,000 Hz reach a half width of only ±20°. These concentrations of sound radiation play an important role in the tonal effect because of the upward direction of the bell.

The shadowing effect of the player is relatively minor for the bassoon. On the one hand, this is the result of the size of the instrument – there is hardly an angle behind the musician from which not a portion of the instrument is visible – on the other hand, the lower tone contributions are more readily refracted around the player. When comparing levels in directions toward the side with those in the direction of sight of the player, differences, as entered in Fig. 4.11, are negligibly small. Not until an angle exceeds 120° does the decrease reach a value of 3 dB. The front-back relation then falls into the order of magnitude of 3–7 dB, it is thus significantly smaller than for the oboe and the clarinet. The relatively broad spread for the level values of the bassoon are determined on the one hand by the directional characteristics, and the lesser shadowing effect on the other hand. As made clear by the principal radiation regions in Fig. 4.12, the tone contributions between 550 and 1,300 Hz are radiated especially strongly toward the rear by reason of the secondary maxima around 240°, while preference is given toward the front of the instrument for components between 400 and 550 Hz, as well as for those above 1,500 Hz.

The statistical directivity factor is relatively small, as is the case for the other reed instruments in the mid range, because of the equatorial principal radiation region. At 350 Hz the shadowing effect by the player is relatively small, so that the statistical directivity factor for the principal direction comes to a value of about 1.4, which corresponds to a directivity index of 3 dB. This is raised to 2.1 (corresponding to 6.5 dB) at 2,000 Hz and after the joining of the sideways maxima at 3,500 Hz to 2.5 (corresponding to 8 dB).

4.4 String Instruments

4.4.1 General Considerations

The directional dependence of sound radiation for string instruments rests on the subdivision of vibrating plates into zones of different amplitudes and phases. Added to that, especially at low frequencies, is the sound radiation by the air space through the f-holes. The directional characteristics, therefore, depend on frequency but also on the structure of the wood so that based on material characteristics, every instrument exhibits individual patterns. Nevertheless, for a large number of instruments, common tendencies are found for the sound radiation. This presents the

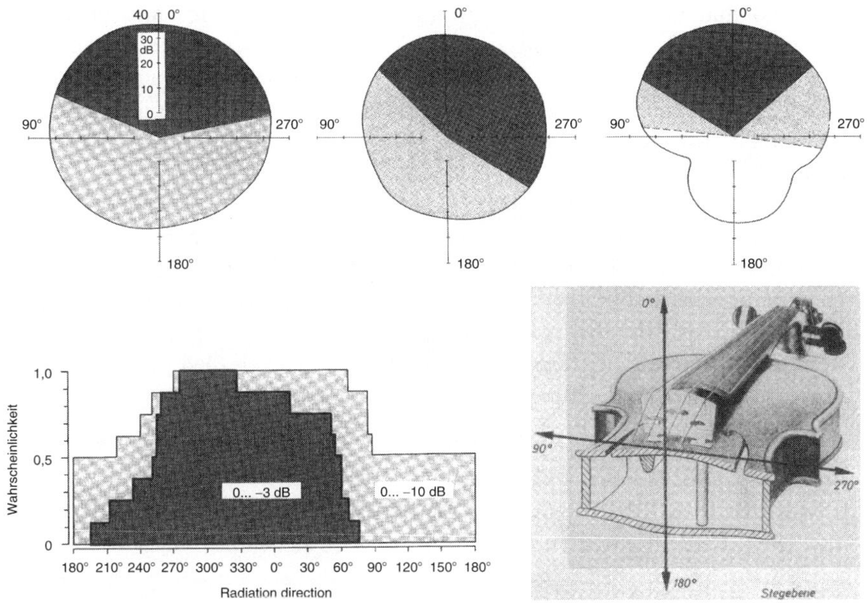

Fig. 4.13 Directional characteristics of the violin at 1,000 Hz. *Top*: polar diagram of three different instruments. *Bottom*: histogram for the principal radiation regions (of eight instruments)

possibility of describing directional characteristics with some generality for the violin, the viola, the cello, and the bass, including essential characteristics for these instrument groups.

As an example, polar diagrams are given in Fig. 4.13 for three different violins. Each case represents a directional characteristic at a frequency of 1,000 Hz. As a comparison of the boundaries of the 3 dB region shows, similarities are recognizable with respect to regions of preferred sound radiation, in spite of individual differences. In order to obtain an average value for the results of all instruments of a family, the number of instruments was determined for which a preferred region (in the sense of a 3 dB and a 10 dB limit) exists in the corresponding angular region. The result for the frequency of 1,000 Hz is also summarized in Fig. 4.13 in the form of a histogram. In this partial picture, the probability that in a given direction a principal radiation region is found, is entered as the abscissa above the angular direction. When the probability reaches the value 1, then all evaluated instruments exhibit a preferential region for this angle. For lesser values there are correspondingly fewer instruments.

In the example pictured, the results for eight violins are summarized as described. One recognizes that between 275 and 325° all instruments have a 3 dB region and between 258 and 15° there are only seven instruments, etc. In contrast, none of the investigated violins have their preferred region of sound radiation in the range of 75–195°. Such an average value formation was also carried out for the 10 dB range. Of the eight instruments, never did more than four drop below the 10 dB limit in any direction so that the probability does not drop below 0.5 for any orientation. Between 220 and 85° a grouping of the 10 dB regions occur, which reaches its maximum between 270 and 65°. With this procedure one consequently obtains a grouping effect in the characteristic radiation regions, while the idiosyncrasies of the individual instruments become less important because of their nonuniform distribution.

An important characteristic of principal radiation regions for string instruments can be seen in the fact that the relationship between the harmonic tone contributions and the unavoidable noise mixtures is especially favorable. Fig. 4.14 shows a comparison of two directional characteristics. One for a steadily vibrating sinusoidal tone of 1,250 Hz and the other for a narrow band rush noise of 1,250 Hz center frequency. Both curves show the same basic shape, however, the typical differences can be recognized, which can be summarized as a generalization:

1. The narrow and deep cuts in the curve for sinusoidal excitation are not as well pronounced for the noise and occur mostly only as shallow depressions.
2. The half width of the principal radiation region is greater for noise components and can certainly assume values of twice that for the sinusoidal tone.

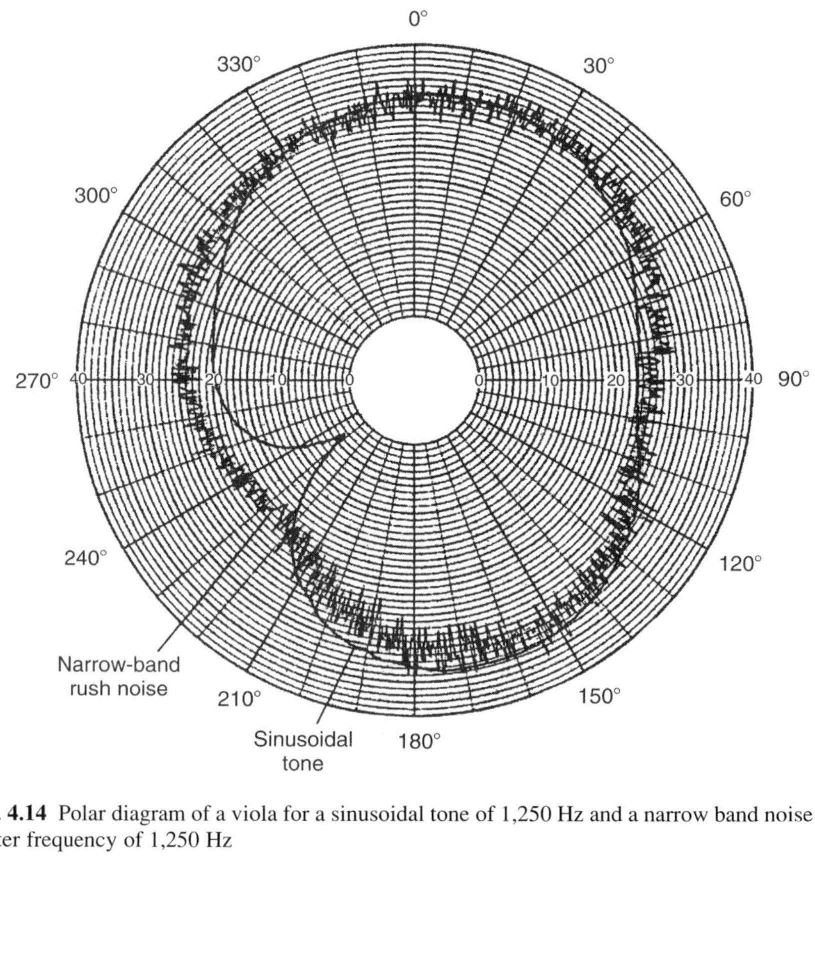

Fig. 4.14 Polar diagram of a viola for a sinusoidal tone of 1,250 Hz and a narrow band noise with center frequency of 1,250 Hz

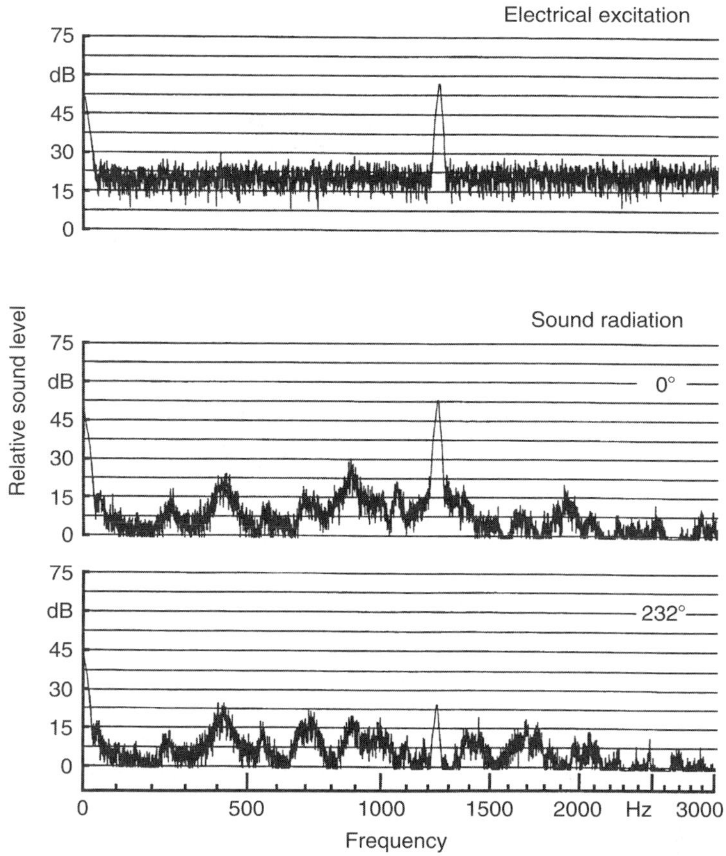

Fig. 4.15 Tonal spectra of a viola under excitation by a mixture of white noise and a sinusoidal tone. The angular indications refer to the polar diagram on Fig. 4.68

3. The noise contributions in the entire frequency region hardly ever fall below the 10 dB limit. Only in the region of middle frequencies a number of depressions occur which, however, are rarely broader than 90°.

Consequently, the noise separation is largest in the principal radiation region, while in those directions for which the sinusoidal directional characteristics show deep indentations it is least favorable. Fig. 4.15 gives an example for this. It contains sound spectra which are radiated by a viola in two directions when the instrument is simultaneously excited by the sinusoidal tone and white noise. The middle diagram corresponds to principal radiation directions, while the bottom diagram shows the relation in the direction of the deep incision of Fig. 4.14. It is clearly evident that the noise components differ in intensity only by an order of 5 dB while the sinusoidal tone changes by approximately 30 dB and consequently hardly exceeds the noise level in the lower diagram.

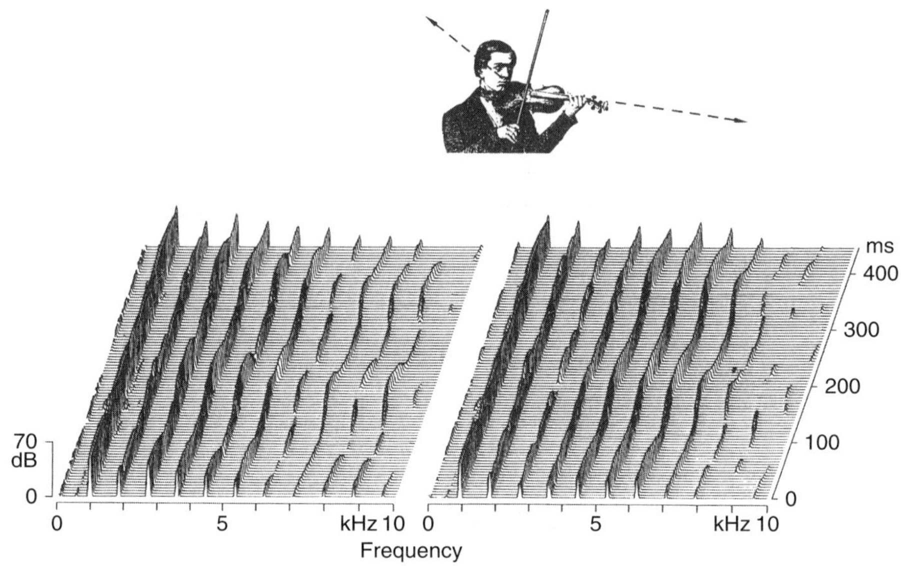

Fig. 4.16 Time variations of the tonal spectrum for a violin vibrato. *Left*: perpendicular to the violin top plate. *Right*: in the direction of the neck

Since in a string instrument each body resonance has its own directional characteristic, this can lead to interesting phenomena. For example: In a vibrato, a partial can excite several overlapping resonances, with one resonance dominating in one direction, and a separate resonance in another. Fig. 4.16 shows an excerpt from a vibrating violin tone which was recorded once in a direction perpendicular to the top plate and once along the direction of the violin neck. In this example the 5th partial varies only slightly in the direction of the neck while the level change perpendicular to the plate is approximately 20 dB. Also the 7th partial varies in the direction of the neck only slightly while in the direction perpendicular to the plate it clearly evidences the double frequency of the vibrato. The 9th partial also varies with a double frequency in both directions, however, in opposite phase. In the left partial image the level goes through minima at the lowest and highest frequency and the maxima are reached at a frequency lying somewhere in the middle. In the right picture the level experiences its maxima at the lowest and highest frequency and the minima at the mid frequencies. This type of directional dependence of sound radiation is a particular characteristic of the string instrument sound. It is also one of the reasons why electronic imitations never, and even good recordings of actual instruments, only very rarely, sound natural in speaker reproductions.

4.4.2 The Violin

The principal radiation regions for the violin are represented in Figs. 4.17–4.19 as histograms for a number of frequencies. They refer to conditions in the plane of the bridge, i.e., in a plane perpendicular to the instrument as well as to sound radiated in

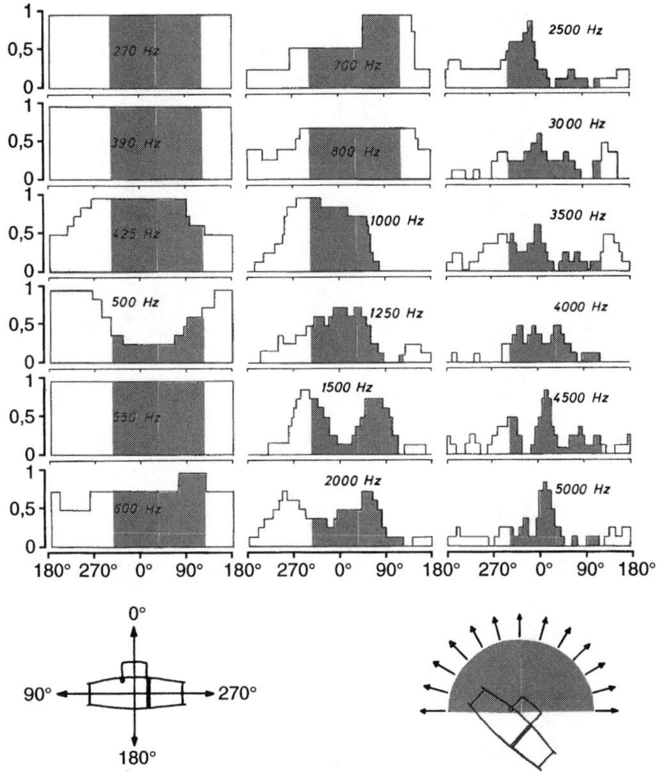

Fig. 4.17 Histogram for the principal radiation directions (0 – 3dB) for violins (bridge plane). *Red areas*: Angular region for first violins oriented toward the audience directly or with one ceiling reflection. *Blue areas*: the corresponding angular region for the second violins with European seating (See Color Plate 2 following p. 178)

a horizontal plane where the angled position of the instrument during performance is already considered. The color coded regions point to the difference in sound radiation between the first and second violins in the orchestra when they are placed opposite each other in the so-called European seating. This question will be considered later. At this point, only the nature of the curves in the individual partial images is of interest.

Already a cursory look makes plain that the directional characteristics of string instruments in their frequency dependence do not exhibit the same steady tendency as those for wind instruments. Rather, frequent changes relative to the direction of preferred sound radiation occur. This becomes very clear when examining the 3 dB regions in Figs. 4.17 and 4.18. Thus, in the plane of the bridge there is practically no directional effect evident below 400 Hz, as the rectangular histograms show. In this diagram probability over the entire angular region maintains a value of 1. Also in the region around 550 Hz, violins radiate uniformly in all directions. In between, a region of preferred top-plate radiation (around 425 Hz) as well as a region of

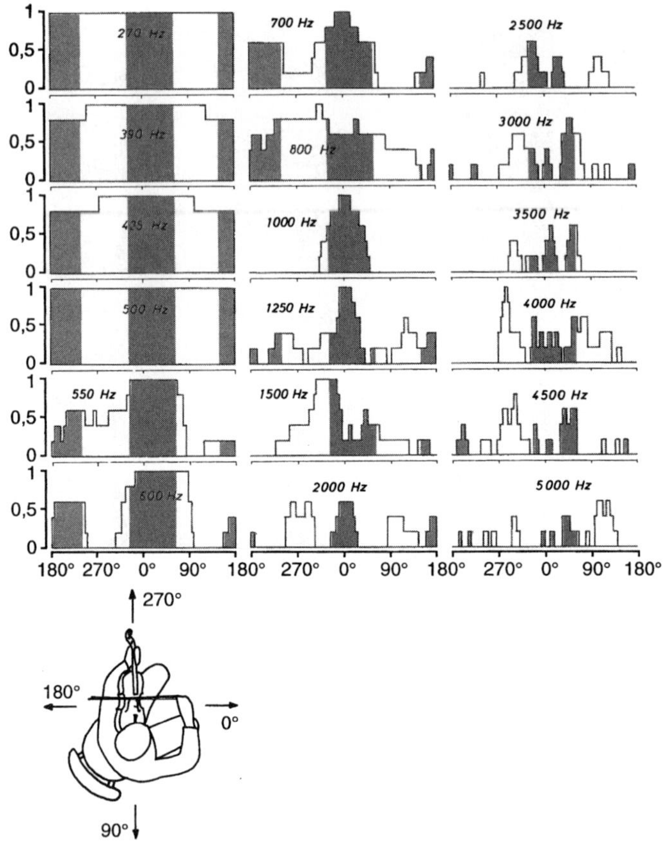

Fig. 4.18 Histogram for the principal radiation directions (0 – 3dB) for violins (*horizontal plane*). *Red areas*: Angular region for first violins oriented directly toward the audience. *Blue areas*: the corresponding angular region for the second violins with European seating (See Color Plate 3 following p. 178)

preferred back-plate radiation (around 500 Hz) is located. In contrast, there is no essential deviation from uniform sound emission in all directions in the horizontal plane, only from 550 Hz on, a concentration near 0° with a secondary maximum toward the back is observed. Correspondingly, there are higher values in the bridge plane near 650 and 700 Hz in the region around 90°, i.e., toward the back of the instrument.

While for 800 Hz no uniform concentration is noted in either plane, in the region of 1,000 and 1,250 Hz again, clear maxima are formed. In the horizontal plane, they lie near 0°, that is, they are directed toward the right from the standpoint of the player, and extend between 270 and 45° in the bridge plane. This results in preferential sound radiation by the top plate. The diagrams for 1,500 Hz and 2,000 Hz show a division in the bridge plane into two maxima, while at higher frequencies the principal radiation zones are concentrated in an angular region around 0° which is perpendicular to the top plate. In the horizontal plane the high frequency maxima are distributed in the

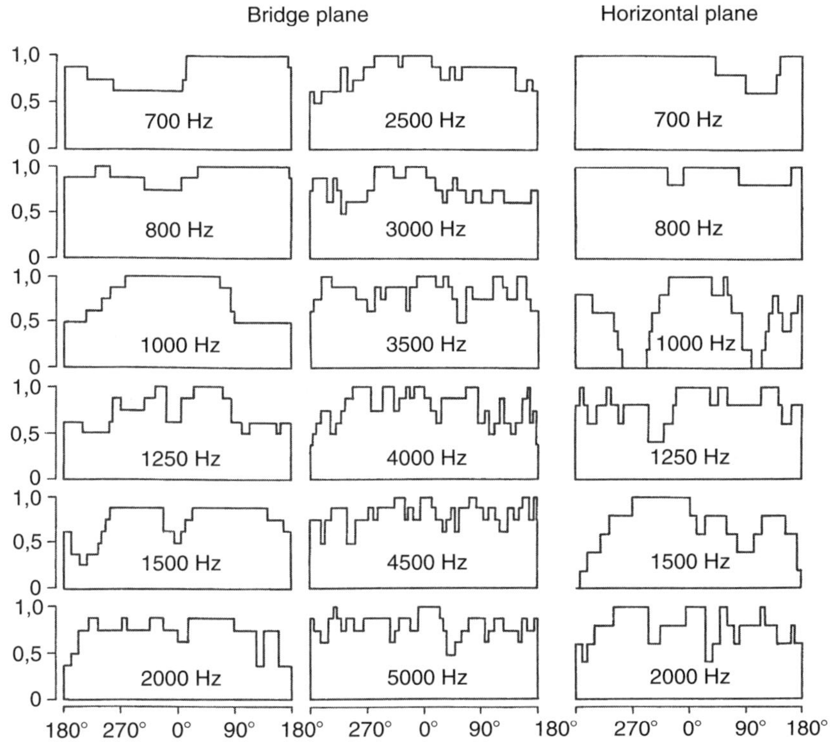

Fig. 4.19 Histograms for directions of principal sound radiation (0–10 dB) for violins

relatively broad region between 270 and 90°. Comparatively strong concentrations for frequencies of 4,000 and 4,500 Hz at around 300° are noteworthy. This nearly approximates the direction of sight for the player. In this context it should be noted that many old Italian violins, valued for their tone quality, possess a directional characteristic for which the principal radiation regions are concentrated into the angular regions which are directed by the top plate of the instrument into the hall (Meyer, 1964b).

For angular regions in which the sound level does not drop by more than 10 dB below the maximum value there are no deviations from isotropic radiation up to 600 Hz. In Fig. 4.19, therefore, only histograms for frequencies above 700 Hz are represented. In the partial pictures for the bridge plane, a relatively uniform distribution is recognized at 700 and 800 Hz, also from 2,000 Hz on upward there is no more pronounced concentration in certain angular regions. A sharper sound concentration, however, is present in the frequency region from about 1,000 to 1,500 Hz. This result also finds support in the diagrams for the horizontal plane where especially at 1,000 Hz near 270° and 100°, that is, in the direction of the fingerboard, and the opposite direction, all instruments exhibit decreases in the directional characteristics, which are deeper than 10 dB.

The particular concentrated sound radiation of the violin around 1,000 Hz also becomes noticeable in the determination of the front – back ratio. In reference to the normal seating arrangement of violins on the podium, this refers to the relationship of the amplitude of the sound radiation to the right (0°), to the value toward the left (180°). A difference of about 7.5 dB is obtained for frequencies around 1,000 Hz. For higher frequencies as well as for lower registers, this front – back ratio again decreases. In a very narrow region around 500 Hz even negative values are obtained because of the preferred floor radiation, however, already at 425 and 600 Hz the radiation toward the right is stronger by about 3 dB. A decrease of the front – back ratio at higher frequencies by about 3 dB should not cover the fact that violins radiate these tonal contributions on the right side predominantly upwards so that by this means additional energy reaches the hall.

The statistical directivity factor for the principal radiation directions moves near values of 1.25 in the neighborhood of the region of top and back plate resonances (425–550 Hz) and in the frequency region of about 1,000–1,250 Hz it reaches its highest value of 2.1. For yet higher frequencies it drops to an average of 1.5 at 3,000 Hz. This, however, does not exclude the possibility that it can be significantly higher for individual, strong resonances. This corresponds to a directivity index of 1.8 dB for 425–550 Hz, 6.2 dB for around 1,000 Hz, and 5 dB around 3,000 Hz.

4.4.3 The Viola

The directional characteristics of the viola have great similarity with the violin, since the size difference between the two instrument groups is relatively small (Meyer 1967a). For very high frequencies the radiation of the viola is, however, concentrated more sharply. In detail, the principal radiation regions of the viola are again given as histograms for the 3 dB regions in Figs. 4.20 and 4.21. Initially it can be noted that the radiation for frequencies up to 500 can be essentially designated as spherical, merely in the region around 200 Hz, many instruments show a directed radiation from the top plate and the F holes. However, exceptions for individual frequencies can occur. These "special cases" are associated with the fact that for violas the differences between different models as far as body measurements and plate tuning are concerned, are greater than for violins.

Above 600 Hz, however, there are clearly preferred regions of radiation and also the gaps in sound radiation are concentrated in the same angular regions for all instruments. In the bridge plane (Fig. 4.20), a particularly strong radiation is observed from the back plate in the region between 600 and 700 Hz, while between 800 and 1,200 Hz maxima of different width are noted in the region of 0°, i.e., perpendicular to the top of the instrument. At 1,500 Hz no clear preference region is recognizable. Here the individual characteristics of each instrument dominate too strongly. In the direction of higher frequencies these diagrams become increasingly complex because the directional characteristics frequently are divided into several maxima. The preferred regions are distributed predominantly over the region of top plate radiation, i.e., between 270 and 90°. At 3,500 and 4,000 Hz they are oriented nearly in a perpendicular direction from the top plate.

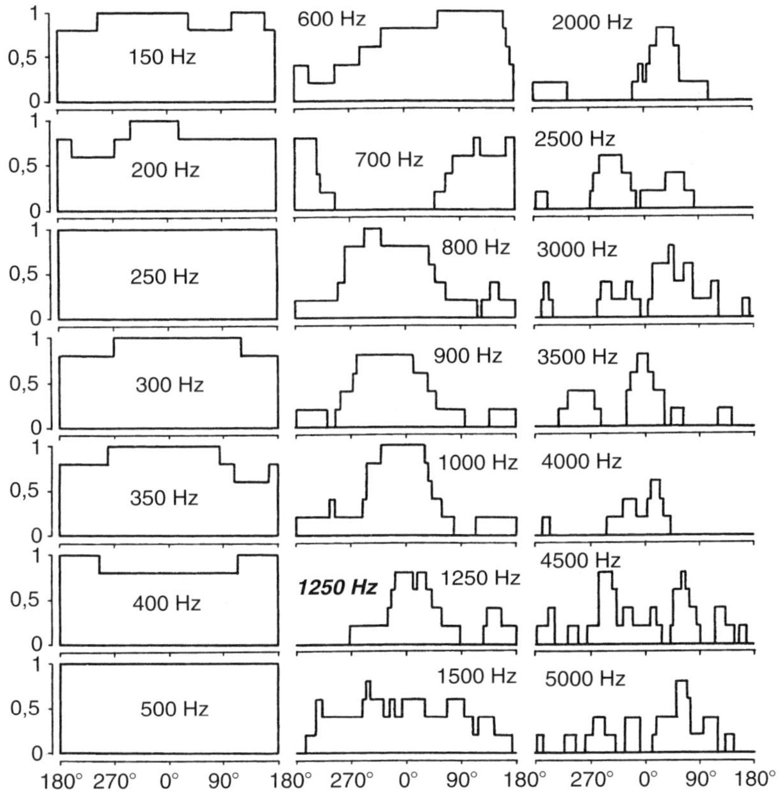

Viola bridge plane

Fig. 4.20 Histograms for the principal radiation directions (0–3 dB) for violas (bridge plane)

The corresponding results for the horizontal plane are given in Fig. 4.21. As mentioned, spherical sound radiation dominates here also for the low tone contributions up to approximately 500 Hz. Only at 200 Hz a pronounced drop is observed toward the left, as seen from the position of the player. The air resonance is located in this frequency region and consequently the F holes strongly participate in the radiation. The partial images from 600 to 1,000 Hz exhibit very sharp concentrations. The contributions around 600 Hz show directional preference toward the left, the others toward the right. At 1,250 Hz a separation into two preferred regions can be recognized. They are oriented toward the direction of the finger board and toward the player. Also at 1,500 Hz the region of strongest amplitude is directed toward the player. For increasingly higher frequencies the concentration regions are distributed into an angular region directed toward the right, especially preferred are orientations around 60°, i.e., approximately in the direction of the extension of the right shoulder of the player.

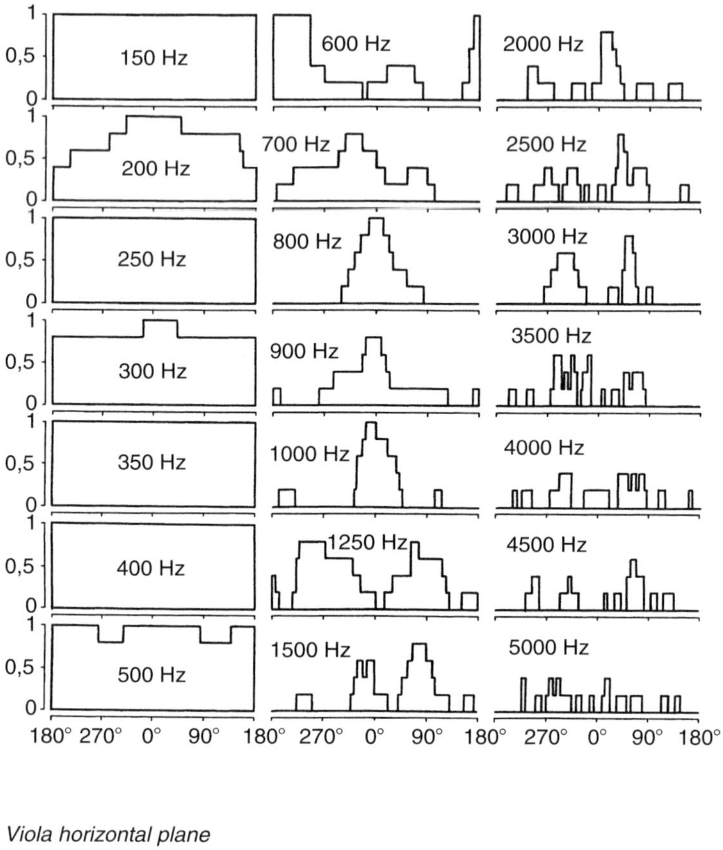

Fig. 4.21 Histograms for the directions of principal sound radiation (0–3 dB) for violas (horizontal plane)

The front-back ratio, again given as difference between the horizontal radiation toward the right (0°) and toward the left (180°) – corresponding to the practice for violins – leads to similar values for the viola as for the violin. It is, however, noteworthy that the region of pronounced sound concentration from the back plate of the instrument extends over a broader frequency range and it lies near higher tones than for the violin. Between 600 and 700 Hz, the intensity toward the left is more than 5 dB stronger than toward the right. The maxima of the concentration toward the right lies, as is the case for violins, with a front-back relationship of approximately 7 dB between 900 and 1,000 Hz.

The statistical directivity factor for the appropriate directions of strongest radiation is still relatively small with a value of 1.15 at 500 Hz. However, it rises to a value of above 1.9 for frequencies around 1,000 Hz and to 2.6 for 3,000 Hz. This corresponds to a directivity index of 1.2 dB at 500 Hz, 5.4 dB at 1,000 Hz, and 7.2 dB at 3,000 Hz.

4.4.4 The Cello

Corresponding to the larger dimensions, the region of the uniform radiation for the cello is shifted toward lower frequencies. As the histograms in Figs. 4.22 and 4.23 clarify, the spherical characteristic can only be expected for frequencies below 200 Hz. In the plane of the bass bar, i.e., a plane located vertically in space, a preferred radiation toward the front is observed in the region around 200 Hz for several instruments. In the region around 250 Hz and also around 300 Hz the effect of pronounced plate resonances is recognizable, in the subsequent frequency region, up to about 500 Hz, the maximum of the sound level lies in an angular region of $\pm 40^\circ$ around the 0° axis. It is, thus, oriented toward the front. In the frequency region between 1,000 and 1,250 Hz, the strongest contributions are radiated in a

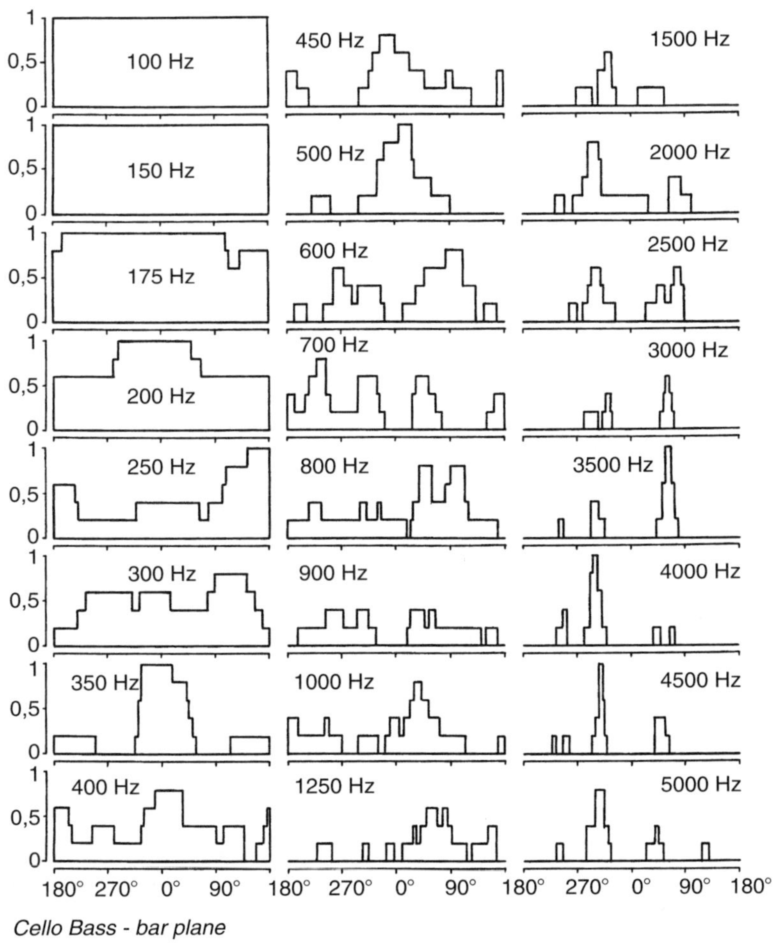

Fig. 4.22 Histogram for the principal radiation directions (0–3 dB) for Cellos (Bas bar plane)

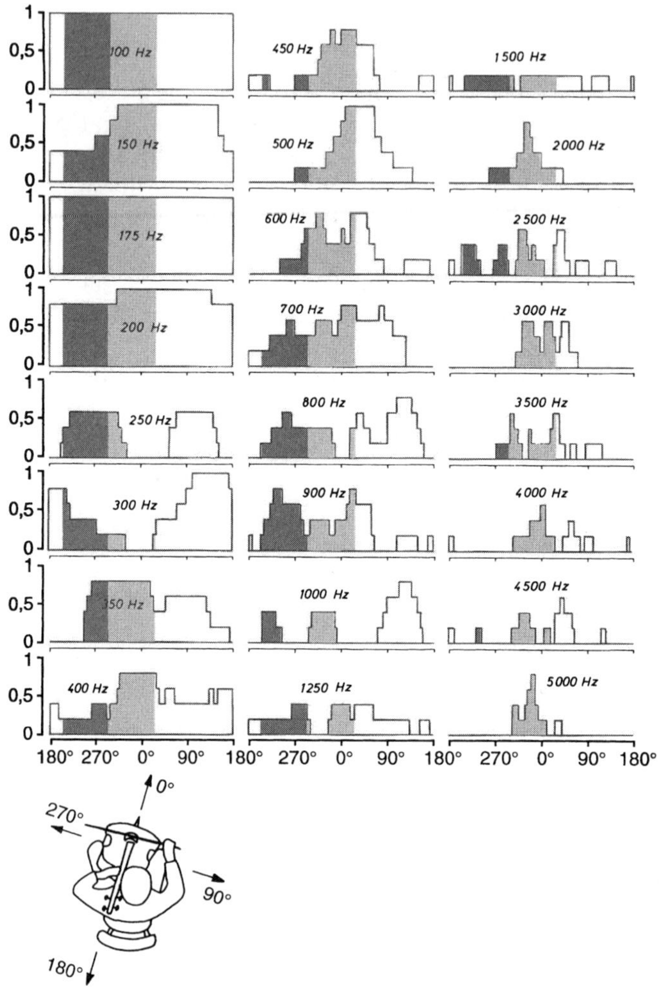

Fig. 4.23 Histogram for the principal radiation directions (0 – 3dB) for Celli (bridge plane). *Yellow areas*: Angular region oriented directly toward the audience for seating position facing the audience. *Blue areas*: The corresponding angular region for a sideways positioning near the edge of the stage (See Color Plate 4 following p. 178)

more upward direction. They are concentrated approximately between 25 and 75°. The directional characteristics at these high frequencies are noteworthy. In contrast to violins, where these components are radiated most strongly in a direction perpendicular to the top plate, in the cello, two extremely narrow preferred regions are formed, which are found with a width of 20° each, at the angular locations around 300 and 60°.

The histograms for the bridge plane in Fig. 4.23 show that already at 150 Hz a certain amount of sound concentration occurs, although over a large angular region.

The right side and the direction of view of the player are preferred over the left side. Furthermore, in this plane the radiation up to about 200 Hz can be considered as spherically symmetric. At 250 Hz a split into two maxima occurs. They are oriented toward the left and the right. In contrast, the radiation near 300 Hz is stronger toward the rear, however, the width of the maxima differs from instrument to instrument. The subsequent frequencies from 350 to 700 Hz show the highest intensities in the region around 0°. This is similar to the situation in the bass-bar plane. The spatial distribution of the principal radiation region at these frequencies thus has the shape of a cone, the axis of which is perpendicular to the top plate of the instrument. Between 800 and 1,000 Hz the preferred directions are predominantly oriented towards the sides. At the high frequencies from 2,000 Hz upwards, the strongest radiation again occurs forward from the top plate, where the left side, as seen from the viewpoint of the player, however, is frequently disadvantaged in comparison to the right side (Meyer, 1965c).

For the cello, the front- back ratio is best related to the direction of sight of the player. In that setting it is noted that already at frequencies near 200 Hz the radiation toward the front is 5 dB stronger than toward the back. In the region around 250 Hz a radiation toward the rear is somewhat stronger than toward the front because of the preferred back plate radiation, in spite of the shadowing by the player. For all frequencies from about 350 Hz on upwards a positive value for the front-back relation is obtained in the direction of sight. A first maximum of more than 10 dB is reached near 500 Hz. Because of the shadowing by the player, the tonal contributions above 1,500 Hz are likewise radiated more weakly toward the rear in comparison to horizontally forward. Additionally consideration needs to be given to the fact that the principal directions of these high components are oriented toward the front downwards and upwards.

The statistical directivity factor for the important resonance region between about 350 and 500 Hz already reaches a value of 2.1 in the principal radiation direction, which corresponds to a directivity index of 6.5 dB. For frequencies around 1,000 Hz, the statistical directivity factor in the principal direction also has a value of 2.1, and rises to 3 for the principal radiation direction at frequencies around 3,000 Hz. This corresponds to a directivity index of 6.5 dB at 1,000 Hz and 9.3 dB at 3,000 Hz.

4.4.5 *The Double Bass*

For the directional characteristics of the double bass it is noteworthy that even for low frequencies a pronounced concentration of the radiated sound is present. As recognized from the histograms in Fig. 4.24, only in the region around 100 Hz, and in a narrow frequency band below the air resonance, a directionally uniform sound radiation can be expected for the majority of instruments. In contrast, in the neighborhood of this resonance, nearly all basses possess a preferred region which in the bridge plane encompasses approximately the front semi-circle. A similar concentration, though slightly wider, results toward the right near 160 Hz.

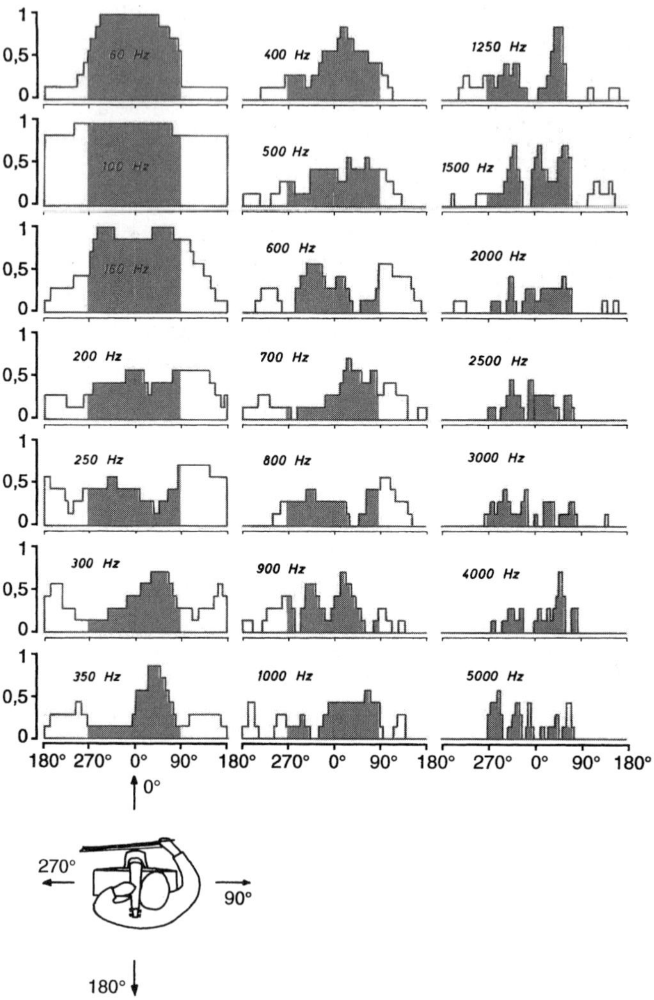

Fig. 4.24 Histogram for the principal radiation regions (0 – 3dB) for double basses (bridge plane). *Red areas*: Angular region oriented directly toward the audience for an instrument location in the left rear corner. *Blue areas*: The corresponding angular region for a right rear corner location (See Color Plate 5 following p. 178)

Such uniform concentrations are rarely encountered at higher frequencies, occasionally a splitting into two or three preferred region occurs. Thus between 200 and 250 Hz two maxima are present, one of which is oriented toward the front and left, while the other one points toward the right rear. Nevertheless, even at these frequencies, the intermediate angular regions show a significant level. Between 300 and 400 Hz a concentration appears in the region between 345 and 90°, also toward the front and the right. Similarly located concentrations, though not quite as sharply clustered are found in the diagrams for 500 and 700 Hz. In contrast, the preferred directions for 600 and 800 Hz are again divided into two regions, which in their angular locations correspond to the circumstances near 200 Hz. The high frequencies from 1,000 Hz on upwards are concentrated in an angular region between 290 and 70°, they are thus oriented toward the front in a rather wide angle (Meyer, 1967a).

This type of directional characteristic leads to a front-back ratio for which the radiation at high frequencies toward the front – similar to the cello – is from 10 to 15 dB stronger than toward the rear. However, relatively large individual differences between instruments occur. The front-back ratio for sound radiation at low frequencies is also informative: While positive values from 5 to 7 dB are observed in the region from 50 to 200 Hz, in a small region around 275 Hz a preferred radiation toward the rear with values up to −10 dB occur. Finally, positive values around 10 dB characterize the front – back relation between 500 and 700 Hz.

The statistical directivity factor for the respective principal radiation directions already has a value of 1.5 in the low register (around 160 Hz). At frequencies between 300 and 1,000 Hz it moves around values of 2.1 and rises to a value of 2.6 at 3,000 Hz. This corresponds to a directivity index of 3.6 dB for low-, 6.5 dB for mid-, and 8.3 dB for high-frequencies.

4.5 The Grand Piano

4.5.1 Lid Open

For the concert grand piano the vibrations of the sound board are primarily responsible for the sound radiation (Grützmacher and Lottermoser, 1936). The directional dependence comes about by the fact that the sound reaches the hall by radiation directly upward and also by reflection from the lid, it is furthermore also radiated downward and influenced in the hall by floor reflections (Meyer, 1965a; Bork, 1993). In addition, at very high frequencies, the strings themselves radiate some energy which in part reaches the room directly and in part by reflection from the lid. The directional characteristics are formed by these components in cooperation, whereby consideration must be given to the fact that the lid as well as the entire instrument have a shadowing effect in certain directions for the direct as well as the reflected sound. This effect increases in importance with rising frequency since the sound waves of low registers are bent around obstacles.

Inasmuch as the vibrational shape of the sound board does not only depend on frequency but also on the location of the excitation, the directional characteristic depends within certain limits also on the fundamental of the excited string. The following directional data are thus separated into the low, middle and upper register of the keyboard. In Fig. 4.25, the directional characteristics of a grand piano are represented for these three regions in the form of polar diagrams in a vertical plane, the angular designation of directions is explained in Fig. 4.26, and is supplemented with relevant data in the horizontal plane.

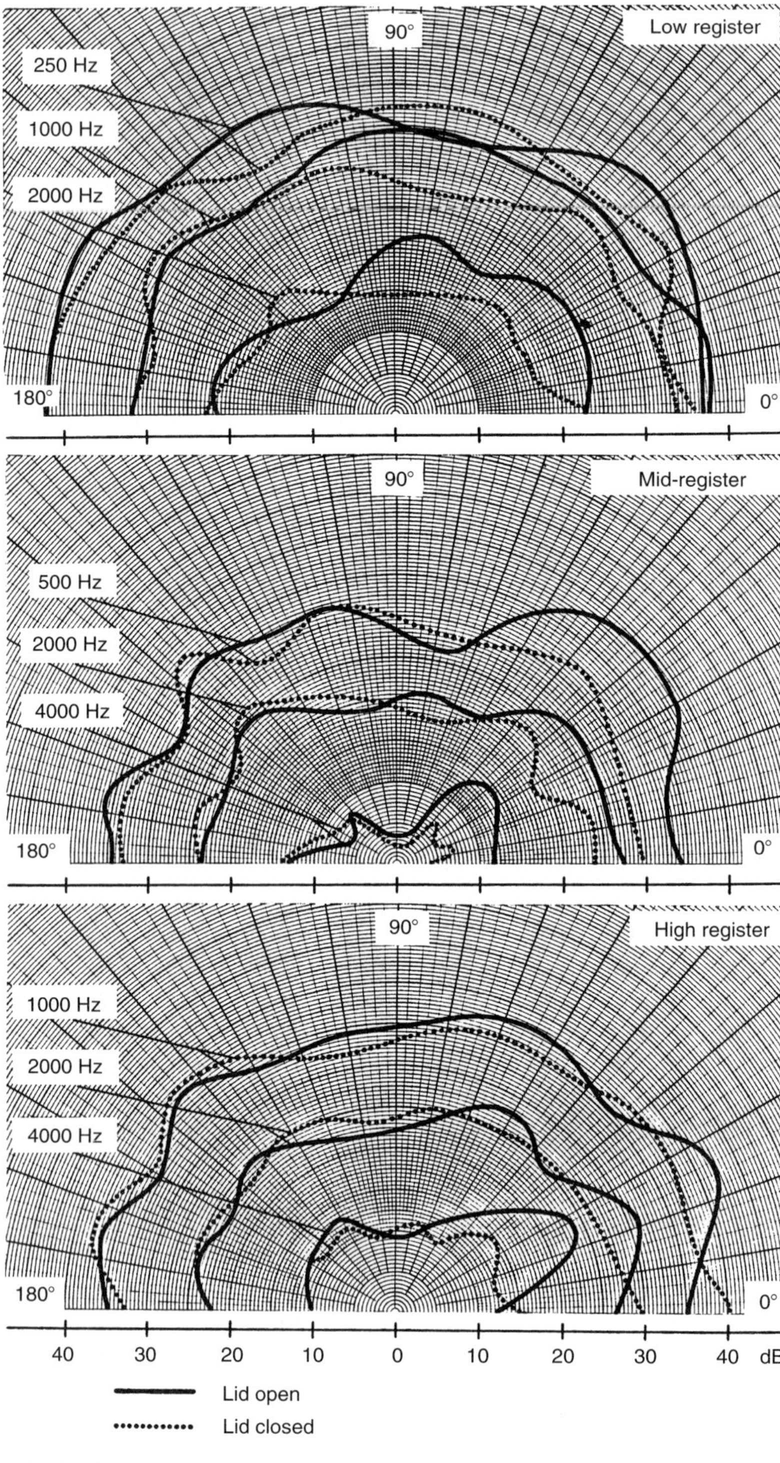

Fig. 4.25 Directional characteristics of a grand piano in a vertical plane

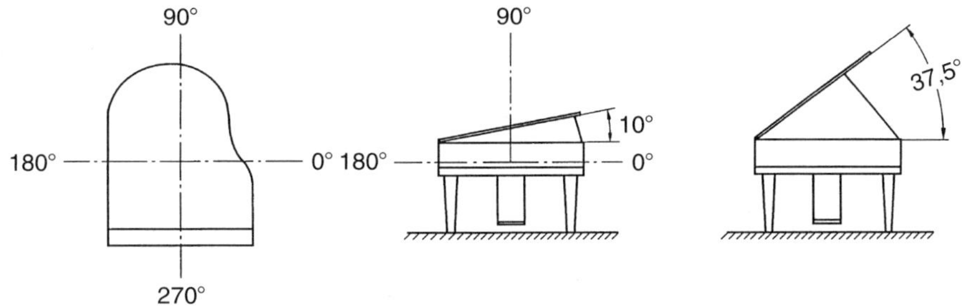

Fig. 4.26 Specification of polar coordinates for the directional characteristics of a grand piano

In the low register, i.e., in the region around C_2, sound radiation in all directions is relatively uniform. In the horizontal plane, up to an angle of about 20°, the fundamentals are only about 4 dB weaker than in the steeper directions. The reason for this can be found in the fact that the sound above and below the sound board is radiated in opposite phase, and thus the sound contributions in the horizontal plane weaken each other; at low frequencies the sound board effectively functions as a dipole. In the frequency region around 250 Hz the level behind the piano is stronger than in front of the open lid by about 5 dB; here the shadowing toward the rear, of the energy radiated in the upward direction by the sound board becomes noticeable as a level increase, since the interference between the sound radiated in opposite phase between upper and lower sound contributions is avoided. Consequently, at some distance behind the piano a listener localizes the sound source below the instrument. The higher frequency sound components from about 1,000 Hz upwards present somewhat more pronounced directional characteristics. In the vertical plane the preferred radiation occurs approximately between 0 and 5°, i.e., out of the open lid. At these frequencies a preference for directions around 0° is noticeable also in the horizontal plane. The half value width of these radiation regions lies at about ±30°.

In the middle register of the keyboard the shadowing effect of the lid toward the top increases, while at the same time the sound from the instrument is reflected in greater concentration. Again, the maximum of high frequency components, as seen from the player, is located toward the right. The highest level is reached for an elevation angle of about 40°, the half value region encompasses the angle between the horizontal and approximately 55°. A secondary maximum is found towards the left between approximately 130 and 150°. Aside from the fundamentals, the directional characteristics are flattened toward the top. The timbre directly above the piano is comparatively dull. In the horizontal plane, the radiation in the mid register of the piano is relatively uniform, only for the fundamentals, the radiation between 90 and 180° is somewhat disadvantaged, however, nowhere do the amplitudes fall below the 10 dB boundary of the maximum value.

In the high register the influence of the lid on directional effects is significantly more characteristic. Already the fundamentals in the region of 1,000 Hz show pronounced clustering between 15 and 35° in the vertical plane i.e., somewhat upward toward the right. The upper limit of this half value region remains steady with increasing frequency, it corresponds approximately to the direction of the angled lid. The lower boundary of the principal radiation region varies somewhat, however it never reaches the horizontal plane. The sharpness of the clustering is particularly acute for components around 4,000 Hz, here the levels in the angular region from 0 to 5° and 60 to 180° lie at more than 10 dB below the maximum value. The brilliance of the sound is thus limited to a relatively narrow angular region. This also appears in the horizontal plane, where (also at 4,000 Hz) the principal maximum at 0° and a secondary maximum at around 30°, each possess a half value width of ±5°, and further regions to the left, as well as the direction of sight of the player remain below the maximum level by more than 10 dB. The fundamentals of the high register do not fall below the 10 dB boundary anywhere, and the octave partials show such a minimum only for small angular regions near 135°. The statistical directivity factor for a grand piano can only be determined with a low degree of accuracy, inasmuch as the spatial directional characteristics are divided into many branches already at mid frequencies. The following values can serve as reference points: for the low register (around C_2) 1.2 at 250 Hz and 1.5 at 1,000 Hz, for the upper register 2.0 at 1,000 Hz and 3.5 at 4,000 Hz. This corresponds to rounded values for the directivity index of 1.5 and 3.5 dB (low register), as well as 6 and 11 dB (upper register).

Even the initial transient noise contributions are not radiated uniformly in all directions, though their directional characteristics are not as strongly pronounced as those of the harmonic partials. The lowest resonance of the sound board, which lies in the region of 90 Hz has a dipole characteristic with maximum radiation upward and a weakening of around 15 dB in the horizontal plane. In contrast, the resonances between 300 and 600 Hz which create the knocking noise, are shadowed by the lid which is turned at an upward angle so that they are radiated more strongly in the horizontal and low angle upward direction by an order of magnitude of 10 dB in comparison to the upward direction. The relative loudness relationship between noise contributions and the actual tone is important for the tonal impression. Therefore in Fig. 4.27, the level of the low tone of the sound board and the knocking

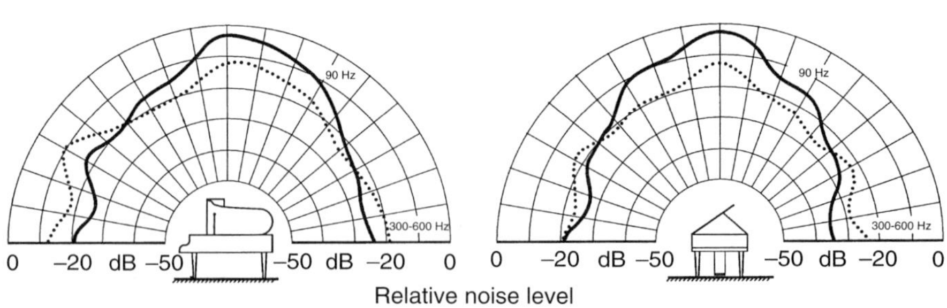

Fig. 4.27 Directional characteristic of the attack noise for a concert grand piano for the range around C_7. The relative level of the sound board tone and the knocking noise in reference to the strongest partial to the piano sound is represented

noise is given in reference to the level of the fundamental, which in the upper register is by far the strongest partial. The directional dependence shows that from the horizontal up to approximately 40° (in the direction of sight of the performer approximately 30°, toward the rear approximately 50°) the knocking noise dominates in comparison to the low sound board tone. However, it also shows, that nevertheless both noise components stand out particularly strongly in the upward direction, because the harmonic partials experience significant shadowing. The indicated angular dependence is valid for lower register with almost no change, however, the relative level of the noise approaches the level of the harmonic partials to within 15–20 dB even in the least favorable directions (Bork et al., 1994).

4.5.2 *Lid Closed*

When the lid is closed, clear differences in the directional characteristic in comparison to the open lid, become apparent in the vertical plane, especially in the angular region between 0 and 90°. In contrast no essential changes are noticed in the direction toward the back of the piano. As a comparison of curves in Fig. 4.25 shows, closing the lid creates configurations responsible for amplitude distributions of equal value for the right and the left halves. In the directions for which the open lid causes strong sound concentrations, the maxima drop out without additional energy being radiated toward the other sides. The intensity decrease for higher frequency sound contributions is especially significant in some areas it amounts to up to 10 dB. Because of this the tone color become duller and loses its brilliance and brightness. In some slightly upward angled directions, even the total intensity of the sound drops noticeably.

Also in the horizontal plane the strong components drop out towards the right side (0°). At higher frequencies the radiation occurs preferentially in the direction of the player. In this angular region the amplitudes mostly correspond to the situation for the open lid. This principal radiation region, which includes the contributions above 2,000 Hz could be caused by the fact that the sound can emanate from the approximately 25 cm wide opening between the lid and the music stand.

4.5.3 *Lid Half Open*

The case of a half open lid naturally takes the place of an intermediate position; for the usual length of the short support generally an angle of about 10° results as indicated in Fig. 4.26. One would expect a preferred region to be formed toward the right side (as seen from the player position) which, corresponding to the lid angle, would be more shallow than for a normally opened lid. It must be noted however, that the principal radiation region for a half open lid extends to a similar angular region as observed for the open piano: the half value region extends from approximately 10 to 60°, however, the amplitudes are somewhat weaker than for an open lid. This difference becomes all the larger as partial frequencies rise; in the upper registers it amounts to approximately 6 dB for partial contributions near 4,000 Hz. At the same time the radiation of the low sound board tone in the upward direction is somewhat restricted so that the minimum of the dipole characteristic is not as clearly pronounced as for the open piano. Consequently the radiation toward the right side of the player is stronger, by up to 10 dB. As a result, the tonal picture in the principal radiation direction as a whole, is only marginally softer, however, it does not possess the brilliance and brightness present for a wide open lid. Toward the back of the piano no essential changes in directional characteristics are observed, in comparison to the open or closed conditions.

4.5.4 Lid Removed

When the lid it totally removed, the piano obtains a rather equalized directional characteristic in the low and mid registers. For most frequency contributions the deviations in the vertical plane are not stronger than ±2 dB, only for the very low tonal contributions below 250 Hz, a decrease in amplitude near the horizontal plane becomes noticeable. At 65 Hz it is about 5 dB. In the upper register, however, a preferred radiation of the highest components in the upward direction is observed, consequently the intensity in the region of 4,000 Hz in this direction is by 3–5 dB greater than for an open or half open lid.

Therefore, seen as a whole, the upward radiation is relatively large after removal of the lid, which naturally must affect a weakening in the horizontal plane. Thus, especially the higher contributions recede strongly in importance in the tonal picture, so that the impression becomes dull. The timbre furthermore becomes flat by the additional lack in very low components. The directional characteristic in the horizontal plane is by no means round. A maximum is found near 300° in the upper register, where it is most pronounced, it is thus directed toward the right past the player, while in the angular region from 0 to 200° the higher overtones are especially weak.

In the middle and lower registers, the radiation is relatively round in the horizontal plane. In the region around C_4, however, components of the frequency region around 100 Hz are directed preferentially toward the left.

4.5.5 The Harpsichord

In comparison to the concert grand piano, the ribs, and above all the lid, are much lighter, less stiff and thus more capable of vibrating. Therefore, they can certainly contribute to sound radiation especially at low frequencies. The region around 120 Hz can be considered as a typical frequency location for the lid resonance at which the vibration amplitude of the lid can be as large as that for the sound board. In addition, the directional characteristic is determined by the subdivision of the vibrating surfaces into smaller vibrating zones which in each case vibrate in opposite phase. It is thus even more strongly structured than for a piano. Furthermore, significant individual differences between neighboring frequency regions, as also between instruments of different construction can occur. This strong frequency and tonal dependence of directional characteristics is an essential factor for the vibrant character of the harpsichord sound, and thus it is at least as important for the sound as the form of the directional characteristic per se (Elfrath, 1992).

For room acoustical considerations the directional characteristics of the harpsichord can in essence be summarized by the fact that at low frequencies, up to about 100 Hz, the radiation is still round. In the region of about 100–150 Hz a dipole characteristic appears, for which the radiation toward the right (as seen by the player) is strongest, toward the left by approximately 5 dB weaker, and in the direction of sight, as well as toward the back by 10–15 dB weaker than in the maximal direction. For increasingly higher frequencies, more or less narrow preference regions of sound radiation are distributed in all directions, where the sound level toward the back of the instrument is by 6–10 dB weaker than from the direction of the open lid.

4.6 The Harp

For the harp, the sound energy is radiated almost exclusively by the cover of the resonance body. The other surfaces of the body, and its rear opening participate only weakly in the sound radiation. In its resonances the cover vibrates with more or less strongly subdivided sections; whereby the nodal lines predominantly run across the cover, however, above approximately 170 Hz, vibrational shapes with an additional nodal line in the long direction occur (Firth, 1977). Accordingly the directional characteristics of the harp at intermediate and high frequencies are subdivided by small structures; in addition they also depend somewhat on the location of the excitation on the cover, i.e., on the played note.

In Fig. 4.28 the directions of preferred sound radiation are presented for three planes as 3 dB regions. Considered here are averaged results for various registers (Bell and Firth, 1989). In the horizontal plane the radiation up to approximately 400 Hz is round; between 400 and 2,000 Hz it is concentrated in two regions toward the front and back, above 2,000 Hz in two regions at an angle toward the front. In the plane of the strings preferred radiations perpendicular to the cover (toward the front as well as toward the back) appear below 400 Hz, additionally there is a region up to approximately 1,000 Hz with predominantly upward radiation which continues, with downward contributions, up to 4,000 Hz. In the third plane, oriented at right angles to the string plane, broad radiation regions are observed toward both sides, in the region up to 1,000 Hz the sound is radiated preferentially upward, and above 1,000 Hz, clover leaf like to the side upward and downward. Since the

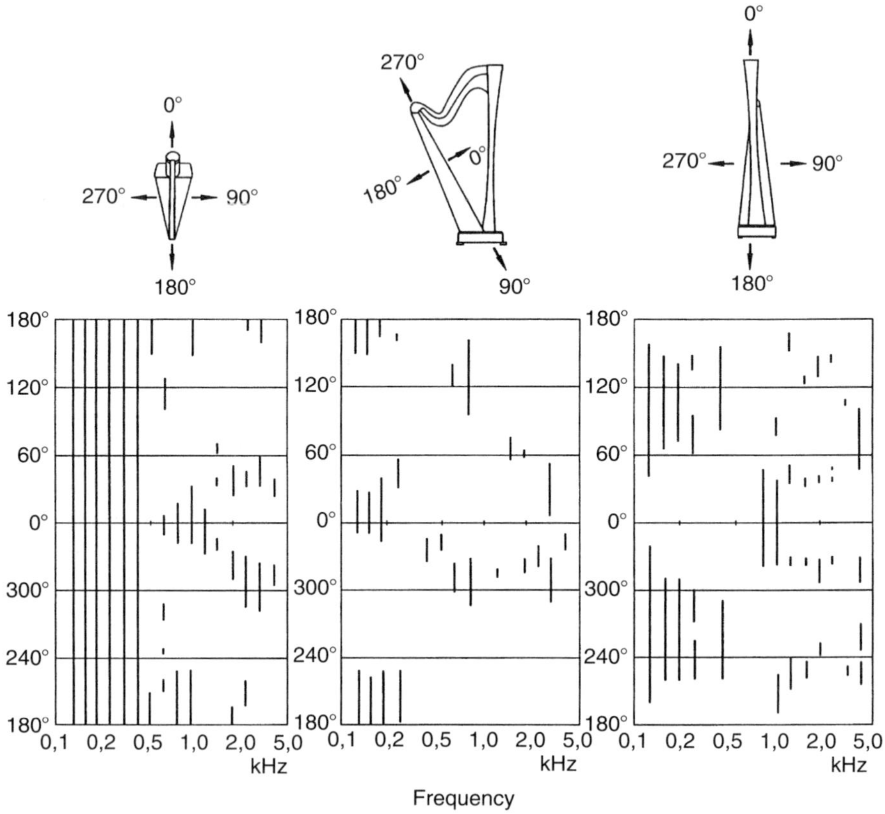

Fig. 4.28 Angular range of principal radiation regions of the harp (0–3 dB after Bell and Firth, 1989)

representation for the three planes are given in reference to the maximum level values in each case, it is not possible to determine values for the statistical directivity factor since each has a difference reference level.

4.7 Percussion Instruments

4.7.1 The Timpani

The directional characteristics of timpani are determined by the sound field radiated upward by the membrane. The sound field below the membrane is shadowed by the kettle and neither the kettle nor its lower opening contribute noticeably toward the sound radiation. The shape of the radiated sound field is therefore only determined by the subdivision of the membrane into differently vibrating zones (Rossing, 1982b; Fleischer, 1992).

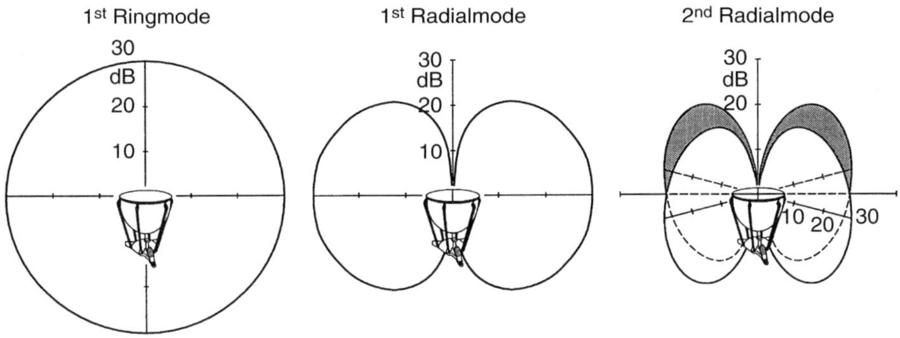

Fig. 4.29 Calculated directional characteristics for the three lowest partials of the timpani sound

For the three lowest vibrational modes of the membrane, calculated directional characteristics are presented in Fig. 4.29. In each case, those vertical sections through the spatial directional characteristic are represented, for which the directions of strongest sound radiation are observed. In each case, the nodal lines are positioned to make the directional characteristics most recognizable. In the first ring mode the entire membrane vibrates in phase, consequently a spherical characteristic is formed. The first radial mode, which is the most important for the tone and pitch impression is characterized by a nodal diameter which separates two regions vibrating in opposite phase: a dipole characteristic results which is rotationally symmetric about the horizontal axis. The direction of maximum sound vibration lies in the horizontal plane, the 3 dB region is calculated at $\pm 45°$, the 10 dB boundaries are crossed at an elevation angle of $\pm 71.5°$. For two nodal diameters, a quadrupole is formed with four maximum regions which show a clover leaf arrangement when viewed from above. Again, the directions of maximum sound radiation lie in the horizontal plane; however, toward the top and the bottom, the level decreases more rapidly than for the dipole: the 3 dB region encompasses only $\pm 31°$, the 10 dB region $\pm 56°$. For higher vibrational modes the number of radiation regions increases, corresponding to the number of radial nodal lines, the individual radiation bulges thus become increasingly smaller.

Measurements of sound radiation in the horizontal plane confirm the basic shape of the directional characteristics. Uniformity of membrane manufacture from a material standpoint and uniformity in stretching enhance the clarity of the basic shape of these directional characteristics. While no variations exceeding 3 dB occur for the first ring mode, depressions of a depth of approximately 18 dB between the principal radiation directions are noted for the first radial mode, and of approximately 16 dB for the higher modes (Fleischer, 1988). The boundaries of the 3 dB region agrees largely with the calculated values; the 3 dB regions are slightly extended since in the level depressions, asymmetries of sound radiation become more strongly noticeable: the first radial mode results in $\pm 73°$. Because of rotational symmetry this can also be transferred to the vertical plane. For the second radial mode, the 10 dB width of the four radiation regions in the horizontal plane amount to

±41° each from the principal radiation direction (ideal quadrupole ±36°). Because of the shape of the directional bulges, this value cannot be transferred to the vertical plane. The orientation of the individual directional characteristics in the horizontal plane is determined by the impact point, which in each case is located in the middle between two nodal diameters.

The statistical directivity factor for the principal radiation directions has a value of about 1.7 for the first radial mode, and about 1.9 for the second radial mode; the corresponding directivity indices are 4.5 dB and 5.5 dB respectively.

4.7.2 The Drum

In contrast to timpani, drums also radiate sound toward the bottom. The directional characteristics are thus formed by the sound field toward the top and toward the bottom. The vibrations of the kettle can be neglected since their amplitudes are only approximately 1% of the vibrational amplitudes of the membranes. The experimental results for directional characteristics of a small drum are represented in Fig. 4.30 for the most important vibration shapes, the shape and phase relationships of the membrane vibrations are also represented (Rossing et al., 1992).

In the first principal vibration (image a), both membranes vibrate in parallel, consequently the sounds radiated toward the top and bottom vibrate with opposite

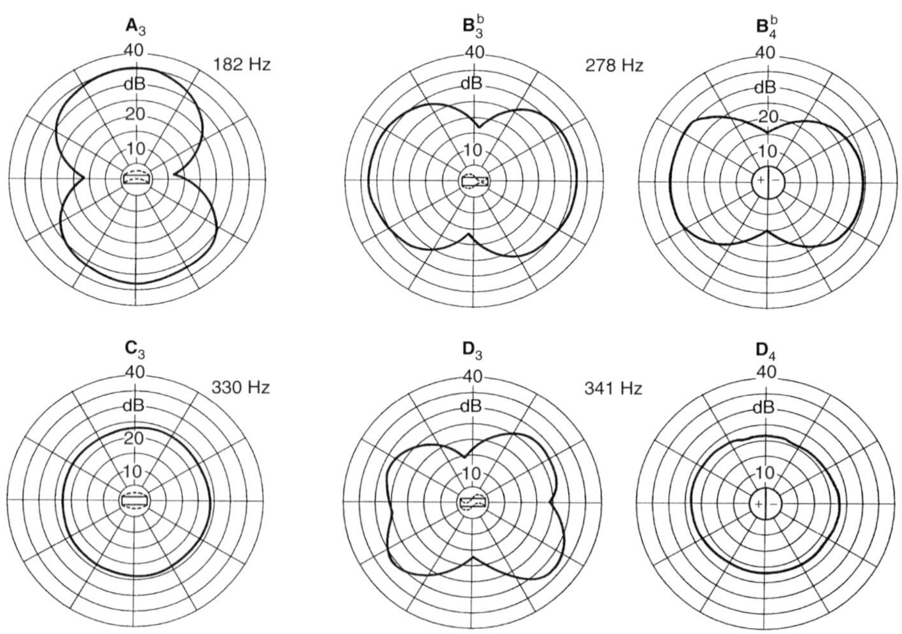

Fig. 4.30 Directional characteristics of a small drum (after Rossing et al., 1992)

phase. The principal radiation regions each have a width of about ±30°. In a range of about ±20°, relative to the horizontal, the level drops by more than 10 dB below its maximum value. The statistical directivity factor for the principal direction comes to about 1.6 (corresponding to 4 dB). This dipole characteristic is also formed when the drum does not have a second membrane and is open toward the bottom. However, when the lower membrane is clearly mistuned in relation to the impact membrane, it only vibrates weakly and a uniform sound radiation in all direction results. For the second principal vibration (image c), both membranes vibrate against each another: "a breathing sphere" is produced with uniform radiation in all directions.

When a nodal diameter is formed on the membranes, the directional characteristic is determined by the two halves of the membranes vibrating in opposite phase. The nodal line runs perpendicular to a line through the center of the drum and impact point. When the upper and lower membranes vibrate against each another (images b and b'), again a dipole characteristic is created, however, this time, with two horizontal principal radiation directions. The 3 dB regions have a width of about ±45° and the dips at a width of ±20° are deeper than 10 dB. The statistical directivity factor for the maximal direction lies at 1.5 (corresponding to 3.5 dB). When both membranes vibrate in parallel (images d and d′) a quadruple results with four preferred regions oriented into the room. They have a 3 dB width of about ±20° and are separated by indents of approximately 8–15 dB depth. In the horizontal plane, the radiation for this vibration form is relatively weak.

For a large drum, Olsen (1967) indicates a relatively uniform radiation, and at 120 Hz indicates a radiation which is by 8 dB stronger toward the front than toward the back in the direction perpendicular to the impact membrane. Four preferred directions are formed near 400 Hz. These occur in pairs, with two in the plane of the membrane and two in the plane perpendicular to it, which can be related to two nodal diameters. This is also expected to lead to a fourfold structuring of the directional characteristic in the plane of the membrane.

4.7.3 Gongs

For gongs, and analogously for tam-tams, the directional characteristics also rest on the superposition of sound fields radiated toward the front and the back. These in turn are divided into zones of differing phase by the sub division of the vibrating surfaces. In Fig. 4.31, the directional characteristics of a D_3 gong in are represented. The upper image shows a fundamental with spatial level structure, details of which change within the first seconds after impact. The direction of strongest radiation is turned by approximately 30° from the plane of the gong. The statistical directivity factor for this direction is about 1.8 (corresponding to 5 dB). The 3 dB region initially covers a region of about 25–65° and is later narrowed to 25–55°. Near the plane of the gong, the sound level drops by more than 10 dB.

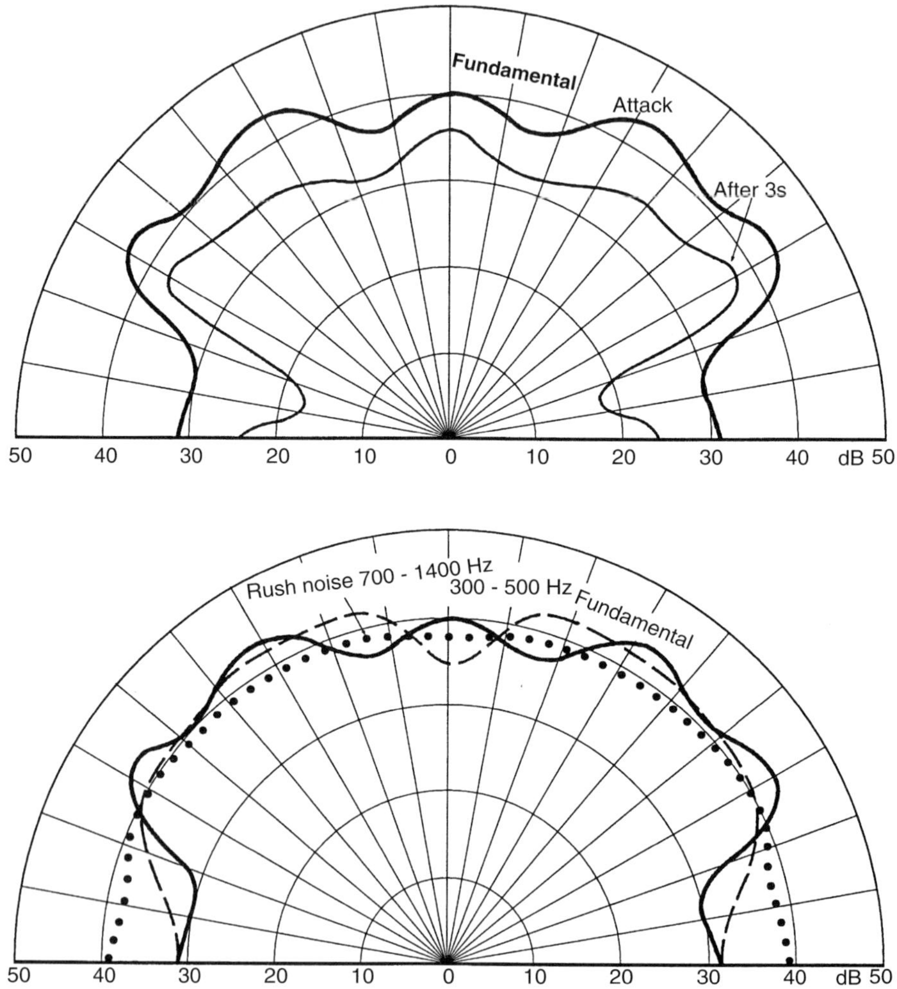

Fig. 4.31 Directional characteristics of a gong. *Dotted line:* rush noise 700–1400 Hz, *Broken line:* 300–500 Hz, *Solid line:* Fundamental

The curve for the overtones between 300 and 500 Hz (lower image) shows a similar form, since structural details are smoothed out by formation of average values. However, the depression at right angles in front of the gong, which can be explained by the symmetry of vibrational regions with opposite phase, is notable. Its depth only amounts to 7 dB, and yet the fundamental is raised by it above the overtones. Otherwise, the overtones dominate over the fundamental for an angular deviation of 15°. In contrast, the noise contributions typical for the gong are radiated with relatively spherical symmetry, since no standing waves are formed on the gong for these components; the fluctuation over all directions do not even exceed a value of 2 dB. This means that the intensity relationship between the tonal sound contribution and the noise depends on the angle: near the plane of the gong, the noise is especially strong.

4.8 The Singing Voice

The directional dependence of sound radiation for a singer is not only determined by the shadowing of the sound by the head, but also by a funnel effect of the mouth, whereby the positioning of the mouth also plays a role, consequently the directional characteristic of the singing voice differs from that of a speaker. According to the most frequently quoted references of Dunn and Farnsworth (1939), as well as Niese (1956), for the speaking voice, a steady decrease in level from front to back in the horizontal plane is indicated, and also in the vertical plane the level in the direction of sight is practically never exceeded.

In contrast, in the case of a singer, the highest sound level is certainly not always radiated in the direction of sight as the example for sound contributions in the frequency range around 2,000 Hz shows in Fig. 4.32. On the one hand, the direction of strongest sound radiation is directed at an angle of 20° downward toward the front. On the other hand, it also slightly increases toward the side in the horizontal plane. The clear definition of these two effects depends on mouth positioning, and especially the vowel which is sung: for "o(oh)" the effect is stronger than for "a (ah)" and for this, it is stronger than for "e(eh)"; furthermore, this also applies to reduction of sound radiation toward the rear (Marshall and Meyer, 1985).

Individual difference between directional characteristics of trained (female and male) singing voices is so minute, that it is certainly possible to derive general directional characteristics. Pitch and dynamics also only play a subordinate role.

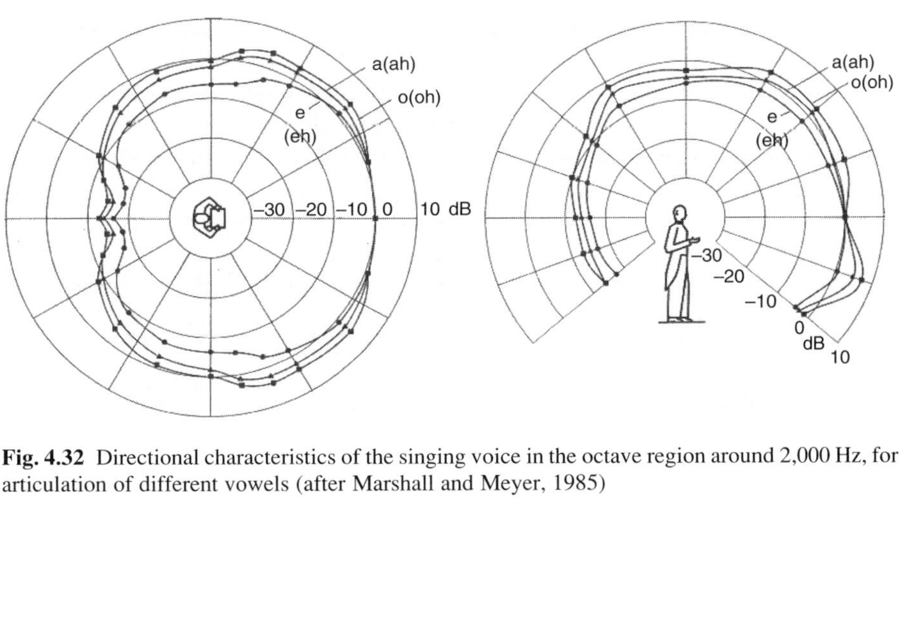

Fig. 4.32 Directional characteristics of the singing voice in the octave region around 2,000 Hz, for articulation of different vowels (after Marshall and Meyer, 1985)

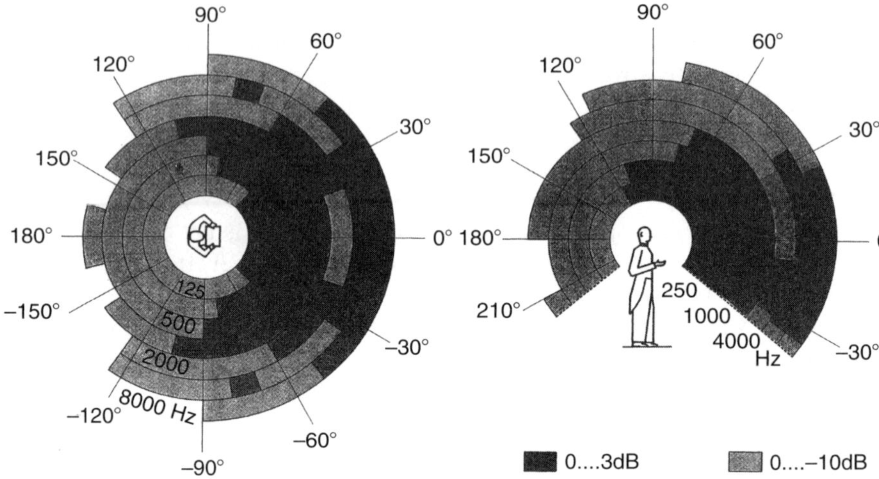

Fig. 4.33 Angular regions of principal radiation regions of the singing voice in octave bands (after Marshall and Meyer, 1985)

When, in addition, results obtained for different vowels are averaged, the principal radiation regions can be assembled in the form represented in the Fig. 4.33.

In the horizontal plane, the 3 dB region broadens from low frequencies up to 1,000 Hz by a factor of about 2, it again becomes narrower from 2,000 Hz on upwards, whereby components around 2,000 Hz experience a dip in the horizontal plane because of the previously mentioned maximum which is directed downward. In the octave of 4,000 Hz, the 3 dB region is particularly narrow with an angle of only ±35°, at 8,000 Hz the radiation again becomes somewhat broader. These numbers, however, refer only to the harmonic overtones of the sung vowels, not to the components of the consonants located in this frequency range. According to Niese (1956), a 3 dB range of only ±30° can be observed around 8,000 Hz.

In the vertical plane a preferred direction pointing slightly downward is clearly recognizable for all frequency regions. In the upper direction the reach of the 3 dB region becomes wider as the frequency decreases. This tendency is also repeated for the 10 dB region, which indicates a stronger level decrease upwards than toward the sides for higher frequencies. Directed toward the back, the level decrease at 3,000 Hz is 15 dB and increases to over 25 dB at 8,000 Hz.

The statistical directivity factor for the singer is of particular interest for the tonal effect within the hall. In Fig. 4.34 it is therefore represented in its frequency dependence, along with the respective directions of strongest sound radiation. For the direction of sight it moves from a value of 1.2 for frequencies up to 2,000 Hz past a value of 1.65 at 4,000 Hz, rising to 1.95 at 8,000 Hz. Corresponding directivity indices were measured at 1.6 dB, 4.4 dB, and 5.8 dB. For the direction of strongest sound radiation, the statistical directivity factor reaches a value of about 2.0 (corresponding to 6 dB) in the frequency region of the singer's formant.

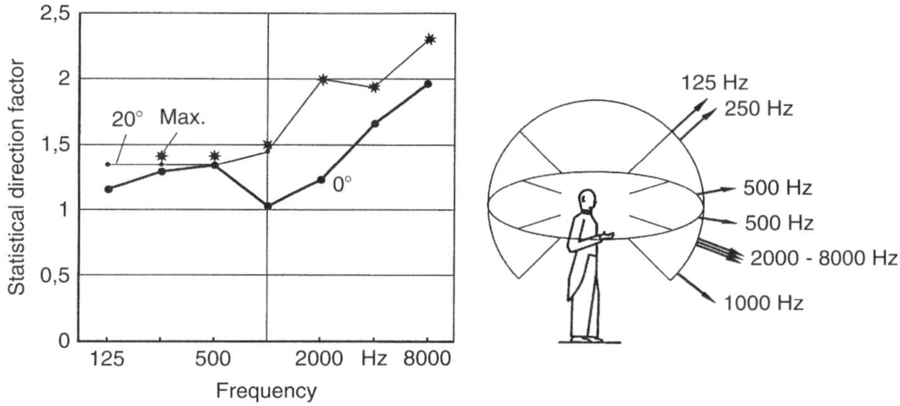

Fig. 4.34 Statistic directivity factor of the singing voice in the direction of sight (0°), inclined downward by 20° as well as in the direction of strongest sound radiation (corresponding to the arrows with frequency indications)

The angular dependence of the statistical directivity factor for the horizontal plane is assembled in the appendix (page 323 in original German version) in tabular form.

The influence of pitch on directional characteristics becomes noticeable only for peak notes in individual voice ranges. For individual vowels it exhibits pronounced differences. In the upward direction, peak tones are weakened by 2–4 dB less than all other tones, toward the back this even ranges from 4 to 5 dB. Above all, this impacts frequency contributions between 500 and 2,000 Hz. In contrast, for peak tones toward the side, components in the region of 4,000 Hz are weakened more by 3–5 dB than all other tones, i.e., the singers formant is less pronounced for peak notes. The only influence of dynamics on directional characteristics worth mentioning consists of a 3 dB larger weakening of frequency contribution between 1,000 and 4,000 Hz in a sideways direction at a piano dynamic level. This tone color change can have the consequence that the dynamic toward the side of the singer can be perceived as somewhat broader than in the direction of sight.

Colour Plate 1

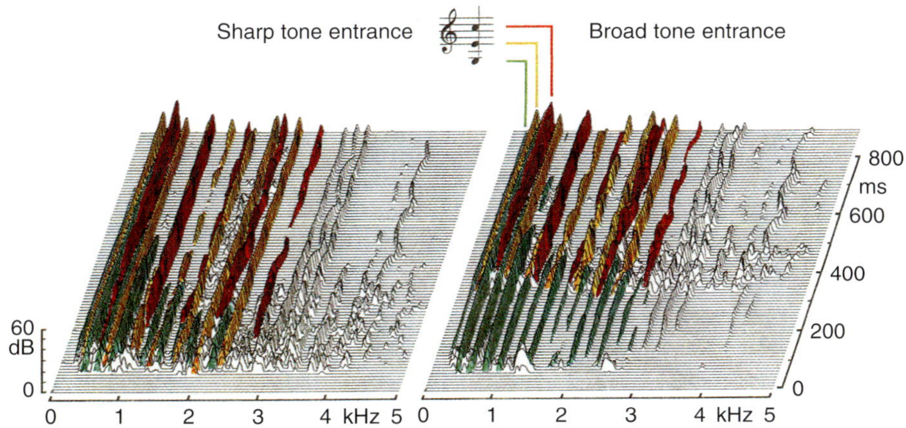

Chapter 3 Fig. 3.19 Time evolution of tone spectrum for a three note violin chord with two different playing approaches

Colour Plate 2

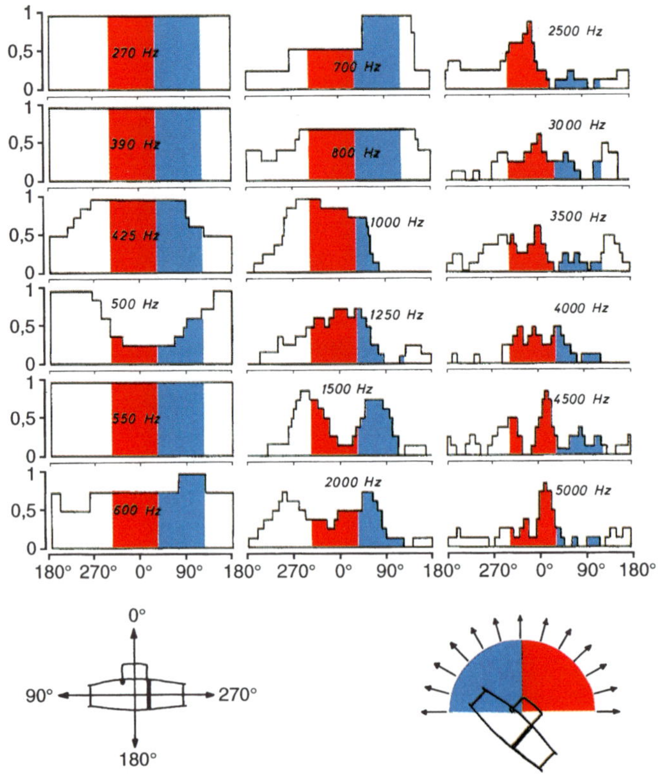

Chapter 4 Fig. 4.17 Histogram for the principal radiation directions (0 – 3dB) for violins (bridge plane). *Red areas*: Angular region for first violins oriented toward the audience directly or with one ceiling reflection. *Blue areas*: the corresponding angular region for the second violins with European seating

Colour Plate 3

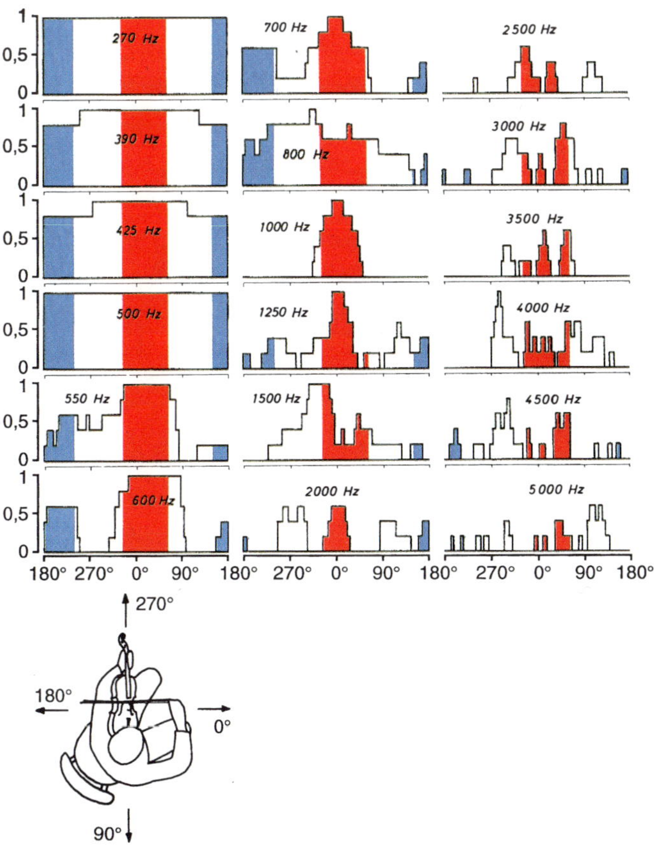

Chapter 4 Fig. 4.18 Histogram for the principal radiation directions (0 – 3dB) for violins (*horizontal plane*). *Red areas*: Angular region for first violins oriented directly toward the audience. *Blue areas*: the corresponding angular region for the second violins with European seating

Colour Plate 4

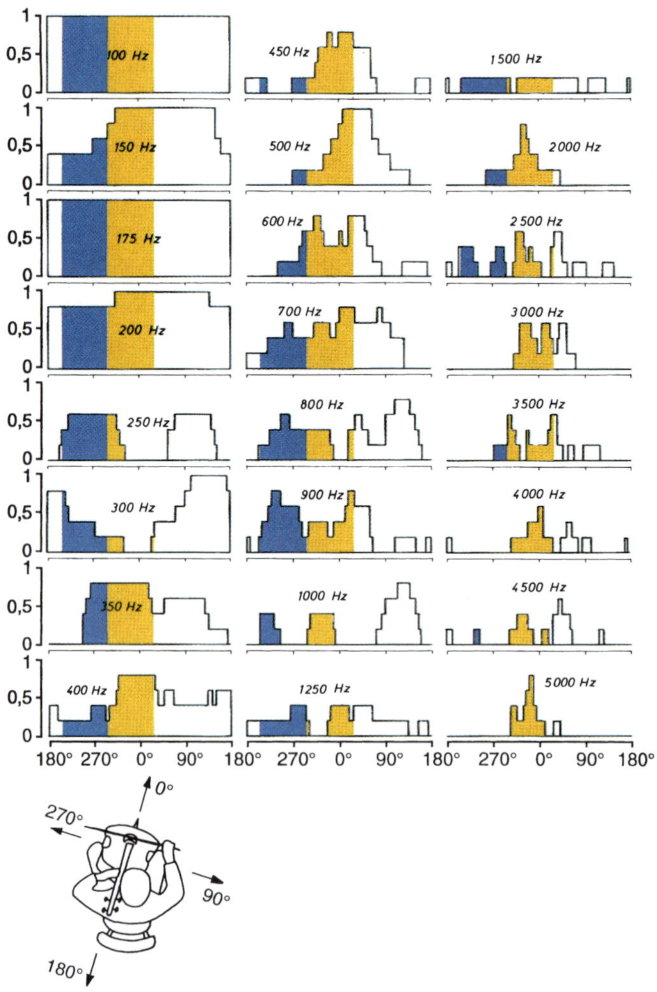

Chapter 4 Fig. 4.23 Histogram for the principal radiation directions (0 – 3dB) for Celli (bridge plane). *Yellow areas*: Angular region oriented directly toward the audience for seating position facing the audience. *Blue areas*: The corresponding angular region for a sideways positioning near the edge of the stage

Colour Plate 5

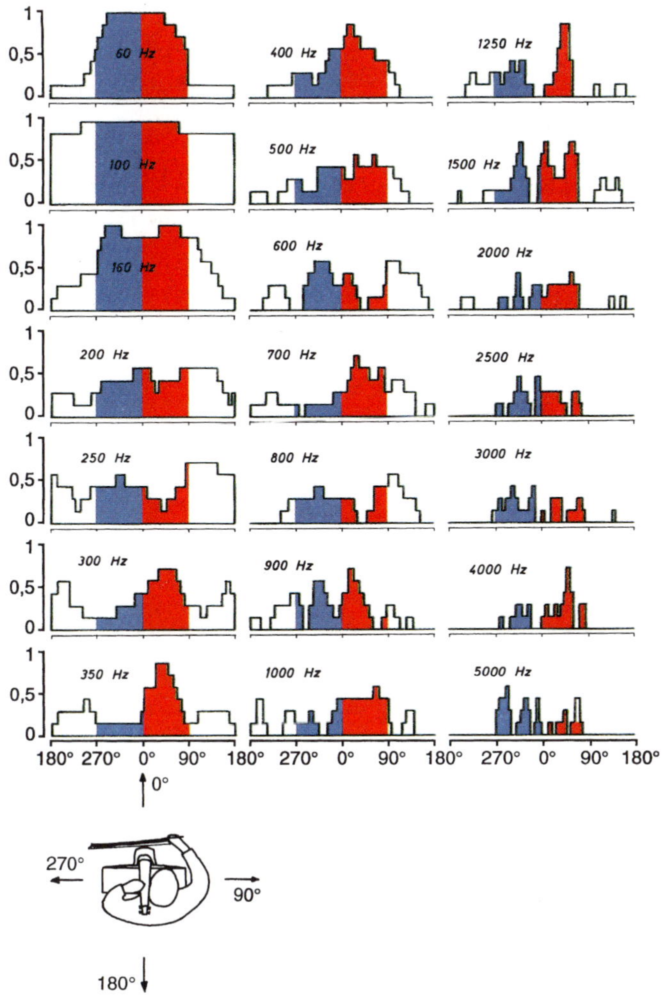

Chapter 4 Fig. 4.24 Histogram for the principal radiation regions (0 – 3dB) for double basses (bridge plane). *Red areas*: Angular region oriented directly toward the audience for an instrument location in the left rear corner. *Blue areas*: The corresponding angular region for a right rear corner location

Colour Plate 6

Chapter 6 Fig. 6.3 Joeph-Keilberth-Hall of the "Sinfonie an der Regnitz" in Bamberg (Stadthallen GmbH, Bamberg)

Colour Plate 7

Chapter 6 Fig. 6.4 Large hall of the Gewandhaus in Leipzig (Gert Mothes, Leipzig)

Colour Plate 8

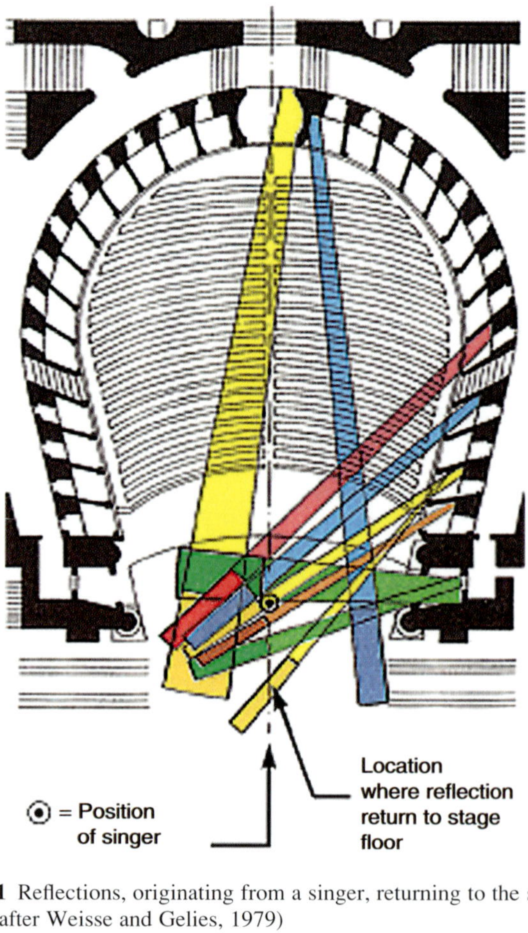

Chapter 6 Fig. 6.21 Reflections, originating from a singer, returning to the stage, in the Teatro San Carlo, Naples (after Weisse and Gelies, 1979)

Chapter 5
Foundations of Room Acoustics

5.1 Reflection and Refraction

5.1.1 Reflection from a Flat Surface

Inasmuch as sound propagates in a straight line in a homogeneous medium, as is the case for air at rest at uniform temperature, the presentation of a sound ray, which connects the sound source and the observer in a straight line, is well suited for a number of acoustical considerations. It forms a visual foundation for a geometrical point of view of sound propagation in a room.

When a sound ray impinges on a sufficiently large flat surface, it is reflected. The well-known law of optics, that the angle of incidence is equal to the angle of reflection, goes into effect. This is a phenomenon which has its cause in the wave nature of sound or light respectively. Furthermore, the incident ray, reflected ray and a line perpendicular to the wall at the contact point lie in the same plane. This reflection process is represented schematically in Fig. 5.1a. This illustrates clearly that the angle between the incident and reflected ray depends on the incident angle, thus, by appropriate orientation of the surface, the incoming ray can be reflected into any desired direction.

When two walls are perpendicular to each other the incident sound is reflected twice. It always leaves the corner in the direction exactly opposite to the incident direction. This case is illustrated in Fig. 5.1b. From the two indicated ray paths it is noted that, in contrast to simple reflection from one plane, the direction of the reflected ray depends only on the direction of incidence of the incoming sound.

If two walls in contrast, form an obtuse angle, a single reflection results for steep incidence. For shallow incidence, double reflection results in the corner. Both possibilities are represented in Fig. 5.1c for the case that the sound initially impinges on the lower wall. The second reflection of the sound ray represented by the broken line leads to a direction which is relatively flat in relation to the left wall. When considering, that this secondary reflection can only become steeper when the incident ray on the right becomes even shallower one recognizes that there is a limiting case for double reflection when the incoming sound ray is parallel to

J. Meyer, *Acoustics and the Performance of Music*.
DOI: 10.1007/978-0-387-09517-2_5,

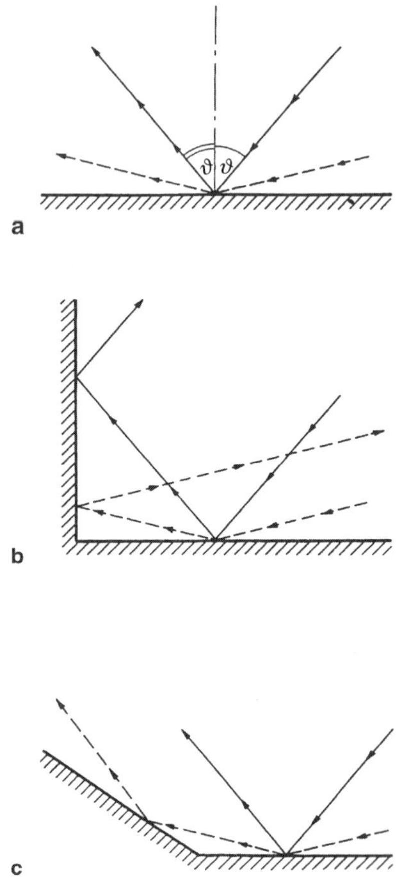

Fig. 5.1 Reflections from a flat surface

the wall. The consequence is that for an obtuse corner, in a certain region near the angular bisector, there are no secondary reflections.

5.1.2 Reflection from Curved Surfaces

When sound falls on a large wall with a curved surface, it will be reflected according to the angle of incidence of the ray, in the plane tangential to the curve at the point of contact, where again the incidence ray, the reflected ray, and the normal to the tangential plane, all lie in the same plane. Depending on the distance of the sound source from the wall and the radius of curvature, focusing or spreading effects can occur in analogy to optical curved mirrors. The most important cases are represented in Fig. 5.2:

(a) If the distance from the sound source to the wall is greater than half the radius of curvature, a focusing point beyond the center of curvature is formed. If the

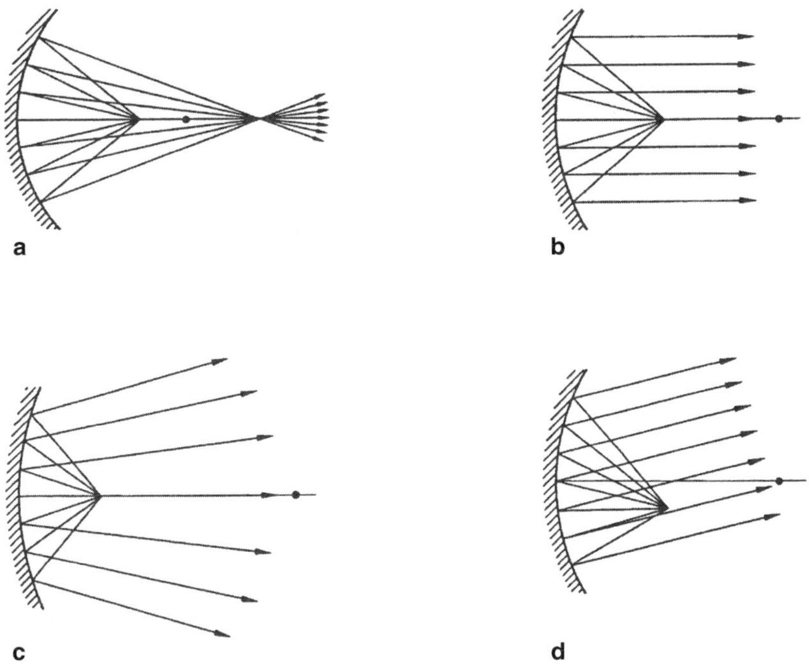

Fig. 5.2 Reflection from curved surfaces

sound source, as drawn, lies between the center of curvature and the wall, then the point of focus is located at a greater distance. Such sound energy concentration is perceived as detrimental when the focusing point falls within a region of the audience. It is, however, perceived as advantageous when the sound is to be focused onto a microphone by a ellipsoidal mirror (for example, such an arrangement has found long time use in the "Marktkirche" in Hannover, prior to the development of technically superior micro-port arrangements).

(b) When the distance of the sound source to the wall is equal to half the radius of curvature, a parallel ray bundle in the direction of the axis of the reflector is generated. A parabolic mirror is the most favorable shape to take advantage of this effect. Fig. 5.2d, however, can also be interpreted in the inverse direction: When a parallel ray bundle, i.e., sound from a far distance source falls on a hollow curve, a focal point is created at a distance corresponding to half the radius of curvature.

(c) When the source to wall distance is smaller than half the radius of curvature, the reflected sound rays spread as though they came from a single point behind the wall. Such a broadening of the sound field can occasionally be advantageous for uniform energy distribution over a certain angular region. Similarly, convex curvatures can lead to a fanning of an incoming ray bundle.

(d) When the sound source is not located on the axis of the reflector, then, for a parabolic mirror, for a source to wall distance equal to half the radius of curvature, a nearly parallel ray bundle results which, however, according to the angle of incidence at the center of the mirror is directed at an angle with respect to the axis. In similar fashion, the focusing point is shifted for a sideways repositioning of the sound source in case (a) to the opposite side, and in addition, it will lose sharpness.

In summary, one can already conclude from these examples that for the positioning for a large number of musicians in front of a concave wall, the danger exists that individual instrument groups will be reflected into different directions, or that for different locations in the hall, their intensity could stand out of the ensemble sound because of the focusing effect.

5.1.3 Influence of the Wavelength

In considering reflection processes so far, the assumption was made that the surfaces impacted by the sound were sufficiently large. Relevant dimensions must be related to the magnitude of the wavelength. The conditions for forming reflections in analogy to optics can be stated more precisely by requiring that the dimensions of sound obstacles must be at least several wavelengths. Otherwise, the sound is bent around the obstacle and hardly reflected. Thus, a wall of dimensions comparable to three wavelengths, presents only a short sound shadow and elements of the size of one wavelength practically do not disturb the sound field at all. Inasmuch as the relationship of the geometric dimensions relative to wavelength depends on frequency, the acoustic effect of reflectors or sound barriers changes with pitch. Low frequency contributions, i.e., components with long wavelengths, can still be heard strongly even when the sound source is hidden from view. High tones, however, are even reflected by small objects.

The shadowing effect of obstacles between sound source and listener can be quantitatively determined on the basis of Fig. 5.3. If a wall is located between sound source and listener, as for example the partition of the orchestra pit in the opera house, then, on the one hand, the angle with which the sound ray is bent at the edge is important. On the other hand, the effective height with reference to the straight-line connection between sound source and listener plays a role, more exactly, the relation of this height to the wavelength. It is noteworthy that already for an angle of 0°, i.e., a just possible sight connection, an attenuation of from 5 to 6 dB occurs. This is understandable when one considers that by the diffraction a part of the sound energy is bent into those regions behind the barrier which would not receive any sound energy without refraction. It is further noteworthy, that the degree of shadowing even for small angles varies strongly, that however, between 30 and 90°, there is relatively little change.

Inasmuch as in practice, frequently individual free-standing or hung reflectors are used, the question naturally becomes of interest: Above which frequencies do

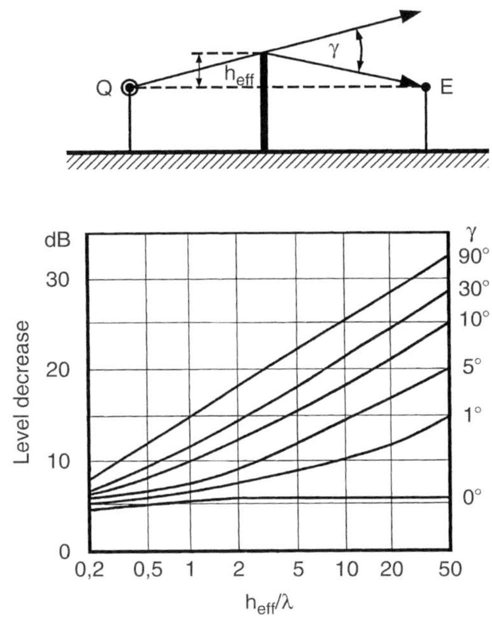

Fig. 5.3 Level decrease for diffraction of direct sound around a barrier (after Redfearn, 1940). S Sound source, R Receiver, λ wavelength of the sound

they reflect effectively? As sketched in Fig. 5.4, if the distance from the sound source to the middle of the reflector is designated as a_1, the distance from listener as a_2, the width of the reflector in the observed plane with b and the angle of incidence as θ, then wave theoretical considerations concerning the effectiveness of reflectors give a lower frequency limit as

$$f_u = \frac{2c}{(\mathrm{b}\ .\ \cos\theta)^2} \cdot \frac{\mathrm{a}_1\ .\ \mathrm{a}_2}{\mathrm{a}_1 + \mathrm{a}_2}$$

In this equation, c is the speed of sound in air. Furthermore, the condition must be satisfied that the distances a_1 and a_2 are larger than the reflector width b (Cremer, 1953). Below this limiting frequency, the level of reflective sound drops with 6 dB per octave (Rindel, 1992).

This formula indicates that the regions of reflector effectiveness reaches increasingly lower frequencies as:

Reflector size increases
Distance to the sound source decreases
Distance to the listener decreases
The sound falls more steeply onto the reflector

This means that seats located far back in the hall will receive fewer low frequency contributions than the front rows when all reflectors are of equal size.

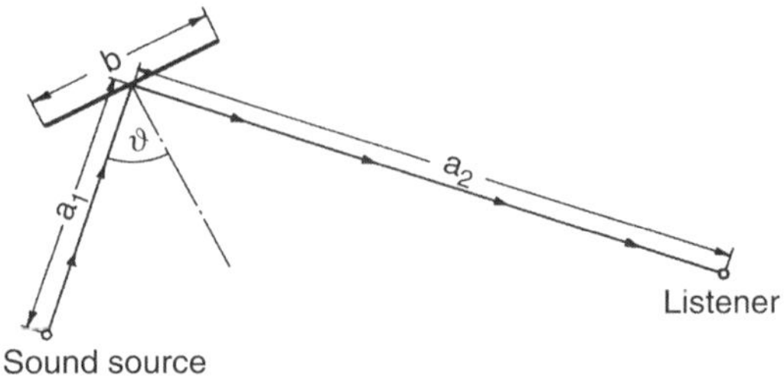

Fig. 5.4 Parameters to calculate the effectiveness region of a reflector

It also results in the requirement to increase the size of those reflectors which are oriented to reach the back of the hall.

In order to simplify the practical application of this equation, Fig. 5.5 presents graphical solutions for the most important variation regions of individual dimensions. If for example, the two distances $a_1 = 10$ m and $a_2 = 20$ m are given, then the lower diagram gives a value of 6.7 for the second term of the formula. From the upper diagram, for a reflector width of $b = 1.5$ m and perpendicular sound incidence ($\theta = 0°$), an upper frequency of about 2,000 Hz is relevant, for a sound incidence of 45°, however, approximately 4,000 Hz. In contrast, if a reflector of 2 × 2 m size is located at a 2 m distance behind the player, the effective region is increased to include frequencies below 300 Hz.

Aside from the size of the reflector, its mass also plays a role for its effectiveness. Plates (or foils) which are too light transmit a portion of the sound energy, or they themselves vibrate too strongly. A desirable area weight of at least 10 kg m^{-2} for reflectors of mid and high-frequencies can be used as an initial point of reference, this corresponds approximately to a 12 mm wood plate. This requirement is especially relevant for reflectors of speech and singing. If frequencies of the bass region also need to be reflected, approximately 40 kg m^{-2} are required.

The previous considerations dealing with reflection relationships were relevant for smooth, i.e., unstructured reflection surfaces. Frequently, however, wall and ceiling surfaces are structured by small protruding or receding surfaces or added profiles. Such surfaces can have a scattering effect, i.e., instead of the previously described geometric reflection of the sound waves (in one direction), the sound energy is reflected diffusely in all directions. This effect is most strongly pronounced when the depth of the structure is of the order of magnitude of one-fourth to one-half of a wavelength. The frequency region for diffuse reflection can be broadened by a differential depth arrangement of the wall structure. Staggering, according to principles of certain probabilistic sequences of whole numbers, according to elementary number theory, or so-called maximal sequences, are particularly advantageous (Schroeder, 1979).

Below the frequency regions for diffuse reflection, i.e., for larger wavelengths, the structured surface behaves like a smooth wall; i.e., geometric reflection occurs. Above the frequency regions for diffuse reflection, the individual surfaces of the structure act as geometric reflection surfaces and angular mirrors. This can certainly result in reflection directions other than those corresponding to the main orientation

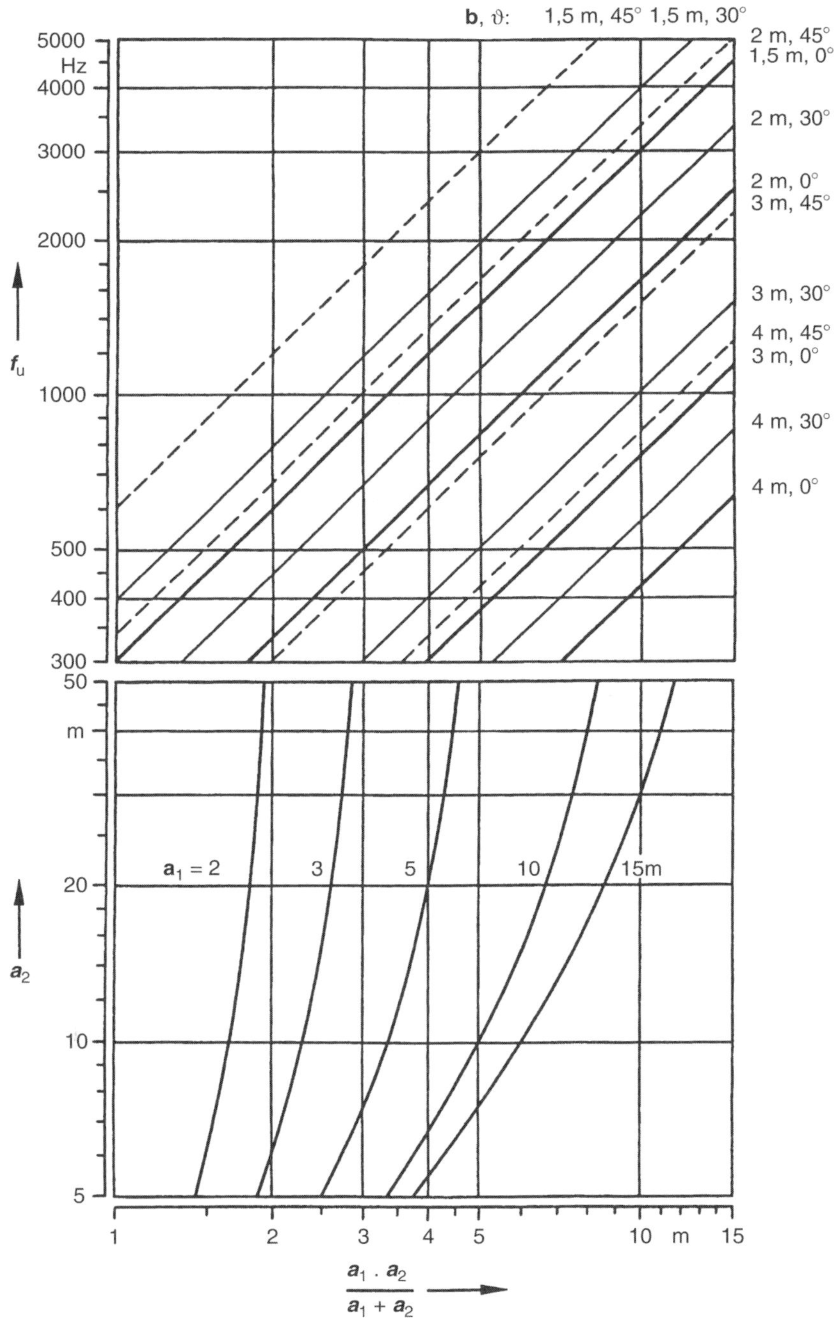

Fig. 5.5 Diagrams for the determination of the lower frequency limit of a reflector. Values for f_u, corresponding to given values for the reflector width (b) and the angle of incidence (θ), are obtained by starting from a given distance a_2 at the left edge on the bottom diagram and going to the curve of known distance a_1, and proceeding vertically upward from that point of intersection to the straight line on the upper diagram corresponding to the given values of b and θ. f_u can then be read from the left edge of the upper diagram

of the wall. Strongly structured surfaces can lead to differing tone colorations of the sound reflected in different directions because of this triple division of the entire frequency region.

For regularly stepped structures, an additional impulse sequence may be formed in certain directions when the higher frequency reflections from the individual steps arrive in regularly spaces sequences. A unique pitch is perceived (mostly in the middle register) corresponding to the frequency for which half a wavelength corresponds to depths of the individual steps.

5.2 Absorption

In the previous section, when considering reflection processes, the only questions considered dealt with directions of incident and reflected sound for different spatial circumstances. Not considered were the amplitude relationships between incident and reflected sound. As in the case of optics, previously cited for comparison reasons, where dark surfaces only reflect a small amount of light, so it is in acoustics, where the portion of reflected energy differs depending on the nature and composition of the walls.

Generally the percentage of energy absorbed during reflection, is indicated when describing material or construction characteristics. This quantity is designated as the "absorption coefficient" and is usually specified by the letter α. On the other hand the "equivalent absorption area" A refers to the sound absorption of a surface of specific size or a room (as the sum of all its surfaces). The following relationship is valid

$$A = \alpha \cdot S$$

where S represents the size of the surface with absorption coefficient α. An important property of the absorption coefficient is its frequency dependence. Sound contributions of different pitch are generally not absorbed or reflected in equal strength by the same material.

There are, for example walls, which by reason of the porous structure only absorb high frequency components, while the low components are nearly totally reflected. Such materials are therefore called high frequency absorbers. The typical frequency dependence of the absorption coefficient for such a case is schematically represented on the left of Fig. 5.6. The height of the absorption coefficient as well as the location of the frequency limit above which significant sound absorption occurs, depends on the composition and thickness of the porous layer: as this layer becomes thinner the frequency limit moves upward.

The audience in a hall can essentially be considered as a high frequency absorber. Thus the audience absorbs all sound contributions from approximately 500 Hz on upwards. In this, the surface occupied by the audience is a determining factor, while the seating density is only of subordinate significance. This means that the same number of persons affect a higher degree of absorption when distributed over a larger surface. For example, 200 persons distributed over an area of 100 m^2

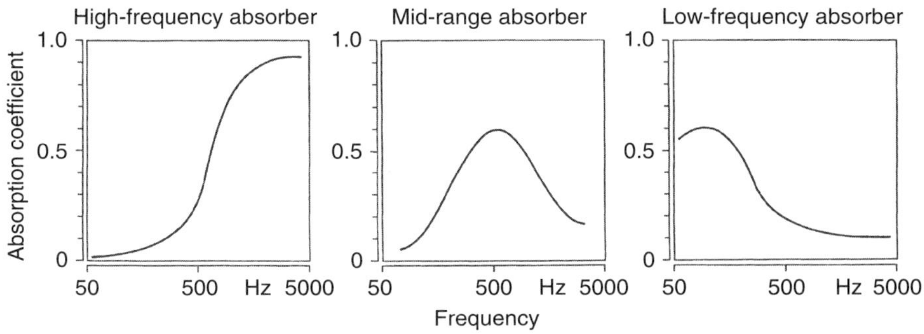

Fig. 5.6 Degree of absorption of materials for differing frequency dependence

possess an equivalent absorption area of 95 m^2 at a frequency of 1,000 Hz. However when they are distributed over 200 or even 300 m^2 their equivalent absorption area rises to above 140 or 165 m^2 respectively.

In contrast, the diagram on the right shows the typical behavior of a low frequency absorber which predominantly absorbs the low tone contributions and reflects the high ones. This effect comes about when plates can vibrate in front of a hollow space as is most often the case for wood paneling for example. As the hollow space becomes deeper and the plate becomes heavier the frequency region of maximum absorption moves toward lower frequencies. For very thin plates and hollow spaces of shallow depth, and above all for constructions where the hollow space opens into the room through slits or holes, the absorption maximum is shifted toward the region of intermediate frequencies, one therefore speaks of midfrequency absorbers. The effective frequency region can thus also be influenced by construction techniques.

There are virtually no materials with frequency independent absorption. The only ones worth mentioning are the "sound – hard" materials such as concrete, marble, or plastered stone walls, which reflect sound of all pitch regions with nearly no attenuation. In this context, however, it should be mentioned that an organ, in reference to its front surface, provides an absorption coefficient between 0.55 and 0.60 in the entire frequency range from 125 to 4,000 Hz. Deviations from this, determined by construction differences are relevant only for low frequencies (Meyer, 1976; Graner, 1988). In concert halls or radio studios it can be meaningful, in individual cases to consider covering the organ while not in use when a reflecting surface rather than an absorbing one is desirable for the orchestra sound at the relevant location.

In practice, curtains are a particularly interesting case, because they can be used without structural changes, or as a temporary proviso. As a porous material, they absorb, as do carpets, preferentially high frequency contributions. However, their absorption regions can be broadened to include lower frequencies when suspended at certain distances from the wall. A somewhat uniform absorption results above the frequency for which the wall distance amounts to one-fourth of the wavelength. From this condition the following formula can be derived

$$f_u = 8{,}500/d \quad f_u \text{ in Hz, } d \text{ in cm,}$$

where f_u is the limiting frequency and d is the wall distance (Cremer, 1961). However the material can not be too light, furthermore the curtain should be hung with folds. As an example, for wall distance of 10 cm a limiting frequency of 850 Hz results, and for a distance of 25 cm this is already 340 Hz.

5.3 Reverberation

The geometric viewpoint of the sound-ray path between source and listener naturally has to be limited to the direct path, as well as to detours with only a few reflections, because otherwise the process becomes too cumbersome and visually complex. In order to include all reflection processes to the point of complete absorption, and at the same time describe, if possible, relationships at all points of the room, statistical methods become necessary. For consideration of sound processes after turning off the sound source, it becomes particularly advantageous to include all reflections: The sound reflections arrive at the listener in an increasingly dense time sequence and thus they shape the reverberation in slowly decreasing intensity.

A typical level progress is represented in Fig. 5.7 for such a case. After turning the sound source off, the level (in logarithmic dB scale) decreases approximately linearly with small variations, until it merges with the ambient noise level in the room. In that context it is irrelevant for the basic shape of the curve whether the concern is with the stopping of an electroacoustic sound source or the termination of an orchestral chord. The listener can follow this reverberation until it is submerged in the noise level. In the graphically represented example this

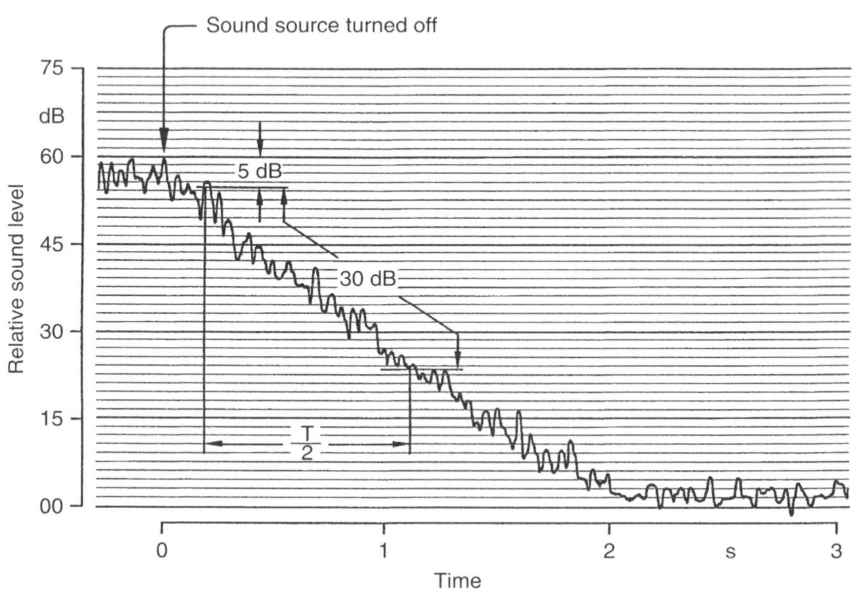

Fig. 5.7 Sound level during a reverberation process

is approximately 2 s. This time is designated as the decay time. It naturally depends on the value of the initial sound level on the one hand, and the value of noise level on the other.

It can be demonstrated that the slope of the linear level drop depends only on the characteristics of the room and not on the sound source, as long as the discontinuation is sudden. Consequently an objective quantity for the acoustic behavior of a room can be derived: the time during which the sound level drop by 60 dB in comparison to its initial value is designated as the reverberation time. This value of 60 dB corresponds approximately to the dynamic range of a large orchestra. Inasmuch as such a dynamic range is not always accessible for measurements, and furthermore, the initial point often is not uniquely recognizable, the following procedure has been specified for the determination of reverberation times: starting with a drop of 30 dB from a value of 5 dB below the steady level, the measured time is doubled. This method is indicated in Fig. 5.7. From the graphically represented level drop a reverberation time of approximately $2 \times 0.9 = 1.8$ s is obtained.

In the course of a musical performance, the temporal note sequence and the actually dynamic structure rarely provides a drop of 60 dB. Consequently, particular attention is given to the beginning portion of the decay curve. A slope of the first 10–20 dB is determined, and from that, the time for a uniform level drop of 60 dB is calculated. For an evaluation of the first 10 dB one speaks of the "Early-Decay-Time" (Jordan, 1968), for the first 15 dB the "Initial-Reverberation-Time" (Atal et al., 1965) and for 20 dB "Beginning-Reverberation-Time" (Kürer and Kurze, 1967).

The reverberation time of a room decreases as individual reflections become weaker, i.e., the stronger the walls, the floor, and the ceiling, etc. absorb sound. In contrast it becomes longer for increased time separations between individual reflection processes; this "free path length" increases with the size of the room. When the total absorption of the room is not too large, these relations are represented by the Sabine reverberation formula:

$$T = 0.163\ V/A \quad T\,\text{in s},\ V\,\text{in m}^3,\ A\ \text{in m}^2,$$

where T is the "Sabine" reverberation time (corresponding to the 60 dB definition), V is the volume of the room and A is its equivalent absorption area which is calculated from the sum of the A values of the individual surfaces and objects.

Inasmuch as absorption is frequency dependent, the reverberation time also shows a frequency dependence, which, depending on the nature and furnishings of the room, can exhibit rather differing characteristics. As an example for a reverberation curve, the reverberation time of the fully occupied Bayreuth Festspielhaus is represented in Fig. 5.8. It is noted that the reverberation time for low frequencies is longest and decreases for the higher registers. It should further be noted that for high frequencies, an attenuation occurs during sound spreading (dissipation loss) in addition to the absorption at the walls as described.

The frequency dependence of the reverberation time is of great importance for the auditory tonal impression, since it causes a tone color change during the decay.

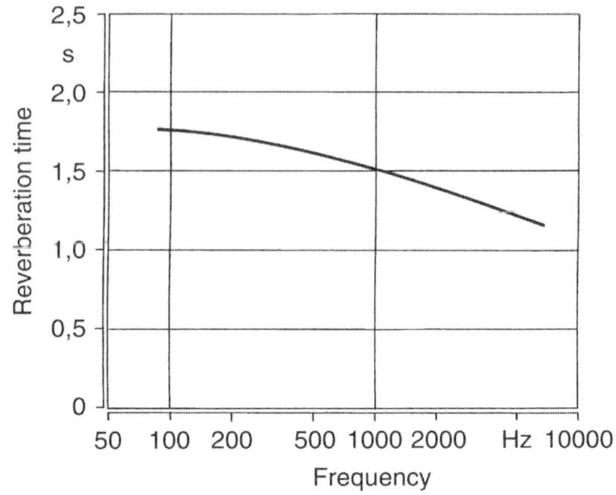

Fig. 5.8 Reverberation curve of the Bayreuth Festspielhaus under fully occupied conditions (after Beranek, 1962)

The faster the high components lose in intensity, the duller the decay process, as caused by the room. However, consideration must be given to the fact that in a certain sense the rise in the reverberation time below 125 Hz finds compensation in the characteristics of the ear. Since the "equal loudness curves" (see Fig. 1.1) are more closely spaced at those frequencies, a region of 60 phons is traversed by a level range of less than 60 dB so that a reverberation process at low frequencies appears shorter than an equally steep level drop at higher frequencies.

The early time decay in the region from 125 to 2,000 Hz has proven particularly suitable for characterizing the influence of a room on tone color (Lehnmann, 1976). This is because the dynamic short-time structure of music occupies a region of relatively narrow level difference. The Early-Decay-Time for most rooms is somewhat shorter than the Sabine reverberation time, particularly at low frequencies, when coupled room sections – e.g., above ceiling reflectors- contribute to late reverberations.

When evaluating reverberation curves of different rooms it becomes of interest to determine which smallest variations in the reverberation time are discernable for the trained ear. In a simplified approach this question can be answered as follows: for short reverberation times below 0.8 s, steps of approximately 0.02 s are noticeable, while the sensitivity above this limit amounts to approximately 3.5% of the relevant reverberation time (Seraphim, 1958). In practice, this means that values need not be specified more accurately than 0.1 s.

5.4 Direct Sound and Diffuse Field

5.4.1 The Energy Density

When a sound source radiates a long tone or a continuing sound, the direct sound and the multiplicity of sound reflections arrive at the listener at the same point and time, all of which have different run times and are attenuated in varying degrees. The combination of all of these components results in the observed sound level.

When an energy balance is established one must assume an equilibrium between the energy swallowed by the absorption surfaces and the energy provided by the source once a stationary state is reached. From this condition the so called energy density, i.e., the sound energy per unit volume, can be calculated. From this, the sound pressure level L_p of the diffuse sound field in the room is given by

$$L_p = L_w - 10\log(V/V_0) + 10\log(T/T_0) + 14\,\mathrm{dB},$$

where L_w represents the sound power level of the source V, the volume of the room ($V_0 = 1\ \mathrm{m}^3$) and T its reverberation time ($T_0 = 1$ s). The term 14 dB is the result of several constants as well as the reference point for sound power-, and sound pressure- levels. This sound pressure level of the diffuse sound field is, at least theoretically, the same everywhere within the room.

It depends on the size of the room and the power of the sound source, it is, however, also influenced by the reverberation time. Since the latter is frequency dependent, a pitch dependence results also for the energy density, and thus for the sound level of the diffuse field. This means, that because of its reverberation characteristics, the room not only influences the sound level but also the tone color. The shorter the reverberation time is in a certain frequency region, the more this region is disadvantaged in the tonal spectrum.

Still, the variation range of the reverberation time is not as large as the range in room volume which from a small chamber music room to a large cathedral encompasses several powers of 10. Consequently, practical experience has taught that it is advantageous when the reverberation time increases somewhat with increasing room size. However, this can only compensate in small measure for the decrease in energy density. On the other hand this presents an advantage since the duration of the audible reverberation is slightly shortened by the reduced sound level, so that, again the extension of the reverberation time finds partial compensation in the subjective reverberation impression. Finally, within limits, it is of course also possible to adapt the power of the sound source to the room acoustical conditions.

In order to make it possible to compare several rooms directly, the option is presented to combine the last three expressions on the right side in the formula for the sound level of the diffuse field into one quantity:

$$L_p = L_w - D_A.$$

The quantity D_A will be designated as "room damping index." Accordingly, the room damping index (which is independent of the sound source) is the numerical difference between the sound power level of the source (also room independent) and the sound pressure level achieved by the source in the room. If the volume and reverberation time of the room are known, the room damping index can be calculated as

$$D_A = 10\log(V/V_0) - 10\log(T/T_0) - 14\,\text{dB}.$$

In practice, the room damping index indicates by how much the average sound level is changed when the same sound source is moved to different rooms. However, it should be emphasized, that in all cases this applies to a spatial average for the diffuse sound field i.e., for the steady state of the room (such as for long notes).

In order to represent the differences between energy densities in different locations in the same hall, as determined by different room reflections and directional characteristics of the sound source, the so called strength factor (also in dB) must be determined. It indicates the difference between the sound power level of the sound source and the sound pressure level as it actually occurs at the relevant location in the hall. This strength factor, however, must be measured in each situation by means of an impulse response, whereby it makes sense to include reflection with up to 80 ms delay relative to the direct sound, because it is also within this time duration that loudness impressions are formed. With some difficulty the strength factor can be calculated from several sets of data for the room (Lehmann, 1976).

5.4.2 *The Direct Sound*

As the sound radiated by the source spreads, its level naturally decreases with increasing distance. Energy conservation mandates that in an undisturbed sound field the sound pressure is inversely proportional to sound source distance, one speaks of the $1/r$ law or $1/r$ decay, where r represents the distance to the sound source. The dependence of the sound pressure level on sound source distance is given by

$$L = L_w - 20\log(r/r_0) - 11\,\text{dB} + D_i,$$

where again L_w is the sound power level of the source, r_0 the reference value of 1 m, and D_i is the directivity index of the source for the radiation direction under consideration. From this it follows for the sound level, that it decreases by 6 dB for each doubling of the distance to the source; a tenfold increase of the distance leads to a decrease by 20 dB.

Figure 5.9 represents this level change of the direct sound for a range of 28 cm to 20 m, where the reference level of 0 dB corresponds to the sound power level of the source radiating equally in all directions. In the immediate neighborhood of the sound source, this curve can deviate somewhat from the line drawn for the omnidirectional sound source, depending on its size and vibration shapes. Furthermore, dissipation losses already mentioned, effect a somewhat steeper level drop at very high frequencies. When the sound passes a strongly absorbing surface at a very shallow angle, the level drops somewhat faster than that which would correspond to a $1/r$ decay, since the sound waves are practically bent into the absorber. This case is of particular interest to the spread of sound over seating rows in an audience.

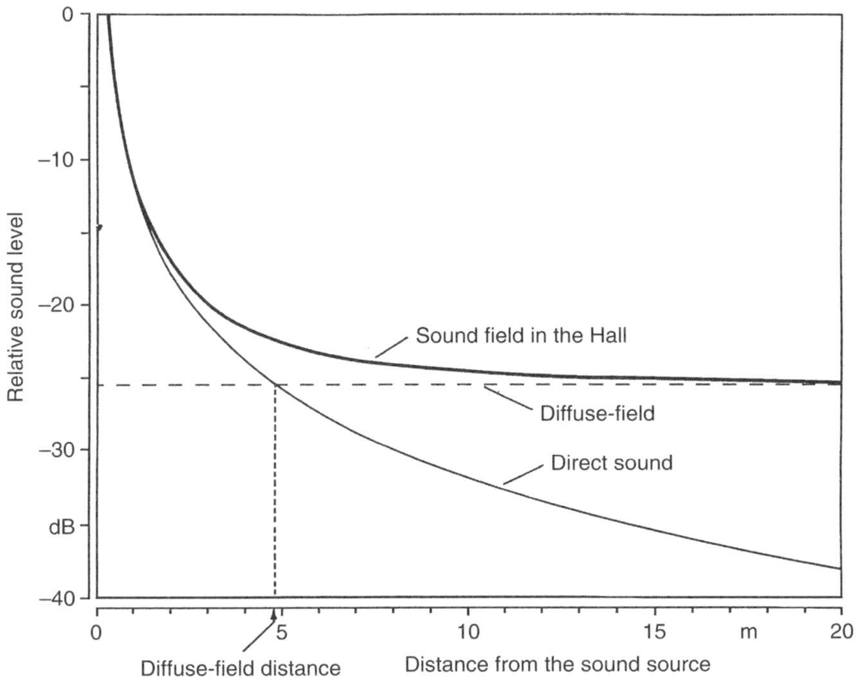

Fig. 5.9 Sound level in dependence on sound source distance

In Fig. 5.10 sound level reduction beyond the $1/r$ decay is represented for a listener position at a 13.5 m distance from the source (12th row) for different sound source heights. A more or less pronounced frequency dependence appears, which is most strongly noted for a sound source at head height. This finds its explanation in the bending of the sound around the head, especially at relatively high frequencies. This weakens the direct sound, especially for those tone contributions important for brilliance and clarity of articulation. This effect is not as strongly pronounced when audience seating is offset than is the case when heads are located exactly behind each other in the direction of sound spreading. Furthermore one notices a clear decrease of the additional damping when raising the sound source on steps. At a height of 2 m the influence of audience rows is nearly negligible: this additional damping therefore can be avoided in large measure with a higher podium and rising audience seating (Mommertz, 1993).

As the left end of the curves already indicates, the seating rows, however, effect an additional level decrease beyond the $1/r$ decay for a certain region of lower frequencies. The maximum of the resonance-like weakening lies in the region of 130 and 170 Hz. This frequency location is independent of seating row occupation and separation, it is only determined by the height of the backrests and their lower connection to the floor. Inasmuch as this additional level drop can take a value of 10–20 dB for a source distance of 20–25 m, it can affect an essential change of the direct sound of the low instruments. It is particularly noticeable in the sound

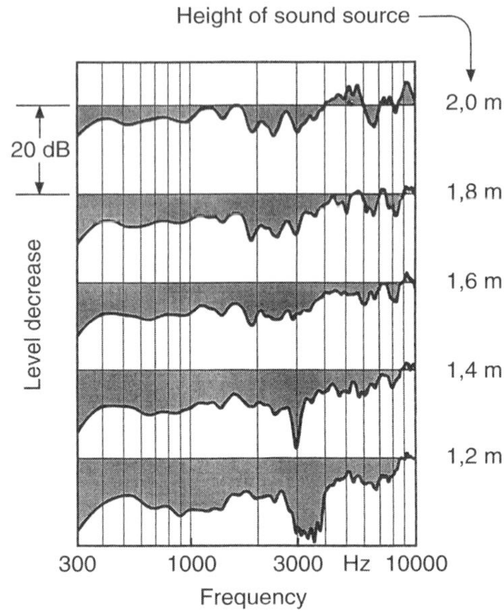

Fig. 5.10 Sound level reduction over audience seating rows which are not offset relative to each other, distance between sound source and observation point 13.5 m (after Mommertz, 1993)

impression when the components are contained only weakly in the first reflections from the ceiling or from hanging reflectors. This effect can be reduced when absorbers for the relevant frequency region are installed on the floor between rows of chairs (Schultz and Watters, 1964; Ando, 1985; Davies and Lam, 1994).

5.4.3 *Diffuse-Field Distance*

In the sound field present in a room, the direct sound, which decreases in intensity with distance from the source, is superimposed onto the diffuse field distance, which has the same level everywhere. Consequently, close to the sound source, the direct sound dominates, while at large distances it only makes a minor contribution to the overall sound level. This connection is represented in Fig. 5.9. In addition to the $1/r$ decay of the direct sound already explained earlier, the level of the diffuse field (for a certain volume and a certain reverberation time), as well as the total sound level resulting from the super-position of these two contributions is entered. One notes, that in consequence of the logarithmic dB scale, for the energy addition of the direct sound and statistical field, a curve results which deviates only a little from the one $1/r$ decay curve in the neighborhood of the sound source, which however, for increasing distance approaches the level of the diffuse field. Additionally, it should be noted, that in large halls the diffuse field is often not formed exactly, but rather its level drops by approximately 0.85 dB per 10 m distance from the sound source. This has the consequence that the energy density level can be by about 3 dB lower in the rear of the hall, which represents an audible loudness loss (Barron and Lee, 1988).

For the description of room acoustical quantities, it has been useful to delineate the region in which the direct sound is stronger in comparison to the region of predominantly diffuse sound contributions. The distance of this boundary from the source is referred to as the "diffuse – field distance." In the case of omnidirectionally uniform radiation by the source one also speaks of a diffuse-field radius (previously the designation "diffuse-field radius" was in general use, even for directional sound sources).The intersection of the two relevant curves in Fig. 5.9 correspond to the condition, that at this point the levels of the direct sound and the diffuse field are equal. From this, the example represented yields a diffuse field distance of 4.8 m for an omni-directional source. The following numerical-value equation applies

$$r_{\mathrm{H}} = 0.057\Gamma_{\mathrm{st}}\sqrt{(V/T)} \quad r_{\mathrm{H}}\ \mathrm{in\,m};\ V\ \mathrm{in\,m^3};\ T\ \mathrm{in\,s},$$

where again V represents the room volume, T the reverberation time and Γ_{st} the statistical directivity factor of the sound source. When the sound source has a pronounced directional effect, the diffuse-field distance in the preferred direction becomes correspondingly large while in other directions, it becomes smaller than for a spherical radiator. It is important to note, that the diffuse-field distance is independent of the power of the sound source (Cremer, 1961).

A graphical solution of the equation for the diffuse-field distance is represented in Fig. 5.11, where it is assumed that the sound source has spherical characteristics,

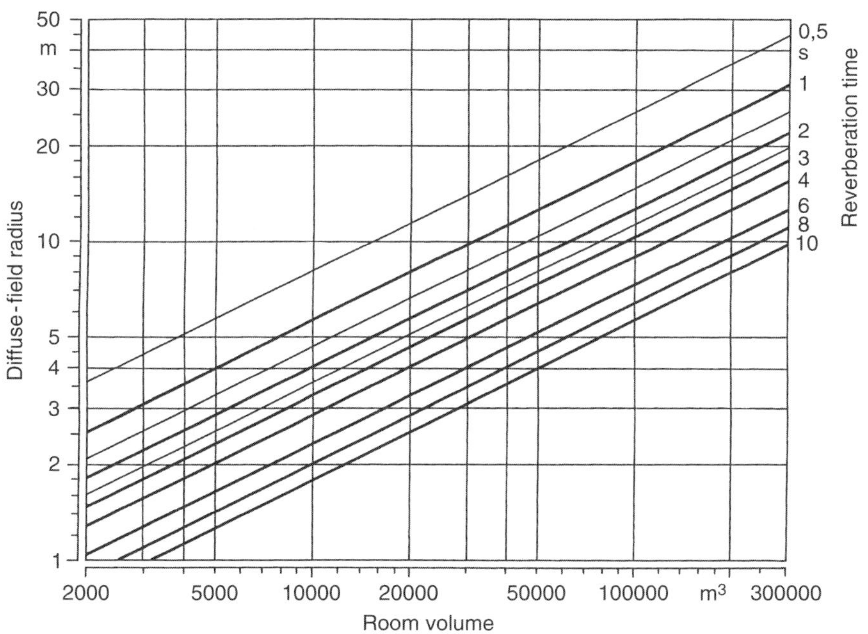

Fig. 5.11 Dependence of diffuse-field distance on room volume and reverberation time (for omnidirectional sound sources)

Table 5.1

Distance from the source	$r_H/2$	r_H	$2\ r_H$	$3\ r_H$
Relative sound level in dB:				
Direct sound	+6	0	−6	−10
Diffuse field	0	0	0	0
Superposition	7	3	1	0.4
Level increase in dB:				
By direct sound	–	3	1	0.4
By diffuse field	1	–	–	–

that is, the statistical directivity factor in all directions is equal to one. It is noticed that the diffuse-field distance increases with increasing room size and constant reverberation time, that, however for a fixed volume, it remains smaller for longer reverberation times. The advantage of a larger energy density and thus greater loudness, which accompanies the increase in reverberation time in large rooms is thus, purchased with a reduction in the diffuse-field distance and consequently, with a decrease in clarity.

In Table 5.1, some levels for the combined values of the sound field at several distances are assembled, using diffuse-field distance as a specific room characteristic for the starting point. The tripled diffuse-field distance proves to be of great importance, since at this point the direct sound lies by 10 dB below the diffuse field, which constitutes a boundary for certain locations of the sound source based on the reaction of the ear to the wave front arriving first. It can furthermore be seen that for the double diffuse-field distance, the direct sound effects an increase on the total level by only 1 dB; at half the diffuse-field distance, the influence of the diffuse field in comparison to the direct sound is correspondingly minute.

5.5 Temporal Structure of the Sound Field

The diffuse sound field can only be formed when sound waves arrive from all sides in a dense time sequence. However, this condition is not satisfied during the initial transient of a sound or noise within a room. The behavior of a room during an initial transient, therefore, does not represent a process to be treated statistically. Rather, the direct sound is followed after a short pause by the first reflection whose delayed time in comparison to the direct sound depends on the length of the detour the sound must traverse. Only after several additional reflections, which already arrive with shorter time separations, the uniform reverberation follows.

This process is represented schematically in Fig. 5.12 for a single sound source, where for the sake of clarity, only a selection of early reflections is shown in the picture of the hall. Among these are the specially important early reflections from the side walls (W_1 and W_2) as well as those formed by the gallery and the side wall barriers (W_1' and W_2'). The strength and the time of arrival of the individual

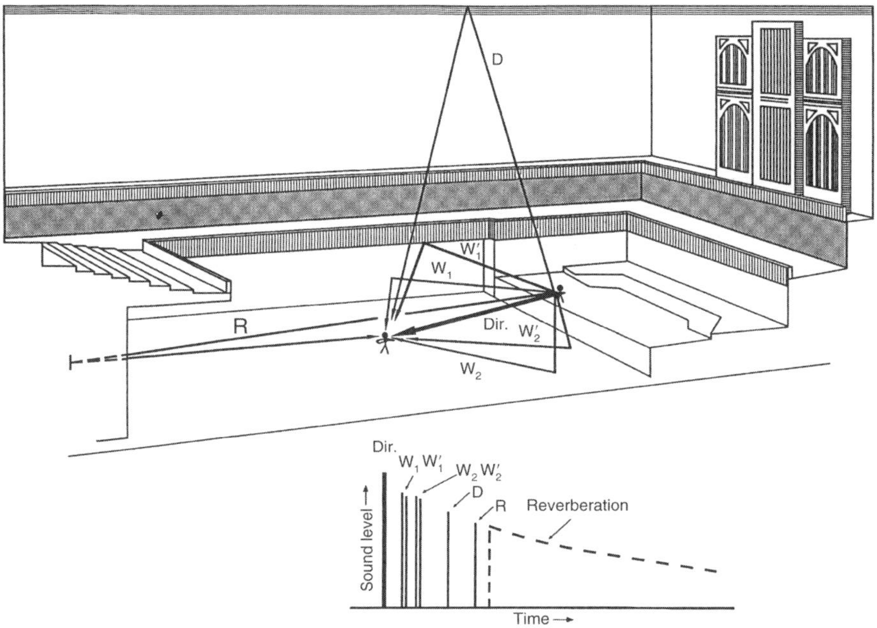

Fig. 5.12 Development of sound field in a concert hall: Sound paths (*top*), and time sequence (*bottom*) of the direct sound and the first reflections

reflections naturally depends in large measure on the shape of the room, as well as the reflection characteristics of the wall and ceiling, furthermore, both characteristics vary for different seating areas within the hall.

These room acoustical processes play an important role in the tonal impression for a musical performance. In this context, the direct sound is responsible for the clarity, in particular for rapid tone sequences, for the transparency of the tonal picture. It also transmits the tonal impression of the spatial arrangement of the performer. Furthermore, the direct sound contributions affect the sensation of the proximity of the orchestra or the stage to the listener. The increasingly delayed reflections, which in their totality shape the reverberation, particularly influence the melting of the individual voices into a complete sound. In similar fashion, they bridge the time sequence of short breaks between individual notes, whereby the melodic line obtains greater uniformity. In addition, the reverberation raises the loudness impression considerably.

Depending on delay times and intensities, the early reflections can influence the tonal impressions in different ways (Kuhl, 1965). Also the directions, from which the reflections reach the listener, play an important role, and finally, the nature of the performed music itself has an influence on the subjective perception of the first reflections (Schubert, 1969). For increasingly impulsive music, i.e., for sharp *staccato* and *pizzicato*, the ear responds more sensitively to reflections from the room. The reflection arriving first carries particular significance. For a delay time of approximately 30 ms, which corresponds to a detour of about 10 m, it affects the orchestral music by way of enhancing the direct sound without detracting from the clarity. This could merely result in minor tone color changes, which are connected to the relative phases of direct and reflected waves. However, this effect is only audible when the first reflection is dominant for the initial room sound, i.e., when further reflections are only weak or follow relatively late, as is the case above all for open air performances or in extremely dry halls. For solo passages, with a strongly pronounced rhythmic dynamic structure, this time span is reduced to approximately 20 ms. Since a long delay naturally causes an impression of a larger room, the time span between direct sound and first reflection is responsible for the degree of intimacy of the spatial sound impression (Beranek, 1962). The shorter the time span, the closer one feels acoustically to the musicians. Beyond that, an impression of the dimensions of the room is transmitted to the listener by the time sequence of the first wall and ceiling reflections, independently of their strengths.

As already mentioned elsewhere, the first reflection can even be 10 dB stronger than the direct sound without having the latter lose its direction-determining characteristic (Haas, 1951). This can occur when the sound source is blocked from the listener as is the case, for example, for instruments in the orchestra pit in relationship to the seats on the main floor in an opera house.

Concerning the minimum intensity necessary to obtain any tonal impression from the reflection, the direction of incidence at the listener plays an important role. While reflections, which reach the listener from the front or above within the time interval of the first 30 ms, depending on the nature of the music, can be by 10–20 dB weaker than the direct sound, reflections arriving from a side can still be perceived when their level is lower by an additional 10 dB (Schubert, 1969).

Under the assumption that the first reflection arrives sufficiently early, one can generally conclude that all reflections which follow the direct sound within 80 ms are useful for the precision and clarity of the musical tone picture, while later reflections and reverberations diminish the clarity. One can therefore define a "clarity factor" (C_{80}), wherein the sound energy (E_{80}) arriving within the first 80 ms is compared to the remaining energy:

$$C_{80} = 10\log(E_{80}/(E_{\infty} - E_{80}))\,\mathrm{dB},$$

where E_{∞} is the total arriving sound energy. This clarity factor should lie between -2 and +4 dB, for distant seats a value of -5 dB is still defensible. For speech – and to some extent for sharply structured music – the boundary between reflections which enhance clarity and those which reduce it lies at 50 ms. Accordingly, the "clarity factor" (C_{50}) is defined as

$$C_{50} = 10\log(E_{50}/(E_{\infty} - E_{50}))\,\mathrm{dB},$$

where E_{50} is the sound energy arriving during the first 50 ms. This clarity factor should lie above 0 dB (Reichardt et al., 1975).

Individual reflections, which arrive with a delay of more than 25 ms can effect very different sensations, depending on their directions. When coming from the median plane, i.e., from the plane standing vertically in the room, which also includes the direction of sight of the listener, they do enhance the sound, however they do not create a spatial effect. The upper limit for this lies at about 80 ms. Side reflections arriving in this time range determine the tonal room impression, as do all reflections arriving later (Barron, 1971). The boundary between side reflections and those not coming from the side is naturally somewhat fluid; The difference between the sound contributions arriving at the two ears increases as the incident direction is increasingly turned toward the side. In fact, the value of side reflections in relation to the room impression increases with the sine of the incident angle (calculated in relation to the direction of sight) (Alrutz and Gottlob, 1978). With that, the "interaural correlation" which determines the spatial impression decreases (Gottlob, 1973). In order to be able to measure this sound effect as "spatial impression measure", a 40° cone about the direction of sight is specified as a boundary (Lehmann, 1975; Reichardt et al., 1975). All sound energy contributions with more than 80 ms delay, as well as all sound reflections arriving from the side (i.e., outside of the 40° cone) with 25–80 ms, delay enhance the spatial sensation. All sound contributions arriving within the first 25 ms, and those arriving from the front (within the 40° cone) with 25–80 ms delay do not contribute to the spatial sensation. When designating the spatially enhancing sound energy as E_{R} and the sound energy which does not contribute to the spatial sensation as E_{NR} the spatial impression measure is given as

$$R = 10\log(E_{\mathrm{R}}/E_{\mathrm{NR}})\,\mathrm{dB}.$$

A range of 1–7 dB is desirable for concert halls. From a standpoint of measurement it is simpler to determine the relationship of the sound energy arriving from the side between 25 and 80 ms to the entire sound energy arriving between 0 and 80 ms ("lateral efficiency") (Jordan, 1979).

The spatial impression measure, as well as the "lateral efficiency", are quantities which are independent of the loudness of the sound impression for the listener. Thus they are not concerned with the sensation that music only assumes spatial characteristics with increasing loudness (Keet, 1968). This effect is represented by the concept of "spatial impression". This is to be understood as the subjective impression that the sound of an orchestra seems to expand into the space between the musicians and the side walls and possibly from the floor to the ceiling without limiting the localization of the instruments (Kuhl, 1978). The sensation that sound in a room also comes from behind occurs only rarely, e.g., when one is very close to the sound source in a large church. As a matter of principle one can assume that the acoustic space impression is limited to the hemisphere in front.

Sufficient loudness is an essential condition for an impression of space, so that the effect, when present at all, only occurs for *forte* passages in music. Precisely therein lies the value of this criterion: even in mediocre halls a *piano* passage can be made to sound well, while a convincing tonal development in *forte* succeeds only in acoustically good halls. This is because both sufficient loudness and the impression of spatial expansion of the sound are of great significance for an emotional experience.

In addition to loudness, the spatial sensation demands an energy relationship between reflections not in the median plane with a delay time of 10–80 ms on the one hand, and the direct sound on the other. The level difference between the sum of these reflection and the direct sound is also designated as the "lateral sound level". For a quantitative description of the spatial sensation, a scale is determined in such a way that an increase in the spatial sensation by one step is caused by an increase in the sound level by 5 dB, provided that a spatial effect is already sensed at the initial level. Below a certain level, no spatial sensation occurs (step 0). Figure 5.13 shows a spatial sensation diagram for a seat in the Hamburg Musikhalle. It is evident that below a level of 76 dB no spatial sensation occurs and that above this level-boundary, three spatial steps are reached up to a *forte* level of 91 dB.

The level-boundary for the onset of spatial impression can be different for different halls. The value of this level-boundary becomes lower as the sound level coming from the side increases. Figure 5.14 shows this relationship. Particularly low values, and thus pronounced spatial impressions, are found in long stretched rectangular rooms. This also makes the requirement understandable that, for sound esthetic reasons, the first side wall reflections should arrive earlier than the first ceiling reflections and reflections from the back wall (Marshall, 1967).

Fundamentally, all spectral components of the early lateral reflections contribute to the subjective spatial impression, and the broader the frequency region contained in these reflections, the more strongly do these reflections enhance the spatial impression. However, the frequency contributions contained in the reflections can evoke a difference in the nature of the spatial impression: if the reflections contain only low and middle frequencies, a predominant sensation arises that the sound is broadened in the depth of the room (to the front and back). Many listeners connect the feeling of being immersed in the sound with that effect. When the spectrum of

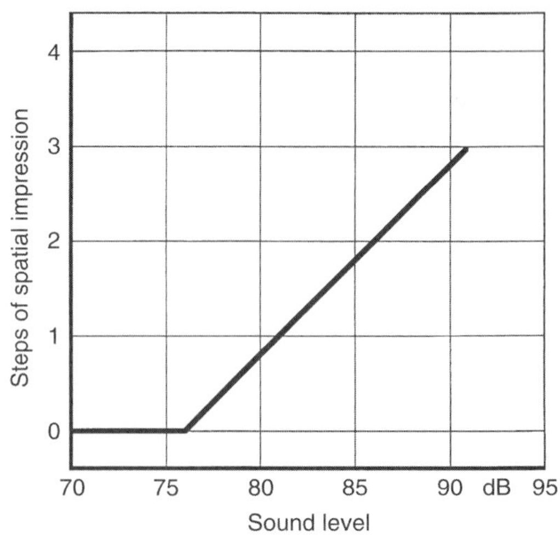

Fig. 5.13 Spatial impression of the orchestral sound in the Hamburg Musikhalle in dependence on sound level (after Kuhl, 1978)

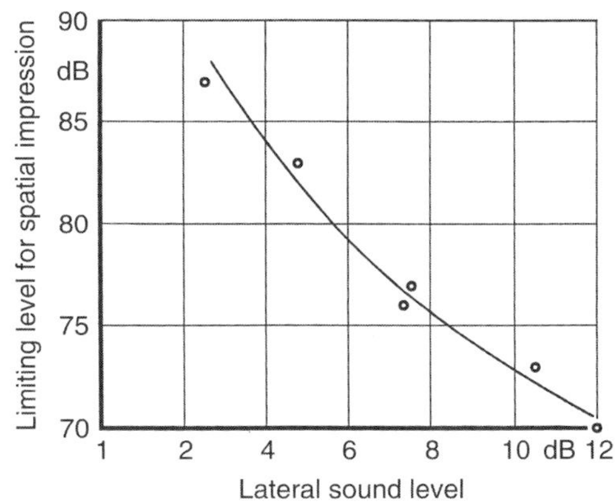

Fig. 5.14 Limiting value of sound levels for onset of spatial impression. Data points for halls (After Kuhl, 1978) from *top* to *bottom*: Stadthalle Hannover, Rheinhalle Düsseldorf, Philharmonie Berlin, Musikhalle Hamburg, Stadthalle Wuppertal. The lowest data point refers to hearing tests with electroacoustic instrumentation

the reflections also includes sufficiently high frequencies (above 3,000 Hz) a preferential sensation of tonal broadening occurs (Blauert and Lindemann, 1986). These differing dimensions of spatial impression can be influenced within certain limits by construction modification of the relevant reflection surfaces. They belong to specific tonal characteristics which differentiate large halls from each other.

By changing reflector arrangements, delay times of individual sound reflections can occasionally be influenced. However, it should be noted, that the ear responds more sensitively to reflections from the side than to those which come from the front, from above or behind (Reichhardt and Schmidt, 1967). Thus for lateral reflections in a test series, changes in delay time of 7 ms were noted; this corresponds to a lengthening of the sound path of approximately 2.4 m. For ceiling reflections, in the same test, changes of approximately 12 ms corresponding to an additional path of 4.1 m were required. It can be assumed, however, that for music with strong rhythmic structure, these values are reduced by approximately one-half (Schubert, 1969). This is also supported by the fact that musicians sense a clear difference in the spatial effect, depending on whether the orchestra is located on the same level as the concert hall floor or on a 1.5 m high podium.

Based on the very different sound field developments at different locations in the hall, specifying a uniform initial transient, as is done for musical instruments, is difficult. If the time required for the sound level to reach a value of only 3 dB below the steady state value is calculated, a relatively short initial transient time is obtained under the assumption of statistical conditions: it amounts to 1/20 of the reverberation time. After approximately 1/14 of the reverberation time, the level lies only 2 dB below the final value. In reality, in such a short time, the requirements for a diffuse sound field are certainly not met. Accurate studies of the spatial sound field development in a concert hall however, have given the result, that for most locations the level of 3 dB below the final value was already reached within three quarters of the time span which would have been calculated from statistical considerations; only in the least favorable locations were these two values the same (Junius, 1959). From this, one can generalize that the initial transient times of a room can be assumed to be somewhat shorter than 1/20 of the reverberation time for the majority of the locations.

Chapter 6
Acoustical Properties of Old and New Performance Spaces

6.1 Concert Halls

6.1.1 Tonal Requirements

Among the spaces used for musical performances, large halls for symphonic concerts play a particular role when considering acoustical properties, because in these, the musical experience of the listener is least influenced by visual impressions. While in the opera, the action on the stage, for chamber music performances, the close visual contact with the performers, and for church concerts the atmosphere of the room can make an essential contribution to the overall impression, for symphonic performances the sound – esthetic evaluation assumes predominant importance. Nevertheless, one should not underestimate the visual relationship between listener and podium, as well as the architectural impression of the room, as they influence the tonal sensation (Winkler, 1992; Opitz, 1993).

An important criterion for the acoustical quality of a concert hall is its reverberation time. As mentioned earlier, reverberation enhances the melting of the individual voices of the orchestra into a closed overall sound and lends a uniform flow to melodic phrases in their time progression. Furthermore, the reverberation time plays a role in the loudness impression, since the resulting energy density and thus, the sound pressure is directly proportional to the reverberation time for tones which are not excessively short. Extremely long reverberation, naturally can also lead to negative effects, when short pauses are no longer clear or when a piano passage after a forte ending gets lost in the reverberation.

Next to the homogeneity of the overall sound, the clarity of passages with strong rhythmic structure or of polyphonic vocal arrangements is essential for favorable sound impressions by the audience. Since this condition must be met by the intensity relationship between the direct sound (including the first, only slightly delayed reflections) and the later reverberant sound, the various seating groups in a hall naturally don't exhibit the same acoustic quality in this respect.

J. Meyer, *Acoustic and the Performance of Music*.

DOI: 10.1007/978-0-387-09517-2_6;

A balanced relationship between the early sound and the reverberant sound is a necessary requirement for a concert hall, to give the audience a satisfying tonal impression. However, there are additional considerations which also play an important role, thus, the listener does not want to be confronted by the orchestra from a tonal standpoint, but rather be included in the musical experience. In addition to adequate loudness, this effect is brought about by the direction of incidence of the first reflections combined with the diffuse reverberations for which sound arrives uniformly from all sides (Meyer, 1965). The opposite of spatial impression is observed when music with reverberation is heard from a loudspeaker: the reverberation can bring about a good overall sound, but it cannot provide spatial inclusion of the listener.

Finally, at different locations in the audience, the intensity relationships between the individual instrument groups, dispersed laterally and in depth over the podium, must be as uniform as possible and should not deviate too much from the sound picture at the position of the conductor. This addresses a particular problem of acoustical relationships during musical performances: the listener should obtain a possibly complete tonal impression, however, the conductor can only shape the sound of the orchestra on the basis of the aural impression from his position. Only during the rehearsal can he gain an impression of the tonal effect in the hall. He must, therefore, be able to obtain an optimal acoustic tonal impression from the podium in order to achieve the best possible artistic result.

The individual players must also be able to hear each other well, otherwise the ensemble performance suffers. With regards to the demands placed by the musician on the acoustical and also the nonacoustical requirements for the platform, three stages can be identified which lead to a difference in quality of performance and thus, to the orchestral sound. At the lowest level we have the minimal demand of a technical ensemble performance, free of errors in relation to intonation, rhythmical precision and adherence to dynamical categories. Already at this stage, difficulties can arise. When the self perception of the musician is too loud and other performers are heard too little, harmonic relations are still perceived and intonation can still be correct, but the precision of rhythmic adaptation and consequently, the certainty of entries and articulations will suffer. In contrast, if the self perception of the performer is inadequate in its rise above the sound of the orchestra, the intonation is compromised, while the insufficient aural impression from the instrument of the performer will still permit a temporal precision in the performance by virtue of the performance mechanics controlled by the feeling of the musician.

On the second level, the individual performer senses optimal conditions for the shaping of the tone of his own voice, including a sure and easy attack of the instrument: the performer has a "feeling of well-being." Within this framework, dynamic and tone color variation possibilities can be exhausted to the technical limits. Included in this limit is also the circumstance that performers of low instruments, especially the bass, should not feel disturbed by sympathetic vibration of their own instruments or the floor, which are excited by other voices, as for example, low percussion instruments. This could make their own intonation more difficult.

The highest level results in a common artistic achievement of all participants. This especially includes an integrated overall sound of the individual string groups based on a high measure of uniformity with respect to the temporal fine structure of the tone formation. To accomplish this, the individual performer must sense a tonal inclusion in the group. This involves the knowledge of reactions of other performers and also, the trained experience based on the traditions of the orchestra. Visual contact and joint breathing play an equal valued role next to acoustical communication (Meyer, 1994a).

Similarly, these conditions of mutual listening can be transferred to the situation of a soloist. On the one hand, the necessary "resonance" in the form of early reflections and appropriate reverberation is expected (which at the same time is sensed as an easier approach to the instrument), on the other hand, the feeling is desirable to be carried by the instrument (in relation to intensity). For singers it has even been noticed that the acoustical feedback of the room is capable of a physiological enhancement of the vocal functions (Husson, 1952). In this context it should be mentioned that instrumental soloists, especially string players, feel very uncomfortable in rooms in which the sound is reflected too strongly by the walls and the ceiling from the area of the podium into the auditorium, because the missing or overly weak first reflections to the podium cause the impression that the tone can no longer be influenced by the instrument, as soon as it is placed in the room. On the other hand, the soloist should be easily heard by the audience, therefore, the position on the podium in relationship to the orchestra should not be acoustically disadvantaged, but rather occupy a preferred position.

Inasmuch as mutual listening by musicians in the concert hall is determined by the spatial arrangement of the region around the podium, while the resonances noticed by the soloist are determined by the acoustical relationships in the entire hall, it is not surprising that hall quality judgments of musicians are very different, depending on their function, and that they in turn can be very different from the judgment of the audience. This discrepancy in evaluation of halls is all the more evident because many musicians rarely take the opportunity to hear concerts in a fully occupied hall from the location of a normal listener.

6.1.2 Reverberation Time and Hall Size

With the multitude of factors which influence the acoustical quality of a concert hall, naturally the question of optimal relationships for symphonic performances arises. In order to be able to draw conclusions about the necessary requirements for excellent acoustics from technical data, it is important to determine which halls are judged to be especially good, both by musicians and audiences.

In the years from 1950 to 1955, F. Winckel conducted a survey of approximately 25 internationally recognized conductors concerning the acoustically best concert halls. The result was the rank-ordered listing of the five best halls given in Table 6.1, for which also the year of dedication, the volume, number of seats, reverberation

Table 6.1 Concert halls before 1940

Concert hall	Constr. year	Volume (m^3)	Seats	Reverberation time (s)	
				Mid	Low-frequency
1. Large Musikvereinssaal Vienna	1870	14,600	1,680	2.05	2.4
2. Theatro Colon Buenos Aires	1908	20,550	2,487	1.8	–
3. Concertgebouw Amsterdam	1887	18,700	2,206	2.0	2.2
4. Symphony Hall Boston	1900	18,740	2,631	1.8	2.2
5. Konzerthus Göteborg	1935	11,900	1,371	1.7	1.9

Fig. 6.1 Grosser Musikvereinssaal in Vienna (Bärenreiter – Verlag Kassel (MGG))

times in the middle and low frequency region (500–1,000 Hz and 125 Hz, respectively) are given under occupied conditions.

This table shows that the optimum of the average reverberation time for halls of that size is evidently in the region between 1.7 and slightly above 2 s. The Musikvereinssaal (Fig. 6.1) has the longest reverberation time, even though it is next to the smallest of the five halls. This hall is famous for its ability to enhance the performance of romantic works, yet the acoustics are also praised for an ability to present classical compositions. In contrast, the hall in Amsterdam is judged by sensitive musicians not to be as reverberant, even though the difference of the reverberation times is only approximately 0.05 s. Finally, the Konzerthus in Göteborg is the driest of the halls considered, it also is nearly the smallest.

The connection of J. Brahms and A. Bruckner to the Vienna Musikvereinssaal is often pointed out. Both performed their own works here, therefore it is considered especially authentic for their music. Nevertheless, one should not overlook the fact that this hall has undergone changes during reconstruction in 1911 which also affect the room acoustics. Among other things, the wood construction of the upper galleries was replaced by steel beams and the load bearing statues, which earlier were located at the front edge of the galleries, were moved to a location in front of the side walls. The entire weight of the ceiling construction was increased by a layer of sand and bricks. The stage surface was widened, and the number of seats was increased. From an acoustical standpoint, this meant an increase in the effective hall volume, a decrease in the absorption of the low frequencies, and a change in the time sequence determined by the widening of the hall. Even though the hall was praised for its good acoustics before the renovation, one can assume that the renovation resulted in a further improvement of the hall acoustics (Clements, 1999).

For newly constructed concert halls, those that were built within the first 20 years after the second world war, experiences with older halls were utilized. However, in the meantime, the body of knowledge related to room acoustics has increased so much, that from today's standpoint, those newer halls, which by now are almost designated as historic halls, are no longer evaluated as equally good. Table 6.2 gives five examples of well known concert halls from the 1950s and 1960s. Their listing order is given according to the building year and in no way reflects a value judgment.

From this listing it is clearly noted that at the time an effort was made to obtain shorter reverberation times. In the case of the Royal Festival Hall, this may be related to the extraordinarily large number of seats. It can also be concluded that the tastes of musicians and listeners has changed with time, as has the interpretation style for classical and romantic music. For example, the Liederhalle in Stuttgart was initially judged as acoustically excellent upon completion (Beranek, 1962), while today it is considered relatively dry. The tendency for this development of room acoustical perceptions in these, now considered historic concert halls shifts from a greater transparency of the tonal picture in the years around the middle of the

Table 6.2 Concert halls after 1950

Concert hall	Constr. year	Volume (m^3)	Seats	Reverberation time (s)	
				Mid	Low-frequency
1. Royal festival Hall London	1951	22,000	3,000	1.45	1.35
2. Liederhalle Stuttgart	1956	16,000	2,000	1.65	1.8
3. Beethovenhalle Bonn	1959	15,700	1,407	1.7	2.0
4. Philharmonie Berlin	1963	26,000	2,218	2.0	2.4
5. Meistersingerhalle Nürnberg	1963	23,000	2,002	2.05	2.2

Source: 1. Parkin et al. (1952); 2. Cremer et al. (1956); Gade (1989b); 3. Meyer and Kuttruff, (1959); 4. Cremer and Müller (1964)

century to a more homogenous richness of sound in the subsequent decades. Among other things, this may be based on the sound expectation of the 1950s related to the larger transparency of radio and record reproductions, while later the sound aesthetic expectation for real concert experiences became clearly separated from that for sound recordings. A concert hall cannot reach the transparency of a recording, nor a recording the spaciousness of a concert hall. Judgments about the acoustic qualities of concert halls can therefore only be interpreted in the context of the times, detailed questions are not necessarily universally applicable.

Thus it becomes understandable that the long-time chief conductor of the Berlin Philharmonic, A. Nikisch, in the years around 1920, considered the Hall of pillars in the "House of Unions" in Moscow as the acoustically best concert hall in Europe: this hall has room for 1,600 people, a volume of 12,500 m^3, and an average reverberation time of 1.75 s when occupied (Lifschitz, 1925). It is particularly suitable for transparent presentations of polyphonic structures, as found in the neo-baroque composition style of that era. For that matter, the "Neue Gewandhaus" (1886) in Leipzig, famous for its room acoustics, had an average reverberation time of only 1.55 s (Kuhl, 1959).

The fact that the actual reverberation time of 2 s of the Berlin Philharmonie (Fig. 6.2) and the Nürnberg Meistersingerhalle for a long time was considered as optimal, suggest a comparison not only with the Vienna Musikvereinssaal but also with the majority of concert halls built in recent years throughout the world. Extensive comparisons are found in Barron (1993) and Beranek (1996). Table 6.3 assembles (again five) examples of concert hall development in Germany in the 1980s and 1990s. While the Schaupielhaus in Berlin and the Bamberg Hall (Fig. 6.3) – with a volume similar to the Musikvereinssaal- today are counted among the smaller new constructions, the Munich Philharmonie, with its size, approaches the limit of the framework acceptable for symphony concerts. Characteristic for the Leipzig Gewandhaus (Fig. 6.4) is a reverberation time which remains uniform for low frequencies while in the old opera house in Frankfurt the reverberation time clearly diminishes.

Table 6.3 Concert halls after 1980

Concert hall	Year of construction	Vol. (m^3)	No of seats	Rev. time (s)	
				Mid	Low fr.
1. Gewandhaus Leipzig	1981	21,560	1,905	2.0	2.0
2. Old Opera Frankfurt	1981	22,500	2,353	1.95	1.55
3. Schauspielhaus Berlin	1984	15,000	1,674	2.0	2.2[a]
4. Philharmonie Munich	1985	30,000	2,400	2.1	2.2
5. Sinfonie a.d. Regnitz Bamberg	1993	14,300	1,420	1.9	2.2

Source: 1. Fasold et al. (1981, 1982); 2. Brückmann (1984); 3. Fasold et al. (1986, 1991); 4. Müller and Opitz (1986); 5. Opitz (1993)

[a]Now 2.6

Fig. 6.2 New Philharmonie in Berlin. In the background left, one of the three small tower stages for spatial sound effects can be recognized (Foto Friedrich, Berlin)

As is the case for the remaining three examples of this table, the majority of other halls of all periods, recognized as having good acoustical properties, show a frequency dependence of the reverberation time in the lower registers with a more or less clear rise, which supports the aural reverberation sensation. As the two curves for occupied halls in Fig. 6.5 make clear, this lengthening of the reverberation time becomes noticeable below a frequency of approximately 200 Hz. For a frequency of 125 Hz, the curves of these two halls even reach a value of 2.4 s. Such an increase in reverberation time is very advantageous for orchestra concerts for energetic reasons, since most low instruments have their strongest sound contributions above 200 Hz and radiate only relatively weakly at lower frequencies. It is therefore advantageous when the fundamental registers of bass voices are strengthened by the hall.

At high frequencies the reverberation curve in all halls drops. This is associated with dissipation losses and surface roughness of walls and ceilings. In general, the reverberation time at 2,000 Hz is by 5–10%, and at 4,000 Hz around 15% less than at 1,000 Hz. There are, however, current tendencies to use technical means, such as covering wood surfaces without pores, in order to maintain the 2,000 Hz reverberation time constant at the same value as for the midfrequencies (Munich Philharmonie and Bamberg). The value at 4,000 Hz lies by 10–15% lower (Müller and Opitz, 1986; Opitz, 1993). This rise of the region around 2,000 Hz gives a particular brilliance to the tonal picture and also supports the singers formant.

Fig. 6.3 Joeph-Keilberth-Hall of the "Sinfonie an der Regnitz" in Bamberg (Stadthallen GmbH, Bamberg) (See Color Plate 6 following p. 178)

The two reverberation curves for halls in Vienna and Berlin show a difference in detail which can be considered typical for these two kinds of concert halls. While the reverberation time in Vienna decreases relatively steadily for higher frequencies, the curve for the Philharmonie is nearly horizontal in the midregion. In fact, the open tonal effect is enhanced by a slight rise in the region around 1,000 Hz, i.e., in the frequency region of the a(ah)-formant, a circumstance which can be considered advantageous, particularly in view of the reflection conditions caused by the shape of the hall, since it enhances the clarity of the tone picture. In contrast, the reverberation behavior of the Musikvereinssaal leads to a more soft and rounded sound.

Figure 6.5 at the same time visualizes the difference which can arise between acoustical conditions in an occupied and an empty hall. If the seating from its inception is not arranged in such a way that it causes similar sound absorption properties as the audience, the absorption in the hall is increased significantly by the audience. Thus, in the Musikvereinssaal, as a result of the un-upholstered wooden chairs, the empty hall has a reverberation time of more than 3.5 s at frequencies near 500 Hz, so that for rehearsals or recordings a far greater reverberation is present than in a concert, therefore, during recording sessions, additional absorption surfaces are brought into the hall (e.g., wool blankets hung over the chairs). In contrast, in the Berlin Philharmonie, the seats are strongly absorbing. The audience and

Fig. 6.4 Large hall of the Gewandhaus in Leipzig (Gert Mothes, Leipzig) (See Color Plate 7 following p. 178)

orchestra reduce the reverberation only by about 0.4 s. In a studio setting, i.e., with orchestra but without audience, the reverberation curve follows a path approximately midway between the two drawn curves (Cremer, 1964). Inasmuch as the audience is responsible for the largest portion of the absorption in an occupied hall, the upholstery in an occupied hall should, if possible, not serve as an additional absorption surface, since otherwise the desired reverberation time of 2 s can hardly be realized. This is accomplished by having the cloth cover of the back only reach to shoulder height, with the additional wood surface behind the head serving as a reflecting surface.

The problem of favorable reverberation time, however, should not only be considered in the context of hall size. The question arises to what extent a depen-

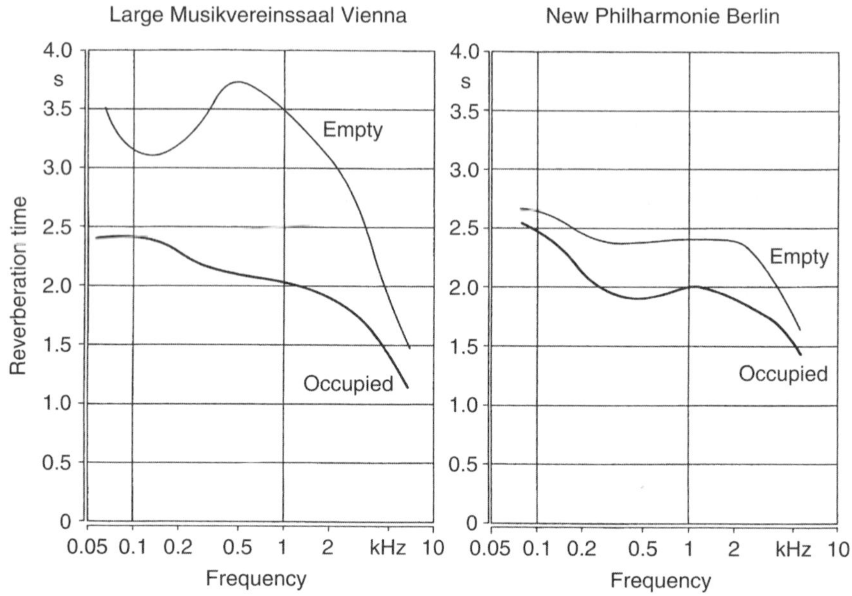

Fig. 6.5 Reverberation curves of two concert halls (after Beranek, 1962, and Cremer, 1964)

dence exists on the character of the music to be preformed in the hall. Should the tonal picture tend more toward a glass-clear transparency or an immersing overall sound, would, within certain limits, likely depend on the style of the work, since composers will be significantly influenced in their sense of tone by the acoustical environment in which they have preformed or heard their pieces (Blaukopf, 1960). Thus, composition styles are influenced by room acoustical conditions, as they are similarly influenced by the technical development of instruments. The performance which is to reproduce the tonal conception of the composer must therefore also be oriented toward the acoustical givens of the "original." It is of course self-evident that a performance under the same conditions as during the time of creation is impossible in most cases, and in most circumstances not even meaningful. It is likely wrong to assume that composers of the great master works considered the acoustical conditions under which they experienced their performances always as optimal.

Reference was already made to the connection between the Vienna Musikvereinssaal and the great masters of the romantic era. Also the other halls which were built at that time have essentially similar characteristics which enhanced the romantic sound ideal; in this context the destroyed hall of the old Philharmonic in Berlin should be mentioned, which, in its acoustics, corresponded to the type of the Vienna Musikvereinssaal. In contrast, the hall of the old Gewandhaus in Leipzig (Fig. 6.6), where premier performances of R. Wagner's Prelude to the Meistersinger and J. Brahms' violin concerto took place, is a relatively small hall. This concert hall was built in 1780 for an audience of approximately 400, and after the addition of the upper balconies provided seating for 570. The size of the hall is 2,100 m^3, the

Fig. 6.6 The concert hall of the old Gewandhaus in Leipzig after remodeling in 1842 (Bärenreiter - Verlag Kassel (MGG))

Table 6.4

Concert hall	Volume (m^3)	Reverberation time (s)	
		Mid freq	Low freq
Schloß Eisenstadt	6,800	1.7	2.8
Schloß Esterháza	1,530	1.2	2.3
Hanover Square Rooms London	1,875	0.95	1.6
King's Theatre London	4,550	1.55	2.4
Festsaal der Alten Universität Vienna	5,250	1.7	2.6

reverberation time for occupied circumstances was calculated as about 1.2 s using old drawings (Bagenal and Wood, 1931). Under such circumstances the music naturally exhibited an essentially greater presence than in the halls of much greater volume dating from the second half of the nineteenth century. The works of the classic composers therefore obtained a far more transparent tonal shape.

This also applies – though in different measure – to the concert halls for which J. Haydn composed his symphonies, and in which he preformed them himself. Fortunately two of these halls, i.e., the ones in Eisenstadt and Esterháza (today "Fertöd") have been preserved in practically original condition so that reverberation measurements could be conducted (Meyer, 1978a). Table 6.4 contains the results, which have been recalculated for occupied conditions, along with recalculated reverberation times of two London halls based on literature data (Elkin, 1955; Robbins Landon, 1976), and the values for the festival hall of the old university in Vienna, in which in 1808 the now legendary performance of Haydn's "Creation" took place in his presence.

All of these halls point, or pointed toward a significant rise in reverberation time at low frequencies. The halls in Eisenstadt and Vienna, with their reverberation of 1.7 s at midfrequencies, correspond to our contemporary conception of the reverberation time of early classical orchestral music. The former King's Theater in London comes quite close to this value. Within this context it should be noted that Leonhard Bernstein considered the Beethoven Hall in Bonn, the reverberation time of which also is 1.7 s, as particularly suitable for performances of Haydn symphonies, while it is considered somewhat too dry for romantic music. The hall in Esterháza and the only, slightly larger concert hall in the Hannover Square Rooms in London, were not only significantly smaller in dimensions than the other ones listed in Table 6.4, but their acoustical conditions also were characterized by significantly shorter reverberation times. In today's terms, these halls, as for example the Eroicasaal in the Palais Lobkowitz in Vienna (see page 252), correspond to relatively large chamber music studios.

And in fact, concerts for orchestral works as well as chamber compositions were performed in these halls. The symphonies of J. Haydn, written specifically for these halls of varying reverberation relationships, are especially appreciated under these circumstances. Thus the symphonies written for Esterháza contain rhythmic finestructures and dynamic jumps from an orchestral *forte* to a *piano* of individual voices, which would have been totally lost in the reverberation of the London halls, and also in Eisenstadt. In contrast, the symphonies written for the king's theater, always have a fermata associated with *forte* breaks, or a following pause. The composition style in these symphonies also frequently utilizes the tonal melting effect of the reverberation (Meyer, 1978a).

The historical development of concert halls, which in this context could only be discussed in a few examples, leads to the conclusion that only a slightly longer reverberation time is appropriate for the works of the classical period in comparison to romantic music. This is especially true since during the classical period orchestral concerts frequently occurred in theaters, which in comparison to the concert halls of that time, had a noticeably shorter reverberation time. Investigations concerning optimal reverberation times have led to similar results. These results will be further considered in the section on studios (Kuhl, 1954a, b).

In two recent concert halls, new approaches have been pursued which, within certain limits are designed to adapt the reverberation time to the compositions to be performed. The halls in Dallas (1989) and Birmingham (1991) not only utilize moveable absorption curtains (not too uncommon), which make it possible to reduce the reverberation time, but also add echo chambers to increase the reverberation time. They are located all around the hall above the highest balcony as well as in the front of the hall at a lower elevation. They can be closed with concrete doors (Forsyth, 1987; Graham, 1992). These, in connection with height-adjustable reflectors weighing more than forty tons, located above the stage, are designed to create acoustical conditions not only appropriate for the standard symphonic works but especially for orchestral works with large choirs written by composers from Berlioz to Mahler which require halls with a cathedral-like character.

A reverberation time of less than 1.7 s in a large hall meets with decided universal rejection, since then even intensity relations become very unfavorable. When, nevertheless concerts under unacceptable acoustical conditions are to be performed in such halls, the utilization of electroacoustic installations becomes an option. This applies especially to halls which are so large that the sound energy of an orchestra or a soloist respectively is no longer sufficient to bring adequate intensity to the distant seats.

If the only concern is to increase reverberation time, then the microphones have to pick up sound contributions of the diffuse sound field, which are then radiated over speakers distributed over the hall. Because of feedback possibilities, limits are set for the degree of amplification. Thus, already during the years 1960–1970, such installations were developed in steps for the Royal Festival Hall in London (Parkin and Morgan, 1965 and 1970). This procedure utilizes a very large number of very narrow band channels and thus a clear limit of reverberation increase can be recognized near 800 Hz. Furthermore, very narrow limits apply to the diffusivity of this artificial reverberation, since each speaker handles only a very narrow tone region.

These difficulties are circumvented by a procedure for which a number of microphones feed each speaker with a broad band signal. Thus, the reverberation time can be uniformly extended, spatially, as well as in respect to frequency, i.e., there is no dependence on the location of the sound source (Kleis, 1979). An ealier example of this procedure is represented by the installation in the Royal Concert Hall in Stockholm, it increases the reverberation time in an occupied room from 1.6 to 2.3 s (Dahlstedt, 1974).

Naturally, it is also possible to limit the individual channels of the electroacoustic instrumentation to an average width without making the band too narrow. Thus a sufficiently large number of channels with the same frequency dependence can result in uniform sound radiation. Such a system, which occupies an intermediate position between the two systems described earlier, is installed in the International Congress Center "ICC Berlin" (Keller and Widmann, 1979).

Finally it is also possible to amplify the direct sound contributions in the framework of an electroacoustic installation. By appropriate choices of delay times for individual speakers, it is possible to achieve the impression that the sound originates with the original source for practically all locations in the audience. This naturally requires a substantial effort, especially for a wide stage. However this can be technically accomplished. For example such an installation is installed in the Palace of Culture in Prague (Ahnert et al., 1986).

6.1.3 Sound Field and Hall Shape

The intensity relationship between the direct sound and the statistical sound field, under given room acoustical conditions, does not only depend on the distance between the listener and the performers, but also on the directional characteristics of the instruments. The more sharply the sound radiation is clustered, the deeper the

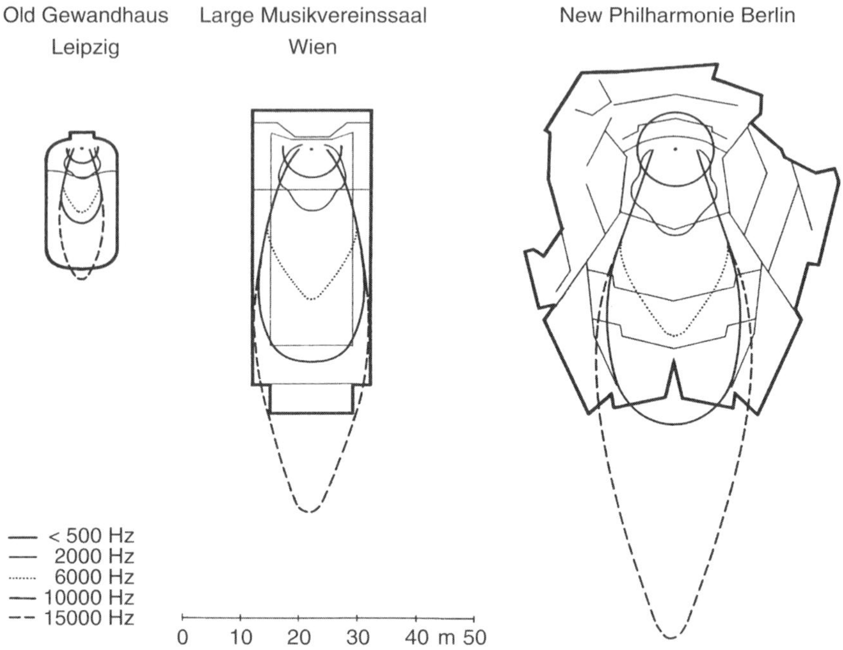

Fig. 6.7 Diffuse-field distance of a trumpet for different frequencies in three concert halls

region of predominantly direct sound reaches into the room in the preferred direction. This effect is all the more important when the diffuse-field distance for omnidirectional sources does not extend significantly beyond the podium. Thus the Vienna Musikvereinssaal has a value of 4.75 m in the midfrequency region. For the significantly larger hall of the Berlin Philharmonie a value of 6.5 m is found, and for the relatively reverberation-poor Royal Festival Hall, a value of 7.0 m. In the old Gewandhaus with its much smaller dimensions, this value amounted to only 2.4 m. Since the reverberation time at very high frequencies is shorter than in the mid-region, this diffuse-field distance is somewhat larger for these tonal contribution; for the highest components it can grow to up to twice this value.

How far this region, for which the direct sound dominates, can be extended for a sound source with pronounced directional characteristics is shown in Fig. 6.7 for several frequency regions in the three concert halls mentioned earlier. This summarizes the effect of the directional characteristics for the reverberation time which decreases for the highest frequencies. Already at 6,300 Hz the trumpet possesses a diffuse-field distance of approximately 27 m in the direction of its axis in the Vienna hall, at 10,000 Hz it increases to 38 m and at 15,000 Hz it reaches a value of 64 m (only theoretical under existing room dimensions). The width of the region enclosed by the diffuse-field distance curve increases only slightly: from 15 m at 6,300 Hz it merely increases to 20 m at 2,000 Hz. As can be recognized from the figure, the Vienna Musikvereinssaal appears to be nearly tailored for the precision of trumpet passages rich in overtones. In contrast, the edge seats in the much smaller old Gewandhaussaal are no longer so advantaged for the brilliance of the trumpet.

Naturally a concert hall design which distributes the audience all around the orchestra leads to a very different tonal effect for trumpets in the hall. The example of the Berlin Philharmonie demonstrates this situation in an extreme manner. However, even for halls with a volume similar to that of the Vienna Musikvereinssaal, whose width exceeds a measure of 20 m significantly, must count on lack of clarity for trumpets-staccato passages.

The different tonal effects created in the audience in front of, to the side of, or behind the orchestra are naturally not limited to trumpets but are more or less pronounced for the other instrument groups. The major considerations are the front to back relationships and the front to side relationship of the directional characteristics. When considering that, for example for the oboe, the front to back relationship for the strongest tone contributions (around 1,000 Hz) amounts to approximately 16 dB, it becomes very clear how difficult it is, to achieve a uniform balance between the individual instrument groups in halls like the Berlin Philharmonie or the Gewandhaus in Liepzig (opened in 1981) (Fasold et al., 1981). It is thus not surprising that for new concert halls the tendency again prevails to move the orchestra from a central position to the proximity of a head-wall, where only choir seats remain behind the orchestra as already utilized in the Gewandhaus.

In addition to the diffuse-field distance, the tonal effect of the direct sound can be supported by reflections with sufficiently short delay i.e., they arrive from not too distant surfaces. A requirement for the effectiveness of such reflection surfaces is that the sound arriving there from the instruments has a sufficiently high level. This condition is particularly met when the relative reflection surface lies within the angular region of principal radiation of the directional characteristic. In the same sense, this naturally also applies for the effectiveness of absorbing surfaces near the orchestra (Meyer, 1976, 1977).

In Fig. 6.8 the principal radiation regions for the group of the first violins and cellos are entered as examples into the schematically represented length-wise section of a hall. The region of the podium in which the relevant instruments are situated is represented as a black bar where a frontal arrangement for the celli in front of the conductor is assumed. The sound rays which delineate the principal radiation region (0–3 dB) emanate from the surface of the instrument. Those parts of the ceiling, rear wall and floor, which lie in the angular region of the strong sound radiation are indicated outside of the room boundaries by bars with specification of relevant frequency regions. In the partial picture for the first violins, the angular region for frequency contributions around 500 Hz are indicated by arrows for clarity. It is noteworthy that this reflection region, important for the brilliance of the violins, is located at the ceiling in front of the podium. In contrast, the reflection surfaces essential for the brilliance of celli are located above the podium.

The multiple divisions of the reflections surfaces for the individual instrument groups show that it is hardly possible to find a comprehensive representation for the entire orchestra, without losing information for the sake of simplification. Figure 6.9

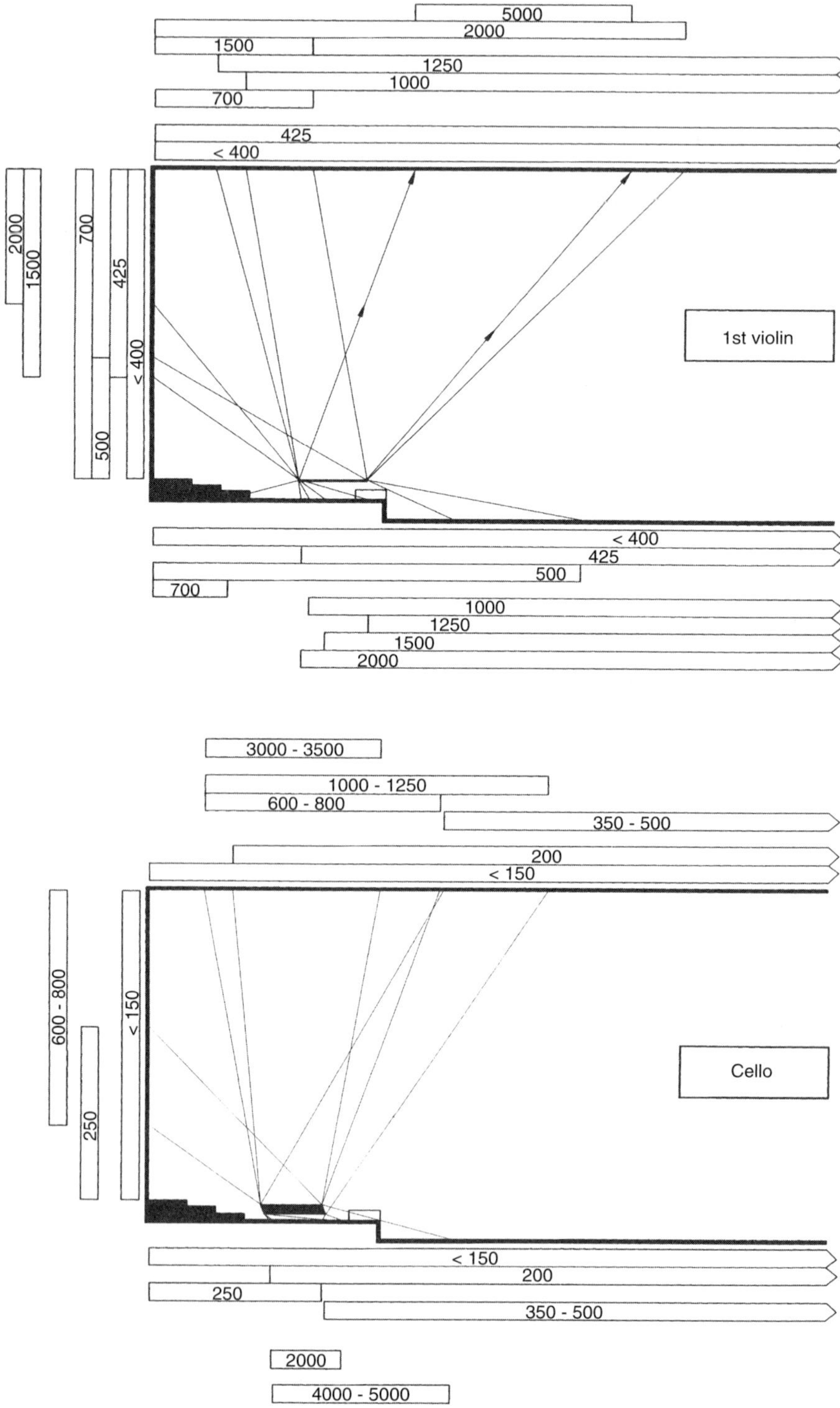

Fig. 6.8 Reflection surfaces on ceiling, rear wall and floor of a concert hall, which lie within the principal radiation regions (0–3 dB) of string instruments. The numbers are frequencies in Hz

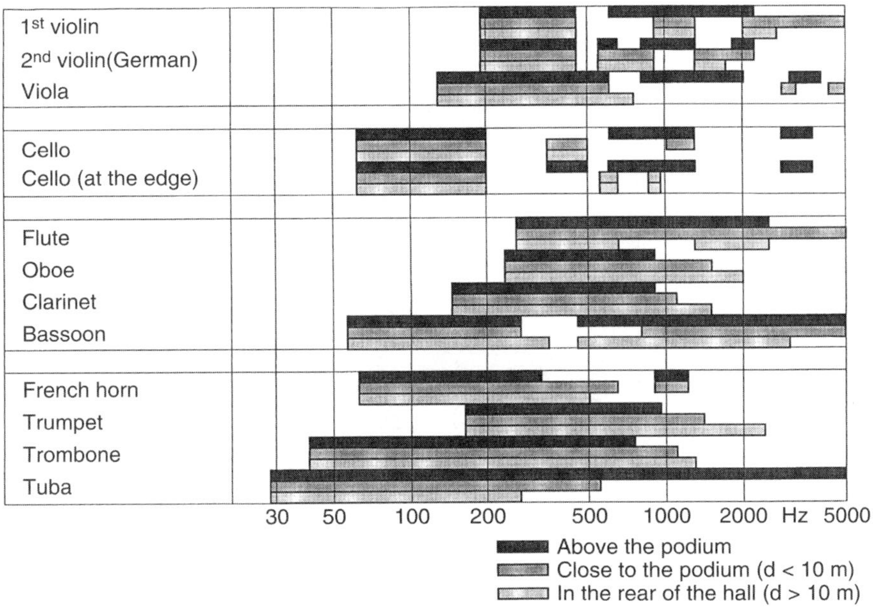

Fig. 6.9 Frequency regions of preferred ceiling reflections for orchestra instruments

shows a schematic representation of the frequency regions for which the ceiling lies either partially or entirely within the principal radiation angle. Here the ceiling is divided into three parts: "above the podium," "in front of the podium" (to a distance of approximately 10 m), "in the distant hall" (a distance greater than 10 m). The spatial relationships correspond to Fig. 6.8; a lower ceiling would bring about changes which would correspond to a shift of the boundary between close and distance audio region in the direction toward the podium. A ceiling rising into the audience would correspond to a shift of this boundary further into the hall.

It is recognized that the ceiling above the podium and also in front of the hall can reflect the sound in relatively broad bands for all instruments and thus it contributes significantly to the loudness impression and also to the brilliance of the tone color. Further into the hall, the high frequencies of most instruments are much weaker as they hit the ceiling, so that here above all the reflections of the lower tone contributions play a role.

In similar fashion Fig. 6.10 visualizes the reflection effect of the wall behind the orchestra, where a distinction is made between the lower region (up to height of 5 m) and the region above that. As can be seen, at low midfrequencies almost all instrument groups utilize the entire height of the wall. Only the lowest strip is not strongly affected for strings because of shadowing by other musicians. At higher frequencies the rear wall is of little significance for winds while the reflections from the upper regions of the wall certainly play a role for the strings.

The side walls lie within the principal radiation region for low and midfrequencies for strings as well as most winds (see also Figure 138 and 150), thus, a priori,

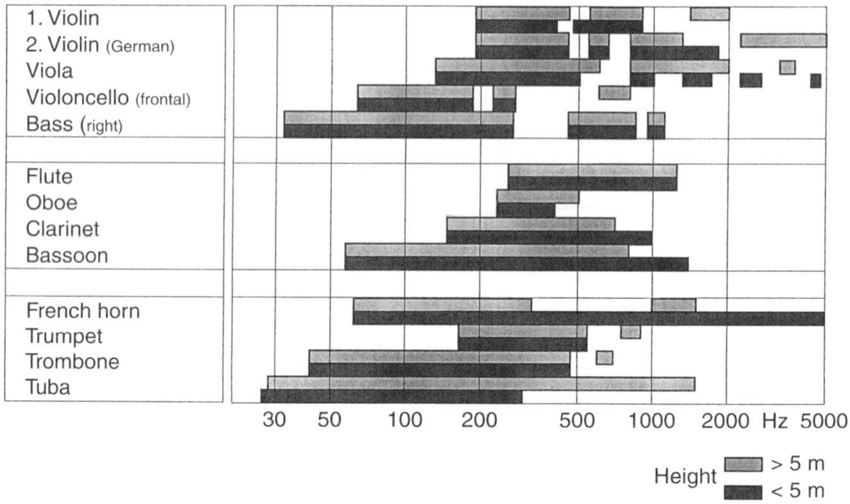

Fig. 6.10 Frequency regions of preferred rear wall reflections for orchestra instruments

they have more influence on the tonal richness than on the brilliance. The fact that the universal validity of this statement has certain limits can be noted for some halls in which large unstructured side walls have such a flat surface that also very high frequencies are reflected with optimal effectiveness. The geometric side wall reflection receives a prominence through the high frequency contributions, which leads to a cutting impression of the violin sound, and the spaciousness of the overall sound obtains a grating component. When the side walls are structured – for example by folding, or the placement of figures – the higher frequencies are reflected diffusely, and this effect becomes inaudible.

The basic hall floor plan and thus the directions of the side walls in the region of the podium and the audience play a decisive role for the acoustical room effects. Thus diverging side walls carry the danger that individual wind groups stand out too strongly in certain regions of the hall because of wall reflections, when the shading by the front rows of performers, and thus that equalizing effect is diminished. Normally an orchestra occupies a rectangular area, so that for strongly divergent room walls, open flanks are created.

The time sequence and the directions of incidence of first reflection are naturally determined by the distances of the reflection surfaces at the ceiling and the walls from the listener. It is therefore notable that three of the five halls identified as optimal in Table 6.1 are rectangular whose height is greater than half the width so that the first reflection in nearly all seats comes from the side and not from above. In Table 6.5, values for the height, the total width of the hall, as well as that width, which is important for first reflections, between boxes or galleries are assembled for several halls with elongated floor plans.

Table 6.5 Dimensions of several concert halls

Concert hall	Height (m)	Overall width (m)	Inner width (m)
Musikvereinssall Vienna	18.5	20	14
Concertgebouw Amsterdam	17.5	28	29.5
Symphony Hall Boston	21	23	17
Meistersingerhall Nürnberg	14	38	27
Sinfonie a. D. Regnitz Bamberg	14	33	22

Among concert halls of the 1960s and 1970s, the Meistersinger Hall approximates conditions for especially good halls relatively closely. For a large number of seats, the reflection from the ceiling arrives after a wall reflection. Many halls of that same period are, however, lower and longer, so that the spatial impression is not created and the width at the top is missing, rendering it less than optimal for Romantic music.

In the Joseph-Keilberth Hall in Bamberg, the side walls form an angle of 10° with respect to the central axis, diverging toward the back, consequently the width measurements are given for the width of the podium where the critical reflection surfaces are located. It is a successful example for the tendency in concert hall construction to achieve sufficiently early side reflections for all seats in the hall. In very wide halls, this can also be achieved by horizontal stepping of the seating regions; a characteristic example for such a design is the Orange County Performing Art Center in Costa Mesa California (Forsyth, 1987).

When calculating the admissible width for a hall, for a condition that the first wall reflection at a location of at least 15 m from the podium should not arrive later than 30 ms after the direct sound, one arrives at a value of 20 m. For larger distances, this condition is also met. The stronger direct sound closer to the orchestra permits a slightly longer delay time. This result is particularly interesting, because the width of 20 m corresponds to the value which is optimal for a sharp trumpet staccato, based on the directional characteristic (see Fig. 6.7).

The shape of the hall, including wall and ceiling structures, crucial for reflection directions, has a decided influence on the level of the sound energy arriving from the side with a delay of 10–80 ms, and thus on the spatial characteristics of the tonal impression. In this context, early sidewall reflections are important, as best realized in rectangular halls or halls with only slightly diverging walls, or by reflections from angular reflection surfaces or correspondingly structured wall features formed by the walls and the underside of galleries (Kuhl, 1978; Wilkens, 1975). A characteristic example is the Gewandhaus in Leipzig, where the wall elements direct the sound partly downward toward the audience and partly up toward the ceiling (see Fig. 6.4). In this way, depending on seating groups, a spatial impression measure of +3.1 to +5.5 dB is achieved (Fasold et al., 1981). For an appropriately shaped ceiling, the radiated sound can be directed to reach the audience from the side after a second reflection (Schroeder, 1979).

This viewpoint, i.e., to create a possibly high measure of spaciousness, can naturally also be taken into account with an electroacoustic reverberation

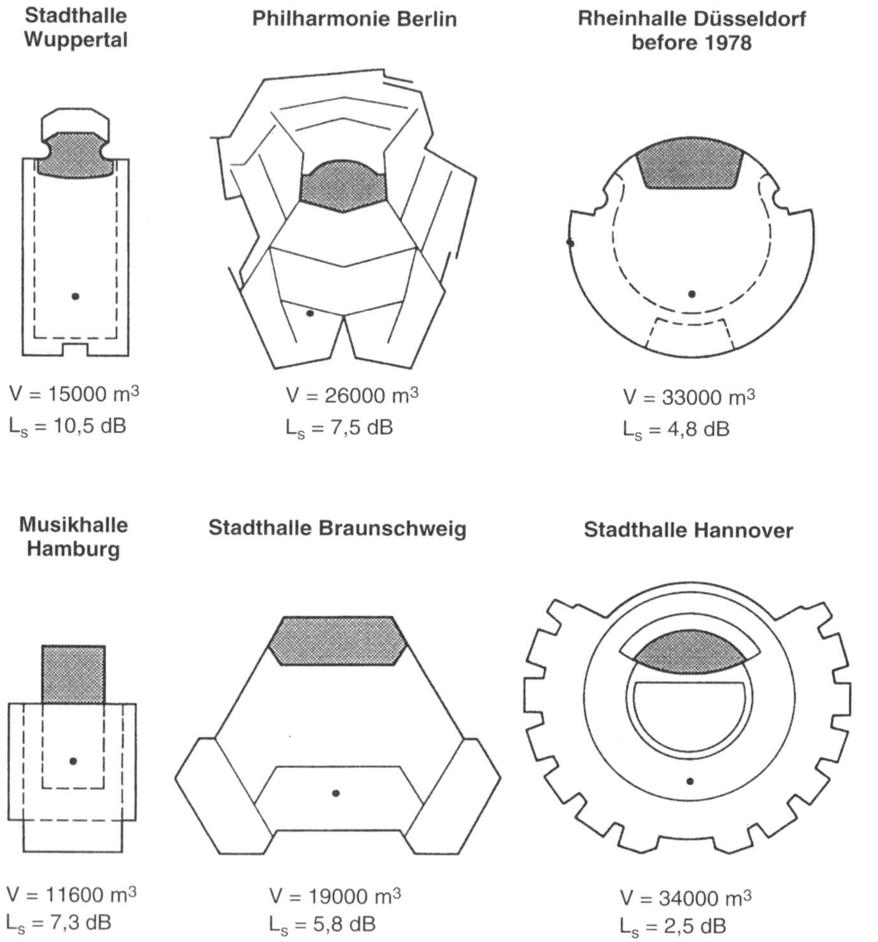

Fig. 6.11 Floor plans of six halls with indication of volume and lateral-sound levels (After Wilkens, 1975 and Kuhl, vor 1978)

installation. However, there is a danger that instruments whose sounds normally reach the audience predominantly in a vertical plane, become estranged by the tonal broadening. This applies particularly to the concert grand piano which ordinarily should distinguish itself from the orchestra sound by spatial precision.

In Fig. 6.11, the floor plans of six concert halls along with hall volumes are assembled. In addition, in each case, a value for the lateral sound level is given. This value is given for a seat in the rear third of the hall with appropriate distance from the wall (Kuhl, 1978; Wilkens, 1975). The advantage of rectangular halls in comparison to long or wide shapes in particularly noticeable. However, one also recognizes that by meaningful structuring of reflection surfaces in the amphitheater-like hall of the Berlin Philharmonie, a high side sound level was achieved for the relevant audience region, while the Braunschweig Stadthalle with its unstructured and angled walls, does not permit a spatial sensation. Thus, it is not surprising that the completely redesigned concert hall in the Lincoln Center in New York was reshaped into a rectangular hall with side walls structured by the presence of gallery boxes (Kuttruff, 1978). This is all the more noteworthy since fan shaped concert hall designs are generally approached skeptically and consideration of strongly structured walls are imperative.

An additional problem for halls with large opening angles or very large width is that a more or less large number of seats on the side are outside the region of optimal direct sound exposure. This is particularly relevant for vocal soloists standing in front of the orchestra: a tonal change is already noticeable outside of a cone of $\pm 40^\circ$, and outside a cone of 80° the tonal impression is unsatisfactory. One should not be deceived by a smaller opening angle of a fan-shaped hall. Such a hall opens from a broad basis, the directional characteristic of a singer or an instrument, however, from a point.

Inasmuch as a high ceiling supports the Romantic tone concept, whereas the transparency of classical music is enhanced by a less delayed reflection, i.e., rooms with lower ceilings, which furthermore do not spread, one option is to hang reflectors above the orchestra and to adjust their height to conform to the program. As already mentioned, this however, requires a change of at least 2 m so that the ear can notice the difference in spatial impression. However, it should not be overlooked that, while lowering the reflectors, the regional effectiveness at lower frequencies is broadened (see Sect 5.1.3), so that the tonal color is changed.

This subjective impression of clarity and spaciousness of the tonal picture can be supported by visual impressions. This is less the difference between very austere halls, and rooms, which radiate a more festive character, though the latter possibly stimulate the emotional expectation of the listener more strongly, rather, the distribution of light and color in the hall play a significant role. Illuminated regions stimulate visual attention and raise the concentration of the listener in that direction. In a somewhat darkened room, it is easier to concentrate on an illuminated podium than in a bright hall, particularly when large (not totally immovable) audience regions are in the field of view.

In this context, a comparison of the Leipzig Gewandhaus and the Bamberg Concert Hall are of interest. In Leipzig (Fig. 6.4), the bright podium with its light edge strip, draws the eye. The dark walls retreat in the optical impression: this can strengthen the subjectively sensed clarity, particularly for the distant listener. In Bamberg (Fig. 6.3), the wood framing of the stage extends to the audience region, the white walls are especially noticeable, thus these reflective surfaces play a particular role in the optical impression, in contrast, the blue ceiling is less noticeable. This also corresponds to the directional sensitivity of the ear, which is more sensitive toward lateral reflections than to those from above (Opitz, 1993). This color and light relationship, which is attuned to acoustic processes in the hall, enhances the sensation of spaciousness.

6.1.4 Acoustic Conditions on the Stage

The reflections from the regions close to the orchestra play an important role for aural interaction between musicians. The fact that in some halls musicians hear themselves as loud, and other musicians as insufficiently loud, or not at all, points to the fact that the direct sound is insufficient for mutual communication and must be supported energetically by reflections. When one considers that the accuracy of a rhythmic group performance, at least for chamber ensembles, permits deviations from exact synchronicity of no more than 35 ms (Rasch, 1979), one can derive an upper limit for useful reflections.

The question of time delays for reflections, which aid or hinder mutual listening between musicians, is considered in Fig. 6.12 where results of several authors are assembled: the horizontal axis gives the distance between musicians, and below that, the associated traveling time of the direct sound is noted. The vertical axis gives the ceiling height above the podium. Since the ears of the musicians are located at about a 1 m level, the distance of the reflecting surface should only be calculated from that height, which is included in the measurement of the wall distance. Furthermore, in this coordinate system, the delay times of reflections are indicated in broken lines. The positive or negative effect of reflections on musicians

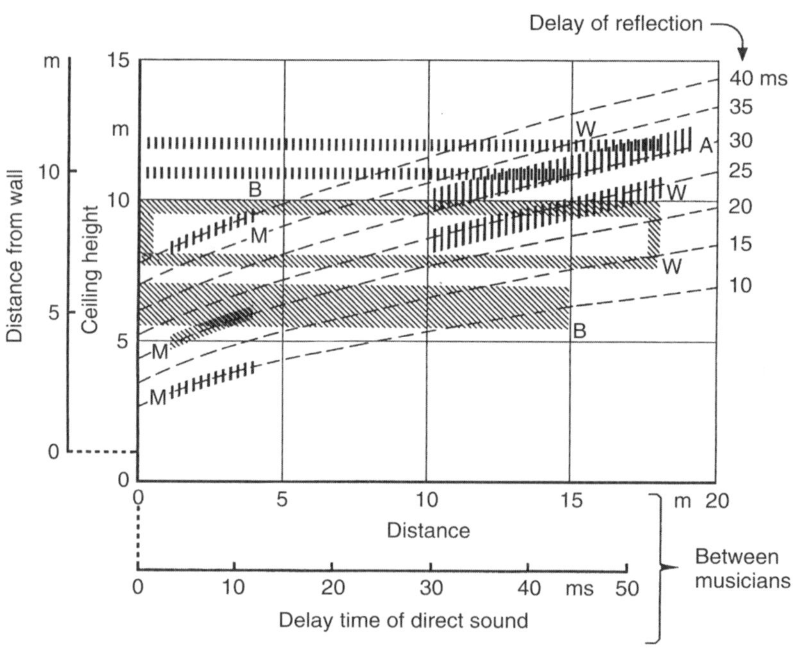

Fig. 6.12 Influence of hard reflection surfaces on mutual hearing (after Allen, 1980; Barron, 1978; Marshall and Meyer, 1978; and Winkler, 1979). *Angled shading*: favorable reflections; *vertical shading*: unfavorable reflections

is indicated by the nature of the shading with references to individual authors (Meyer, 1982b).

Arrival times of approximately 10 ms are sensed as disturbing. This is connected to the fact that the level of reflections becomes too strong and performers can no longer hear their own instruments clearly. Reflections are judged supportive, when the reflecting surfaces are so close to the orchestra, that the delay relative to the direct sound assumes values between 17 and 35 ms (Marshall et al., 1978). This corresponds to a detour of the reflected sound of about 6–12 m. For reflectors above the orchestra, a height of about 8 m is considered optimal. Heights of 12 m are considered as still marginally useful (Winkler, 1979).

For musicians seated only a few meters apart, even ceilings of 8 m heights can bring about disturbing reflections with excessive delay times, which are sensed as separate sounds from the room. These disturbing reflections, however, can be avoided when vertically upward radiated sounds are prevented from being reflected along the shortest path to the neighborhood sound source by folding the reflecting surface. For a very large distance between musicians, the danger of excessively delayed reflections for greater ceiling heights is not a problem, however, for energy reasons the limit of 10 m should not be exceeded when reflections from above are intended to aid mutual hearing.

The lower limit for favorable ceiling height, for small distances between musicians, is between 4 and 5 m. Lesser values lead to reflections which are too loud. This can be counteracted by partial absorption or by folding of the ceiling which affects a reflection of vertically upward radiated sound to distant parts of the stage. Such low ceiling heights, however, are only to be expected in small halls.

Considerations for ceiling heights can also be related to favorable distances for reflecting walls. For chamber music ensembles, distances from 3 to 6 m from the wall are advantageous, for orchestras on the other hand, a surface extension appears to be meaningful, so that the walls enclose the orchestra directly, since otherwise the wall distances become too large for those musicians sitting further in. Excessively large orchestra stages with surrounding reflecting surfaces at distances which are too large lead to situations for which the desks of the front strings and the location of the conductor are acoustically unsatisfactory, because the rear string desks are too weak and many wind groups can only be heard imprecisely. In contrast to that is the undesirable situation for the rear string desks, when they have no rear wall, but instead are located in front of the region of the stage enclosed by the side walls. The minimum required height for the surrounding wall surfaces is indicated by various authors between 1.8 and 3 m, where in the latter case the lower frequency contributions are also strongly reflected, which is of little significance in the rhythmic context. Lateral reflections are also supported by appropriately angled surfaces of gallery enclosures or by angular double reflectors formed by the bottom of the galleries and the wall below them.

The sum of the reflections arriving at the musician should result in a sound level relationship for which the extraneous sound at the ear of the performer should be from 15 dB below to 5 dB above the sound level of the player's instrument; within these limits the performance of the musician is not unduly influenced

(Naylor, 1987). However, independent of that, it should not be overlooked that especially in large halls, strong reflections returning to the orchestra bring the danger that the musicians overestimate the loudness impression in the audience, and consequently do not play a *forte* as strongly as the room acoustical conditions demand, keeping in mind the tonal impression of the listener.

For interactive listening of singers in a choir, considerations similar to those in an orchestra apply, however, the reverberation plays a stronger role than for instrumentalists (Marshall and Meyer, 1985). Reflections with a delay of about 15–35 ms are considered especially favorable. Since the reflected energy should be between −15 and 5 dB in relationship to the level of the voice at the singers ear (Ternström and Sundberg, 1983), reflections rich in energy are necessary: therefore, early reflections are better than later ones, and lateral reflections, by reason of the directional characteristics of the voice, are more effective than reflections from above or behind. An unfavorable region for first reflections appears to lie at delayed times near 40 ms, a phenomenon which has not yet been explained. In contrast, first reflections are again evaluated as better when they only arrive after 60 ms: for such long delay times it is actually better when lateral reflections arrive first.

For a solo vocalist, reverberation dominates even more strongly than for singers in an ensemble in respect to voice control and ease of singing. Early reflections (up to about 25 ms) can affect additional improvement, when coming from the direction of sight, from which also strong reverberation contributions arrive. Reflections from above and behind should only arrive after that (Marshall and Meyer, 1985). In similar fashion, instrumental soloists perceive reflections in a relatively long time range as supportive: the favorable region includes reflections from 20 to 100 or even 200 ms (Gade, 1989a).

Figure 6.13 assembles several floor plans of concert halls, which show different ways of including the orchestra in the total hall, in order to illustrate the aspects of interactive hearing. Furthermore, for these, and two additional halls (included in Fig. 6.11), dimensions are given in Table 6.6 which are important for the acoustical conditions in the area of locations of musicians: concerns are the width of the stage at the front edge, the depth of the stage, the height of the ceiling above the front edge of the stage, or the height of individual reflector hanging at that location, the angles of the side walls of the stage in relationship to the long axis of the hall, and the width of the hall at the end of the stage. See table in text

Overall it is noted from these examples that the ceiling above the stage in most cases is higher than expected from demands of musician-friendly reflections. Reference to tonal advantages of high ceilings for the audience has already been made several times; a further consideration is the tonal impression at the location of the conductor (see Sect. 6.1.5.). The following details for stage design of the individual halls need to be pointed out:

In the Musikvereinssaal in Vienna, musicians receive reflections of short delay from the rear wall and the side walls as well as from the double reflection formed by the surrounding gallery and walls.

In the Symphony Hall in Boston, the orchestra is narrowly enclosed by rear and side walls by the somewhat constricted stage space. The deeply structured ceiling in

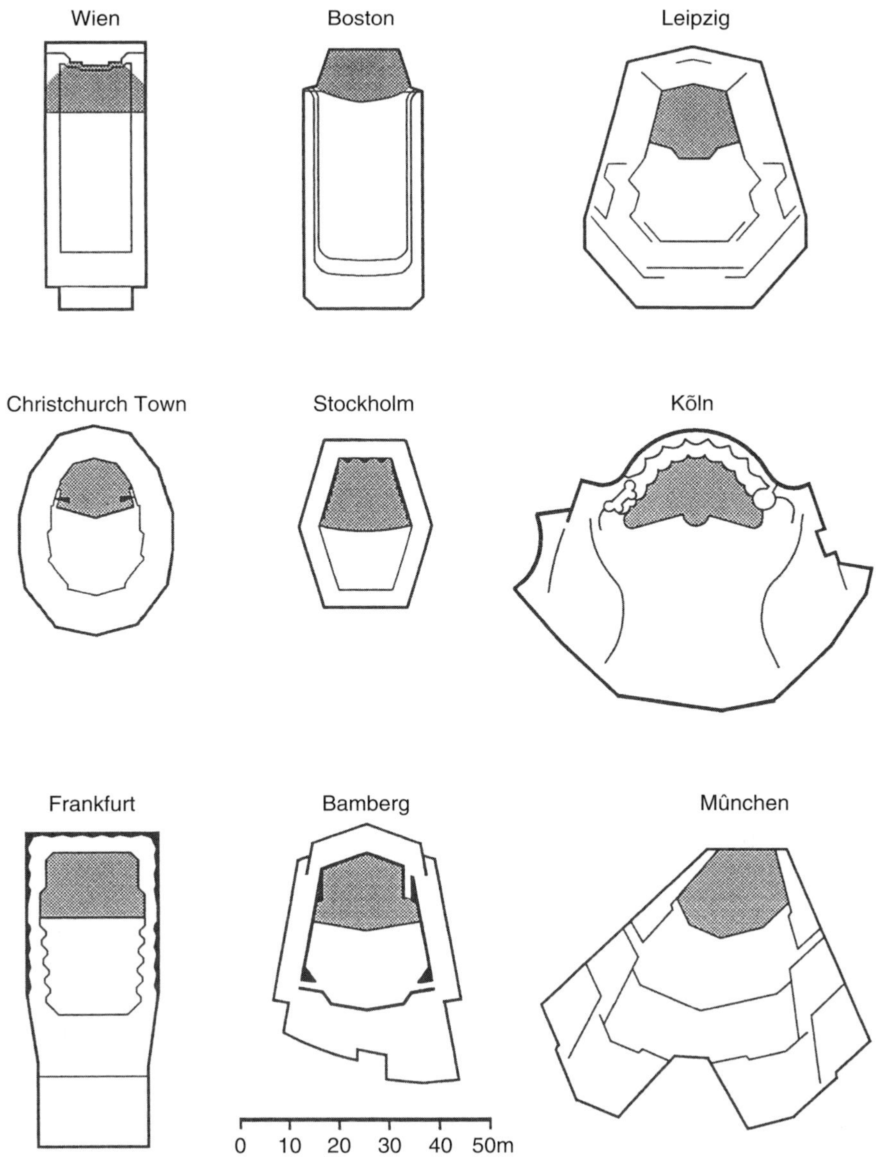

Fig. 6.13 Floor plans of several concert halls. The *shaded areas* identify the stage

the main hall presents a sequence of high frequency reflection into the orchestra which assures a time connection to the subsequent reverberation.

The Philharmonie in Berlin dispenses with a closed rear wall behind the orchestra, which is enclosed by 3 m high walls diverging slightly toward the front. Reflectors hang above the stage, which partially reflect the sound back to the stage.

Table 6.6 Orchestra stages for several concert halls

Concert hall	Year of Constr.	Width depth of podium m	m	Ceiling height above podium	Angle of sidewalls	Hall width in front of podium
Wien	1870	15	9	16	0	20
Boston	1900	17	10	13	18	22,5
Berlin	1963	17	12.5	11	4,5	45/20**
Braunschweig	1965	17	14	10	30	38
Christchurch Town	1972	15	11*	15	–	30
Stockholm	1980	17	13	14.5	17,5	27
Frankfurt	1881	21	13	14	0	25
Leipzig	1981	18	13	15	15	37/21**
München	1985	21	16	14	35	35/25**
Köln	1986	21	10.5	11.5*	20	49/31**
Bamberg	1993	20	17	13	10	31/20**

*Reflector height, **in upper/lower region See table in text.

In the Stadthalle in Braunschweig, early reflections reach the stage area only from the wall behind the orchestra and from the relatively low, folded ceiling, however not from the side walls.

In the Christchurch Town Hall, four large-area unstructured reflection walls of 3 m height are located behind the orchestra stage. Above the front stage area, the front gallery walls are angled, such that they reflect the sound toward the musician. In addition, a reflector hangs above the orchestra, which, however, directs the sound mostly toward the audience.

In the Berwald Hall in Stockholm, rear and side walls of the orchestra stage are designed to absorb low frequencies but reflect middle and high frequencies diffusely, in order to create a clear musical picture for the musicians. Furthermore, the somewhat diverging side walls are folded in such a way that the reflection surfaces are parallel to the room.

In the concert hall of the Alte Oper in Frankfurt, the orchestra is enclosed by parallel wall surfaces extending to the edge of the narrow circumferential gallery. Higher frequencies are also reflected back to the stage by the double-angle reflection surfaces formed between gallery and wall structure.

In the Gewandhaus in Leipzig, the orchestra is surrounded in the back and the sides by walls of approximately 2 m height. Above these is an additional 1 m high railing, which is angled toward the stage. Furthermore, there are additional high frequency reflections from the wall structures above the gallery. In spite of the large room height, reflectors above the musicians are deliberately avoided.

In the Munich Philharmonie, the surrounding walls of the orchestra stage diverge towards the hall. Reflections from these surfaces, therefore, support especially the musicians located relatively closely. Curved reflectors at the side walls throw the sound back to all regions of the stage, thus delay times for the middle of the stage are relatively long. Recently, additional reflectors were suspended below the reflecting ceiling.

In the Cologne Philharmonie the round rear wall of the orchestra stage is structured such that the sound is distributed largely over the entire orchestra. Since the individual surfaces of the wall, as also the railings of the balcony behind the orchestra are inclined, a relatively large amount of sound energy comes back to the orchestra. Furthermore, a nearly horizontal reflector is located above the stage.

In the Joseph-Keilberth Hall in Bamberg, the orchestra is enclosed by walls from the rear and sides, the latter are parallel to the rear of the stage and diverge only slightly toward the front, so that reflections also reach distant regions of the stage. Interactive listening is furthermore supported by the level ceiling as well as angled reflection surfaces at the side walls.

When considering reflections which are effective for mutual hearing, one needs to be aware, that above all, the frequencies in the midrange and possibly the base register are important for intonation. However, for rhythmic ensemble performance the higher components up to a region of 2,000–3,000 Hz are still significant. How strongly these individual contributions to the extraneous sound reaching the performer are perceived, does not only depend on the impinging intensity, but also on the masking of the performer's own instrument, which in turn is directionally dependent (See Sect. 1.2.6). In this context, those reflections are especially important which come vertically from above, while contributions arriving at an angle from above, especially at high frequencies, play a lesser role, particularly when they arrive from the front or the rear. Wind players hear mostly those reflections which they receive from lateral directions; for this, the side walls of the stage are relevant. For strings, these reflections are less useful. For them, reflections from above and from the ceiling in the front of the hall and the rear wall of the stage are more valuable.

The differing directional dependencies, and the threshold of interactive listening for players of different instruments, in addition to the fact that the radiated energy of wind instruments is higher than that for string instruments, is the reason why in many concert halls the balance of the tonal impression for individual player is very poor. Thus, at high dynamic levels, most brass player are themselves relatively loud and are also surrounded by the very loud sound of other brass players, while they hear the sound of the string instruments only weakly or not at all. The strings, in contrast, are located in the angular region of strong direct radiation of the winds, so that they can hear these as very loud, their own instrument very weakly, and the other instruments in their own group even more weakly. The masking by the wind sound, by reason of its spectral composition, goes so far that the violinists can only hear the higher frequency components of there own instrumental sound, and consequently force these contributions by increased bow pressure.

It is precisely this phenomenon which points to the importance of the higher frequency tone contributions for the mutual listening by string players within the individual listening groups. Therefore, ceiling surfaces above the string players should be structured or shaped so that they reflect the sound of the strings somewhat diffusely to the region of the strings. This diffusion characteristic refers predomi-

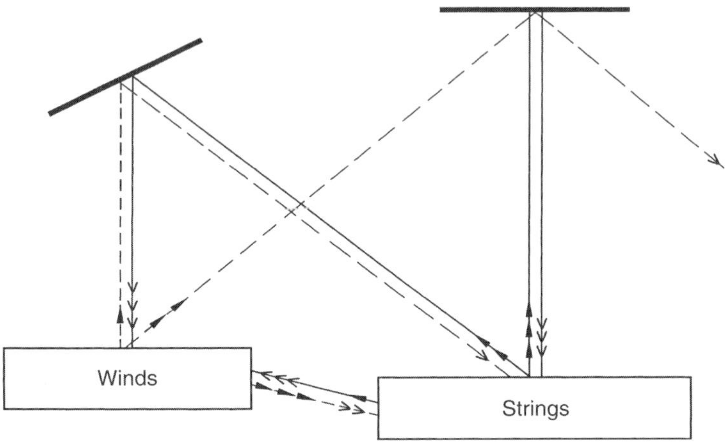

Fig. 6.14 Direct sound and ceiling reflections between winds and strings. *Solid arrows* refer to the intensity of the sound radiation, broken arrows to the sensitivity of the ear for extraneous sound

nantly to the sound contribution over the breadth of the stage, while the reflection surfaces in the direction of the depth of the stage should be practically horizontal. As indicated schematically in Fig. 6.14, these reflectors include directions of strong sound radiation of the instruments, and reflect the sound in such a way that it reaches the players from directions of highest hearing sensitivity. This improves the mutual listening of the string players. In contrast, the sound of the winds is directed toward the audience by these reflecting surfaces.

If, on the other hand, the ceiling surfaces above the wind players are oriented so that reflections from the winds are directed toward the strings, and naturally also in the opposite direction, then the reflections of the string sounds arrive at the winds in a very sensitive direction, which is advantageous to the tonal impression by the wind players. The opposite direction, however, combines a direction of weak radiation by the winds with a direction of low sensitivity for the strings, so that these reflections do not amplify the wind sound for the strings too much. A reflector arrangement, installed with this in mind, has proven very successful in the Hans-Rosbaud-Studio in Baden-Baden, for example.

For large orchestras, the higher dynamic level can lead to phenomena which can influence the clear recognition of the far distant voices, or even make it impossible: the sound not only becomes louder – above all for horns, trombones, and even for the celeste - but also "more dense," and thus masks the clear contour of other instruments (Schultz, 1981). This effect occurs especially in relatively narrow and low orchestra shells, or when in higher rooms the ceiling reflectors are hung at locations which make them too low or too narrow. A clear explanation for this is not yet found in the literature; however, nonlinear effects in the ear can not be excluded.

It is, however, possible that a special tone impression is created by a large number of slightly delayed wall and ceiling reflections, which prevents locating individual sound sources, and thus a separation in the sense of the "cocktail party effect." As already mentioned, spatial localization of soft sound sources is no longer possible when the sound level at the ear is by less than 10 or 15 dB above the mutual hearing threshold as determined by louder sound sources. This limit is certainly shifted to the detriment of softer sources, particularly in situations which make it increasingly difficult to separate them directionally from louder sources.

The fact that in such cases the addition of sound absorbing surfaces behind the winds (up to head height) can bring effective help without reducing the loudness impression for other players (Schultz, 1981), speaks for this explanation for the intensity of the direct sound, which gives strong dominance over the reflected tonal contributions so that spatial location becomes easier and the transparency of the entire ensemble tone is raised.

Inasmuch as mutual hearing by the musicians depends strongly on the direct sound and the first reflections, players also notice differences in the reverberation distance for different room occupancy. The physical distance between players is precisely in the order of magnitude where already small distance changes change the intensity relationship between the direct sound and the statistical sound field noticeably (see Fig. 5.9). Thus, players in an occupied hall frequently find themselves within the diffuse-field distance of other instruments, while during rehearsal in an empty hall, they are seated outside that distance, because of the narrower diffuse-field distances. Consequently, musicians on stage sense the influence of the audience on the reverberation time (and thus the diffuse-field distance), more strongly than the audience in the hall.

6.1.5 The Location of the Conductor

The conductor differs in several points from the orchestra musicians, not only in the musical task but also from the standpoint of the acoustical environment. The conductor carries the overall responsibility for the tonal shape received by the audience, where the fact should not be ignored that this tonal shape is certainly not always the same for different seats. Technically this means that the conductor has to watch over the tempi and the rhythmic connections, furthermore the conductor must shape a dynamic development as a whole, as well as the balance between the individual musical groups. On the other hand, for practical purposes, during the performance the conductor no longer influences the intonation.

The slightly elevated location of the conductor is already sufficient to reduce the additional attenuation of the sound which occurs for flat spreading above the heads, as seen from Fig. 5.10. The audibility of distant musicians, in comparison to a close player, such as for example the concert master, is therefore better for the conductor. In addition, the masking of sound by the players own instrument, which otherwise is sensitively close to the ear, is not present. A further difference in relation to communications with other musicians consists in the fact that, while the conductor receives acoustical information, information to musicians – at least during the performance – is communicated only optically. Thus information coming from the conductor reaches all musicians without time delay.

While instrumental musicians and singers rely on the correct balance between their own voice and the other ensemble members, the conductor requires a balanced relationship between the direct sound of the orchestra (and the soloist) and the sound in the hall. The optimum for these relationships can be shifted within certain limits depending on the quality of the orchestra: a high degree of rhythmic accuracy in ensemble playing (and thus also better intonation) will allow a stronger hall sound to aid the conductor, and it is precisely this communication with the acoustic room response which enables the conductor to make adjustments for a particular hall. This is certainly very important for an optimal performance.

As already mentioned, the spatial tonal effect generally consists in the fact that the space in front of a listener appears to become filled in a larger or smaller degree with sound; only in rare cases, when the largest part of the hall is located behind the listener, a weak impression of the hall sound is created from behind (Kuhl, 1978). Since the conductor's back is turned toward the main part of the hall only a relatively small spatial tone development becomes noticeable – unless the hall has adequate free space in front and above the conductor.

In this context it is interesting to note that Herbert von Karajan in the 1960s considered the Massey Hall in Toronto, which has a height about 30 m above the orchestra, as one of the most outstanding halls in the world from the standpoint of spatial tone impressions (Winckel, 1962b). Based on previous considerations about mutual hearing within the orchestra this hall must present musicians significant difficulties to maintain rhythmic cooperation. Thus Karl Böhm reported (Winckel, 1974) that he required approximately 10 min in this hall before the orchestra came to a precise ensemble playing after initial floundering; however thereafter he judged the hall as one of the most outstanding from a tonal standpoint. From Table 6.5 it is already evident that halls highly valued for acoustical characteristics have a ceiling height which exceeds by far the measure, which has been found to be favorable from a standpoint of reflections useful for mutual hearing.

Evidently the large hall height conveys to the conductor the spaciousness of the sound from above, which it can not develop from the front by reason of the strong direct sound. This effect is very helpful to the conductor for the overall acoustical impression and thus for the interpretation. A feeling for the spatial tone development evokes within the conductor a sense of the dynamics as determined by the increase of the tonal volume, and prevents forced entrances and dynamic excesses. From this standpoint it is not surprising that regular attenders at Berlin Philharmonic concerts made the observation that reflectors above the orchestras in concerts with H. von Karajan apparently were always located in their top-most position. This consideration also explains why conductors are relatively unanimous in their opposition to placing audience seas behind and partly to the side of the orchestra.

While the hall response gives important information to the conductor about the balance between the individual instrument groups, the effect of the direct sound is also essential. However, differing distances between individual players and the conductor add some difficulties. Aside from the fact, that for most listeners the distance relationships to the instrument groups is different than for the conductor (the differences to the middle string desk and to the brass players is approximately 1:2 or 1:3 for the conductor, however for the listener something like 1:1.2 or 1:1.5), a partial masking occurs for the conductor as a result of the high level of neighboring strings, i.e., a weakening of the sound of the further distantly seated players.

This applies not only to winds, which are sensed by the conductor as increasingly soft in halls with less reverberation, making it difficult to find a correct balance. An example of this is a dry TV studio with large free surfaces within the orchestra (providing free access to TV cameras). This also applies to string desks in the back, so that the conductor predominately hears the front desks, for, at the ear of the conductor, the level from the third desk is already decreased by 7 dB relative to the sound from the first desk, and by 10 to 11 dB weaker from the fifth desk, assuming equal sound power from all players. If there is a wall located directly behind the strings, which, because of its structure sends at least two reflections back to the conductor, the fifth desk (with a level difference from the first of about 7 dB) becomes much more clearly audible at the location of the conductor (Meyer, 1994a). In this connection it should also be mentioned that orchestra musicians value a second reflection of inclined surfaces of the stage edge, since it improves the mutual listening between instrument groups. Particularly woodwinds profit (Winkler and Tennhardt, 1994).

In contrast, when the stage wall is only located 4 m behind the last desk, the equalizing effect is no longer present for the conductor. In this context, reflections from above also have very little effect. On a limited scale stepwise raising of the outer string desks can support a more closed tonal impression of the group for the conductor. However, this is not connected with advantages to the sound in the audience.

The differing demands placed on the acoustical conditions at the podium by the conductor and orchestra musicians has the consequence that there is no absolute optimum for a concert hall. Rather, acoustically good halls will always differ in the fact that they will either favor the intention of the conductor or the tonal impression for the orchestra musicians. The question thus remains open to what extent limitations can be imposed on musicians with regards to mutual hearing in order to make the tonal impression for the audience as good as possible. Driven to an extreme, this leads to the question: must the conductor have the best place in the hall or is the conductor capable of projecting a subjective tonal experience to the perspective of the audience without causing the interpretation to suffer? To this is added the further question: to what degree can the conductor demand additional efforts on the part of the members of the orchestra when acoustical conditions at the podium are tailored to strongly to the position of the conductor? At this point it is certainly relevant to note that the quality of the orchestra plays a decisive role.

6.2 Opera Houses

6.2.1 *Reverberation Time and Room Size*

The acoustical demands on rooms in which musical stage works are to be performed are significantly more complex than the demands in concert halls. Without loss of significance of orchestra sounds, the effect of the singers is intended to move to the foreground. For this, *bel-canto* passages require sufficient reverberation, possibly even a certain measure of spaciousness, so that the voice achieves a luminous fullness, and the melodic line a tonal continuity. Furthermore, singers want to feel a certain resonance of the hall which transmits a sense of security in development of their own voice, but also security of adjustment to the ensemble. On the other hand, a high degree of clarity is demanded in order to guarantee sufficient understanding of the text, so that the audience can follow the stage action. Therefore, the reverberation time in opera houses must be shorter than in concert halls.

These requirements, however, cannot be generalized by one optimal value depending only on room size. Even more so than in symphonic music, the composition style plays a role. Since the clarity of understanding largely depends on the tempo of sound, it is also influenced by the technique of instrumentation. In fast *parlando* passages, and for spoken texts, the danger exits that the articulation drowns when the reverberation time is too long. This is also relevant for very rapid recitatives, which mostly contain the significant process of the action, while the arias often have a static character in relation to the stage action. The operas of Mozart and Rossini, therefore demand a shorter reverberation than for example, the majority of works by Verdi, Wagner, or R. Strauss.

Thus, in these latter, large "fully composed" stage works, in contrast, the problem of text understanding essentially lies in the fact that singers must not be covered in intensity by the orchestra. This task however, cannot only be mastered by a short reverberation time, but must be accomplished by appropriate reflection surfaces which concentrate the sound energy of the singer into the audience, and by measured dynamics of the orchestra. For energetic reasons, a slightly longer reverberation for the singers than for play-operas is advantageous. Table 6.7 contains a survey of existing reverberation conditions in several opera houses, which in each case can be considered as representative for their time period. In several cases, the year of construction refers to the year of last renovation and the indication of the room volume, relate in each case, only to the audience space. Specifically, this table contains the reverberation time (occupied hall) for midfrequencies as well as for the octave region around 4,000 Hz, since this frequency region includes an important contribution to the singer's formant, which is relevant to the ability to carry the singing voice. In addition, the values for the diffuse-field distance (for omnidirectional sound sources) at midfrequencies and in the 4,000 Hz octave regions are given.

Among the older opera houses, the Festspielhaus in Bayreuth stands out because of its relatively long reverberation time, it is therefore, especially designed for the great flow of Wagnerian music. The frequency dependence of the reverberation

Table 6.7 Acoustical data for several opera houses

Opera house	Year	Volume (m^3)	Number of seats/ standing places	T_m (s)	T_4 (s)	r_{Hm} (m)	r_{H4} (m)
1. GroBe Oper Paris	1875	9960	2131/ 200	1,1	0,9	5,4	6,0
2. Festspielhaus Bayreuth	1976	10300	1800	1,55	1,3	4,6	5,1
3. Scala Milano	1946	11250	2289/ 400	1.2	1,0[a]	5,5	6,0
4. Staatsoper Wien	1955	10660	1658/ 580	1.3	1,1	5,6	6,3
5. Festspielhaus Salzburg	1960	14000	2158	1,5	1,3	5,5	5,9
6. Metropolitan Opera New York	1966	30500	5000	1,8	1,3*	7,4	8,7
7. Semper–Oper Dresden	1985	12500	1290	1,85	1,3	4,7	5,6
8. Opéra National de Paris	1989	21000	2700	1,55	1,25* *estimated	6,6	7,4

Source: 1.–5. Beranek (1962); 6. Tarnoóczy (1991); 7. Schmidt (1985); 8. Müller and Vian (1989)

time was already represented in Fig. 5.8; the rise in the low register favors a round and sonorous tone coloring, which furthermore supports the instrumental sound by the peculiar covering of the orchestra pit. An attempt was made to approach a similar reverberation time in the new Festspielhaus in Salzburg, whose audience space, however, is larger than in Bayreuth. Thus, from an acoustical standpoint, this house also appears to be created for the large opera. Thus, it forms an interesting alternative to Bayreuth as a performance space for Wagnerian music dramas, where the most significant difference is seen in the open orchestra pit.

Opera houses in Paris, Milano and Vienna, built in the typical style of the nineteenth century, with their reverberation times of slightly more than 1 s, represent a compromise between the Festspielhauses designed primarily for tonal fullness, and the demands of good speech understandability. Yet, even these houses are more suitable for the large opera than for the transparent tonal events of Rokoko-works. Already from these examples it becomes clear, that with the passage of time evidently the perception of an optimal reverberation time has undergone slight changes in the direction of more reverberant halls.

While the long reverberation time in the New York Metropolitan Opera is justified by the extremely large audience space, the long reverberation time in the reconstructed Semper-Opera in Dresden is all the more noteworthy, since it is comparable to values of concert halls of corresponding size, and the acoustical characteristics are evaluated as very positive – at least for today's sound impressions (Schmidt, 1985). However, long reverberation times at low frequencies with large orchestras and full instrumentation lead to sounds for singers, or even choirs, which easily become excessively "massive" or "heavy." As an aside, the opera house in Göteborg, (opened in 1994), has a similarly long reverberation time. In contrast, the Opéra National de Paris (formerly Opéra de la Bastille) in Paris corresponds more to the traditional line, as can also be seen from Fig. 6.15.

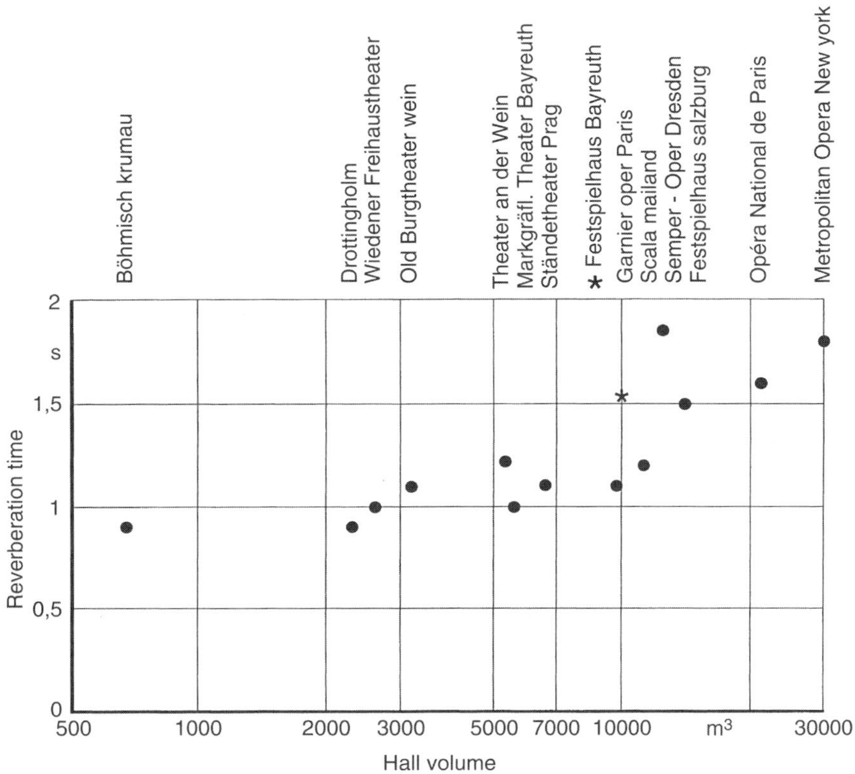

Fig. 6.15 Reverberation times for several opera houses (midfrequencies, occupied hall)

Table 6.8 Acoustical data of historic opera houses (see Table 6.7)

Opernhaus	Year	Volume (m^3)	Number of seats	T_m (s)	T_4 (s)	r_{Hm} (m)	r_{H4} (m)
1. Böhmisch Krumau	1591	670	270	1,0	0,8	1,5	1,6
2. Markgr. Opernhaus Bayreuth	1748	5500	550	1,0	0,8	4,2	4,7
3. Drottningholm	1766	2300	400	0.9	0,8	2,9	3,1
4. Burgtheater Wien	1779	3100	1100	1.1	0,9	3,0	3,4
5. Ständetheater Prag	1783	6600	1100	1,1	0,9	4,4	4,9
6. Wiedener Freihaustheater	1788	2600	800	1,0	0,8	2,9	3,3
7. Theater an der Wien	–	5200	1060	1,15	0,9	3,8	4,3

Source: 1.&5. Januska (1969); 3. Stensson (1968); 4. Singer (1958)

Since many of the great opera houses were built in the same time period in which also the composition of "the great opera" occurred, naturally, the question arises: What was the nature of the halls in which the musical stage works were performed

in Mozart's time? Fortunately, a number of theaters from that time still exist. On the basis of existing drawings it is possible to draw conclusions about acoustical characteristics of buildings no longer standing. Reverberation times have been calculated for the old Burgtheater in Vienna, for which Mozart wrote his operas "The Abduction from the Seraglio," "The Marriage of Figaro," and "Cosi fan tutte." Depending on the level of occupancy, these lie between 1.0 and 1.3 s (Singer, 1959). For the Wiedener Freihaustheater, in which the premier performance of the "Magic Flute" occurred, a reverberation time of 1.0 s can be estimated (Meyer, 1986). These calculated values, along with some measured results from several existing theaters from that time are given in Table 6.8. Even though the measured values for the Ständetheater (Stavovské divadlo) in Prague, in which "Don Giovanni" saw its premier and for the "Theater an der Wien" which was an important performance location for Beethoven, in each case the values apply to remodeled conditions, yet approximate original conditions can be estimated. While data of Table 6.8 represents structural and acoustic parameters of that time, it should be mentioned that certainly not all theaters, which in those days were used for performances, were significantly smaller than the average opera houses built in the past century in cities of average size. The opera house built in 1742 in Mannheim has a volume of 7,700 m^3, and the opera house in Esterháza from 1769 with its volume of 9,500 m^3, nearly had the dimensions of the Garnier-Opera in Paris.

A comparison of reverberation curves – however, this time in unoccupied halls – for five opera houses from different time periods is given in Fig. 6.16. The frequency dependence of these curves is noteworthy. The Ständetheater especially emphasizes the low tonal contributions, consequently it does not permit emphasis of the desired brilliance of the orchestra. The Markgrave opera house in Bayreuth, in contrast, with its reverberation maximum at midfrequencies, presents a very light tonal atmosphere, which makes for a nearly ideal performance of Mozart operas. In that context the orchestra does not even need to hold back, but can develop its full brilliance without influencing the effect of the singers. In contrast to concert halls, a drop in the reverberation curves, or at least a flat continuation toward lower frequencies, is advantageous for opera houses.

The stage opening, and the nature of the stage, have an essential influence on the reverberation time. A scene largely devoid of decorations can prolong reverberation significantly, because the room resonances of the stage area are also excited. Sound reflecting props (such as large plywood structures) also increase the reverberation, while curtains and cloth props strongly absorb the sound. Figure 6.17 shows the reverberation curves for two extreme cases of stage scenery structure with the main curtain closed, for the National Theater in Munich (after opening in 1963). The latter is important for the tonal impression during performance of overtures and interludes. This range of variations, which could be used by stage designers to shape the room acoustics, suggests the possibility of utilizing the structure of the stage design to meet the differing demands discussed earlier of a Mozart opera on the one hand or a large music drama on the other.

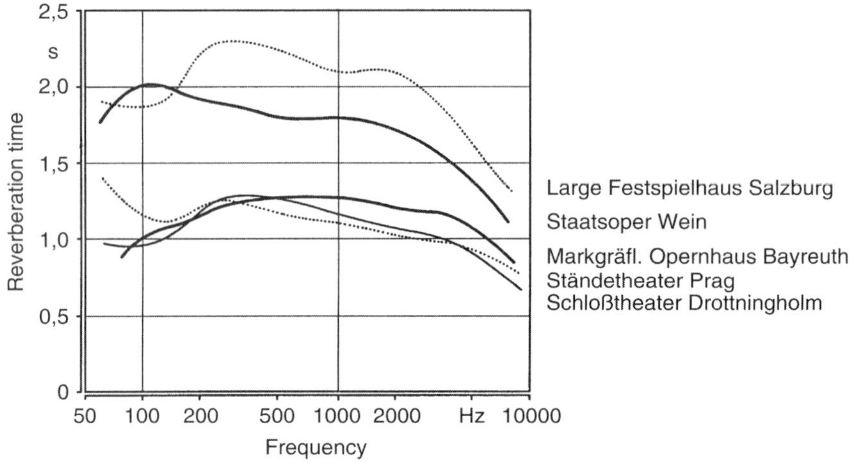

Fig. 6.16 Reverberation curves of several opera houses under unoccupied conditions. Salzburg and Vienna (after Beranek, 1962,) prague (after Januschka, 1969), Drottningholm (after Stensson, 1968)

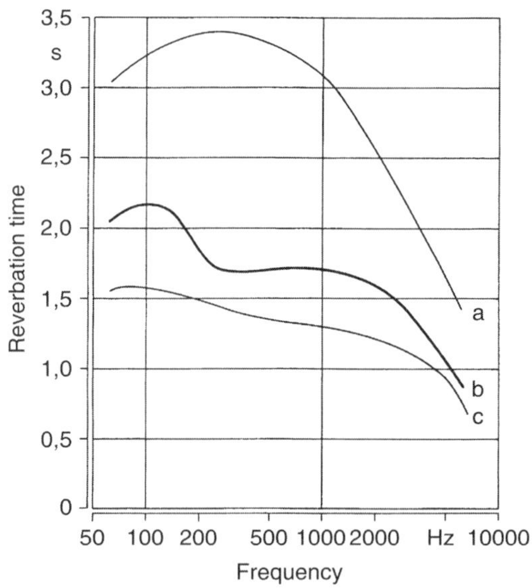

Fig. 6.17 Reverberation curves for the National Theater in Munich (after Müller, 1969). (**a**) for very reverberant scenery, (**b**) with closed main curtain, (**c**) for strongly absorbing scenery

6.2.2 Direct Sound and Early Reflections

The correct intensity relationship between orchestra and stage is a particular problem which will be considered in more detail in Sect. 9.3. In that context the

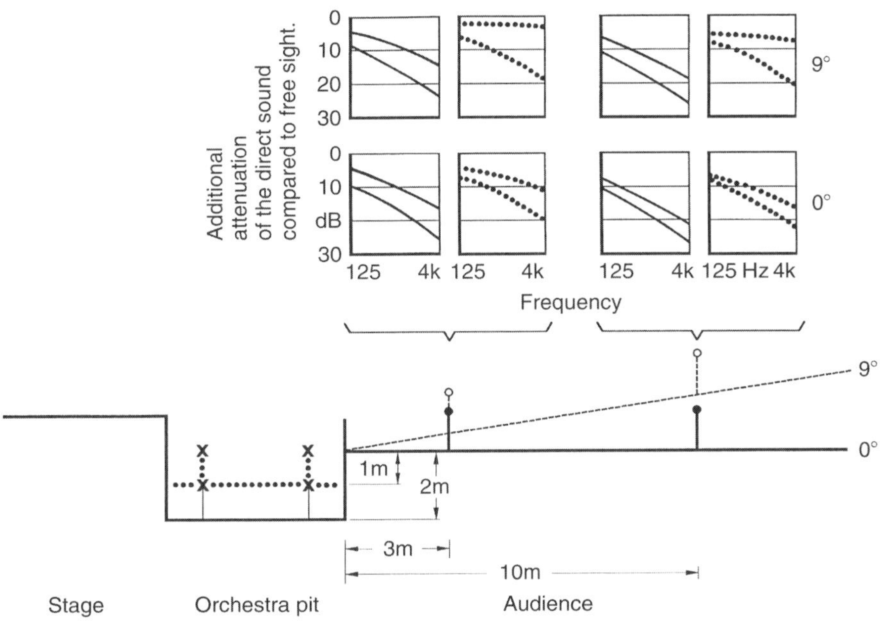

Fig. 6.18 Shading of the direct sound from the orchestra pit into the audience. In the individual diagrams the *upper curve* relates to musicians close to the stage, the lower ones to musicians close to the audience. *Broken*: orchestra location- high. *Solid*: orchestra location – low

nature of the orchestra pit plays an important role. In nearly all old theaters the orchestra pit rarely was located below the level of the front audience rows. Consequently, relatively strong direct sound contributions reach the audience. The sound, therefore, was brilliant and transparent because of the nearly unattenuated higher frequencies. In this respect, the Markgrave Opera House in Bayreuth again represents an optimum. Here the enclosure of the orchestra space is divided by individual pillars, and thus permits the sound to pass through it. The small Baroque Theater in Böhmisch Krumau (Cesky Krumlov) belongs to the few rare exceptions, where the orchestra is located at a level of 2.7 m below the floor of the main hall (Januska, 1969). When the orchestra pit is lowered to some degree, the direct sound is blocked by the enclosure for the audience seated on the first floor of the theater, and is weakened by diffraction around the enclosure.

This effect is represented in Fig. 6.18 for a typical situation; the values are calculated using the curves in Fig. 5.3, and present the frequency dependent level drop which occurs in addition to the weakening with distance (Meyer, 1986). The upper row of diagrams refers to rising chair rows, the lower one to flat chair rows. The left diagrams refer to seats at a 3 m distance from the orchestra pit, the right ones to a 10 m distance. The solid and broken lines refer to an orchestra pit sunk by 2 or 1 m respectively, below the audience floor. In the individual partial pictures, the upper curve relates to a musician sitting at a distance of 4 m from the pit enclosure, and the lower one at a distance of 1 m. As can be seen, the shading

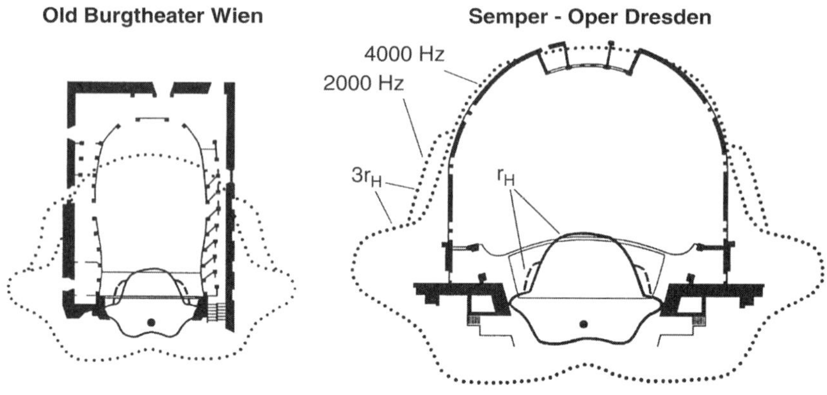

Fig. 6.19 Curves for the simple and threefold diffuse-field distance in the singer's formant

increases strongly with rising frequency. For the lower position of the orchestra, the attenuation at high frequencies can be more than 20 dB. When the orchestra is not lowered that far, the shading can be weaker by 5–10 dB, which is especially noticeable for higher frequencies. The clear sound of a higher seated orchestra effects a particularly advantageous and transparent production of classical operas, or also of many modern works of similar structure, however, it also gives a brilliance to large operas which can not be achieved with deep orchestra pits. Thus, for example, the level of the orchestra pit in the Vienna State Opera is so high that the contra basses in the last row must be lowered by one step so that they don't intrude on the edge of the stage.

While it is possible to keep the overall loudness somewhat lower with deep orchestra pits, particularly for the wind group, this is bought at the expense of a relatively flat string sound on the main floor. Yet, the depth of the orchestra pit has less of an influence on the impression of loudness than on the softness of the attack, because the main part of the energy is shaped for the audience by the strongly delayed reflections. In fact, delays of more than 1/10 s can occur (Reichardt et al., 1972), which naturally makes it easier for the singer to rise above the orchestra through articulation. The total or partial blockage naturally has the strongest influence on the damping of the orchestral sound. The well known blockage in the Bayreuth Festspielhaus (see also Fig. 9.4), does homogenize the tonal impression of the instrument groups, on the other hand, however, it also diminishes the plastic affect of the orchestra. It therefore, supports the classical Wagner sound, as it is most pronounced in the "Ring" and "Parsifal," on the other hand, for example, the "Meistersinger" score, with all its delicacies, is very difficult to realize from a tonal standpoint.

As already represented in Fig. 3.30, it is very important for the singer to rise above the orchestra with the singer's formant. In relationship to the desired clarity, and also the ability to pinpoint location on the stage, it is of fundamental importance for the direct sound to reach the audience with sufficient strength. The determining factor for this is the threefold reverberation distance, this is the distance for which the direct sound drops to 10 dB below the statistical sound field in the hall. Therefore, in Fig. 6.19, the reverberation distances relevant for the singer's formant are given in relationship to the floor plan of two opera houses, where the location of the singer is assumed to be 2.5 m behind the edge of the stage. The basis for the curves is the statistical direction factor, which for 4,000 Hz within an angle of ±20°, remains at a value of 1.6, and for 2,000 Hz, rises above 1.6 in the angular region adjacent to that on each side, so that within an angular region of ±40° from the direction of sight, a pronounced concentrating effect results. It is surprising how well this curve for the threefold reverberation distance fits the Semper-Oper, while in the elongated hall of the former Burgtheater, the rear seats lie outside that region.

As the numbers from Table 6.7 for diffuse-field distance (for omnidirectional sources) at 4,000 Hz show, theaters of current sizes reach values of around 6 m; thus, for a statistic directivity factor of 1.65, a value of 30 m follows for the threefold diffuse-field distance, which should be used as an outside limit for a defensible listening distance. However, this limit should not be considered too dogmatically, particularly when it is possible to reinforce the direct sound with some reflections of only slight delay. The floor, at a distance of 2–5 m in front of the singer, offers an important surface for this purpose, because of the particularly strong tonal contributions of the sound maximum, which are directed downward from the singer. These relatively flat reflections are directed into the hall at an angle of somewhat less than 45°. This floor reflection compensates somewhat for the position of the singer further toward the back of the stage, so that one can defend calculating the "audience distance" from the front edge of the stage.

Inasmuch as the side-walls at the front of the stage are only about one reverberation distance away from the singer, they can be extremely effective as reflection surfaces. To serve in this manner, it is essential however, that they are unstructured and contain no openings for stage lighting or box-seating. Reflections from surfaces near the audience are also effective, because very short delay times can be achieved. This has special application for the under side of the galleries. It is especially relevant, because the ear is particularly sensitive to additional reflections with delays of less than 10 ms, since such reflections lie below the direct sound by a level of 12–20 dB (Kihlman and Kleiner, 1980).

The ceiling above the orchestra can contribute significantly to the balance between singer and orchestra in spite of the relatively long delay times of the reflections. For this, the angle is important, as shown for the example of the Staatsoper Berlin (Fig. 6.20). In the years 1955–1983, the ceiling above the orchestra made an angle of 30° relative to the horizontal. This reflected the sound from the stage mostly into the upper galleries which were already served adequately by the ceiling reflections from the main hall. The sound of the orchestra was reflected toward the main floor so that the sound of the instruments dominated above the singers. Since 1986, the ceiling above the orchestra is angled at only 8°, which reflects the sound from the stage to the main floor, and the sound from the orchestra predominately back into the pit. This results in a more balanced sound level of the singers on stage as perceived on the main floor and in the galleries. It also

Fig. 6.20 Ceiling above the orchestra in the Staatsoper Berlin (after Marx and Tennhardt, 1991)

improves the balance between the singers and the orchestra on the main floor (Marx and Tennhardt, 1991).

The shape and the arrangement in the audience hall can also have an influence on the balance between singers and orchestra. In theaters with box-seating, the possibility exits that the sound from the stage is reflected by the walls and the ceilings of the boxes in such a way, that it will reach the audience and even the stage. This occurs when the boxes are sufficiently high, and not too deep, so that the singer can still see a small strip of the rear wall. If in addition, the boxes at their full height are separated by sidewalls, so that the reflections return to the stage because of the double angled mirror effect caused by the two angles which approximate 90°, then the singer senses a particularly pleasant resonance of the room in relation to his/her voice.

Using the example of Teatro San Carlo in Naples, Fig. 6.21 shows the sound reflections returning from the boxes, along with the relevant surfaces where they arrive on the stage floor, represented for the right half of the room. The long extended arrival surfaces are created by the superposition of reflections from the rows of boxes located above each another (after Weisse and Gelies, 1979). The many reflections returning to the singer are a help in voice control. The fact that delay times of 80–140 ms are involved clearly should be considered an advantage, since less delayed reflections are covered by the masking due to the singer's own voice (Nakamura, 1992). To a lesser degree, a similar effect is achieved in the Semper Opera by relatively high overhangs below the boxes. This is true because the low frequencies naturally play a lesser role in evaluating tone color variations, consequently, the size of the reflection surfaces only need to be adequate for the middle and high frequencies (Fry, 1978). On the other hand, the sound from the orchestra reaches the boxes at a steeper angle, and is thus reflected more strongly against the front box brim. Therefore, only a small portion is returned to the audience, which is very advantageous for the balance between singers and orchestra.

Open galleries, which are separated only by very low sidewalls, reflect the sound more into the rear portion of the hall. In new theater construction, this effect is

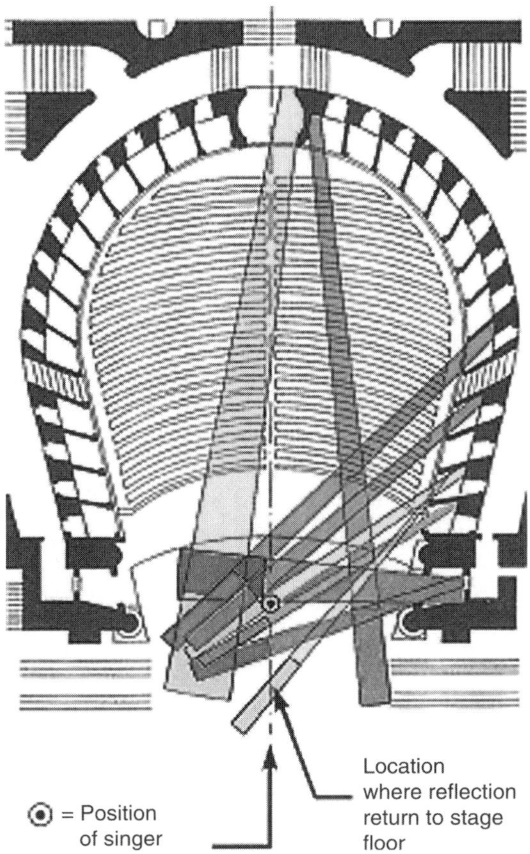

Fig. 6.21 Reflections, originating from a singer, returning to the stage, in the Teatro San Carlo, Naples (after Weisse and Gelies, 1979) (See Color Plate 8 following p. 178)

utilized frequently by appropriate shape and angle of gallery floors and rear walls in order to raise the sound level in the rear portions of the audience (Cremery and Müller, 1978; Fasold et al., 1987). In contrast, very low galleries function as absorbers since even the sound coming from the stage is trapped behind the front rim of the box.

Finally, the stage floor also reflects the sound, which primarily helps the audience in the galleries. As a result, the front seats near the edge of the orchestra pit are not necessarily the acoustically best, from the standpoint of the singer, when a tonal fullness is desired. Corresponding to directional characteristics of the voice, the most effective reflection surface lies in the region of 2–5 m in front of the singer. Of course, proximity to the audience increases clarity of articulation, not even considering the fact that for critical passages (for example, with an accompanying solo instrument), the contact with the orchestra is better.

The depth of the orchestra pit in combination with the angle of the partition separating orchestra from audience plays an important role for auditory communication between singer and orchestra. The reflection from the orchestra to the singers is represented in Figs. 6.22 and 6.23 for several characteristic cases (Meyer, 1988b). If the partition is vertical, instruments close, as well as far from the stage, can be heard from well over 10 m into the audience for a shallow orchestra pit. If the partition is angled toward the inside, the sound transmission from the instruments far from the stage are improved for the singer, the sound of the instruments close to the stage, however, are reflected toward the players, which increases their own loudness impression. If the partition is angled toward the outside, the instruments far from the stage are still easily heard, the sound of the instruments close to the stage are reflected upward above the heads of the singers; however, the singer receives a reflection with relatively short delay of his/her own voice, which can be important since otherwise few reflections are received from the audience.

For a very deep orchestra pit and vertical partition, reflections run more steeply upward and reach the singer only in the front area of the stage; and thus especially the instruments far from the stage are at a disadvantage. Relief can be found in appropriately folded partitions, however, the region of favorable reflections for the singer, don't reach as deeply into the stage as for a less deep orchestra pit. Another option for supporting the singer, under conditions of insufficient reflections of the orchestra, consists in the use of electroacoustic monitors in the regions of the stage. This however, can be an annoyance to the singers because of the unaccustomed directions of the sound source.

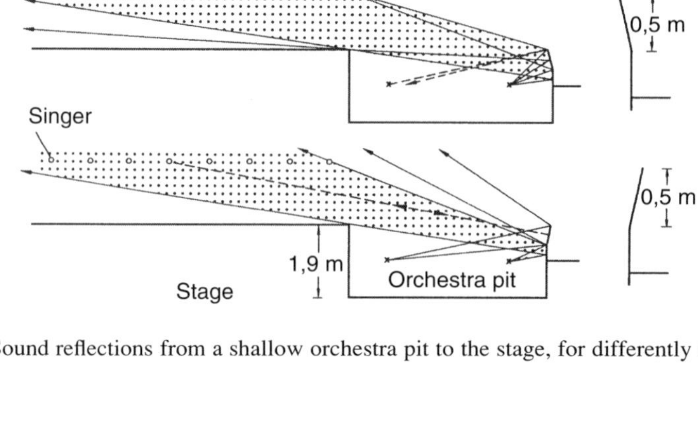

Fig. 6.22 Sound reflections from a shallow orchestra pit to the stage, for differently shaped balustrades

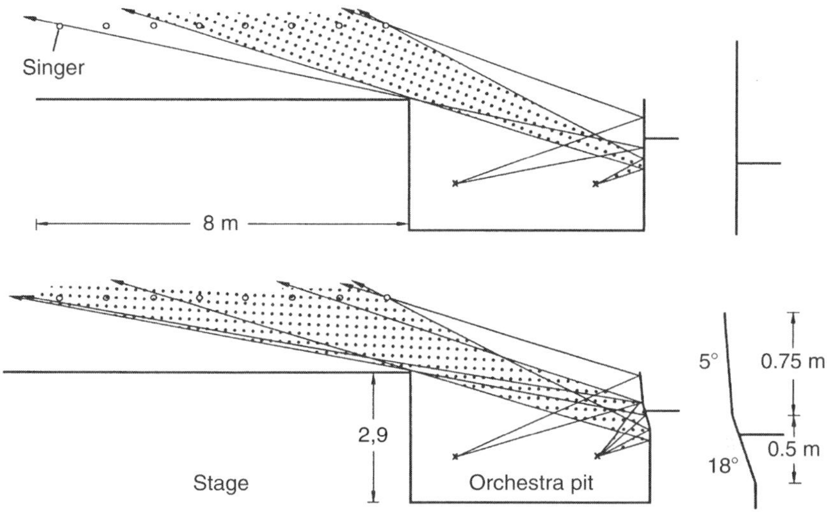

Fig. 6.23 Sound reflections from a deep orchestra pit to the stage, for differently shaped balustrades

6.3 Churches

In many cases, acoustical conditions in churches are characterized by long reverberation times. There are two reasons for this. On the one hand, the walls and the ceilings are frequently highly reflective, on the other hand, in most cases, the ratio of room volume to the volume of persons present is very large, so that absorption effects are not strongly noticeable. Inasmuch as the size of the volume, from approximately 2,000 m^3 in small congregational churches to over 100,000 m^3 for large cathedrals, varies far more than for concert halls and opera houses, the reverberation time also varies in a much greater measure. In addition, the frequency dependence of the reverberation time is closely tied to the architecture style in which the church was built based on the construction materials used (Lottermoser, 1952; Venzke, 1959; Thienhaus, 1962; Meyer, 1977). Thus, acoustical characteristics can be derived, which are typical for church construction styles which not only include the reverberation time, but also the time and directional structure of early reflections (Meyer, 2003).

Three typical reverberation curves for unoccupied church spaces are assembled in Fig. 6.24. Considered here are the Gothic Münster in Ulm, the Baroque St. Michaelis Church in Hamburg, and the small renaissance church of the Darmstadt castle. The Ulm Münster shows a maximum reverberation time at 75 Hz of 12 s. This low frequency of the maximum is characteristic for Gothic churches, as is also shown for other examples in Table 6.9. This is caused by the reflectivity of masonry, which can absorb only higher sound portions because of its roughness; with the exception of leaded glass windows, there are no low frequency absorbers

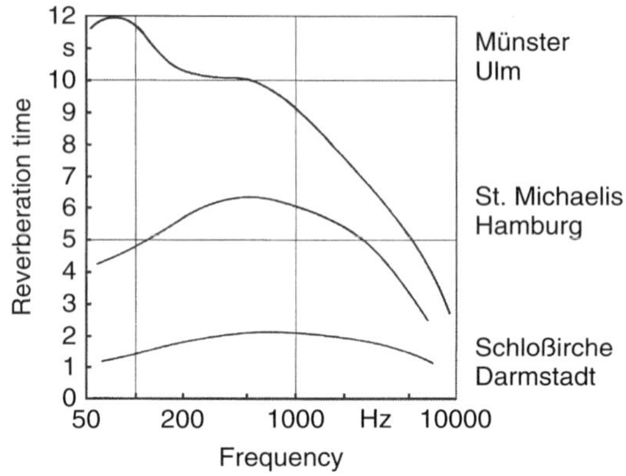

Fig. 6.24 Reverberation curves of several churches

Table 6.9

Church	Style	Volume (m^3)	Maximum reverberation time (s)	At frequency (Hz)
1. Cologne Dome	Gothic	230,000	13.0	100
2. Ulm Münster	Gothic	105,000	12.0	75
3. Freiburg Münster	Gothic	45,000	7.5	90
4. Abtei Church Weingarten	Baroque	53,600	8.5	270
5. St. Michaelis Hamburg	Baroque	32,000	6.3	500
6. Abtei Church Ettal	Baroque	15,000	7.5	750

Source: 1. Winckel (1963); 3.–6. Lottermoser and Meyer (1965)

present. Thus the increase in reverberation time, when comparing midfrequency values to low frequencies becomes greater as the contribution of window area to the totality of walls, ceilings and floors decreases: while a window contribution of 10% results in a ratio of at most 1.1 between low- and midfrequencies, a decrease of window contributions to 2% raises this ratio to1.3 or even more. The reverberation time for Gothic churches in the middle frequencies depends essentially on the volume of the space: A value of 4 s is typical (for an unoccupied room) with a volume of 5,000 m^3, 5 s at 10,000 m^3 and, 7 s at 30,000 m^3. For even larger churches, the reverberation time does not increase significantly and almost never exceeds the value of 11 s (for middle frequencies).

In exceptional cases, Gothic churches have a reverberation time of up to one and a half times the typical values mentioned above. This is the case when the hall in the connection of a "purifying" restoration is too sparsely equipped and above all, the walls and the ceiling are covered with a layer of plaster or paint, which closes the pores. In contrast, the relatively high porosity of materials used in churches of the brick Gothic is very noticeable, particularly when portions of the brick surface are not covered; their reverberation time amounts to only 2/3 of the value typical for natural stone churches with covered walls (Meyer, 2002). The St. Mary church in Lübeck is noted as an example with a volume of 100,000 m^3, it has a reverberation time of only about 6 s. Romanesque churches exhibit reverberation characteristics similar to gothic structures in cases where they possess stone vaulting. However, with a wooden ceiling, the low frequency contributions are dampened somewhat more quickly, so that the maximum of the reverberation time moves to a higher frequency.

In Romanesque and Gothic churches, the pillars or posts play an important role. On the one hand, they block large areas of the side nave. One the other hand, however, they function as valuable reflection surfaces for the audience in the middle nave, provided the sound source is also located in the middle nave. Based on their dimensions, however, they need to be considered as limited reflectors so that their effectiveness becomes significant only above a certain limiting frequency (see Sect. 5.1.3). Depending on the thickness of the pillars or posts, this limiting frequency lies mostly between about 1,000 and 1,500 Hz: complete bending around the pillars can correspondingly be expected only below 200–300 Hz. This is particularly important for the sound of the organ. The high frequency contribution, so important for clarity, including the articulation noise, reaches the listener significantly earlier by diffused reflection from several pillars than the low frequency reflections from side walls and ceiling. This enhances the effectiveness for the direct sound of higher frequencies while the low contributions are perceived as softer than would be expected from their objective intensities, since the initial transient in the hall is slower (Meyer, 2000).

While the reverberation time of middle and low frequencies rises in Gothic churches as a matter of principle, the reverberation time maximum in Baroque churches often shifts into the region of the middle frequencies, since the many wooden structures such as galleries and raised wooden floors underneath the pews, side altars, etc., in most cases largely absorb the low components. This effect is particular strong when the ceiling or the vaulting involves wood. As shown by the example of the St Michaelis Church in Fig. 6.24, a relatively symmetric reverberation curve is created when the maximum is in the region of 500 Hz. Inasmuch as the surface of the walls and additional structures in Baroque churches often exhibit relatively little roughness, the reverberation time at higher frequencies can be somewhat longer in Baroque than in Gothic churches of the same size. Very large Baroque churches with stone vaulting have their reverberation maximum frequently in the region of 250–400 Hz (Lottermoser, 1983), and the difference between Gothic and Romanesque cathedrals is then no longer strongly pronounced. Occasionally however, these large Baroque churches have a second maximum in the reverberation curve, an example of this is the monastery church in Ottobeuren (also known for festival performances), the reverberation characteristics of which are shaped by two maxima at 275 and 1,000 Hz (Lottermoser, 1952). The values for the midfrequency reverberation time typical for this construction style in Baroque churches with volumes of 15,000 m^3, at midfrequencies, lie around 4 s, with 50,000 m^3 around 7 s. This means, that above approximately 10,000 m^3 the reverberation time of Baroque churches is shorter than for Gothic churches with plaster walls, however, longer than for brick churches.

The relatively high frequency location of the reverberation maximum in Baroque Churches leads to the circumstance that the contribution to tone picture in the formant region of the vowels “o(oh)” and “a(ah)” are especially emphasized, and that a clear and brilliant color is created by the fact that the higher components are not very much attenuated, this also corresponds to the optically bright character of the visual impression. This brighter coloration has the additional advantage that the tonal balance between mid and low-frequency contributions is preserved during the reverberation process. Thus, the subjective pitch impression during the decay is not changed for the listener. In contrast, a strong rise in the reverberation time at low frequencies means that the tone coloring during the reverberation becomes increasingly dark. This can lead to the perception of a lowering of the pitch during the decay. The damping of the lower frequency components also avoids the masking effects of the lower voices, so that performance of polyphonic works are possible with sufficient transparency and clarity.

The room acoustic clarity has naturally influenced the instrumental composition style, the development of a polyphonic structure for organ works by Bach can hardly be imagined under acoustic conditions of Gothic cathedrals (Bagenal and Bursar, 1930). These churches especially emphasized the lower registers in their reverberation characteristics and lead to a very dark tonal coloring, which also finds expression in the visual impact made by the somewhat dark rooms on the observer. The duration of reverberations in Gothic churches, therefore, does not permit excessively rapid modulations, and also sets more or less narrow limits on the ability to recognize rapid figures, in spite of the reflections from posts or pillars. The acoustic conditions in these halls are better suited for the slower line sequences of liturgical songs or Gregorian chants. The fact that the Thomas Church in Leipzig, originally a Gothic structure, was suitable for tonal representation of J.S. Bach lies in the circumstance that in its time a multiplicity of additional wooden structures led to acoustic circumstances typical for Baroque churches. The reverberation time in the occupied room was not significantly over 2 s (Keibs and Kuhl, 1959). The harmonically most complicated organ works were created by Bach during his tenure at the so-called Bach-church in Arnstadt, which since Bach’s time has been characterized by a reverberation curve with a maximum above 1,000 Hz and a reverberation time at midfrequencies of around 2 s, when occupied.

When churches are fully occupied, as is frequently the case during concert performances, the difference between Gothic and Baroque reverberation characteristics becomes even more strongly noticeably. Since audience absorption is only effective at higher frequencies, the maximum of the reverberation time is lowered and often shifted to a lower register in Baroque churches. A relatively balanced reverberation curve results, which frequently is also not excessively high. An example of this is given in Fig. 6.25 for the reverberation curve of the Frauenkirche in Dresden, which was destroyed during the war and has since been restored. A curve for the unoccupied church was prepared, based on historic tape-recordings form 1943/1944; the curve for the occupied church was subsequently calculated

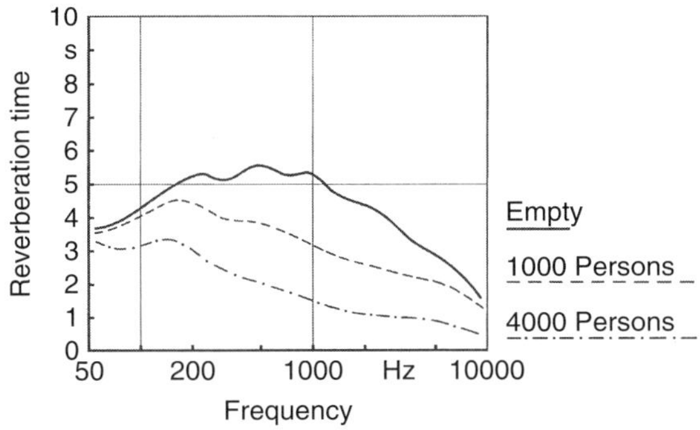

Fig. 6.25 Reverberation curve of the Frauenkirche in Dresden (after Lottermoser, 1960). *Empty*: measured from historic tape-recordings. *Occupied*: calculated from the curve for the empty church (Mozarteum Salzburg)

(Lottermoser, 1960). St. Peters in Rome serves as an additional example, for which at full occupancy values at only 3.5 s were measured (Shankland and Shankland, 1971). In contrast, the maximum for the reverberation time in Gothic churches remains unchanged both in level and frequency location, the drop of the curve for higher frequency regions, however, is steeper. Thus, the coloration of the tone picture has even more emphasis in the low register, however, the clarity of the articulation only suffers slightly since the first reflections from the pillars and posts are not affected by the audience.

Understanding advantages of the Baroque room acoustics has led to an attempt to achieve similar tonal relations by construction measures for numerous newly constructed churches. Alternatively, Many new churches use an architectural conception, which even for relatively small volume, create an excessively large reverberation with preferred low frequency emphasis. This danger can hardly be avoided when using concrete; such rooms are therefore mostly unsuitable for musical performances.

An example of a small church, which lends itself well for musical performances is the Schlosskirche in Darmstadt after restoration in 1969. As shown in Fig. 6.24, with a volume of 1,800 m^3, it has a reverberation time of slightly above 2 s at the midfrequencies and distinguishes itself by a drop toward the lower registers. The tonal effect in this hall is therefore, clear and bright. Many of the small churches constructed in recent yeas, exhibit similar reverberation characteristics. They meet the demands for congregational singing and musical performances of small ensembles and choir (Meyer, 2003).

Long reverberation times in large churches also result in slow initial transients. Since a time period of 1/20 of the reverberation time must pass for the sound level of a sustained note to reach a value of 3 dB below the final value, for a room with a reverberation time of 10 s an initial transient of ½ s can be expected. This naturally leads to a certain "inertia" of the room, and furthermore means that short notes will never reach their final intensity. Thus, they give a softer impression than equally strongly played long notes.

This effect becomes even more significant when the direct sound reaches the audience relatively softly. This is because the diffuse-field distance, by reason of the long reverberation time, is not much larger than for concert halls and opera houses, even though church spaces are usually wider. Thus, even in the Cologne Cathedral, the diffuse-field distance (for an omnidirectional source) for low and midfrequencies is only about 8 m. While first reflections, because of the many pillars and posts in these large churches follow with only very slight delay, the energy density of diffuse reverberations is so high, that it is often impossible to localize the position of the sound source. This applies particularly when the direct sound path of the high frequencies is blocked by pillars. Under such acoustical conditions, a nearly mystical tonal effect is created, which furthermore supports the visual impact of the spatial atmosphere without permitting recognition of precise details of the musical sequences.

6.4 Chamber Music Halls

Today, chamber music is associated with musical performances by small ensembles. From a tonal standpoint, transparent musical performances are expected in which all individual voices are evident. In this context, proximity of audience to the performers fosters immediate contact. These acoustical demands are met by a relatively short reverberation times in rooms which are not too large. This especially emphasizes an intimacy and presence of the tonal impression through very short delays between the arrival of the direct sound and the first reflections. Furthermore, a “chamber- music” tone picture includes the requirement to separate the individual instruments spatially, so that in spite of achieving the homogeneity associated with seamless ensemble performance, a stereophonic effect is perceived.

When considering the historical development of chamber music in the context of the time period during which the works of today’s repertoire were composed, a fundamental change has occurred in the character of the works, as well as in performance practice (Wirth, 1958). In the Baroque era, chamber music was the only alternative to church music. Consequently, instrumentation for those compositions range from one or several instruments for sonatas, clear up to chamber orchestras. These small ensembles consisted predominantly of amateur performers. In approximately 1800, the first chamber music ensembles were organized by professional musicians. These performed in public concerts for a larger audience. For example, from 1803 on we find “the” Gewandhausquartett in Leipzig (Borris, 1969). Naturally, increased performance technical demands were connected with this as well as an evolution of musical content. The expression of both of these is represented especially well in the development of the Beethoven string quartets. At the same time, domestic musical performances developed their own independent form. As determined by the limitations of technical possibilities for amateur performers, this became increasingly removed from concert chamber music.

The chamber music of the Baroque and the early classical era was performed largely in the palaces of the nobility. Thus it was mostly in large rooms or possibly even small halls where a group of performers played for a relatively small number of listeners or occasionally for an audience which filled the hall. According to contemporary artworks, such rooms had a ceiling height of around 4–8 m and a volume between about 200 and 1,000 m^3. Figure 6.26 shows an example of a Paris hall around the middle of the eighteenth century. In such halls, depending on size, the reverberation time should be near 1 s when occupied. Thus the sound should attain the necessary brilliance without losing the individuality of separate voices. In addition, the temporal separation of reflections was sufficiently short for a direct tonal impression.

A hall which played a special role in music history is the so-called Eroica-Hall in the Palais Lobkowitz in Vienna. This hall not only saw the premier performance of the 3rd and 4th symphony by L. van Beethoven but also of many chamber music works by him and other contemporaries. This is a rectangular hall of 8.25 m height and a volume of 950 m^3. In addition to the chamber music podium, it offers seating for approximately 160 people. Its reverberation time is given in Fig. 6.27 both for empty and occupied conditions. When fully occupied, as would have been the case for most of Beethoven's concerts according to contemporary reports, a reverberation time of 1.45 s is determined at midfrequencies, with a steep increase toward

Fig. 6.26 Mozart at the piano in the Palais of Count Conti in Paris, 1763. Oilpainting by M. B. Ollivier (Mozarteum, Salzburg)

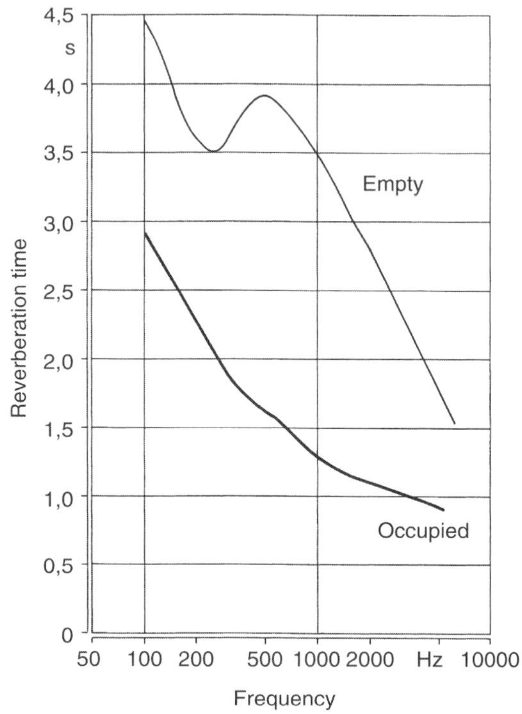

Fig. 6.27 Reverberation curves of the Eroica Hall in the Palais Lobkowitz in Vienna

low frequencies, as a result of the use of marble in the structures. In the empty hall, the reverberation time is significantly longer, thus the number of persons in the audience has a strong influence on the acoustical conditions. The dip in reverberation time at frequencies around 250 Hz can be associated with the wooden podium used today and thus may not reflect the original condition during Beethoven's time.

In this context it is interesting, that in his autobiography, C. Ditters von Dittersdorf (1801) praises a hall of oval design in the castle Johannesberg near Breslau for its especially good acoustics (elliptical halls usually suffer from a disturbing focal point). In this hall, which has a length of 13.5 m, and consequently can not have been much wider than 10 m, the sound concentration evidently must have functioned positively which can only be explained on the basis of a delay time difference between the focusing echo and the direct sound. The direct sound and focus reflections would thus still appear fused to the ear, resulting in a particularly pregnant sound picture. Dittersdorf emphasizes that this construction characteristic contributes much to the amplification of the music without glaring echoes.

The development of house music in common homes, naturally led to the use of smaller rooms in which, determined by the fashion of the times, plush furniture and curtains absorbed the middle and high frequencies strongly. The tone picture, thus lost brilliance, but because of the very short reverberation time, obtained an extremely direct character, which engaged the audience seated close to the performers with a particular immediacy in the tonal procedures. Merely the height of the room, which amounted to at least 3 m, assured that the sound did have some spatial character.

This intimacy of small rooms naturally cannot be transferred to halls for public chamber performances. One can assume that professional quartets use concert halls like the one in the old Gewandhaus for their performances in order to reach sufficiently large audiences. As mentioned, this hall had approximately 400 seats. This already reaches the limit of a meaningful framework for a string quartet; halls, which are larger than the Brahms-hall of the Vienna Musikverein building (with 679 seats and 95 standing places), demand an intensity from the player, which no longer permits a chamber music style and thus already approach a symphonic character. Such halls are only suitable for large ensembles or chambers orchestras, for particularly in chamber music, sufficient loudness is required so that the audience does not sense a disturbing distance.

Modern chamber music halls are therefore usually designed for an audience of 500–600, when acoustics conditions are to be somewhat appropriate for this musical genre. Thus, for example, the small hall of the Meistersingerhalle in Nürnberg, with 500 seats, has a volume of 4,000 m^3. It is thus approximately twice as large as the old Gewandhaus. Under occupied conditions it has a reverberation time of 1.1 s at midfrequencies (Cremer and Müller, 1964). The chamber music hall in the Schauspielhaus in Berlin offers seats for 440 at a volume of 2,150 m^3 and has a reverberation time of 1.3 s. The small hall in the Leipzig Gewandhaus, with a volume of 4,300 m^3 occupied by 450 people has a reverberation time of 1.7 s, with a pronounced rise at the low end, which, at least for small ensembles, is already perceived as too long (Fasold et al., 1981; Fasold et al., 1986). In all three halls, the distance between the center of the stage and the most distant audience row lies between 20 and 25 m. The clarity factor in the Schauspielhaus lies between +1.3 and +3.5 dB, in the Gewandhaus between +0.5 and 1.3 dB. In the latter case, the clarity is already perceived as bordering on insufficient. While the right degree of intensity and, essentially also transparency have been met, the sense of immediate participation is available only in the front portion of the hall, since the distance of 20 m to the last rows already creates a certain separation between the performers and the audience, even when the tone leaves nothing to be desired in brilliance. This does not exclude the possibility that the musicians on the stage have an excellent impression of the acoustics and thus find optimal performance conditions.

The proximity of the audience to the podium is also an essential design criterion for the chamber music hall of the Berlin Philharmonic for which the audience is arranged to surround the stage: 1,064 seats are located within 23 m of the stage. The volume amounts to 12,500 m^3, and the reverberation time at midfrequencies is 1.8 s. This relatively long reverberation time is one of the reasons that the clarity measure lies between −1.3 dB near the podium and −1.6 dB for the distant seats. Numerous reflection surfaces are positioned in the hall to equalize the balance of individual instruments, the surfaces inclined toward the podium reflect the sound toward the player, and thus do not serve to balance the directional characteristics and the blocking by the players for the seats close to the stage (Fütterer, 1988).

As an example of a chamber music hall with a more intimate character, the presentation room of the Schimmel piano Co. in Braunschweig should be mentioned. With a volume of 800 m^3, it offers seats for 80–100, as well as for several upright and grand pianos. Because of its reverberation time of 0.9 s (at midfrequencies in the occupied hall), and a slight drop toward the low frequencies, it provides optimal conditions for the performance of piano and string music for the performer as well as the audience. This was clearly shown, at a comparative demonstration during which the same audience was offered the same program in three halls in immediate succession. The two other halls had a midfrequency reverberation time of 0.7 and 2.7 s respectively, each with a clear rise at low frequencies, for room volumes of 1,750 and 2,750 m^3 (Meyer, 1988a).

6.5 Studios

In a living room, radio transmissions or CD recordings clearly cannot evoke the tonal impression of a public performance. This is particularly true for symphonic music and operas, however, even for chamber music works, certain differences relating to room effects are unavoidable. The reason for this lies primarily in the fact that the reverberation, in essence, is already contained in the recording, while the listening space only adds very damped reflections. In addition, the small size also plays an important role. The spatial representation of a whole orchestra presents great difficulties. Finally, in radio transmission, dynamics are frequently compressed to avoid exceeding loudness levels permissible in residential houses. Recording studios therefore need to take into account the modified sound esthetic demands of such an environment.

The possibility to amplify individual instrument groups selectively, and also control the overall loudness levels, obviates the problem of meeting the listener expectation of conditions for appropriate intensities with suitable combinations of room volume and reverberation times. Thus, the reverberation time can be chosen alone on the basis of tone esthetic view points. For symphonic music, an experiment by W. Kuhl (1954a) with a series of orchestras in 20 different studios had the result that, at least for mono recordings, the optimal reverberation time for orchestra studios, does not depend on their volume, but only on the nature of the music.

The subjective evaluations by over 100 test persons, naturally led to a certain spread of the results, based on the personal tonal conceptions, nevertheless, rather good average values were obtainable. Consequently, the first movement of the Jupiter Symphony, as an example of the classic era gave an optimum of 1.5 s, and for the first movement for the 4th Symphony of Brahms, a value of 2.1 s. Romantic music demanded a significantly longer reverberation; recordings from studios with 1.5 s, almost without exception, were evaluated as too dry. It should be noted that Kuhl, in a discussion, mentioned that for stereo reproduction the possibility of localizing individual voices exists, thus increasing transparency. This allows longer reverberation time in concert halls without diminishing transparency excessively (Kuhl, 1954b).

It is interesting to note, that for an excerpt from the "Sacre du printemps," a reverberation time of 1.5 s was perceived as optimal. In that context, evidently, the transparency of the structure was dominant in the tonal perception. Especially in nonstereophonic recordings dissonant cords can easily obtain excessive hardness, when a separation of voices is no longer audibly possible (the experiment was limited to mono recordings). In contrast, there are, however, conductors, which demand long reverberation times for Stravinsky orchestra works with large ensembles. Thus, Stokowski even speaks of a value of about 4 s for a performance of the "Sacre" (Blaukopf, 1957).

Recordings of symphonic works are usually made in halls, which in size are not much smaller than standard concert halls. These are also used for public performances and also use seating from about 800 to 1,200 people. Accordingly, acoustic characteristics of these large recording halls, at least as far as the interests of the musicians go, are not significantly different from those of normal concert halls. Two examples, which by now have almost become historic, for this type of studio are the larger recording hall in Frankfurt am Main and in Hannover. Both originally had 1,200 seats and had volumes of 12,000 and 15,700 m^3, respectively. The average reverberation time in Frankfurt, (prior to adding the organ) was 1.85 s and in Hannover, 2.0 s; furthermore, the hall in Frankfurt had a low frequency rise up to 2.2 s near 65 Hz (Schreiber, 1958; Kuhl and Kath, 1963). While the radio hall in Hannover has not been subjected to significant changeovers, the hall in Frankfurt was remodeled in the 1980s according to new conceptions: with a volume of around 9,000 m^3 and reduced seating in unoccupied conditions, it offers the same reverberation conditions when occupied, as the concert hall of the old opera in which the public concerts of the radio symphony orchestra are held (Lamparter and Brückmann, 1989).

As far as musicians are concerned, it is advantageous for orchestra recordings in small studios to have the reverberation time not too long and if necessary, add additional reverberation after the recording. An extreme example of this technique is given by a series of performances by A. Toscanini for which he ordered preparation of a studio in New York with extremely short reverberation time. These recordings distinguish themselves by an extreme sharpness of this rhythmically very precise performance. However, they present a very difficult task to the participating musicians relative to the tonal quality.

A reverberation time is viewed as optimal for orchestra studios at 1 s, for a room volume of 1,000 m^3, which increases by 0.2 s for each doubling of the volume (Gilford, 1972). These values can also be considered as advantageous for orchestra practice rooms. Furthermore, the reverberation time at low frequencies should not increase at all, or increase at least significantly less than in large concert halls, since otherwise the bass instruments dominate, which does not correspond to performance conditions in large halls.

Smaller orchestra studios also bring the danger that the weaker instrument groups will be overpowered by those richer in energy. For those cases, one possible solution is not to make the rear wall behind the podium as well as the ceiling above it, fully sound reflecting, as is usually the case, but rather to furnish some surfaces with absorbing elements in order to weaken the intensity of the stronger instrument groups. This must be done with due consideration of the seating of the orchestra and the directional characteristics of individual instruments. It is particularly recommended to place sound absorbing arrangements with appropriate frequency dependence behind the basses and timpani.

Studios for small ensembles, both for entertainment or chamber music, are also partly furnished for public performances and partly for recordings without audiences. The number of listeners is kept small in consideration of the desired room effect on the tonal reproduction. There are rarely more than 300 seats available. The reverberation time for room volumes between 1,000 and 3,000 m^3, for rooms with audience lie in the region of 1.25–1.5 s (Reichardt et al., 1955). The acoustic conditions are thus ideal for chamber orchestras as well as ensembles of several winds and strings. For quartets, trios and sonatas, one obtains a greater intimacy of the tonal picture in studios without audiences which only have a size of 400–1,000 m^3, with reverberation times between 0.8 and 1.0 s. This comes closer to the original atmosphere of producing house music, even though the brilliance no longer reaches the degree desired by some high-fidelity-fans.

6.6 Special Purpose Rooms

In addition to various types of rooms, which serve for artistic performances of musical works for an audience, or for recordings intended for general distribution, there are circumstances for which not the music itself, but the performer or the instrument is the principle interest. This includes rooms for instruction, or for instrumental practice, or for example, tuning cabinets, or intonation rooms in piano factories, as well as performance rooms in which instrument quality is to be judged either directly or in comparison to other instruments. For all of these rooms, the most important acoustic demand is the requirement of a high degree of isolation against sounds from the exterior.

Studios, which are to serve for teaching or practice are usually so small that the internal resonances between parallel walls become particularly disturbing. Thus, wall structure is of great importance. In addition, the reverberation time in such rooms must be kept low, where particularly at low frequencies, a drop in the reverberation curve is desirable. The teaching studios of the music academy in Budapest should be mentioned as examples, the larger ones of these have a volume of 182 m^3 and a reverberation time of 0.9 s at the midfrequencies, at 125 Hz, their reverberation time has dropped to 0.5 s. The smaller studios have a volume of 105 m^3, reverberation values are 0.8 and 0.4 s respectively (Karsai, 1974).

Japanese recommendations give slightly lower reverberation times (Nagata, 1989). As Fig. 6.28 shows, a slight rise is tolerated at least near low frequencies.

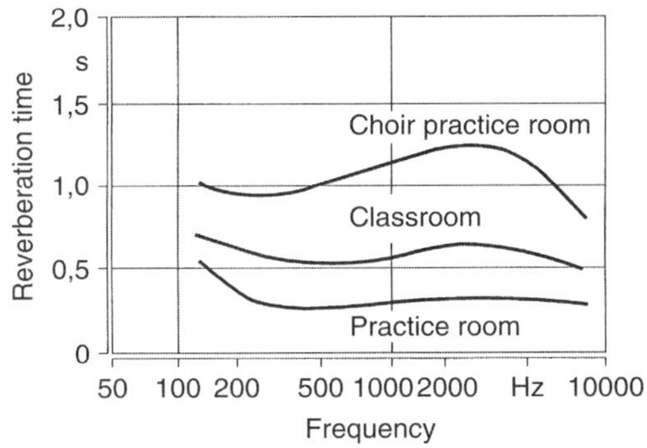

Fig. 6.28 Recommended reverberation times for rehearsal and practice rooms (after Nagata, 1989)

The reason for that may lie in the fact that students do not yet have well shaped sensitivity for tonal changes influenced by the room, on the other hand, performance technical inaccuracies are more clearly recognized in rooms with less reverberation.

Special rooms, which are to serve individuals for practice, usually only have a volume of the order of 30–40 m^3. Based on the fact that in such rooms, tonal impressions comparable to the tonal development in large halls can never develop for the performer, the listening impression should predominately be concentrated on the tonal fine structure. For this, a reverberation time of 0.3–0.4 s is appropriate, where no rise at low frequencies is present, even though the Japanese recommendation (Fig. 6.28) intends that. Wind players prefer slightly longer reverberations than string players, (Cohen, 1992). For percussionist, practice rooms should be largely absorbing. In contrast, experiments in England have shown that in rooms with reverberation below 0.5 s, practice becomes stressful and that rooms with reverberation times of about 0.75 s are preferred for extended practice (Creighton, 1978; Lamberty, 1978).

Reverberation values of the order of 0.3–0.4 s have become accepted for tuning cabinets for pianos. For intonation rooms, the optimum value is slightly higher, i.e., for uprights it is between 0.4 and 0.5 s at a volume of up to 1,000 m^3, and for concert grands, between 0.5 and 0.6 s at approximately the same room size. In order to judge a concert grand accurately, a larger room is needed, which should have a volume between 400 and 600 m^3. However, a reverberation time of 0.6 s for this size is noticeably too short. The optimum lies in the region from 0.8 to 0.9 s. Values above 1 s already increase the difficulty of judgment. Furthermore, these rooms should have no rise at low frequencies.

Practice rooms for orchestras pose a particular problem. If the volume is too small, unacceptable loudness occurs for normal playing intensities; if they are damped excessively, the articulation and sharpness of staccatos cannot be balanced accurately. Therefore, in general, a "tonally appropriate" reverberation time cannot be achieved. Thus, in a room of a volume of 1,200 m^3, a reverberation time of 1.2 s at midfrequencies with a slight low-frequency rise, is perceived as too long for an orchestra with 90 musicians. In such a room, mutual listening can only be assured by strong damping behind the winds and by lowering the reverberation time at low frequencies below the value for the midrange. Völker (1988), therefore, recommends a volume of 50 m^3 per musician (!) with a reverberation time of 1.3–1.5 s. Similar values are included in Japanese recommendations for choral practice rooms (see Fig. 6.28).

Tennhardt and Winkler (1994) recommend a room size of 25–30 m^3 per musician, where for average orchestra size the volume should not be below 2,000 m^3. To the degree that practice rooms are also used for warm-up in production recordings, the average reverberation time in an occupied room should be between 0.5 and 0.7 s, and in other practice rooms, between 0.8 and 1.1 s. A low frequency reverberation rise up to about 1.3 s is still defensible, the reverberation time around 4,000 Hz, should not sink to below 0.6–1.0 s. For smaller orchestras, or section rehearsals, variable sound absorbers, as for example curtains at a distance of 20–30 cm from the wall, are recommended to match the reverberation time to the value of the full occupied room. Sidewalls should be fitted with low frequency absorbers at a level up to 3 m. The wall behind the orchestra is particularly suited for surfaces, which absorb also the middle and high frequencies; absorption surfaces on the ceiling should only cover 30%, since otherwise, at lower dynamic levels, performers tend to compensate for an uncertain tone onset by playing too loudly.

6.7 Open Air Stages

For performance in the open, acoustic conditions are largely determined by the direct sound. Generally this is followed only by a few individual reflections. Since sound energy can escape in nearly all directions, in contrast to closed rooms, a diffuse sound field is not formed, thus there is no reverberation in the usual sense. It is however, easily possible, that individual reflections from distant reflection surfaces are perceived as a discrete echo, which influences the otherwise extremely pronounced clarity of the tonal picture.

To this day, the partly well preserved Greek amphitheaters are admired because of their exceptional ability to make speech understood to the last row. These theaters, which offer seating for many thousands, were frequently built into a mountain with rounded rows of seating, and in those days also had a rear wall behind the stage. The so-called "orchestra" was located between the stage and the audience, which, as a free surface, furnished an energetically important ground reflection. The rise of the seating rows was determined in such a way that each viewer had a clear view of the "orchestra" and thus could receive these reflections without obstructions. A further reflection from the rear wall of the stage, as well as a double reflection from the rear wall and stage floor, completed this package of sound reflections (Canac, 1967).

Measurements on the amphitheater of Epidaurus, which has an audience potential of 14,000 people, showed that all these reflections had a delay of less than 30 ms in comparison to the direct sound, which explains the extreme clarity. The fact that the sound energy of a single voice is sufficient for such a wide auditorium is related to the fact that the low level of extraneous noise, due to its location far from common traffic, makes it possible to hear even very soft tones (Cremer et al., 1968). Such acoustics however, cannot necessarily be designated as optimal for musical performances. Certainly, the extreme transparency of the orchestra's sound can present a unique experience for the listener, which is furthermore enhanced by the extraordinary atmosphere of the location. However, frequent attendance at such concerts can quickly diminish the novelty of this effect.

In contrast, conditions for an opera performance in a Roman arena, as for example, during the festivals in the well known structures in Verona, are significantly different. Due to the fact that the room, which was certainly not originally intended as a theater, forms a closed circle open at the top, a large number of reflections result. This certainly can give the impression of reverberation. Thus the level drop after the first 160 ms corresponds to a reverberation time of 0.85 s at midfrequencies and 1.25 s at low-frequencies (Kurtovic and Gurganov, 1979). Furthermore, for operas, particular clarity in relations to singing is advantageous so that the arena meets the acoustic demands very well. However, one should not overlook that the critical intensity question is solved in Verona by contracting only first class singers and by using large orchestras. The level a singer can create in the upper ranges of the arena lies by 15 dB lower than the value which can be achieved in an opera house with a reverberation time of approximately 0.8 s (Pravica, 1979).

A different type of open air performance is presented by the types of concerts which are performed from a more or less closed orchestra shell to a public sitting in the open. The walls and roof of the music pavilion ensure sufficient loudness for mutual listening by performers and that furthermore, the energy radiated in the outward direction is concentrated in the direction of the audience (Furrer, 1972). In health resorts a variety of shapes for such concert shells are found.

Relative to acoustical effects, structures which have a foundation layout and cross section approximating a parabolic mirror are particularly unsatisfactory since they present varying directions of concentration for the individual instrument groups as already explained in Sect. 5.2.1., and thus, do not present a balanced tone picture to the audience. An example of this shape is given by the music shell at the beach promenade in Westerland on the island of Sylt (Fig. 6.29).

In contrast, pavilions, which have an outwardly directed trapezoidal foundation with a roof directed slightly upward toward the public, are advantageous. When the walls, furthermore, create necessary diffusion by subdivision into small elements, the acoustical conditions are good for both the performer and the listener. The concert pavilions on the island Norderney for example, correspond to this type (Fig. 6.30).

When the region in which the audience sits or walks is spread over a large area, frequently loud speakers are mounted in the frame or on the roof of the music shell in order to achieve sufficient intensity even for large distances. Thus, either a central speaker group can radiate the entire sound, or by using several speakers distributed over the width of the pavilion, which correspond to the seating arrange-

Fig. 6.29 Concert shell in Westerland on the island of Sylt

Fig. 6.30 Concert shell in health resort facilities of the island Norderney (Norderneyer Badezeitung (Foto Kühnemann))

ments of the instrument groups, their sound can be stereophonically projected from the stage. For highly directional speaker amplification, however, it is important that there be no house walls in the neighborhood, which could reflect a disturbing echo, since, for transmission times of an order of magnitude directly related to the musical tempo, very undesirable effects can occur. In this context, facades with contain a number of balconies or other protrusions are especially undesirable since the sound is mirrored back to the source by double reflections from two mutually perpendicular surfaces (see Fig. 5.1b).

The danger of disturbing wall echoes also exists for serenade concerts in interior courtyards or similar spatial situations. For excessive distances between reflecting walls (more than 20 m) so that sound reflections are no longer joined into a relatively well contained reverberation, a solution can be given by having the audience rows rise relatively steeply. The sound which reaches the rear wall above the heads of the audience is consequently reflected upward due to its angle of incidence, and thus no longer influences the perceived sound.

For opera performances and concerts, with an audience of several thousand, increasingly electroacoustic sound installations are utilized, which are intended to coordinate the tonal picture at the location of the audience with a corresponding optical directional impression. For this, numerous microphones and speakers are employed, which are interconnected by delay circuits to ensure that for all audience locations the sound signal of the respective speaker group, which has the shortest distance from the connecting line between the listener to the original sound source arrives prior to the sound from other speakers, however, later than the original sound. In this way, both directional and distance impressions are preserved. Typical examples include the opera performances on the 50 × 60 m Lake Stage in Bregenz (Ahnert et al., 1986) or the Berlin Forest Stage (Schlosser and Krieger, 1993). In these cases, some microphones are permanently positioned and some are affixed to the singers as micro-portable systems. In open air concerts, occasionally microphones are used, which are introduced into the corpus of the string instrument itself at the point where the strings are connected to the body (Winkler and Kaetel, 1990).

Chapter 7
Seating Arrangement in the Concert Hall

7.1 Customary Positioning of Instrument Groups

There is no uniform rule for seating arrangements within an orchestra; the placement of individual instrument groups is handled in many different ways. Thus, tonal as well as performance technical reasons play a role, and finally, the shape and size of the available floor space also are important factors. The latter naturally is especially relevant in orchestra pits of opera houses; however, also for concert stages, which generally provide adequate expansion possibilities for ensembles there is no seating arrangement universally recognized as optimal.

The reason for this can be seen in the fact that acoustical conditions differ from hall to hall and that there will always be an attempt to adjust the orchestra arrangement to the given room acoustic circumstances. Naturally this is the case for the local concert hall, which is mainly used by one orchestra on a routine basis, however, it is certainly not ignored during tours. Thus, for example, it is well known that W. Furtwängler insisted on corrective changes in seating arrangements in various halls during rehearsals, while on tour with the Berlin Philharmonic (Furtwängler, 1965). Other conductors, however, for sound esthetic reasons, insisted on the same seating arrangement, which appeared optimal to them, for each orchestra with which they performed, even when the relevant orchestra was accustomed to a different arrangement (Boult, 1963). The seating arrangement preferred, naturally depends on the personal interpretation style and the subjective tonal perception of the conductor.

The importance ascribed to the grouping and arrangements of performers by composers and contemporary musicians is shown by a look at the history of performance practice (Schreiber, 1938; Hoffmann, 1949; Becker, 1962; Paumgartner, 1966). Thus, already J. Mattheson in the publication "Der vollkommene Capellmeister," (1739) addresses this question and J. J. Quantz, 1752, demands of an "Anführer der Musik" that he "knows how to group, position, and locate the instrumentalists," he also then gives exact directions for positioning of performers, where naturally the location of the harpsichord of the conductor as the central location is understandable for the end of the era of the *basso continuo*. R. Wagner, had a new podium constructed for a performance of the 9th Symphony by

J. Meyer, *Acoustics and the performance of music*.
DOI: 10.1007/978-0-387-09517-2_7,

L. van Beethoven in Paris, in order to arrange the orchestra in steeply rising steps, according to his own system (Wagner, 1911). In his writings about conductors, with many details, Hector Berlioz considers the effectiveness of raised podiums and sound reflecting surfaces, and demands sufficient distance between similar instrument groups which are to enter into a dialogue-like interchange (Berlioz, 1864). A. Melichar (1981) considers the arrangement of the strings in great detail, especially the situation with the second violins, and in doing so quotes A. Toscanini, who considers the two violin groups as a pair of shoulders: "Like shoulders, they must be equally strong and equivalent."

For the seating arrangement within the orchestra, which is currently found in concert halls, the strings are almost always located in front of, and to the side of the conductor and occupy the full width of the podium. The woodwinds and horns are located directly behind, them usually a step up; the final row is formed generally by the heavy brass and the percussions in front of the rear wall of the hall. Within the strings and the wind groups the arrangement of the individual voices can vary considerably.

The currently customary arrangement of the strings is schematically represented in Fig. 7.1. For the sake of clarity, all instruments within one group are drawn as parallel to each other; in practice of course all desks are oriented so that the players have the conductor in their field of view. The so-called German or European seating of the strings, as indicated, represents the longest tradition. Within that tradition, the placing of instrument groups is characterized by the fact that the first and second violins are seated opposite each other on the podium, while the celli are located directly in front of the conductor and slightly to the left of the first violins. The violas fill the space between the second violins and the celli, the basses continue on the left wing toward the rear of the celli. The earliest historical precursors of this seating arrangement can be found in the Mannheim church orchestra under Abbé Vogler (around 1777), who introduced the separation of the two violin groups to the right and left side. In his arrangement the general bass group forms the center. Locating the two violin voices on both sides of the conductor was subsequently adopted relatively quickly in a series of other orchestras. However, the European seating arrangement of the form represented in Fig. 7.1 was not spread universally until the second half of the nineteenth century.

From a purely performance technical standpoint this arrangement has the advantage that the contact between the concert master and the solo cello is very good because they are so close. The contact between celli and basses is also very favorable, which is particularly advantageous for critical common entrances (e.g., at the final presto of the 4. symphony by R. Schumann, in measure 211). To a certain extent the two violin groups are seated as partners opposite from each other, the difference between these two voices can be recognized through a certain "stereo effect" in the tone picture which lends much greater transparency especially to passages within which the motif is exchanged. A further advantage of this seating arrangement is noticed in passages where the first violins and the celli play in unison as for example in the last movement of the first symphony by J. Brahms (score example 11). Placing these two groups as neighbors merges these two voices into a total sound better than if they were separated from each other.

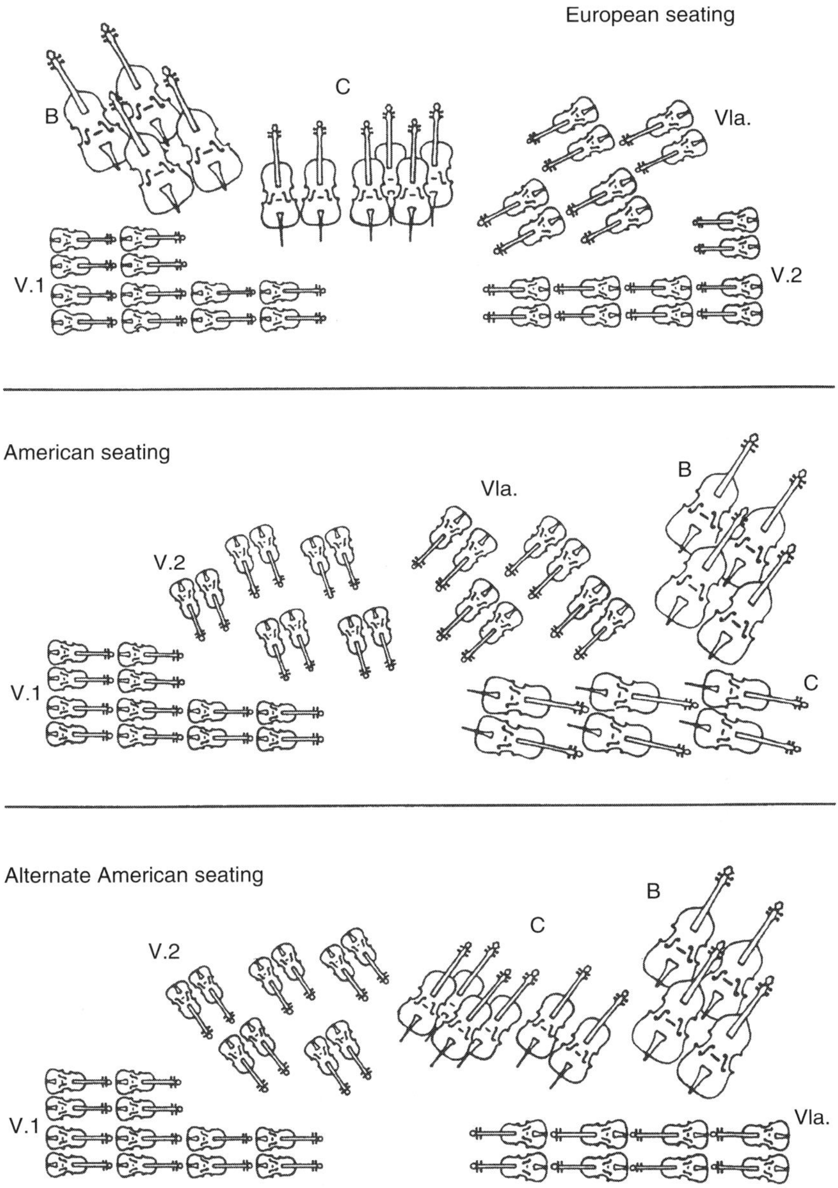

Fig. 7.1 Schematic representation of the currently used seating arrangement of string players in a symphony orchestra. For the sake of clarity all instruments belonging to one voice are represented in parallel fashion

Score example 11 J. Brahms, 1st symphony, 4th movement measure 185 ff. Score excerpt without winds and percussion

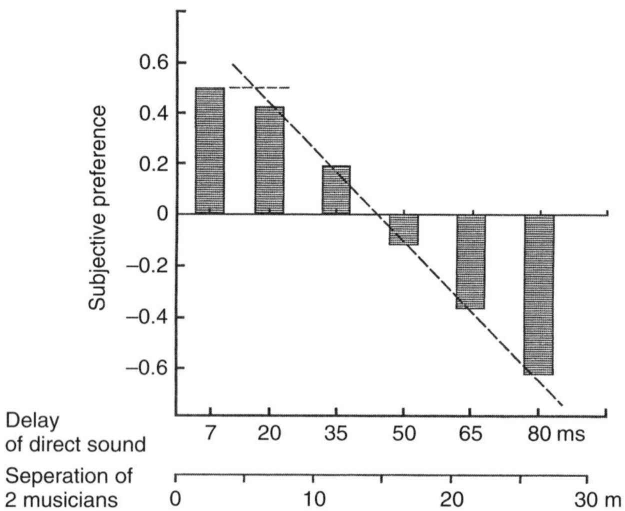

Fig. 7.2 Ease of playing together without visual contact, in dependence on distance between performers (after Gade, 1989a)

With increasing demands for precision in playing together, the requirements for mutual contact increase for the spatially separated arrangement of the two violin groups: acoustically this is supported when the strings are located beneath a nearly horizontal reflection surface (see Fig. 6.14); optically it is supported if the rear desks are slightly raised. The importance of visual contact for players is shown in an experiment by Gade (1989a). He had two musicians perform in separate nearly anechoic rooms. Mutual contact between the two players consisted only in having the sound performed for each by means of a microphone and a speaker; in the experiment, varying delay times could be introduced into the electroacoustic connection. The result is shown in Fig. 7.2: the subjective ease of performing jointly, as judged by the performer, is entered above the delay time; furthermore, the horizontal axis indicates the separation between performers which corresponds to the delay time, or the traveling time between performers when located in the same room.

As evident, performing together becomes increasingly difficult when the traveling time between players becomes longer than approximately 20 ms; this corresponds to a distance between players of about 7 m. Considering, that for an orchestra, a width of about 17 m and a depth of about 12 m is not uncommon, it becomes clear how indispensable visual contact is, especially between players located at the outer edge, as well as with the concert master. In this context it should be mentioned that in a discussion about questions of orchestra seating arrangements P. Boulez represented the standpoint that the difficulties of playing together only arise because of the large orchestra size currently customary. It can easily be overcome with seating limited to eight players within a section (Meyer, 1987).

During the second half of the twentieth century, the so-called American seating was found predominantly. It was initially introduced on a trial basis in the 1920s to make concessions to the initially very primitive recording techniques for radio and recordings. It was then propagated by L. Stokowski, and rapidly entered international concert halls (Melichar, 1981). In this seating, the celli are located to the right side of the podium, while the second violins follow the first violins toward the middle, the violas and the basses take up the space between the second violins and the celli; occasionally the basses are pulled toward the front of the podium following the last desk of the celli. This seating arrangement which rapidly began to become important in the 1950s in Europe, corresponds to the arrangement of the string quartet in a linear voice sequence. However, one should not overlook the essential difference, that in a quartet the first and second violins receive individual tone formation both in regards to performance technique and especially through the vibrato, while these characteristics totally lose their differences.

An advantage of this seating arrangement is of course the good contact between the two string groups which supports critical common passages or entrances. The linear arrangement of strings also appears especially logical in such places where a motif is repeated through all four voices in a rising or falling line, because the time sequence now also corresponds to a spatial sequence and sequence in register. As typical of such passages the score example 12 gives an excerpt from the Scherzo of the 7th symphony by L. van Beethoven; the last movement of this work contains similar locations, where, however, the violas and the celli move together. For this seating arrangement, however, the less than optimal contact between the concert master and the solo celli is a disadvantage, the contact between the celli and the basses also suffers, especially when the latter are not drawn toward the front of the podium.

This deficiency is largely equalized with the variation of the American seating arrangement which was introduced by W. Furtwängler and has since been adopted by many orchestras. The celli and the violas exchanged positions, so that again a closer connection exists between the celli and the basses. The contact between solo cello and the concert master is also improved without having to abandon the performance technical advantages of seating the two violin groups together. This

Score example 12 L. van Beethoven, 7th symphony, 3rd movement, measure 24 ff. And excerpts from the 4th movement, measure 307 ff. (without winds and percussion)

seating arrangement, however, has a disadvantage, especially when due to placement across the entire width of the orchestra, common accompanying figures played by the second violins and violas cover the first violins, particularly when these do not play a closed melody line (e.g., F. Schubert, 7th symphony, 2nd movement, measure 267 ff) (Creuzburg, 1953). Thus the conductor faces a difficult task, at least in the dynamic layering, which for spatial voice separation would not be so critical.

The woodwinds are usually distributed in the two rows behind the strings in a manner such that the flutes and the oboes are in front, and the clarinets and bassoons behind, at least for sufficiently large numbers of players. In this arrangement, the first voices of each group are placed toward the middle so that the quartet of the four solo performers is located with correspondingly good mutual contact in a limited space. Specifically the flutes are generally located toward the left as seen from the position of the conductor, the oboes to the right, while the arrangement of clarinets and bassoons varies.

For a small number of winds, in most cases all musicians can be accommodated in one row, where the highest instruments usually are placed to the left and the additional woodwinds down to the bassoon follow toward the right; the horns are then located at the end. Depending on instrumentation, it sometimes makes sense for individual compositions to depart from this steady rise in pitch. A typical example for this is the symphony in B flat minor by Joh. Chr. Fr. Bach, in which the flute and bassoon both play in opposition to the two clarinets in all movements;

two horns complete the winds (score example 13). In this case it is desirable to place the flute and the bassoon together and not separate them by the clarinets. It thus remains a question of interpretation, to decide whether the "stereo effect" between the woodwind pairs is to be reinforced so that sometimes even the horns are placed in the middle.

A good connection between the bassoon and the cello group is of course important, however this also demands seamless mutual performance with the violins which occurs in many classical works with a so-called Vienna unisono with its octave parallels between violins and bassoons (see score example 14). An excessive separation between the two instrument groups can lead to unpleasant sound effects, since it not only influences mutual listening, but can also lead to a stereophonic separation. In works with smaller numbers of winds, as for example in the early symphonies of J. Haydn and W. A. Mozart this problem is already solved by the fact that all winds are sitting in a row next to each other and the distance between the bassoon and the strings is correspondingly diminished.

The horns are located to the right or left of the woodwinds. The performance technical advantages and disadvantages of both possibilities are difficult to assess, since on the one hand the woodwind group is shifted to one or the other side by the choice of location of the horns, and on the other hand the tonal balance of the remaining brass instruments and the mutual performance with them must be considered. In particular performance technical difficulties arise when the horns are located directly in front of the trombones because the players in both groups will hear the other instruments louder than their own and thus the danger always exists that the loudness level will be raised. Finally the horns, or at least the solo horn must have good contact with the celli, since frequently unisono-instrumentation of these two groups are found, and this often in melodic passages. Contact with the concert master, which is particularly important at the conclusion of the slow movement of the 1st symphony by J. Brahms, will at any rate be only possible visually, so that in this case, positioning the horns to the right side will unquestionably be more advantageous.

In the presence of two to four horns, the heavy brass is generally positioned on the same step with the second row of the woodwinds. When there are eight horns, the trumpets, trombones, and tuba are moved further to the rear, and they are then found in the same line with the timpani and the remaining percussion instruments. For reasons of tonal symmetry a position near the central axis of the orchestra is often preferred for the timpani however there are also seating arrangements for which the timpani are located in a rear corner of the podium. This is the case for example when the basses are positioned in the last row in front of the rear wall.

When positioning percussion instruments it is especially important not to overlook the fact that the traveling time of the direct sound to the more distant musicians of the orchestra can certainly be of the order of 30 ms, so that playing together with desired precision based solely on hearing is no longer possible and visual contact and optical connections through the conductor alone meet rhythmic needs. For difficult joint passages of individual musicians, the choice of seating arrangement with view toward least separation and good visual connection is of the utmost

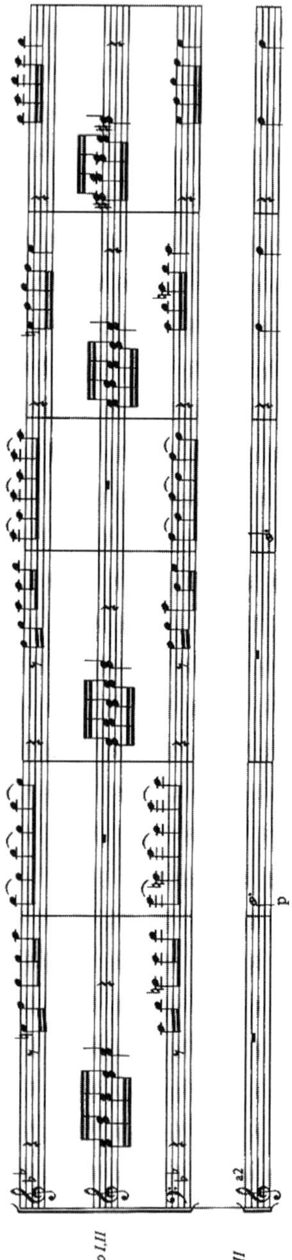

Score example 13 J. Chr. Fr. Bach, symphony in B^b major (1794), 1st movement, measure 73 ff. (without strings)

Score example 14 J. Haydn, symphony No. 94, first part of the trio in the minuet

importance. In this context those percussion instruments are especially important which have their essential tone contribution at higher frequencies as for example small drums, xylophones, or the wooden block. If, for example, the direct sound during its path past the musicians of the orchestra is additionally damped because of its high frequency, it can be attenuated so strongly in comparison to the ceiling reflection that the sound from the percussion instrument is perceived as coming from the ceiling with its corresponding delay determined by the additional path length. This effect particularly adds to the difficulty when a soloist (e.g., clarinet concerto by C. Nielsen) or the concert master is expected to play simultaneously with the corresponding percussion instrument. The score example 15 shows a typical passage for this from the 15th symphony by D. Shostakovitch. For such cases positioning the percussion instruments closer to the front within the orchestra can be an advantage. Also, in the first parts of the Bolero by M. Ravel such a positioning of the small drum is advantageous for the rhythmic precision and dynamic shaping.

The choice of seating arrangement also plays an important role from an acoustical standpoint since the individual instrument groups, depending on their orientation, radiate their sound more strongly or weakly into the hall and thus can also change the tone color. The relationship between the intensity of the direct sound, the early reflections, and the diffuse field which is essential for the clarity of the tonal picture can be influenced by the orchestral seating arrangement because of the connection with the directional characteristics of the musical instruments. This opens valuable possibilities for adapting the tonal body to the room acoustical environment when a musical concept is to be realized in the performance. At the location of the conductor the tonal effects of different seating arrangements are hardly noticeable, since all orchestra musicians are oriented to face the conductor, and thus in all cases the radiation directions remain the same relative to the conductor. However, they are clearly noticeable to the audience.

As a measure to indicate how pronounced the directional dependence of the sound radiation for individual instrument groups is, the range of the statistical directivity factors for three representative frequency ranges is represented in Fig. 7.3. These regions encompass the values for all directions; the maximum value which applies to the direction of strongest sound radiation is of particular

Score example 15 D. Shostakovitch, symphony No. 15, 1st movement, measure 337 ff

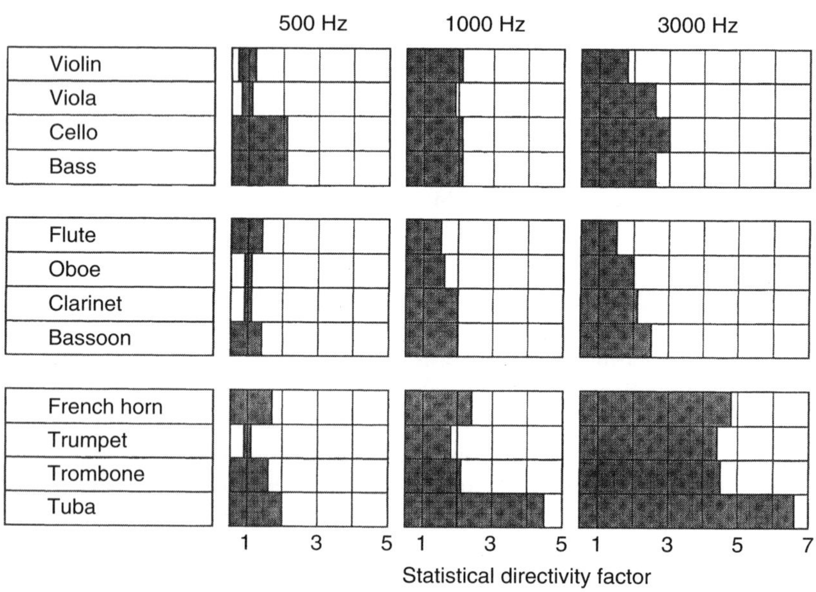

Fig. 7.3 Variation range of the statistic directivity factors of orchestra instruments at three frequencies

importance. In this representation it should be noted that the directional effect for string instruments at higher frequency is more pronounced than for woodwind instruments – it is naturally still far surpassed by brass instruments. The influence of these directed sound radiations on the tonal effect in the hall will be treated subsequently in detail, with its consequences related to performance practice.

7.2 The Tonal Effect in the Hall

7.2.1 String Instruments

7.2.1.1 Violins

The angular regions of preferred sound radiation for violins were already represented in Figs. 4.17–4.19 in the form of histograms. In Figs. 7.4 and 7.5 the essential characteristics of directional effects are once again collected in a schematic overview. They represent those angular regions for which the sound level does not drop by more than 3 dB below its maximum value in a horizontal plane and in a vertical

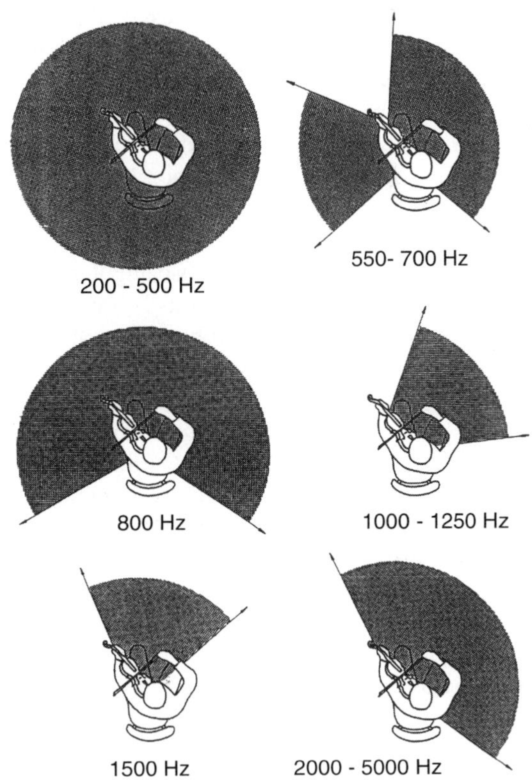

Fig. 7.4 Principal radiation directions (0. . .−3 dB) of violins in the horizontal plane

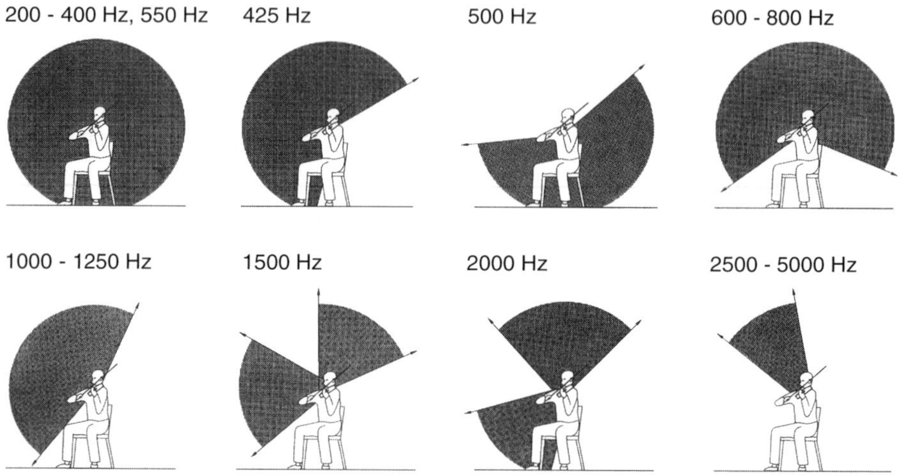

Fig. 7.5 Principal radiation directions (0. . . −3 dB) of violins in the plane of the bridge

plane, with a figure representation of the performer indicating the spatial orientation. These representations are designed to form the foundation for a comparison of tonal effects of the two violin groups under different spatial conditions for varying orchestral seating.

For the first violins, which are seated to the left of the conductor, the angular region which is directed toward the audience is marked in red in Fig. 4.18, in order to emphasize the tonal difference between the two violin voices in the German seating. Correspondingly the region in Fig. 4.17 is marked in red for which the sound radiated from the first violins reaches the audience directly or by a reflection from the ceiling. In contrast those angular regions are colored in blue for which the direct sound of the second violins is important in the German seating arrangement. A comparison of the red and blue areas thus clearly illustrates the tonal difference between the two string groups.

A very pronounced difference appears in the frequency region between 1,000 and 1,200 Hz, which belongs to the high a(ah)-formant and is thus responsible for a strong, bright tone color. In the German seating arrangement, these components are very advantageous for first violins, in contrast, they are radiated significantly more poorly by the second violins. Also, the higher frequencies from 2,500 Hz on upwards are more effective for the first violins, and the tone color receives its brightness and brilliance. Furthermore, it is precisely these high frequencies which in large measure contribute to the clarity and precision of fast passages.

In contrast, sound radiated by second violins has an advantage in the range of o(oh)- and å(aw)-formants, as well as for the nasal contributions (around 1,500 Hz) as perceived by the audience. Thus the tonal effect of the second violins in contrast to the first violins is darker and more covered; the timbre is shifted in the direction toward the violas. This difference in tone color is observed especially clearly in two passages of the 1st movement of the 6th symphony by L. van Beethoven (measure 155 ff and 201 ff), where within four measures, sustained notes of equal pitch are exchanged between the two violins. The duration of these tones is sufficient to enable an audible distinction between the details of the differing tonal structures.

While all this primarily includes the direct sound and the ceiling reflections, Fig. 7.6 makes it possible to include the first side wall reflections in these considerations. In analogy to Fig. 6.9 those frequency regions are represented, for which the side walls of a rectangular hall (as seen to the left or right of the audience) fall within the main radiation region of the string instruments. With reference to the violins it is recognized that for the German seating arrangement strong reflections from the left side wall are to be expected from the second violins, which melt the two violin groups more strongly to a single whole for the audience than apparent at the location of the conductor. At the same time, these representations clarify the significance of the left, and especially the right side wall, for the sound of the first violins, which contributes to the spaciousness of the orchestral sound by appropriate reflections, provided the walls do not diverge excessively.

As can already be seen from Figs. 4.17–4.19 and 7.4 to 7.5 the tonal difference between the two violin groups in the American seating is relatively small. For this seating arrangement relatively good radiation conditions can be expected, even for the high-frequency contributions around 4,000 Hz, provided there is nothing to prevent them from spreading freely. This assumption is confirmed by measurements (from the year 1967) in the Beethoven hall in Bonn, where measurements were performed to determine by how much the sound level of the violins in various locations in the hall is less than the level at the location of the conductor. The result is shown in the diagrams of Fig. 7.7. It is noted, that in most locations the drop for the second violins is not significantly different from that of the first violins. Only in the left portion of the front on the main floor, the performance of the first violins effect some shadowing of the second violins which are seated further back. The relatively even frequency dependence which occurs for the first violins is particu-

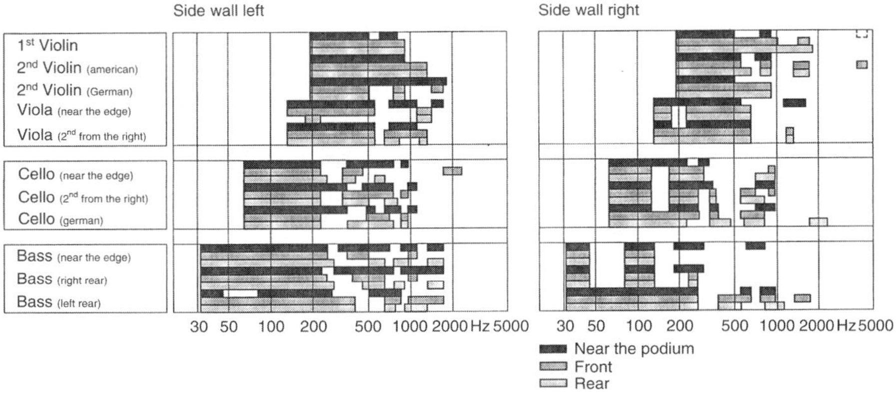

Fig. 7.6 Frequency regions of preferred side wall reflection for strings

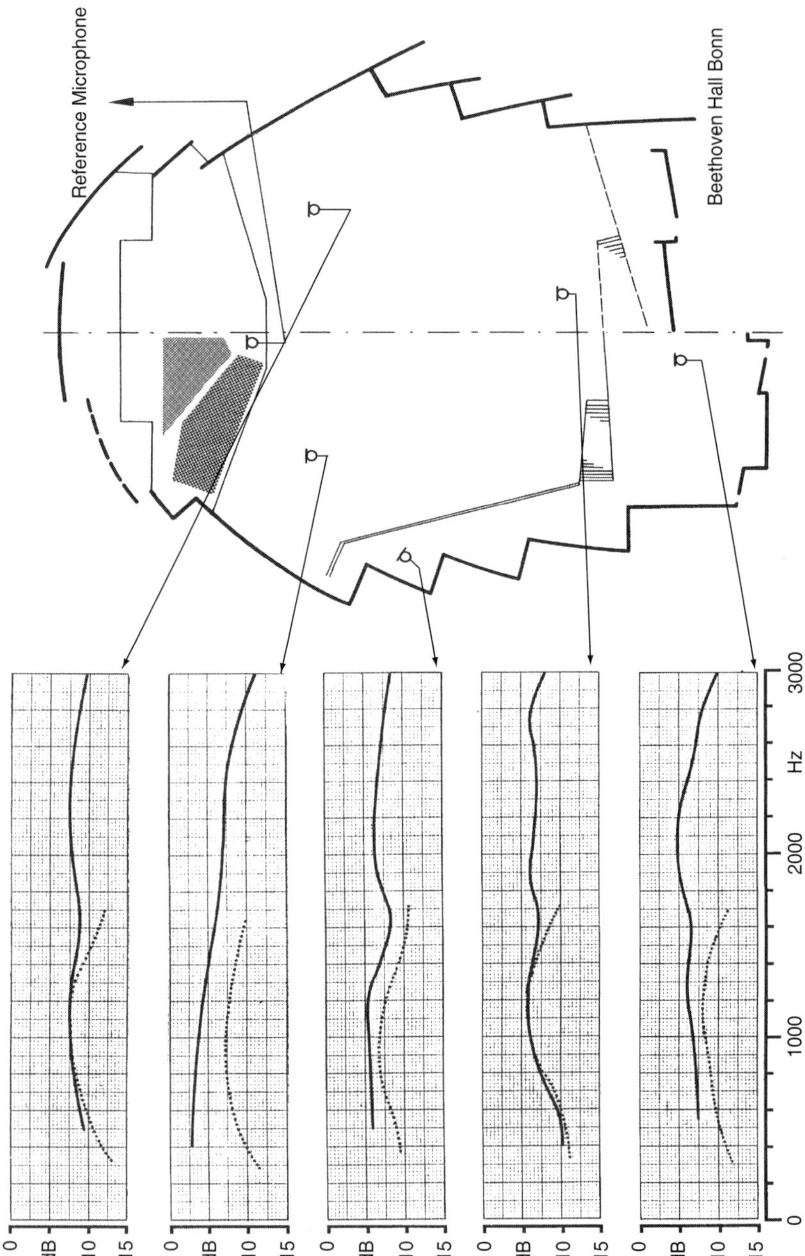

Fig. 7.7 Sound level difference of violins at different locations in the hall compared to the level at the conductor

larly noticeable in these results. Only in the front left of the main floor, are the directional characteristics dominated by the lower components, while the higher frequency components, which are preferably radiated in an upward direction, are not reflected adequately to the front seats of the audience by the ceiling. From an energy standpoint this is an advantage for the rear of the hall. For this reason, the podium in concert halls should not be too high, so that even the front rows receive some of the high portions of the violin sound.

This objectively measured result suggests the conclusion that the tonal effect of the second violins – considered as an individual group – should be significantly better when located next to and behind the first violins, than is the case for the German system. This approach, however, ignores an important psycho-acoustic consideration, which makes these relationships more relative, where the connection to the combined sound of the entire string group has yet to be considered. On the other hand, the tonal difference between the two sections can in many cases have a positive influence on the total impression and should also help the clarity of the voice mixtures. This is particularly the case when both violins have the same, yet temporally displaced motif, which is exchanged, back and forth between the two voices. Score example 16 includes two excerpts from symphonies by J. Haydn and L. van Beethoven, where in each case only the two violins are represented. The upper passage is concerned with the simultaneous performance of the two motifs, which are exchanged between the two voices. When the violin groups sit next to each other the listener has the impression that the one voice only plays eighth notes and the other repeats the motif several times, which consists of quarter notes and one-half note. This certainly does not correspond to the tonal impression intended by the composer. Similarly, the figure grouping of the lower example would become washed out in the American seating arrangement. Furthermore, many tone combinations for two simultaneous melody lines appear less dissonant when the violins are distributed on both sides and both voices can be followed independently, as for example in the first movement of the Jupiter Symphony by W. A. Mozart (measure 220 ff). A missing voice separation in the introduction of the last movement of the first symphony by J. Brahms becomes especially noticeable, where the difference between the alternating 30-s note figures in *forte* and the continuing *piano* parts lose their dramatic effect (score example 17).

Score example 16 *Top:* J. Haydn, symphony No. 103, 4th movement, measure 368 ff *Bottom:* L. van Beethoven, 4th symphony, 4th movement, measure 293 ff. Violin parts

Score example 17 J Brahms, 1st symphony, 4th movement, measure 23 ff. Violin part

The tonal differences between the two string sections under given room-acoustic circumstances of different concert halls naturally depend not only on the direct sound radiation and the reflection of the hall ceiling above the audience but also on the reflections by the rear wall of the stage and the stage portion of the ceiling. For the first violins the reflections from the rear portion of the stage play a role, especially at frequencies between 500 and 700 Hz as well as 1,500 and 2,000 Hz. While the former contributions are partially absorbed by the players behind them, they are still reflected by the rear wall and the ceiling, and thus they, along with the components between 1,500 and 2,000 Hz, can reach the hall after double reflections.

For the second violins it is particularly the important contribution around 1,000 Hz, as well as the very high frequencies, which in the German seating arrangement are radiated upward toward the rear. To what extent they are still effective in the hall depends on the height of the hall (in relationship to the distance of the player from the rear wall). Inasmuch as these components are radiated in relatively narrow bundles, below about 45° upwards, very different reflection effects result, depending on hall height. As the schematic representation in Fig. 7.8 shows for high halls, the strongest sound portions initially hit the rear wall and are subsequently reflected into the hall by an additional ceiling reflection. A slight tilt of the reflection surface is particularly advantageous. If in contrast, the distance of the player from the rear wall is greater than the height of the hall, the frequencies under consideration come back to the orchestra, and even a tilted ceiling cannot avoid this. In such cases, therefore, in the German seating, special attention needs to be paid to the dynamic balance of the two violin groups. In contrast, it is advantageous in high halls to keep the distance of the strings from the rear wall at a minimum when seated in a symmetric arrangement, at the very least it needs to be smaller than the height of the ceiling above the stage.

At first glance the previously named advantages of the German seating, however are placed in question when the tonal effect of audience seating behind the orchestra is considered. The tonal difference between the two violin groups could be perceived as illogical when the second violins have a more brilliant sound than the first violins and thus become unintentionally dominant. In this case the overall impression for the rear seats would be better if the lack of high frequency tonal components becomes as evident for the second violins as for the first, whereby nevertheless the seats behind

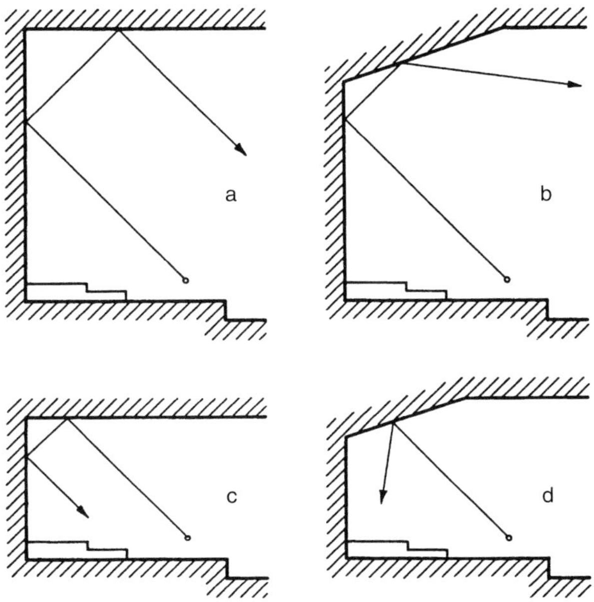

Fig. 7.8 Reflection relations for the very narrowly bundled radiation of high frequencies of the second violins (for German seating) for different hall conditions

the orchestra would not be equivalent to those in the normal direction. At very best, certain compensation can be achieved through appropriately angled reflectors hung from the ceiling. In view of the main radiation angle at high frequencies these must not be located directly above the violins, but rather slightly in front of the orchestra (i.e., in the 0° direction of the bridge plane of the violin). Under such conditions the German seating appears justified even in halls with an audience behind the orchestra, particularly when the points to be considered in Sect 7.2.1.5 are included in the overall sound of the strings.

7.2.1.2 Violas

There are few options in relationship to the normal seating of violas on the concert stage. Locating this section at the right wing of the orchestra leads to the fact that in addition to the components below 500 Hz, which are uniformly radiated in all directions, the components between 600 and 700 Hz are radiated favorably into the audience. These components determine the sonority of the instrument, which also corresponds to the vowel color of a dark å(aw). The frequencies around 1,000 Hz which are responsible for the brilliance of the sound and a certain brightening of the tone, i.e., in the range of the a(ah)-formant, on the other hand, are directed in preference toward the right of the player. The higher partials around 4,000 Hz exhibit a similar characteristic. This means that these components reach the main portion of the hall predominantly by reflections from the rear wall of the hall and the ceiling above the podium. The degree of their effectiveness thus depends on the hall height and the distance of the player from the rear wall in line with the representation for second violins in Fig. 7.8.

For listeners behind the orchestra the directional characteristics of violas result in a tonal change opposite to that of the violins (from the German seating arrangement only for first violins). The comparatively bright viola sounds and the relatively dull violin sounds approach each other in character, while the contrast to the lower cello and bass group increases. It is, however, unmistakable that the violas considered as an individual group develop a very clear and beautiful tone from this direction as can be observed, for example, in certain seats in the Berlin Philharmonie.

The angular region, which, from a tonal standpoint, is evidently very favorable, is directed toward the audience only for special orchestral arrangements without violins. Thus, for example, in the 6th Brandenburg concerto by J. S. Bach, the violas occupy the podium toward the left of the conductor. Also in the A major serenade op.16 by J. Brahms, the violas are often located there. However, in this work winds frequently occupy the position of the first violins. Likewise, the soloist in a viola concerto naturally takes the position at that side so that the instrument radiates a clear tone into the hall.

In contrast, solo viola players within the orchestra most often need to settle for a position in the right wing. For symphonic works with viola soloists (like, for example, the Haydn symphony No. 15, or "Don Quixote" by R. Strauss), however, it does not appear to be a disadvantage to arrange the strings in such a way that the soloist, if possible, is located at the left side. Thus, occasionally a seating arrangement is used, in which the two violin groups find themselves opposite each other on the podium with the violas, – in contrast to the German system – located behind the first violins.

7.2.1.3 Celli

The principal radiation regions for celli are represented in Figs. 7.9 and 7.10 for a horizontal as well as vertical plane. Results for the plane located vertically in the room clearly show the importance of the ceiling reflections for the tonal effect of this instrument group, since the angular regions of especially strong intensities include the ceiling of the hall for all frequencies. This is also one of the reasons why the reflection ceiling in the Philharmonic Hall in New York in its original form, which largely absorbed low frequencies, were such a disadvantage for the cello in the orchestral sound (E. Meyer and Kuttruff, 1963; Beranek et al., 1964).

In detail, these figures show that the radiation at 200 Hz, and in the region from 350 to 500 Hz, occurs predominantly toward the front, so that both floor and ceiling reflections become effective. From 800 Hz on upwards the preferential regions are so steep that the audience is no longer included. Since the upward directed cone becomes narrower with increasing frequency, the tilt of the ceiling, or possibly reflectors hanging above the orchestra, play an important role for these components. This means that in contrast to violins, for which the most important reflection regions of the ceiling are located deeper in the hall, the celli depend on that part of the ceiling directly above the players for reflections. The very high frequencies

Fig. 7.9 Principal radiation directions (0. . .−3 dB) of celli in the horizontal plane

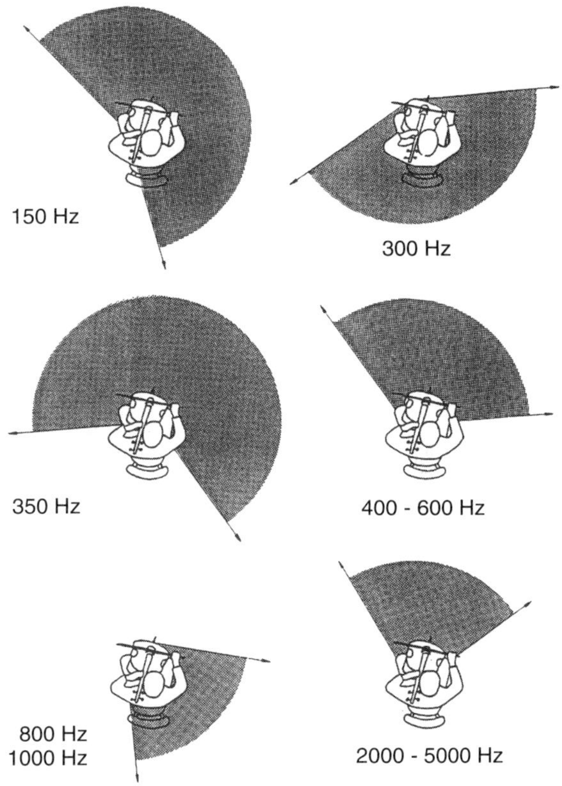

are furthermore radiated in a very shallow direction toward the floor and the front where they are also reflected into the hall unless they are blocked by other players.

The question, concerning the seating of the celli directly in front of the conductor or toward the right on the edge of the podium is of great significance for the tonal effect in the hall. To illustrate the different tonal effects, the angles in Fig. 4.23 (page 319) which in the German and the American seating arrangement are oriented directly toward the audience, are indicated in different colors. A comparison of the yellow and the blue areas clearly indicates, that for the German seating, the high frequencies, which are important for the clarity of fast passages are radiated more favorably into the hall. However, also in the frequency region of 350–600 Hz which includes the region of the o(oh) – formant, which therefore contributes strongly to the sonority of the sound, the frontal radiation is more favorable than the radiation toward the side. It should also be noted that the high register of the A string, used frequently with the fundamental notes of *cantabile* passages, falls into this frequency region. A further advantage of the frontal seating arrangement can be seen in the fact that strong reflections can be expected from both sides for the low frequency contributions below 200 Hz and the tonal contributions in the region from 400 to 500 Hz which are important for the sonority. This means on the one hand, that the tonal effect of the celli is more equalized for

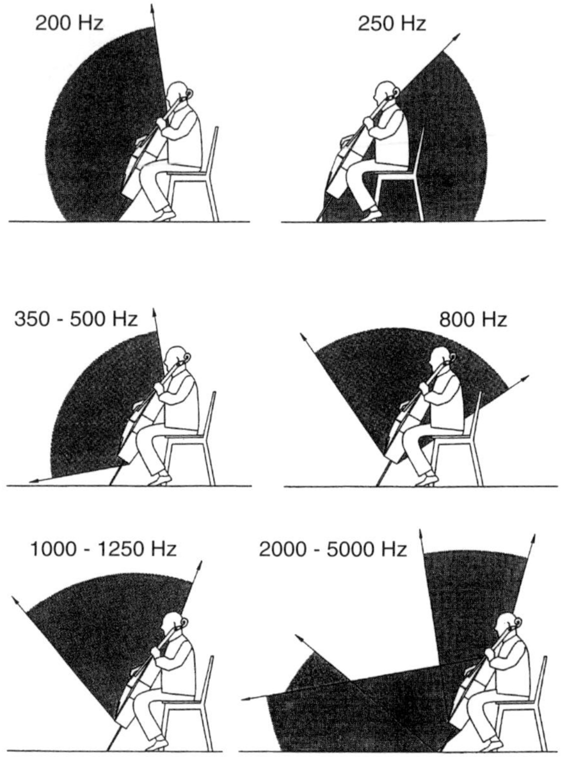

Fig. 7.10 Principal radiation directions (0. . . –3 dB) of celli in the vertical plane

different seats in the hall, and on the other hand, that the spatial broadening of the tone for a *forte* is more pronounced. As the representation of the principle radiation regions and the reflection surfaces lying within these angles show in Fig. 7.11, this symmetry is not achieved by placing the celli toward the edge, so that the sound in the hall is not as full and voluminous.

As is seen in Fig. 7.9 the seating arrangement toward the side leads to stronger contributions in the regions from 250 to 300 Hz and from 700 to 900 Hz; the tone color, therefore, becomes darker. However, it should be emphasized that placing the celli toward the edge certainly does not lead to a louder impression in the hall. This is again supported by recent investigations by Winkler and Tennhardt (1993).

In this context, recent investigations by Prince and Talaske (1994) are very interesting. They measured impulse responses for various positions of a sound source on the stage for sound sources with various directional characteristics in a hall for an audience of 1,200. These results show, among other things, that at least in the middle frequency region the direct sound of the celli is attenuated by 6 dB on the average, when they are placed to the right of the conductor, rather than directly in front. In both cases, the strongest reflections from the stage region have the same intensity, however for frontal arrangement they are slightly delayed. They are clearly weaker than the direct sound (toward the right by about 3 dB, in the center by about 9 dB because of the stronger

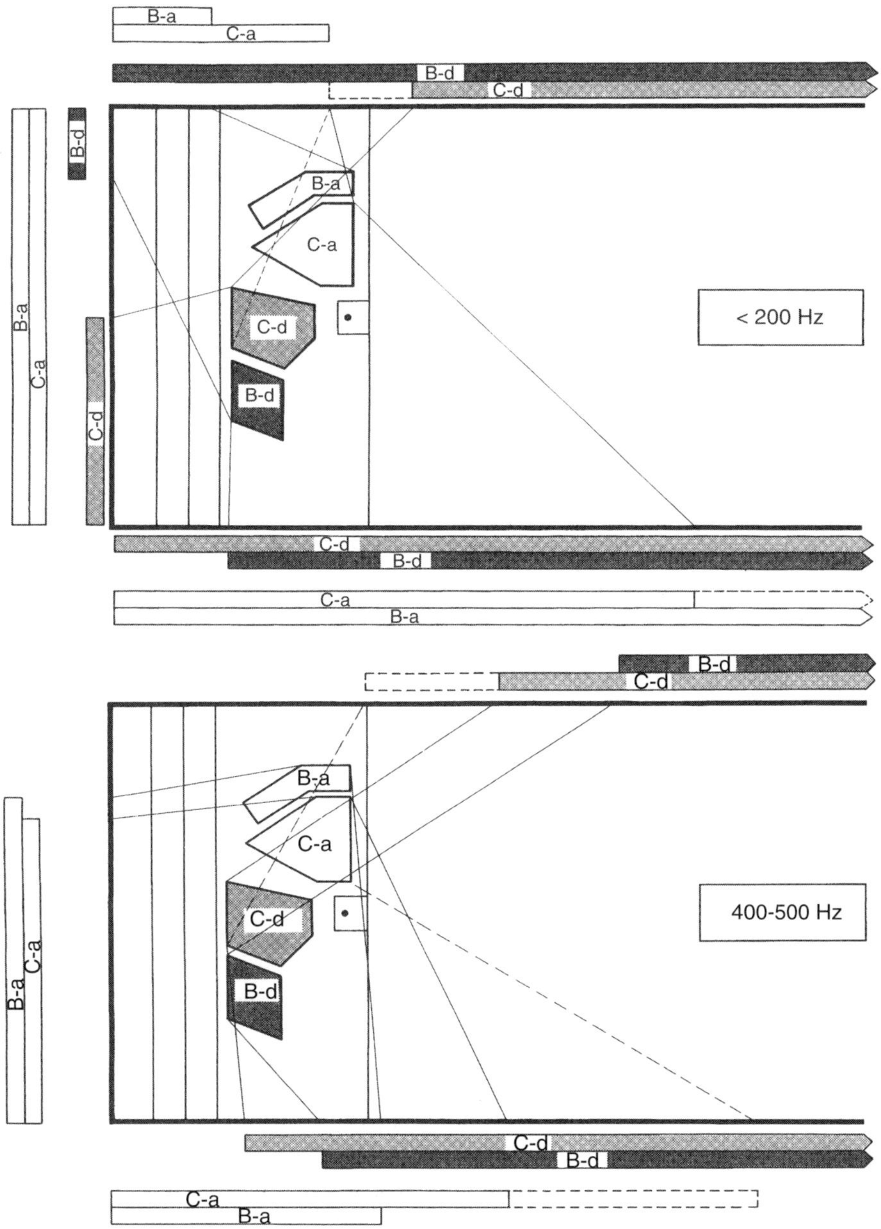

Fig. 7.11 Preferred reflection regions for celli and basses for those frequencies below 200 Hz for which the radiation is not spherical, as well as for the frequency region around 400–500 Hz for two different seatings

direct sound). These results again show the advantages of the frontal arrangement both by reasons of energetic considerations as also the precision. In comparison, the direct sound of the second violins, when placed to the right of the conductor, are lowered only by 4 dB when compared to the frontal arrangement, which is also compensated by the fact that the strongest reflections from the region of the stage (with a delay of about 25 ms) are by 2 dB stronger, and thus exceed the direct sound slightly.

A further problem for celli seating at the side in wide halls lies in the fact that the strongest sound intensity is directed toward the left wall of the hall as seen from the audience, and thus a clear echo is audible when there is no direct sound coming from the left wing of the orchestra. A typical example for this is given in score example 18 where the forte entrances of the celli and the basses are totally open.

The tonal effect at different locations in the hall for American seating is represented in Fig. 7.12 in several diagrams, which again represent the level difference in relation to the position of the conductor. In these results it is particularly noticeable that the celli on the right side of the main floor, i.e, relatively close to the players, are most strongly attenuated in the important frequency region around 400 Hz and are thus strongly influenced in their sonority. In contrast, on the left side of the main floor, i.e., in the direction of sight for the players, the higher tonal contributions are especially pronounced. In the gallery portion located immediately in front of the celli, the tonal effect is very balanced, while in the rear of the main floor a drop in the region of the o(oh)-formant is present which must be judged negatively. These examples demonstrate the advantages of the seats located directly in front of the players, in contrast to those located toward the side. Considerations of the uniform tonal effect for violins, as shown in Fig. 7.7, indicate that the reason for this rests with the sound radiation of the celli and not the nonuniformities of the room acoustics.

The fact, that the components which are responsible for the sonority and the richness of sound are radiated better into the hall for a frontal seating, effects not only the tone of the cello group as a single instrument, but this also is an advantage when playing in octaves or in unison with violins. The theme in the last movement of the 1st symphony of J. Brahms (score example 11), which is initially introduced by the violins only (measure 61 ff), but then appears in unison with the first violins and celli, receives a more closed tonal character for the German seating arrangement than when the celli are located toward the side. The celli enhance the sound of the violins through the strong components in the region of the o(oh)-formant, a frequency range which violins characteristically do not radiate strongly. The overall impression thus gains in tonal richness and expression without darkening the timbre by overemphasizing the low frequency contributions. Reference was already made to the effect that, for the American seating arrangement, the two instrument groups are not blended as well as a result of the stereo effect.

The preferential radiation of high frequencies for the frontal seating is particularly advantageous for fast passages, which therefore gain clarity and precision. However, even in slower passages the plasticity of the melodic line increases, while the initial transients are more easily recognized. The higher frequency com-

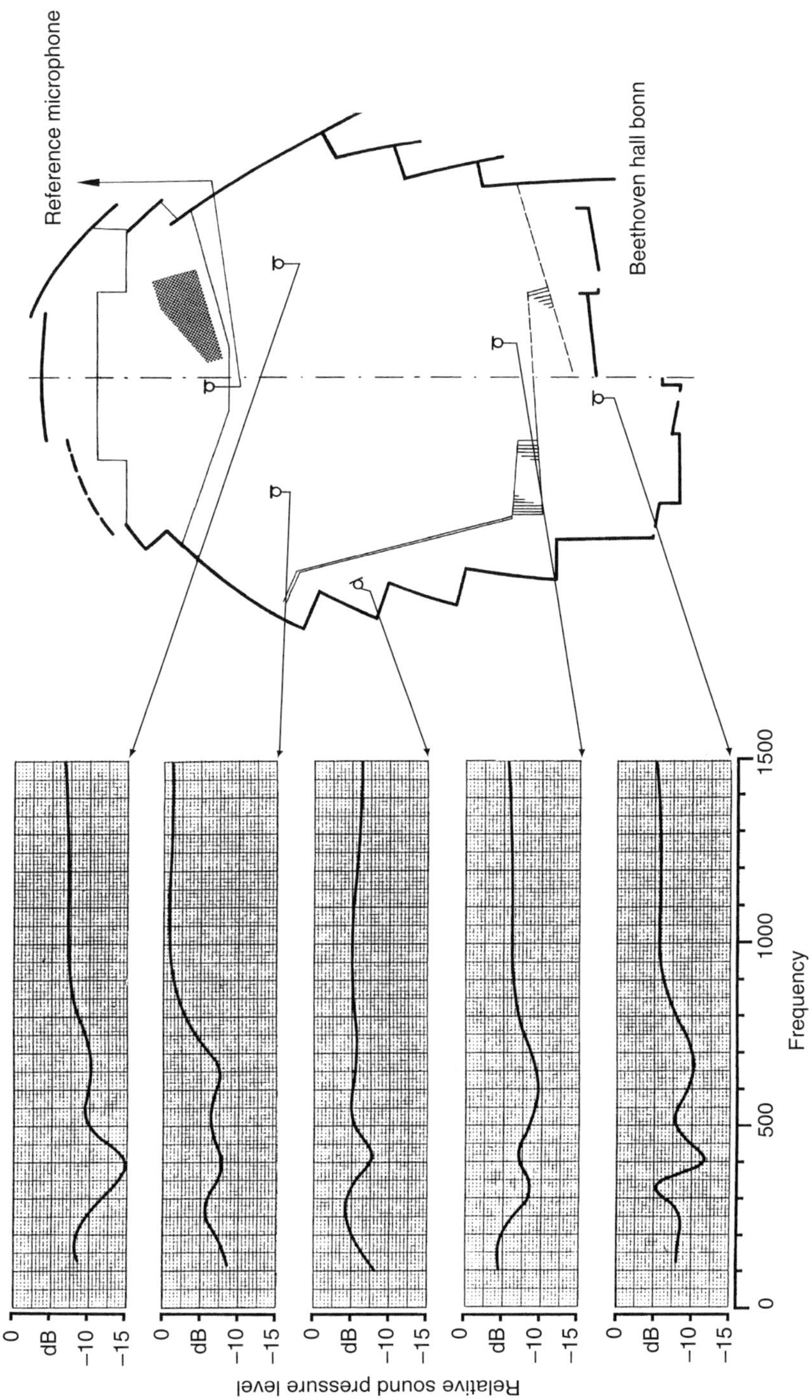

Fig. 7.12 Level differences of Celli at various locations in the hall in comparison to the level at the location of the conductor.

ponents are of course especially important for rapid repetitions of the same note, since for them the rhythmic structure only becomes audible through the articulation of the attack.

As an example for such a passage, several measures from the 2nd movement of the 5th symphony by L. van Beethoven are given in score example 19. (According to the metronome notations, added by the composer later, the 1/32 notes are to follow each other with a separation of 1/6 s) It should also be noted, that in a previous parallel passage (measure 38) for the same execution by the upper strings, the celli play the E as a continuing note; on occasion, it is very difficult to make this difference audible, particularly in large halls.

The Furtwängler variation of the American seating offers similar tonal conditions for the celli in comparison to the German seating, since generally the first or the second chairs sit nearly directly in front of the conductor. For the rear desks, high frequency absorption by players sitting in front, increases at any rate. Consequently, a stepwise rise of the individual rows is not a disadvantage, however, many concert stages do not provide that option.

Score example 18 L. van Beethoven 8th symphony, 4th movement, measure 277 ff

Score example 19 L. van Beethoven, 5th symphony, 2nd movement, measure 87 ff. Score excerpt without horns and bassoons

7.2.1.4 Double Basses

Situating the double basses toward the left or right rear corner of the orchestra means a change of the angular region directed toward the audience by approximately 90°. The difference between these two positions, expected by reason of the directional characteristics, can be noted from the representations of Fig. 7.13; it can also be noted from the recorded angular regions of preferred sound radiation, that rear wall reflections can enhance the sound from the bass especially in the frequency regions between 200 and 250 Hz as well as 500 and 800 Hz.

An additional strengthening effect is noted for very low frequencies, when the instrument is less than a wavelength away from the wall. Under those circumstances the reflection interacts with the instrument itself so that for certain frequencies the radiated energy increases and for others, decreases (Skudrzyk, 1954). In Fig. 7.14 the increase or decrease of this sound power level, in dependence on frequency is represented. The individual curves refer to different wall constructions; an instrument to wall distance of 75 cm is assumed. The broken curve shows that a "hard" stonewall has the effect of strengthening the sound for

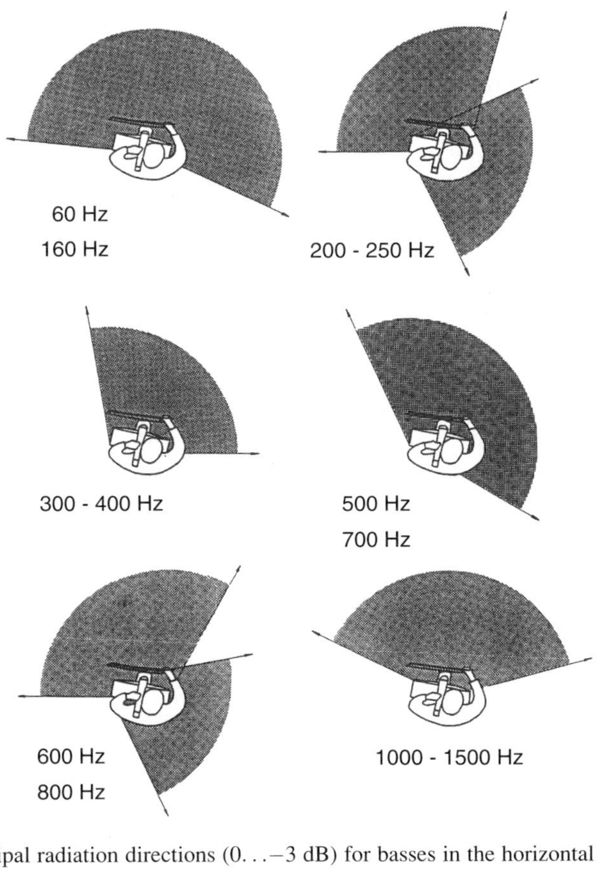

Fig. 7.13 Principal radiation directions (0. . .−3 dB) for basses in the horizontal plane

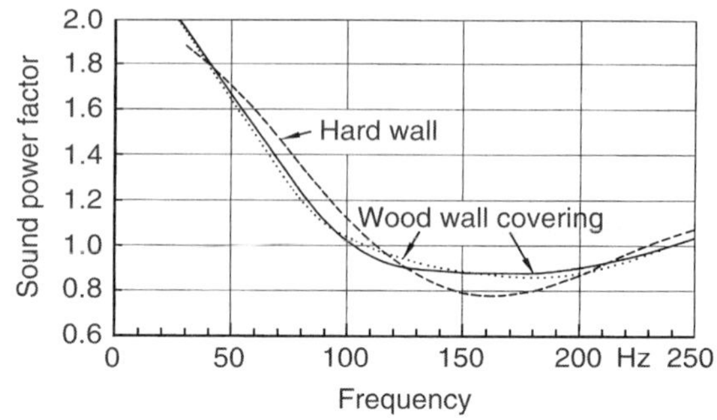

Fig. 7.14 Change of radiated sound power for a source located 0.75 m in front of a wall. *Dashed curve*: hard wall. *Dotted curve*: wood paneling ($\alpha = 0.7$ at 125 Hz), *Continuous curve*: wood paneling ($\alpha = 0.7$ at 160 Hz)

frequencies below 100 Hz, that, however, in the region of 150 Hz the sound power drops. If the rear wall is covered with wood paneling, which has an absorbing effect at these frequencies, a curve is flattened without decreasing the strengthening at lowest frequencies. Thus the bass gains intensity at the low end, where evidently the wood paneling is more favorable than the hard wall.

The effect decreases as the bass is moved further from the wall; at the same time the drop is shifted toward lower frequencies. At a distance of 1.5 m it is moved to the region of the air resonance of the instrument, and thus the energy decrease is equalized. For a distance of 3 m, it moves to 40 Hz where precisely those notes which are to be enhanced are weakened. For all this to be effective, the wall naturally has to be sufficiently high to reflect low frequencies. In order to achieve effectiveness down to 250 Hz, a height of at least 3 m is required, depending on the distance from the bass. For frequencies to 30 Hz, 4–5 m are required (Meyer, 1975). The result of all this is, that, as a matter of principle, from a standpoint of tonal quality, it is not desirable to place the basses so that they do not have a rear wall.

In this connection, the question arises concerning the importance of platforms or supports, capable of vibrating, placed below celli or especially basses. While early measurements (E. Meyer and Cremer, 1933) suggest that a sympathetically vibrating floor has no effect on the sound level in the hall, because the contribution of the radiated energy is too small, more recent investigations (Askenfelt, 1986) certainly recognize the possibility of sound enhancement. It should be noted that these experiments were carried out with extremely thin floor material (12.6 mm plywood). As shown in Fig. 7.15, a level rise (in the hall) as compared to a rigid floor by about 3 dB near 50 Hz and about 5 dB at higher frequencies was measured. Inasmuch as the ear is particularly sensitive to level differences at lower frequencies (see Fig. 1.1), these differences in tonal impressions are clearly noticeable. The psychological feedback for the player should also not be underestimated since the performer senses the floor vibrations as room resonances-even when the floor does not contribute significantly to the radiated sound.

It is thus not surprising when in recent times – as e.g., in the Morton H. Meyerson Hall in Dallas - a floor, capable of vibrating is installed above a special hollow room in the region of the podium, where usually the celli and basses are found. Naturally the acoustic effects of such a construction are tied to a very specific orchestral seating arrangement. This reduces flexibility in relation to seating arrangements.

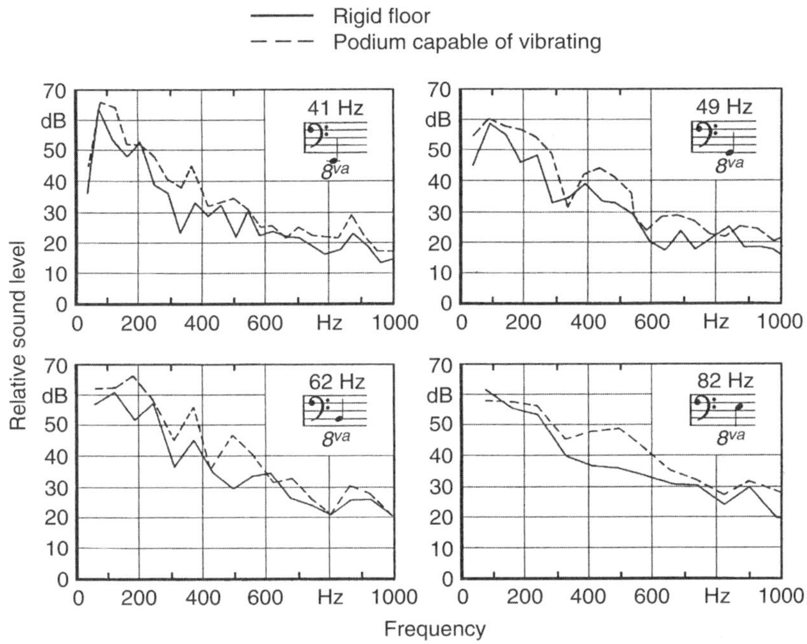

Fig. 7.15 Influence of a podium capable of vibrating on the sound radiation of a bass (after Askenfelt, 1986)

From another standpoint it should not be overlooked that a level rise in the hall by 3–5 dB means that at least the same amount of energy, as that radiated into the air is transmitted through the tail pin into the floor. This means an additional energy loss for the instrument for which bowing technique can compensate, which in turn leads to a more rapid decay for *pizzicato* playing and thus a dryer sound. This could certainly be perceived as a disadvantage for a vibrating podium.

The angular regions oriented toward the audience in the diagrams (Fig. 4.24, page 319) are again emphasized by red and blue coloring in order to clarify the tonal effects of basses for differing seating arrangements. It can be noted that the colored areas are relatively equivalent for wide frequency regions, especially for the high frequencies nearly equal portions fall into both angular regions. Only between 300 and 500 Hz and in the regions around 700 and 1,000 Hz the red areas dominate. Thus the direct radiation is slightly preferred for seating on the left side at these frequencies. The region from 300 to 500 Hz includes parts of the u(oo)- as also of the o(oh)- formants and therefore contributes characteristically to the dark and full nature of the bass sound.

The advantage of locating the basses on the left side, however, becomes very clear when considering reflection conditions in Fig. 7.11. Similar to the case with the celli, utilizing both sidewalls for relatively strong reflections at low frequencies contributes to the spatial fullness of the sound.

The higher frequency contributions are generally not as important for the bass, since basses frequently double the celli and occasionally bassoons. In these cases, the tonal picture is at any rate supplemented by those instruments which are rich in overtones and most often are played an octave higher. For passages which do not provide enhancement by other instruments, good radiation possibilities for the entire bass spectrum is important. This refers either to independent bass voices, as for example the introduction of the 1st movement of the 7th symphony by F. Schubert (measure 17 *ff*, here the melody is shared by celli and violas) or to thematic changes, as for example in certain places in the fugue of the Jupiter symphony. In this example, the difficulty lies in letting the entrance of the basses (e.g., measure 50) stand out sufficiently clearly against the four upper voices where the higher frequency tonal contributions play an essential role. Naturally, for such passages the somewhat greater intensity of the lower components, with an arrangement of the basses toward the left, becomes an advantage. Reference should also be made to the fact that the position of the basses near the left wing is an advantage for the solo bass player, when, for example, in several Haydn symphonies (No. 6, 7, 8, 31, 72), solo passages or an entire solo variation is to be performed. Finally, it is also an advantage for the solo in the 3rd movement of the 1st symphony of G. Mahler, when the solo bassist is not located at the outer edge of the orchestra, but more toward the middle, otherwise the performance with the timpani falls apart.

Also, the way in which the Vienna Philharmonic positions the basses next to each other in front of the rear wall of the podium (see Fig. 6.1) is advantageous for the radiation of the entire spectrum, particularly since the reflection possibilities of the rear wall are optimally utilized. However, this increases the distance to the celli so that the contact with this group is somewhat more difficult. In particular, when there is no wall located at the side of the podium this arrangement provides a very effective adaptation for the basses in acoustically unfavorable halls.

If, however, the basses are placed on the right side behind the violas, or possibly the celli, clear up to the front edge of the podium, a loss of high frequencies in the direct sound results, as seen from Figs. 4.24 and 7.12. The intensities also drop slightly in some of the lower registers, which becomes all the more significant since these frequency contributions are already weakened when considering the propagation of the sound into the audience. In addition, these intensity reductions are especially noticeable because of the characteristics of the ear (as a result of the narrow spacing of curves for equal loudness). Since for this seating arrangement the instruments, including especially the first desk, are unfavorably turned inward in relation to the direction of sight, a shallow tone effect cannot be avoided.

7.2.1.5 Combined Sound of the String Sections

Up to this point the predominant concern was the tonal effect of the individual instrument groups. Yet, the distribution of voices on the podium is also important for the combined sound of the strings. The German seating, with violins distributed equally on both sides and the celli oriented toward the middle, comes close to a symmetric arrangement, especially since the basses in the left wing can be brought closer to the conductor than in the American seating. This symmetry naturally relates predominantly to the direct sound of the individual instrument groups relative to the audience; it is, however, also supported by the early wall reflections, which effect the spatial broadening of the tone. On the other hand, this means that halls of lesser acoustic quality (with lower spaciousness) are more critical in relation to the geometric tone balance.

Especially during the Romantic period, there was no lack of attempts to perfect this symmetry. Thus, H. Richter located the eight basses in two groups of four in the even divided both the celli and the basses into two halves, which were located on both sides of the conductor behind the violins (Becker, 1962). This division of the low voices was also frequently observed in early classical orchestras. Even today several Swiss orchestras, as well as the Vienna Philharmonic, locate their basses in front of the rear wall of the podium over the entire width of the orchestra.

The division of the two violin groups onto both wings of the orchestra provides a pronounced octave performance, or a unison of the two groups in a characteristic fashion, since, at least in the region of low frequencies, very dense and temporally stretched reflection sequences from both side walls supplement the radiated direct sound, which is already very broad (see Fig. 8.2). It is this spatial emphasis in the tonal effect, which Gustav Mahler likely intended when he stated: "Permitting the bright and dominant passages of the second violin to run in unison with the first violins leads to a significant improvement of the tonal effect and to the achievement of a convincing tone brilliance of the violins. This cannot be explained only by the increase of the numbers, but must rest on the acoustic principle which creates such a vital tone and brilliance effect resulting from the two sound waves which meet from both sides" (Paumgartner, 1966). Thus, in Mahler's 6th Symphony there is a passage in which the two violin sections are playing in unison, however are annotated with opposite crescendi and decrescendi so that the tone receives a spatial motion (shortly before the conclusion of the 3rd movement; "always with a feeling of motion, increasing and decreasing.")

The instrumentation of the 14th part of the Bolero by M. Ravel (score example 20) suggests an effort to obtain a spatially broad and uniform tonal effect. Here the violins play the theme in a four-voiced parallel passage beginning with a major triad doubled in octaves. It is particularly noteworthy that the division is not intended to have the first

Score example 20 M. Ravel Bolero measure 219 ff. excerpts: strings

and second violins each take two voices but instead both groups are divided into four sections. When all violins are seated together this division loses its significance, while, when the violins are spread over the entire width of the podium, a closed tonal effect of great spaciousness is achieved. Similar instrumentation is also found in the first movement of the *Symphonie Fantastique* by H. Berlioz (measure 410 ff and 428 ff).

In contrast, the linear American seating arrangement leads to a strong bass emphasis of the right wing and a concentration of the high voices on the left side. While this means greater clarity in relation to the distinction of the instrument groups for the listener, on the whole it leads to geometrically not very even tonal balance and consequently to a less fused complete sound. This division of the high and low string voices proves especially crucial in passages such as the excerpt from the beginning of the Finale of the 5th symphony by P. Tchaikovsky shown in score example 21 where, after the full sounding string entrance, the sound of the E in the second measure suddenly only comes from the right wing even though this note is an integral component of the motif.

From the standpoint of radio recording techniques, there are those who reject this linear seating arrangement, for a number of reasons, among them the artistic argument, because a true stereophonic sound is not fully utilized, but rather an impression is given which approaches a pseudostereophonic effect achievable by frequency differences (Briner-Aimo, 1966).

Furthermore, arranging the bass group toward the side leaves some seats in the audience near the orchestra with an unsatisfying tonal picture, because of the emphasis on the bass. The proximity of the second violins in the German seating order can also lead to excessive emphasis on accompanying figures in many compositions; however, this disadvantage should generally be less important than the excessive emphasis on the bass.

Beyond these objectively measurable differences in relative to sound radiation for different seating arrangements, a psycho-acoustic effect also plays an important role for the tonal impressions of these two string groups for the audience. The brilliance of the first violins does depend, among other things, also on other voices

Score example 21 P. Tchaikovsky 5th symphony, 4th movement. Beginning (without winds)

located in close proximity: when the first violins are located in front of the celli and the basses, their sound is perceived as clear and more brilliant than when they are located in front of the second violins, even though nothing has changed in their sound from a purely physical standpoint. The relevant concept is the more or less strong masking of the higher frequency components for the direct sound of the first violins by the sound of the other instruments (Meyer, 1987). This phenomenon is illustrated by a number of sound power spectra of *forte*-sounds represented in Fig. 7.16. Since the power spectra of string instruments are essentially not changed with intensity, but practically are shifted only up and down, the following considerations can also be transferred to different dynamic levels.

Considering, for example, the overtones in the region of 3,000 Hz for notes in the middle register (see the arrow in Fig. 7.16), one observes that the level of the cello is approximately 20 dB below that of the violin. In addition, it is noted that for the violin, the level below the 3,000 Hz region rises by 5 dB per octave, or alternatively sinks, when one moves to a higher, or correspondingly lower register. In contrast, the level for the 3,000 Hz region in the cello practically does not change between the low and mid registers; in the high register, however, it rises by nearly 10 dB. The level of the bass still lies clearly below the curve for the cello at these high frequencies. With the assumption that the number of first violins is approximately equal to the sum of cellos and basses, the first violins dominate against the low strings by an order of 15–20 dB at the high frequencies in the direct sound of the orchestra.

When the first violins are seated in front of the second violins, a dominance of only about 5 dB results for the first violins at the 3,000 Hz components when the two voices are separated by approximately an octave, which becomes even less for narrower voicing. This causes a partial mutual masking in the ear for the high tone

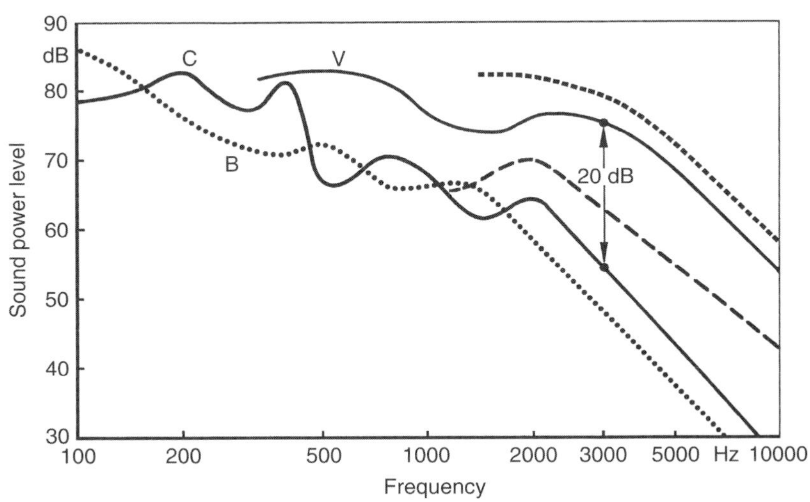

Fig. 7.16 Envelope of sound power spectra of string instruments for *forte*. *Continuous*: midrange (Vl and Vc). *Dashed*: high register (Vl and Vc)

contributions in the direct sound: they are perceived as weaker than when heard from the first violins alone. This means a reduction in brilliance. When in contrast the first violins are seated in front of the low strings, such masking for the sound of the violins does not occur because of the larger level difference of 15–20 dB, and the brilliance of the first violins is preserved. On the other hand, even the high tonal contributions of the second violins are masked in the audible sound impression when both violin groups are seated together. In this the more advantageous sound radiation of the second violins in comparison to the German seating is compensated at least in part. And in fact in a direct comparison of the American and German seating arrangement it is astonishing to experience how unexpectedly clear the second violins sound for an audience in the hall. In this context it is also noteworthy, that the second violins – as emphasized by Herbert Blomsted – are forced to be more independent, and use more initiative (Blomstedt, 2000).

Finally, it is naturally always a question of personal conception, which seating arrangement is preferred. In all this, certainly, the fact plays a role, that, since the changes in the 50 s, a new generation of conductors (and music critics) have come, who have only known the seating arrangements of that period. Since then, increasingly opportunities have arisen to hear orchestras, particularly with internationally renowned conductors, which place the two violin sections on opposite sides. Thus, the beginning of the twenty-first century is characterized by a clear change in the trend towards spatial tone balance as represented by the German seating.

7.2.2 *Woodwind Instruments*

7.2.2.1 Flutes

Since the sound radiation of flutes does not only depend on frequency but also on the order of partials, it is not possible to represent the angular regions of preferred sound radiation as simply as for other instruments of the orchestra. Thus in Fig. 7.17 the principal radiation regions of the first four partials of the not overblown notes, i.e., for the positions of C_5 to D_6, are recorded and are supplemented by the preferred regions at frequencies 3,000 Hz and 8,000 Hz; where the two images on the right in the upper and middle row apply for the first and second partial of the overblown notes E_6^b to D_7.

Since the directional characteristics of the flute are essentially rotationally symmetric, conclusions about the radiation upward can be drawn from the schematic representation of the individual frequency regions of the horizontal plane, when the shadowing effect of the head of the performer is considered. As noted from the pictures, this effect is observable in the horizontal plane approximately from 1,000 Hz on upwards toward the rear and toward the left side of the musician. One must also count on a level decrease in the upward and rearward directions in comparison to the free radiation toward the front. However, there is no noticeable shadowing effect vertically upward.

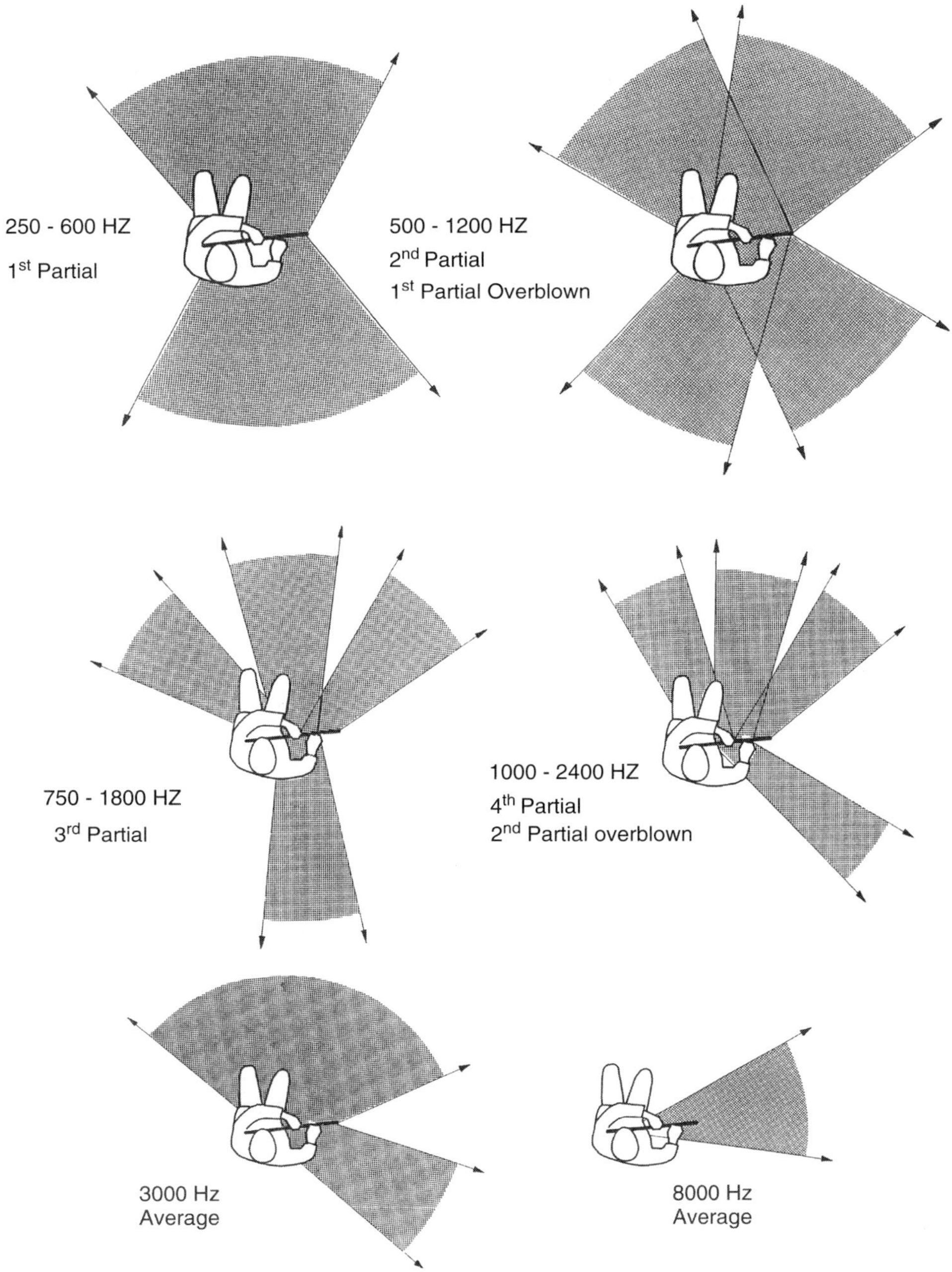

Fig. 7.17 Principal radiation directions (0. . . –3 dB) of flutes

A review of the images of Fig. 7.17 shows that the sound of the flute is radiated very strongly into a relatively broad angle toward the front (naturally also into the corresponding upward directions). While a closed region of high amplitude is present only for the fundamental, the separation into several maxima for the overtones is not clearly noticeable in the tonal effect in large halls. Only for very high frequencies are the harmonic partials preferentially radiated toward the right side, while all other tone contributions are weaker in this direction than in the direction of sight. Since the highest frequencies are most pronounced at great loudness levels, the *forte* easily receives a certain sharpness toward the right side of the performer. In contrast, the very high frequency noise contributions, since they predominantly come from the mouth-hole, are radiated more strongly in the direction of sight, than toward the side, so that the articulation becomes more clear toward the front than toward the side (see Fig. 4.9).

Because of these directional characteristics, the generally customary seating arrangement of flutes on the concert stage with the viewing direction turned toward the audience is considered acoustically favorable. The free radiation toward the front can, however, be impeded more or less by the string groups sitting in front, when the stage is high and the stepping of the rows of winds is flat, or the flutes possibly even sit at the same level as the strings. However, even in these cases, reflections from the hall ceiling are still effective, since it lies within the angular region of preferred sound radiation. In contrast, placement to the left of the conductor as it is occasionally found in works without violins (i.e., Serenade Op. 16 by J. Brahms), or for pure wind compositions, is not advantageous since the flute sound loses in substance.

Sound reflections from the rear wall behind the orchestra do play a role in the standard frontal seating arrangement for flutes, especially at low frequencies; they can reinforce the fundamentals and the first overtones in the low register, for higher passages this reinforcement is essentially limited to the fundamental. Consequently, the rear wall reflections not only effect an increase in loudness, but also support the round tone of the flutes to the extent that they are not shadowed by the winds sitting behind them. As the summary in Fig. 7.18 shows, sidewall reflections for flutes

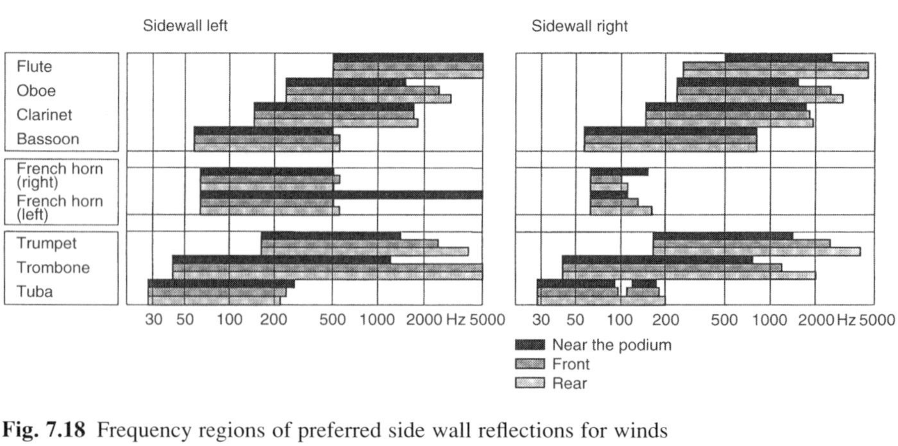

Fig. 7.18 Frequency regions of preferred side wall reflections for winds

include that frequency region which for all woodwind instruments mostly extends upwards. Thus, strong reflections from the left side (as viewed from the audience) reflect the less desirable very high components into the hall. However, this occurs only when the sound is not absorbed by a performer sitting in front or toward the side, as is the case for very wide stage dimensions where sidewalls also diverge in the direction of the hall.

For a soloist, standing in front of the strings, rear wall reflections naturally only play a minor role, the soloist therefore relies especially on the favorable radiation of the direct sound. With reference to the directional characteristic, it therefore appears advantageous when the soloist faces the audience, provided contact with orchestra and conductor is not compromised. The sound is stronger and fuller than when the open end of the flute is turned toward the audience. An additional gain can be achieved through the reflecting floor surface in front of the performer, provided there is at least an expanse of 2 m in the direction of sight.

7.2.2.2 Oboes

The principal radiation directions of the oboe are assembled in Fig. 7.19 for several frequencies. These representations are based on the directional characteristics of the instrument as given in Fig. 4.10, yet at the same time they take into consideration the shadowing of the sound by the performer as expressed in the diagram of Fig. 4.11. As can be seen, the tonal components in the region of the principal formant, which lies near 1,000 Hz, are quite advantaged as far as radiation into the hall is concerned. The principal region extends approximately from the horizontal to an angle of 50° upward so that the audience is reached directly as well as by a reflection from the ceiling. Also at 2,000 Hz, the direct sounds proceed horizontally and in a slightly angled direction toward the audience, however, the reflections from the ceiling depend very much on its inclination. If the ceiling is horizontal, the reflections primarily return to the orchestra. However, if there are angled reflectors, sound reflection toward the audience is also possible.

At higher frequencies, sound radiation occurs primarily toward the floor so that the intensity reaching the audience depends largely on the reflectivity of the floor covering. Within the orchestra these tonal contributions are also rather strongly absorbed by the musicians sitting in front. The resulting unavoidable weakening of the highest tonal components, however, is certainly aesthetically pleasing, since the oboe sound otherwise becomes too shrill and cutting. For frequencies near 4,000 Hz, a special reference needs to be made concerning the distance of the reflection surface in front of the player. In this frequency region, a very steeply reflected sound bundle occurs, which reaches the audience only after a further reflection from the ceiling, provided the ceiling has the appropriate angle.

The increasing shadow effect toward the rear by the performer not only affects the weaker sound for the audience possibly located there – even for a rising angle of 30° the front-back ratio retains a value of approximately 15 dB for the important frequency contributions – but it also means that the reflection wall

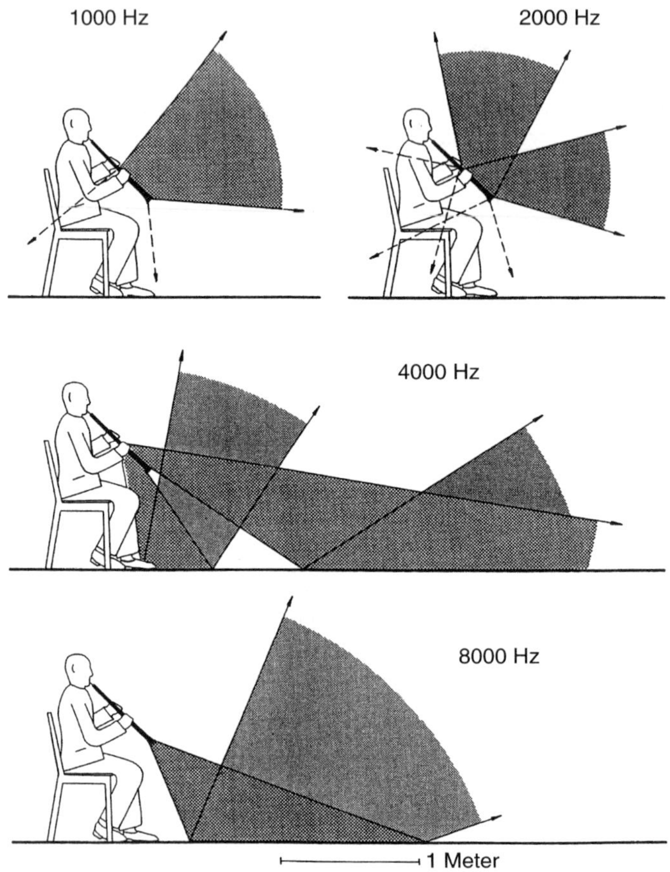

Fig. 7.19 Principal radiation directions (0. . .−3 dB) for oboes

located behind the player does not have very great significance for the oboe. Consequently, the unimpeded sound spreading toward the front and upwards is essential for a good tone development.

The weak radiation of the oboe into the angular region behind the player, with particular damping of the high frequency components, offers the possibility in concert halls, where there is no separate and acoustically usable location for the oboe from the distance (*"lontano"*), as in the 3rd movement of the *Symphonie Fantastique* by H. Berlioz, to perform these passages from the orchestra stage. When the oboist is seated with his back facing the conductor – which, because of the free location of this duet passage, should not influence the ability to perform with the orchestra – the reflections from the podium rear wall and the ceiling dominate strongly for the audience, so that the tonal picture obtains the desired spatial indifference, without localizing the instrument within the orchestra. This effect, however, can only be achieved because the oboe has a much weaker sound radiation toward the back than in the direction of sight of the player; for example, for a flute this would not be possible, furthermore the audience seated behind the orchestra would observe an opposite effect.

Since the radiation of the oboe toward the side is by nearly 5 dB weaker, reflections from the sidewall next to the orchestra are only marginally significant, while in long stretched halls, reflection surfaces which lie deeper in the hall, have a reinforcing effect. Locating the oboes next to the conductor, as might be suggested, for example, for the Serenade for Winds, Celli, and Basses by A. Dvorak, the decreasing intensity of the instruments toward the side would also become noticeable. A frontal placement for the oboes (highest section in this work) should therefore be preferred, especially in large halls. Corresponding considerations naturally are valid also for compositions for winds only.

7.2.2.3 Clarinets

As shown in Fig. 7.20, the angular regions of strongest sound radiation for the clarinet are similar to those for the oboe; only in the frequency region near 2,000 Hz, the upward directed sound bundle is missing, furthermore, the radiation in this region is directed in a slightly more flat direction downward, by reason of the slightly more steep orientation of the instrument. When the clarinets are seated in the orchestra facing the audience, the essential tone contributions up to above 1,000 Hz, as is the case for the oboe, are radiated favorably directly toward the audience as well as by a reflection from the ceiling. Since this involves the components in the region of the o(oh) and a(ah) formants, this directional effect partly explains the full round tone color of clarinets in large halls. Naturally, for this, free radiation without excessive shadowing by performers seated in front is necessary, since otherwise the intensity relationship between the direct sound and the sound reflected by the ceiling is shifted in favor of the latter so that the instrument is no longer located with the desired spatial precision. On the other hand, as is the case for the oboe, the shadowing of the sound toward the rear can also be used to advantage when an echo effect is desired as indicated in the 3rd Movement of the Symphonie Fantastique by Berlioz for the clarinet, which is merely specified as *pppp* and not intended for an additional performer at greater distances.

Rear wall reflections are of relatively little importance for the tone of the clarinet as a whole. They are limited essentially to the fundamentals of the low register, the reinforcing effect is here slightly more pronounced than for the oboe. Furthermore, sidewall reflections by surfaces near the stage can contribute to the overall sound picture slightly more effectively than for oboes. On the other hand, audience seating behind the orchestra is equally disadvantaged for both instrument groups.

The juxtaposition of two-tone spectra in Fig. 7.21 clearly shows the strong influence of floor reflections on the tonal effect for clarinets. Both were recorded in an anechoic chamber with a distance of 1.7 m of the microphone in front of the performer and a height of 1 m. The upper picture shows the tonal relations without

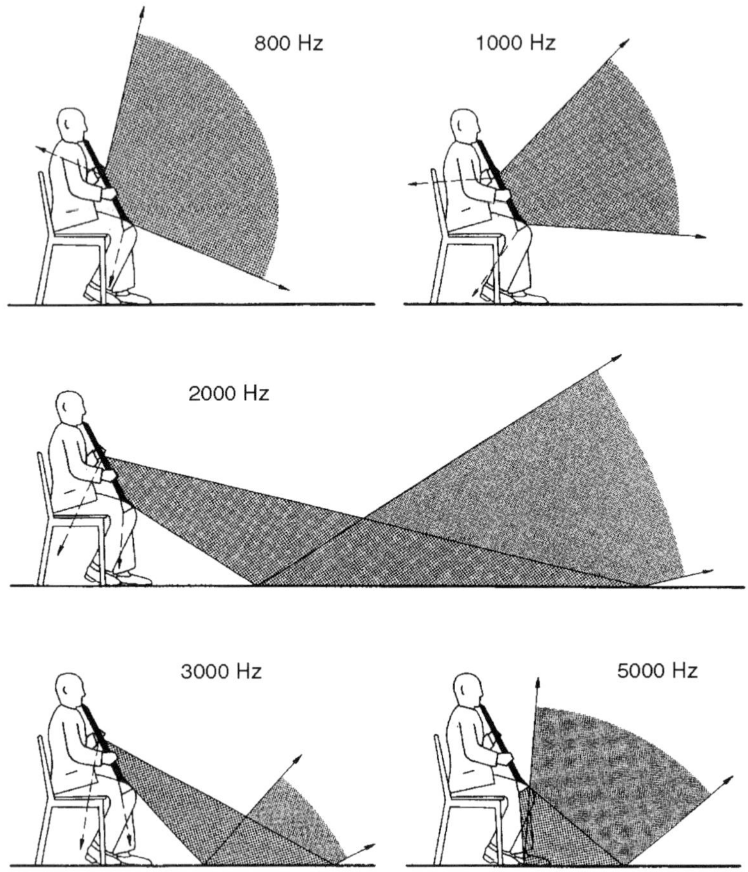

Fig. 7.20 Principal radiation directions (0. . .−3 dB) for clarinets

reflecting floor, the lower one with reflecting floor. Ignoring the fact that, because of the relationship of the microphone distance to the wavelength of the note played, the third partial is weakened by the out of phase reflected wave, one can see a significant intensity increase of the components above approximately 1,500 Hz. The level of the attack noise (in the region of 2,000 Hz) has also increased. Therefore, the tone of the clarinet has increased in brilliance as well as precision; thus, the timbre has become crisper without raising the overall loudness significantly.

The absorption of high frequencies in the hall does not permit this effect to become equivalent for all listeners, however, for microphone recordings of oboes and clarinets, one needs to be aware that even relatively small position changes lead to noticeable changes in the tonal picture because of the variations in radiation angle for the different regions at higher frequencies. In this context, reference again needs to be made to the fact that in the axial direction as well as the direction

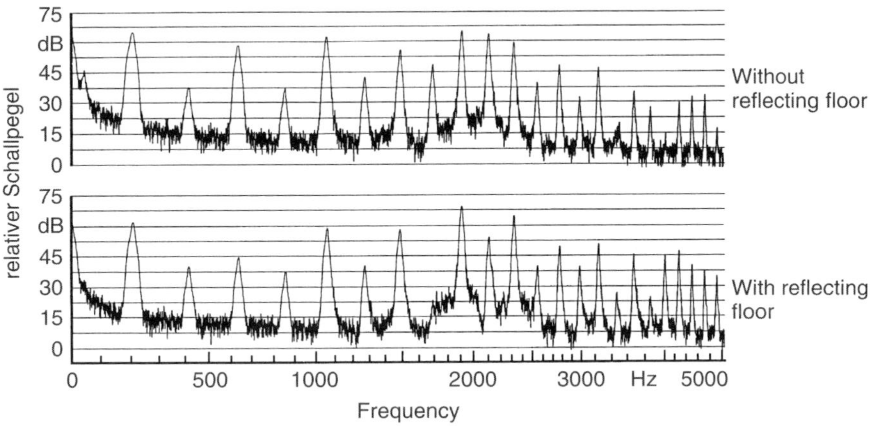

Fig. 7.21 Influence of floor reflections on the tonal spectrum of the clarinet (tone played: A_4) recorded in anechoic chamber, microphone distance 1.7 m, elevation 1 m

perpendicular to that, dips in the directional characteristics which lie by more than 10 dB below the maximum (See Fig. 4.12) are present in the frequency region between 1,500 and 2,000 Hz.

7.2.2.4 Bassoons

By reason of the upwardly directed orientation of the instrument, and its size, different radiation conditions than for the remaining woodwind instruments are given. As seen from Fig. 4.12, the intensity up to approximately 250 Hz is equal in all directions. The preferred directions for higher frequencies are represented schematically in Fig. 7.22. As clearly apparent, the radiation in the region near 300 Hz is very strong in the horizontal plane in all directions, so that reinforcing reflections are to be expected from a rear wall behind the player as well as from side walls of the hall (see Figs. 7.12 and 7.17), reflections from the floor are also possible. In the frequency region of the strongest components of the bassoon (near 500 Hz), the sound is also oriented directly toward the audience, however, the energy also reaches the hall by side wall reflections. In contrast, the rear walls play a lesser role. However, with appropriate angles of the ceiling, reflections from the region above and behind the player can also be effective, reflections from ceiling portions located further in the rear of the hall near the audience can also be expected. The tonal components in the region of the principal formant are thus radiated into the hall with particular advantage.

The structuring of the preferred angular regions into four radiation bundles in the subsequent frequency region up to above 1,000 Hz also leads to strong ceiling and rear wall reflections and also contains contributions directed toward the audience, provided the audience floor rises appropriately. For a level audience floor, these components reach the audience mostly by reflections and will therefore be some-

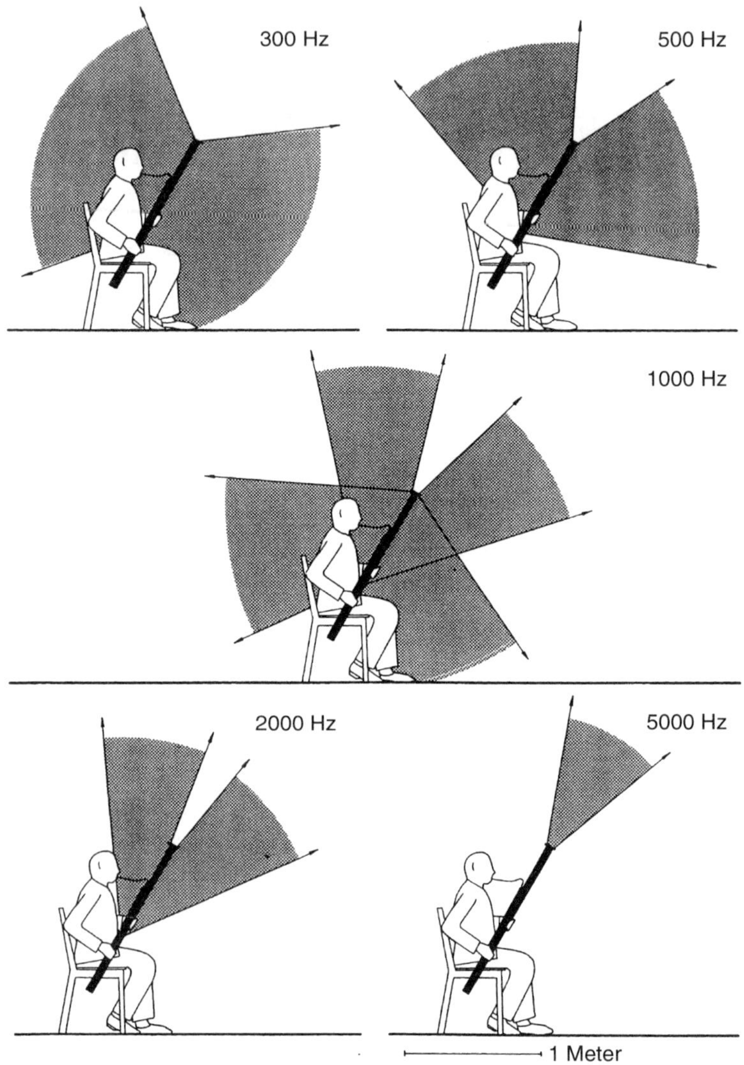

Fig. 7.22 Principal radiation directions (0. . .−3 dB) for bassoons

what weaker. Thus, the sound of the bassoon loses somewhat in strength and brightness. This effect can certainly be desirable for Baroque music, in view of the timbre of contemporary bassoons, in contrast, for musicians of the Classic period, a possibly bright tone color is desired (Meyer, 1968). Consequently, the portions of the frequency region around 1,000 Hz which are directed toward the rear, should be reflected into the hall if possible, while it is certainly not harmful if the low components around 300 Hz are absorbed by the rear wall. A lowering of the intensity at these frequencies would shift the main formant slightly higher so that it more nearly approximates the historical tonal impression.

At the highest frequencies, which occur mostly for very loud tones, the strongest components are concentrated in a relatively narrow angular region upward. However, an amplitude dip is formed above 2,000 Hz in the neighborhood of the instrument axis. At times this dip even exceeds a value of 10 dB in depth (see Fig. 4.12).

With increasing height of overtones the radiation cone closes. At 5,000 Hz the bundle only includes an angle of about ±20° relative to the axis. These very high components, which for tone aesthetic reasons should not reach the audience with excessive intensity, are mostly conducted by ceiling reflections. Thus, it can become dangerous when the bassoon addresses certain reflecting surfaces which are concentrated toward the public because the tonal balance between the high and low contributions is disturbed and the timbre receives an unpleasant hardness and crispness.

Locating the microphone along the extension of the instrument axis is especially undesirable because of the directional characteristics. The position of the microphone should at least be far enough from the player so that also the 500 Hz cone is included, (thus flatter than 45°). In order to avoid these difficulties for recording purposes, a bassoon with a bell bent at the bottom is used for the solo parts in concerts. For an instrument of such construction the high, noise-like contributions, do not stand out.

7.2.2.5 Combined Sound of the Woodwind Sections

The different angular regions for preferred sound radiation by individual woodwind instruments and the difference in the effectiveness of wall and ceiling reflections lead to difficulties in achieving a good balance between the wind voices for all seats in the hall under complicated room acoustic conditions. An example for the tonal effect of two oboes and two bassoons for a very flat seating arrangement, with winds behind the strings, is given for two different seats in the concert hall in Fig. 7.23.

At a location in the upper main floor (see Fig. 7.41), wall and ceiling reflections can arrive from all sides without obstructions, while only few reflections are expected, or can be expected in the gallery, and thus the direct sound contributes more strongly to the overall impression. As recognized from the octave diagrams, the level value for individual chords of this passage is larger in the upper main floor in the lower frequency regions and around 4,000 Hz, while around 1,000 and 2,000 Hz larger values are found in the gallery. This means that in the upper main floor the bassoon can be heard more loudly. In contrast, in the gallery it is the oboe, as can easily be deduced from the tonal characteristics of the two instrument groups.

The reason for this variation in intensity relations lies in the directional characteristics. The bassoon stands out wherever wall and ceiling reflections play an important role for the room acoustical picture. In contrast, the oboe is heard better wherever the direct sound can arrive without obstruction and is not attenuated along the path to the audience. Under given hall conditions, a better

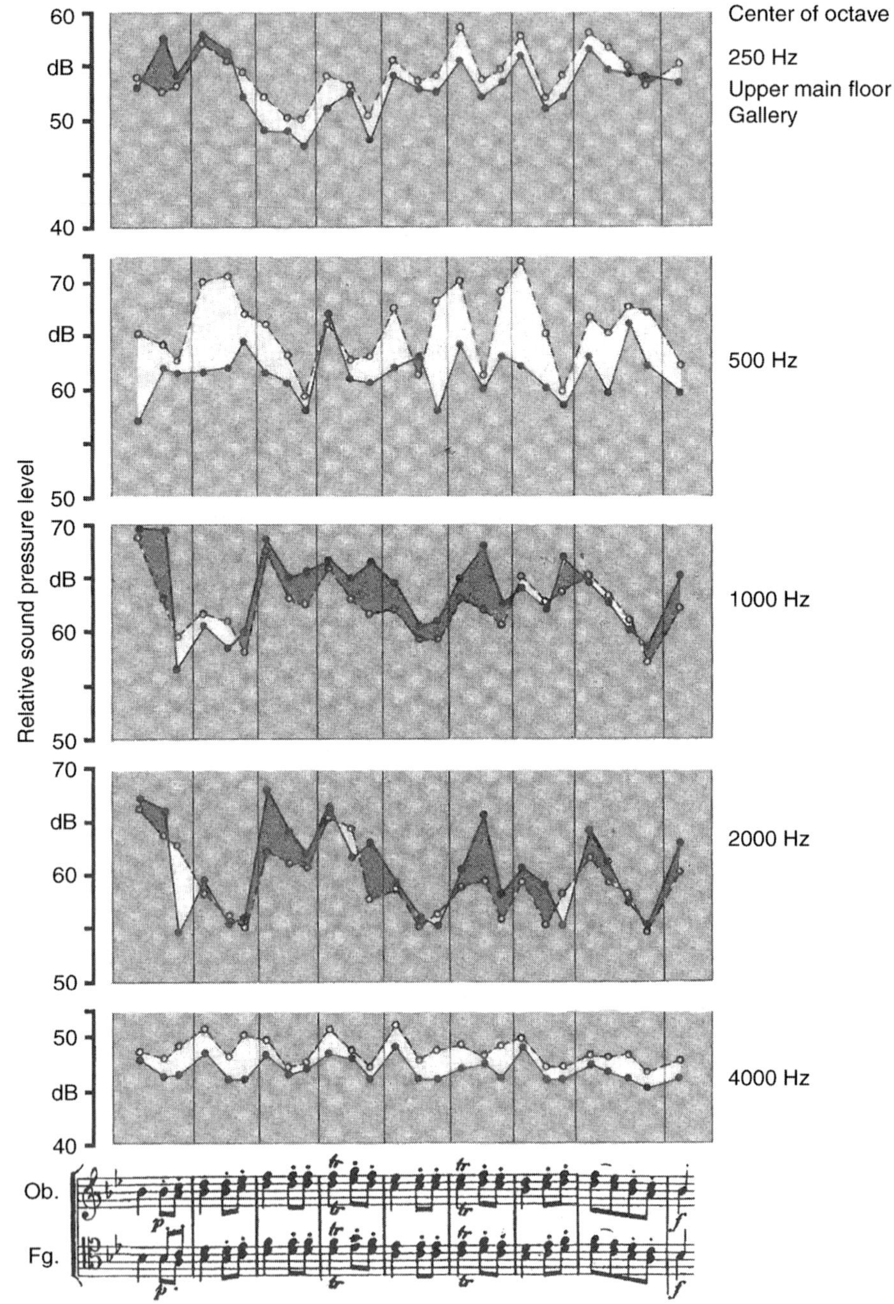

Fig. 7.23 Octave level analysis of a passage for two oboes and two bassoons (W.A. Mozart, Symphony B^{b} major, K319, 4th Movement, measure 336–344) in two locations in the large hall of the Stadthalle Braunschweig

balance can be achieved by placing the oboes appropriately higher to permit free radiation to the main floor.

With regards to the positioning of the four woodwind groups, this example shows that it is advantageous for a balanced overall sound, to flank the bassoons by clarinets and horns, or possibly trumpets, instead of locating them near the edge, in order to reduce wall reflections. This is particularly important when the orchestra does not occupy the entire width of the podium, leaving a free space between the performers and the side walls. In this case the wall reflections of the bassoon are not attenuated by the string players for a portion of the audience, and thus the bassoon appears louder in some parts of the hall than in the remaining seats.

Since free sound radiation is important for clarinets, two levels for woodwind rows are acoustically better than having all players seated at the same level. The question whether clarinets should be located to the left or right of the bassoons is immaterial for the tonal effect of woodwinds. Locating the clarinets is thus primarily determined by the choice of seating the horns. In most halls, even the seating arrangement of oboes and flutes in the front row is of no importance for the acoustic effect in the hall. It could merely happen that a piercing piccolo can influence the tonal symmetry of the orchestra when seated near the outer edge: in such a case switching flute and oboe group positions is occasionally advisable, however, it is very important to consider the character of the composition.

7.2.3 Brass Instruments

7.2.3.1 Horns

The simplified sound radiation relationships for horns are represented in Figs. 7.24 and 7.25, for the horizontal plane as well as for two vertical planes. In addition it should be remembered that the intensity for frequencies below 100 Hz is uniform in all directions. The tone contributions between 150 and 250 Hz, as well as those between 300 and 500 Hz are radiated in the horizontal plane (Fig. 7.24) approximately in the width of a semicircle, while the frequencies located between those two regions are concentrated in a slightly narrower angle. Above 600 Hz the largest amplitudes occur only in a more or less narrow region oriented toward the right rear. These components reach the audience predominantly by wall reflections.

The general preference of the right side, as well as the rear facing direction, is clearly expressed in the figure representations for the vertical plane (Fig. 7.25). Beyond that, these characteristics show a pronounced radiation concentration toward the front and the upper direction: this can reach the audience in the hall through ceiling reflections. Below 600 Hz, this preferential region reaches at least down to the horizontal plane, so that these components, which also include the principal formant of the horn, are radiated directly toward the audience. Furthermore, an additional maximum toward the back of the performer in the region from 1,000 to 1,300 Hz is noteworthy. This can become effective through double

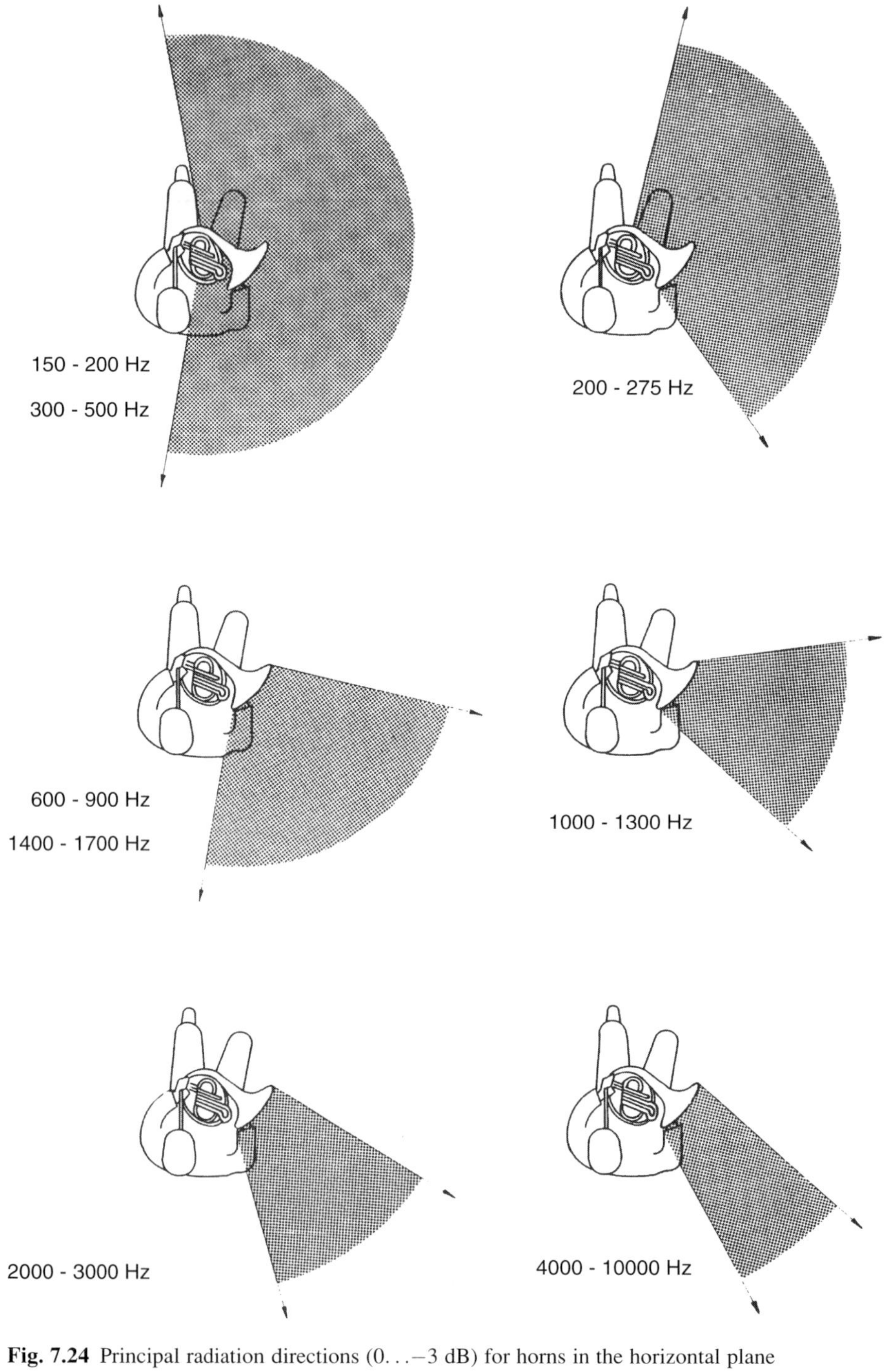

Fig. 7.24 Principal radiation directions (0. . . –3 dB) for horns in the horizontal plane

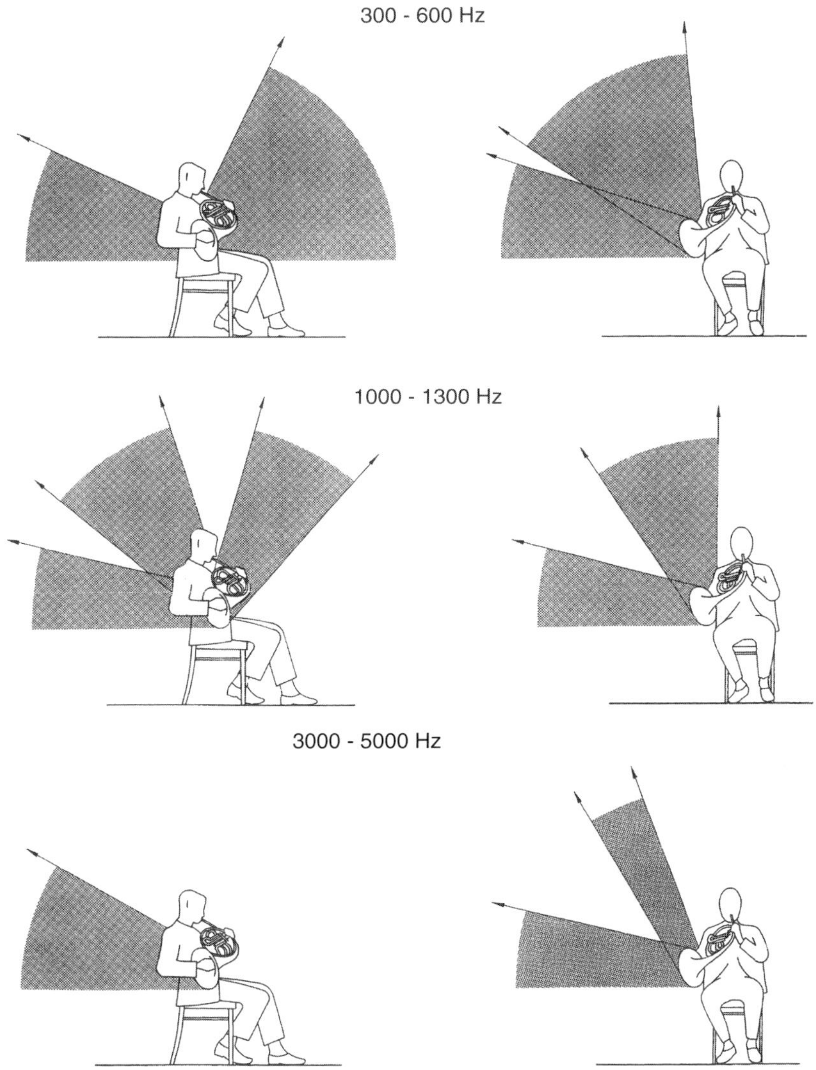

Fig. 7.25 Principal radiation directions (0. . . −3 dB) for horns in two vertical planes

reflection from ceiling and rear wall. Corresponding to its location near the high a (ah)-formant it enhances the forceful tonal characteristic of the instrument.

Toward the right side, an increasing separation into two preferred regions occurs as frequency rises. These in turn also become narrower. Consequently, in addition to sidewall reflections, sound reflections from the ceiling also are significant. These regions of ceiling reflection need to be located at an angled orientation above the player. The dynamics of the performance can increase significantly when the highest frequencies, which occur only for increased loudness are advantageously directed toward the audience, otherwise a *forte* can remain too dull. For unfavorable room acoustic conditions, forced blowing can compensate for this phenomenon, at least partially.

The extraordinarily different tonal effect of the horn is also expressed in the juxtaposition of two sonagrams in Fig. 7.26, where one gives the tonal picture in front of the player, the other one toward the right next to the player. In this representation, the frequency of the overtones is recorded from bottom to top, the intensity is given by the degree of darkening, where the individual grey steps represent level differences of 6 dB. As even a cursory glance reveals, a tone picture much richer in overtones is formed to the right of the player than toward the front. When considering the initial transients of tongued *staccato* eighth notes, a very dense sequence of contours is noted, represented by a very rapid rise in amplitude.

The highest intensity at frequencies below 1,000 Hz is reached within approximately 20 ms. For some tones, the higher contributions have a longer transient than the lower ones. This phenomenon is very pronounced for some of the *staccato* eighth notes (e.g., F_6 and $F_6^{\#}$ in measure 4), especially for notes of longer duration. This *crescendo* within the note is only weakly perceived by the player, since the overtones only reach up to 1,500 Hz.

A tonal difference also results for the connected E_6, which appears more strongly separated toward the right hand of the performer, while the connection (this case deals with a valve connection) results in a softer impression in front of the player. For the last three notes, the loudness is raised to a metallic timbre: toward the side of the player the overtone contributions reach up to about 8,000 Hz, while in front it barely reaches 4,000 Hz. In this context it is worth noting how the player attacks those notes, which by overtone content are related to the *staccato* eighth notes, and thus, thereafter, rather suddenly again achieves the metallic sound. Certainly the

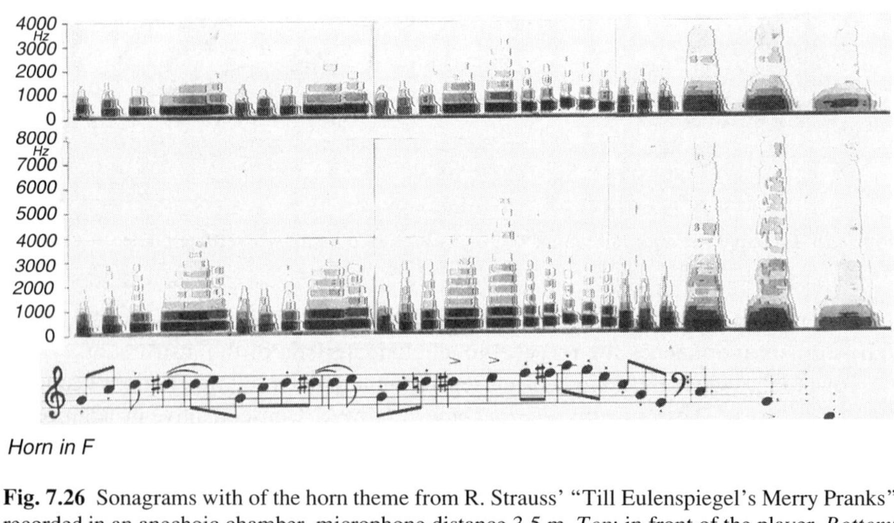

Fig. 7.26 Sonagrams with of the horn theme from R. Strauss' "Till Eulenspiegel's Merry Pranks" recorded in an anechoic chamber, microphone distance 3.5 m. *Top*: in front of the player. *Bottom*: to the right, next to the player

same dynamic tendency can be recognized in both directions; however, the expressivity toward the front is far less. Finally, a valve noise (between C_6 and G_5 in the 5th measure) is noted, which is observed more clearly toward the right of the player in several frequency regions, while it is not observed in front of the player.

Naturally, rear-wall reflections play a particularly important role for the sound of the horn. Consequently, the front-back ratio is represented in Fig. 7.26, above the level increase for comparison, with a hard, sound reflecting wall behind the player. Ignoring variations through interference phenomena at some low frequencies, one clearly recognizes the reinforcing effect of the rear wall in the lower diagram. In the low frequency region, the level rise amounts to approximately 6 dB, and has a maximum of about 15 dB in the region from 1,500 to 2,000 Hz; at still higher frequencies the effect again becomes less. Thus, the sound of the horn generally gains intensity through the reflection effect of the rear wall, where the lower region of the nasal components is strengthened least. The brightening tone contributions in the region of the vowel color "e(eh)" are especially brought out. For all commonly used horns (except for the horn in high F) these form a characteristic side formant.

The spectral broadening is also recognized in the sonagram of Fig. 7.27. The *staccato* motif not only contains more overtone contributions in the presence of rear wall reflections, but even at lower frequencies presents larger amplitudes, as is noted by the total number of darkness steps. Even the particular intensity increase in the region around 2,000 Hz is clearly visible for several of the notes. Inasmuch as in this case, as for the sonagram of Fig. 7.25, the clarity of the attacked notes is more determined by the greater frequency range than by the special processes of tone

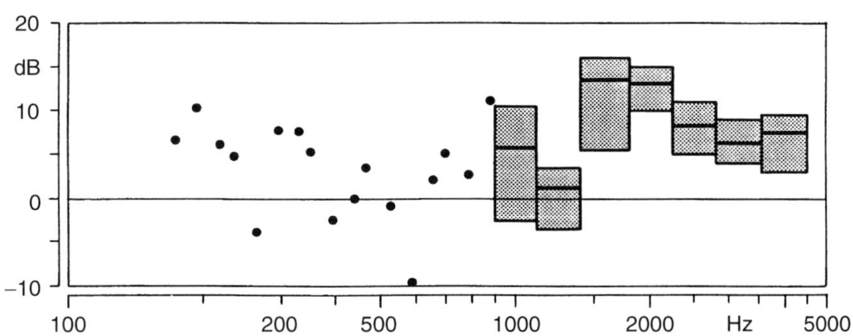

Fig. 7.27 *Top*: front-back ratio of the directional characteristic of the horn. *Bottom*: level rise in front of the player caused by a reflecting rear wall with 1 m distance from the bell

development, it certainly can be said that the rear wall gains in significance, especially for a larger orchestra. When the spectrum is excessively limited, a pronounced soft sound results, however the danger increases that the horns are covered by other instruments in the context of the entire sound. On the other hand, a stylistic comparison with the sound of valveless concert horns (German: Invention Horn) of Mozart's time leads to a conclusion, that especially for compositions of this time period, the brightening effect of the rear wall is advantageous, provided players make allowances for excessively hard attacks in an attempt to equalize the dynamics relative to other instrument groups.

Generally, horns are seated toward the right or left next to the woodwinds; when there are six or eight horns, the entire group is occasionally located as a third wind row behind the clarinets and bassoons. The higher the horns are seated above the strings, and the closer they thus come to the rear wall of the stage, the brighter their tone color becomes, and the more precise they become in their attacks. In contrast, locating them next to the oboes or bassoons is more favorable for a softer tone presentation, since the shadowing by other players especially effects the higher frequencies. It is thus recommended to surround the horn group by other players, in order to achieve the round and sonorous sound of the horns, as demanded, for example, in the symphonies of Brahms and Bruckner, while smaller horn sections in classical symphonies already lead to the fact that horns obtain a more free, and thus brighter tone radiation.

The question, of locating horns preferably in the right or left side of the orchestra is also related to the spatial conditions of the podium. In rectangular halls or also for a nearly rectangular stage, both seating arrangements are possible, since the side wall reflections partly return to the orchestra and are partly attenuated by the string players sitting before them. However, it is important that the first horn is located at the very edge when placed on the left side as seen from the conductor so that that instrument presents the most brilliant sound within the section. Within the framework of meaningful possibilities, this arrangement leads to the brightest sound of the horns. It thus appears to be predestined for classical compositions. It is interesting to note that especially in France this arrangement is frequently used, where also horn designs move in a direction to achieve similar tonal effects.

When the horn section is placed on the right wing, the first horn can sit either inside or outside. On the inside, it is located more closely to the solo players of the other wind groups, which supports mutual contact, and a unison with the cello also works very well. On the other hand, when located on the outside, the first horn can be heard well by the other sections and its influence on intonation and uniform tone development of lower voices is made easier. From a performance technical standpoint, placing the horns toward the right side is also advantageous in cases when the first horn plays together with the solo violin as is the case for example in the slow movement of the 1st symphony by Brahms, since visual connection with the left side is not possible.

When the stage opens cone-like toward the hall, locating the horns on the left side of the orchestra is not advisable. In similar fashion, as already explained for the bassoon, the pronounced side wall reflections reach the hall without obstruction because of the opened flank of the instrument. For a portion of the audience, the horns thus sound significantly louder and more cutting than expected by the conductor, and as perceived in other seats of the house. It thus is certainly possible that the higher frequency tone contributions reflected by the wall – in consequence of directional characteristics – are stronger than the direct sound and therefore the sound source is perceived as coming from a reflecting wall and not from the player. Thus, with increasing loudness, the sound from the horns is spatially separated from the remaining orchestra. This uneven spatial effect of horns in the room is significantly less when the horns are seated toward the right since they then turn their side of least sound radiation toward the open flank of the orchestra.

For a soloist standing in front of the strings, naturally the tonal contributions radiated toward the rear are strongly absorbed, so that the direct sound becomes of increased importance. Some players utilize the possibility of influencing the tone by turning the bell toward the audience during fast passages, i.e., passages of a lively and rhythmic strongly accentuated character, while they adjust their orientation for cantabile portions of compositions, in such a way, that the timbre becomes rounder and softer. This technique, however, can not be used for a horn quartet functioning

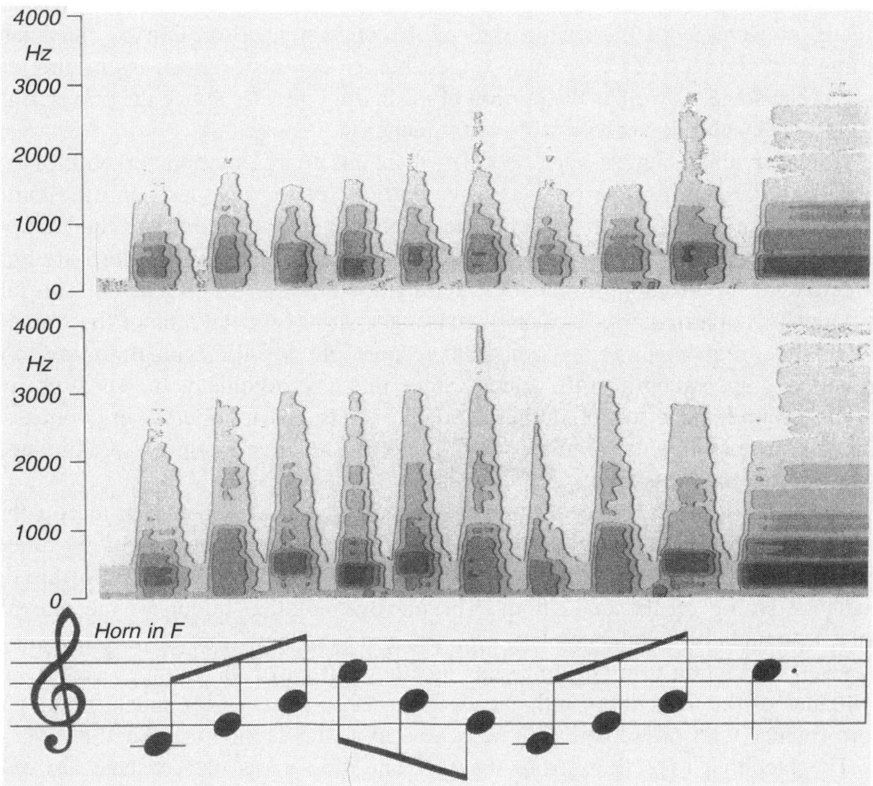

Fig. 7.28 The sonagram of the *staccato* tone sequence recorded at 3.5 m distance in front of the layer in an anechoic chamber. *Top*: without reflecting rear wall. *Bottom*: with reflecting rear wall

as a solo instrument. It is thus more favorable, and this not only in view of the dynamic balance, for works such as the Concert Piece for 4 Horns and Orchestra by R. Schumann, to place the soloists behind the orchestra, though slightly to the side, but somewhat raised. A reflecting wall behind the players can also be used to raise the brilliance.

7.2.3.2 Trumpets

Sound radiation patterns for trumpets are significantly less structured than for horns, as the two representations for the vertical and horizontal planes in Figs. 7.28 and 7.29 show. Following the frequency region of the spherically uniform radiation, which in the low register includes frequencies up to approximately 500 Hz, the stronger energy contributions are concentrated toward the front in an angular region, which approximately fills the frontal hemisphere. Accordingly, at these frequencies the sound reaches the audience not only directly but also by reflections from large sections of the ceiling and the side walls, provided they do not diverge excessively in the region of the audience. Floor reflections are also possible, where, however with increasing size of the orchestra, absorption by musicians sitting in front becomes ever more noticeable. In contrast, reflections by the rear wall of the stage, as well as the portion of the ceiling directly above the player, only plays a very subordinate role at these frequencies.

However, this changes near 800 Hz where the representation for both planes visualizes the significance of the higher portions of the rear wall and the entire ceiling. From 1,000 Hz on upward, the increasing concentration for the trumpet leads to the circumstance that in addition to the direct sound, reflections are only effective for far distant portions of the ceiling or side walls (in long halls). This fact is noteworthy particularly because the strongest tone contributions of the trumpet lie in this frequency region, which is responsible for the typical timbre, and primarily gives strength to the sound. Since this also dominates the spectrum for a *piano* sound, the directional characteristic means that trumpets – in contrast to many other instruments – radiate their sound in a relatively concentrated manner even for low loudness levels.

Above 1,500 Hz the concentration of the sound radiation becomes so narrow that only little value can be attributed to side wall reflections and even ceiling reflections. From 4,000 Hz on upwards, the principal intensity is only directed toward the audience sitting on the main floor. The narrow radiation region of the relevant partial pictures also indicates how important it is to have the player sufficiently elevated in relation to orchestral seats located in front of the trumpet, so that the brilliance of the *forte* is not influenced. The related effect of the music stand, and performance with raised bell, will be treated in a later section (see Sect. 8.3.3).

The fact that reflections from the rear and side walls of the stage are only important for the lowest tone contributions becomes noticeable, especially when large choral works are performed. While most instruments are influenced in inten-

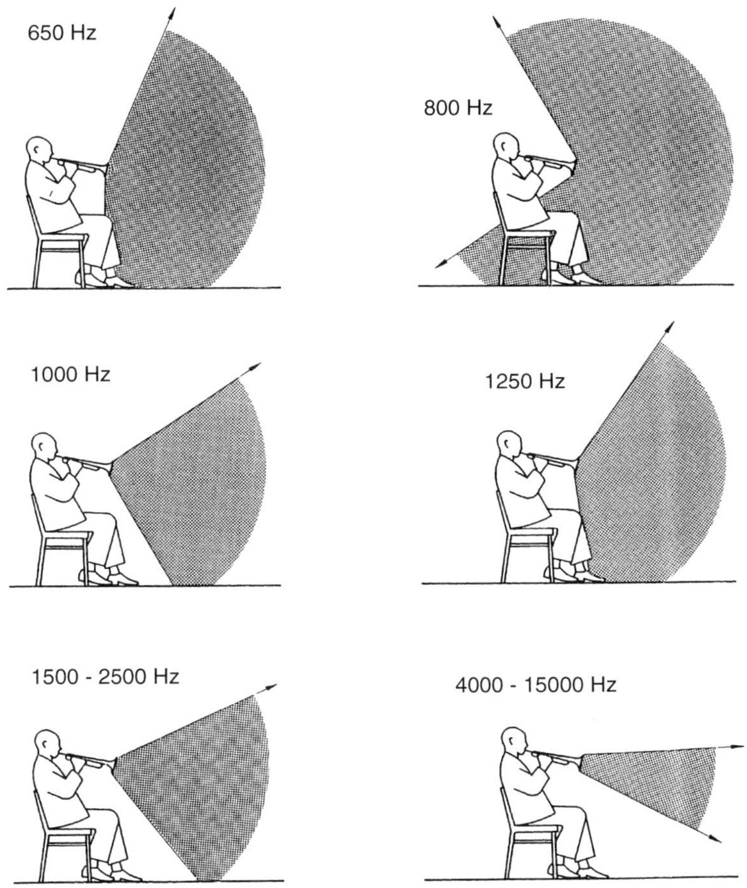

Fig. 7.29 Principal radiation directions (0. . .−3 dB) for trumpets in the vertical plane

sity and especially brilliance by the absorption effects of the choir, the timbre of trumpets are essentially preserved for the audience in the hall.

The especially sharp directional characteristics of trumpets naturally results in the fact that in wide concert halls, or halls for which the audience is also seated behind the orchestra, the sound is not perceived with the same brilliance and clarity in seats toward the side and the rear, as it is for seats in the direction of sight of the performer. This is less noticeable in definite solo passages than in passages for which the trumpets are to stand out with a rhythmic motif from the total orchestral sound. As an example for this, a tonal analysis of an excerpt from the 9th symphony by A. Bruckner is given in Figs. 7.30 and 7.31.

Figure 7.30 shows two models of sound levels in octaves for measures 51–71 of the 1st movement; the time sequence is shown in the associated score. The center frequencies of the octave band filters used are noted at the left edge. Accordingly, the low frequencies are located at the rear, the high frequencies at the front. The

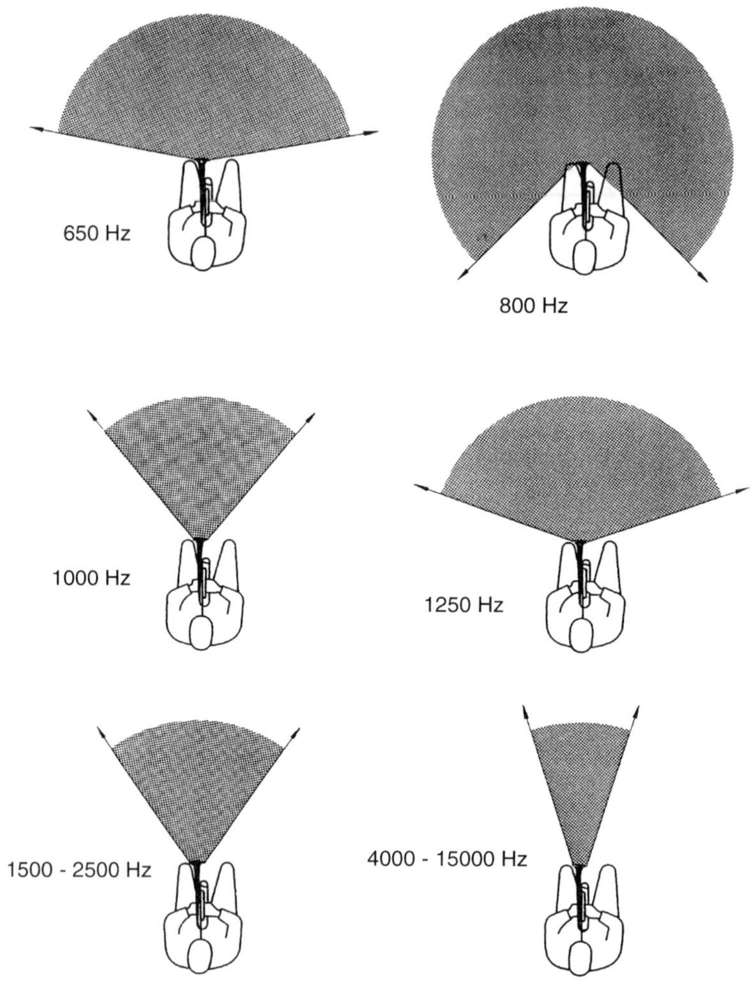

Fig. 7.30 Principal radiation directions (0. . .−3 dB) for trumpets in the horizontal plane

intensity of the individual filter regions is represented on the vertical axis. While the lower model represents the tonal picture at an audience location in the 5th row from the front, approximately in the middle of the hall, the upper model represents the sound relations at the location of the conductor.

In both examples, the same structure of the tonal process is recognized, from the exchange between *piano* and *pianissimo* (twice) at the beginning, over the large *crescendo* with the entrance of the low instruments, to the rhythmically marked theme in the right portion of the picture. However, certain differences are also noted between the models, especially the greater richness in overtones at the position of the conductor. In the context of the previous discussion, the last four measures of the *crescendo*, where the trumpets have to play *staccato* 1/8 notes, deserve particu-

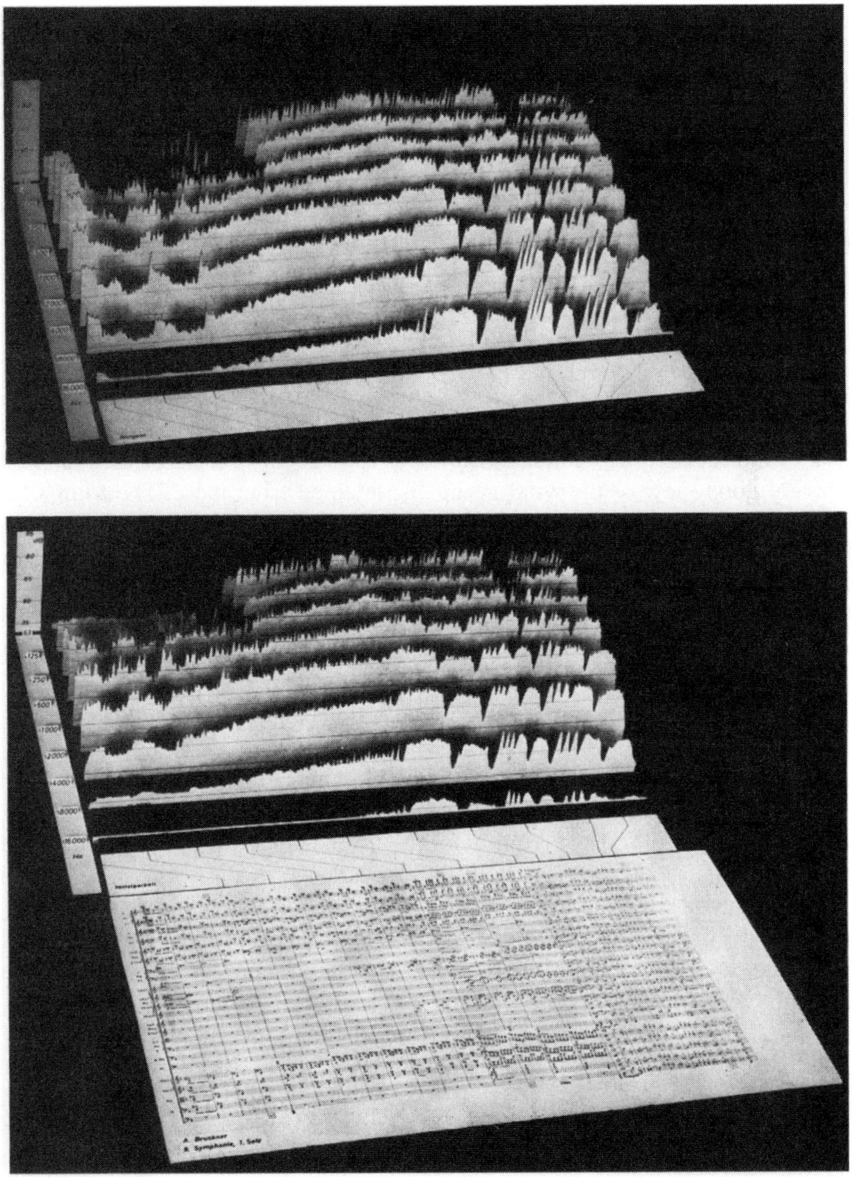

Fig. 7.31 Octave level models of the orchestral sound in the Stadthalle Braumschweig (A. Buckner, 9th symphony, 1st movement, measure 51–71). *Top*: sound at conductor location. *Bottom*: sound in the 5th row of the main floor

lar attention. In the upper model, these figures are not clearly recognized in any filter region (as equally spaced sharp peaks), which is connected to the fact that at the location of the conductor the strings located in close proximity radiate strong components at high frequencies with their tremolo. In the lower model, however, the trumpet 1/8 in the 8,000 Hz region are noticed as regular peaks even though the trumpets are slightly shadowed by musicians sitting before them for this audience location. This is because the high tone contributions of violins are received only weakly in these front audience seats as shown in Fig. 7.7 (different hall, however).

The difference in the effectiveness of trumpets becomes especially clear when a location, which is still within a diffuse-field distance, is compared with one toward the side. For this purpose, the interesting measures, are given, with sound levels in octaves for an 8,000 Hz filter, while identifying the recording locations in the floor plan. In this frequency range, trumpets stand out most clearly (if at all) from the orchestra sound. The score excerpt includes only those voices which can deliver sufficiently strong tonal contributions at those high frequencies. The time base on the recording strips is given by connections to the measure bars in the score, thus one can clearly recognize that the individual level peaks in the lower diagram correspond to sharply attacked 1/8 notes.

Since the seats in the audience rise relatively steeply in front of the rear wall, a relatively good view of the wind groups is obtained from there. The trumpets are not shadowed by other performers, and thus stand out sharply from the total sound. This is not only the case for the *staccato* 1/8 before letter C but also in the clearly separated entrance in the 2nd measure after letter C. At the location on the side, beyond the diffused-field distance, the individual 1/8 notes, however, are no longer uniquely identifiable, even though the distance to the player is only half as large, and furthermore, the other high voices arrive with relatively low overtone content because of the placement of the two violin sections and the flute on the left side of the podium.

It should, however, be especially emphasized, that a maximum in intensity in the region of highest frequencies is not necessarily to be considered as desirable for trumpets in principle. This requirement is primarily only valid for the higher dynamic levels, as can be seen from the dependence of the spectral composition on the loudness. For smaller orchestras, a less cutting trumpet tone also merges better into the overall tone. Nevertheless, it will always contribute less to the spatial representation of the total orchestra sound than all other instruments, and will stand out from the overall tone because of the precision of its spatial location.

7.2.3.3 Trombones

Concern for equalized tonal effects at high frequencies is even more important for trombones than for trumpets since this instrument group easily has a harsh sound in slow passages and misses the desired sonority. As seen from schematic representations in Figs. 7.32 and 7.33 the concentration of the radiated sound of a trombone plays an important role. As is the case for trumpets, the radiation is directed mostly toward the front and the ceiling which contributes to an intensity increase by reflections for frequencies below approximately 1,100 Hz, at least at far distances. Since the components directed toward the floor experience strong absorption by the

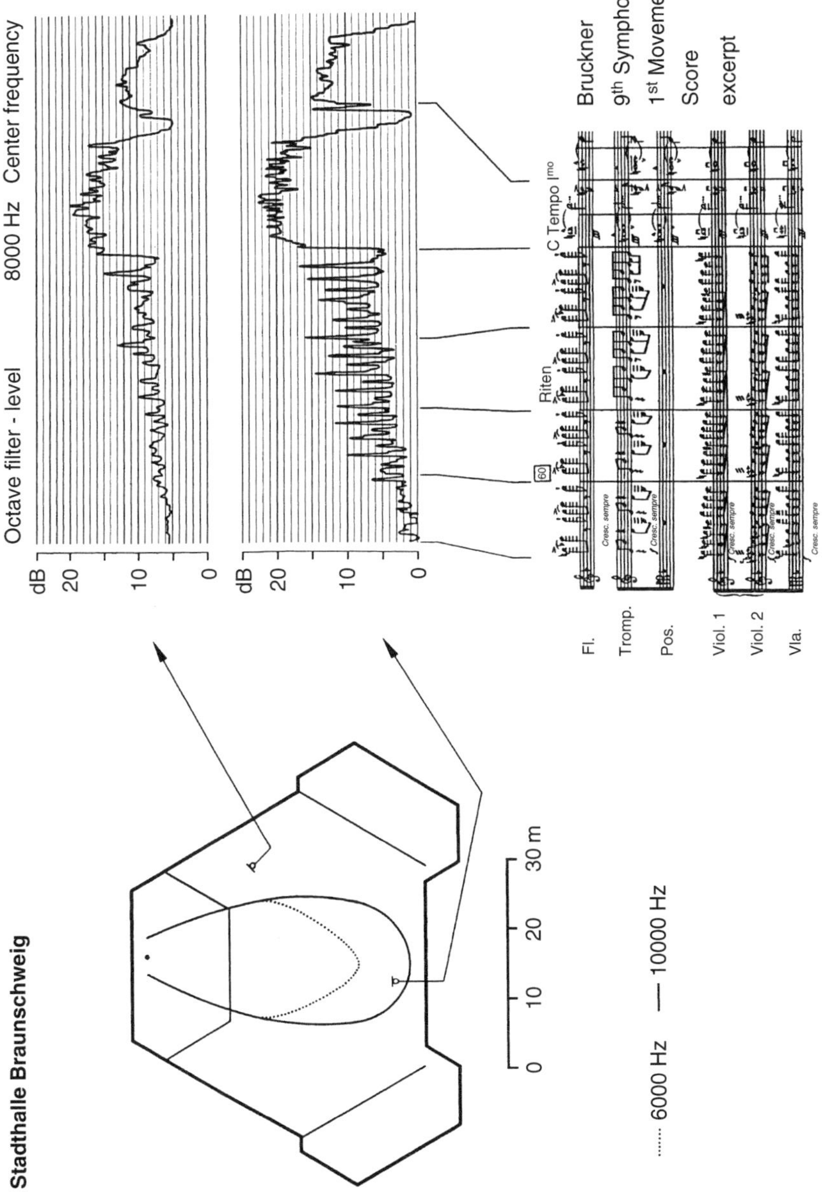

Fig. 7.32 Tone effect of the trumpets at two locations in the Stadthalle Braumschweig

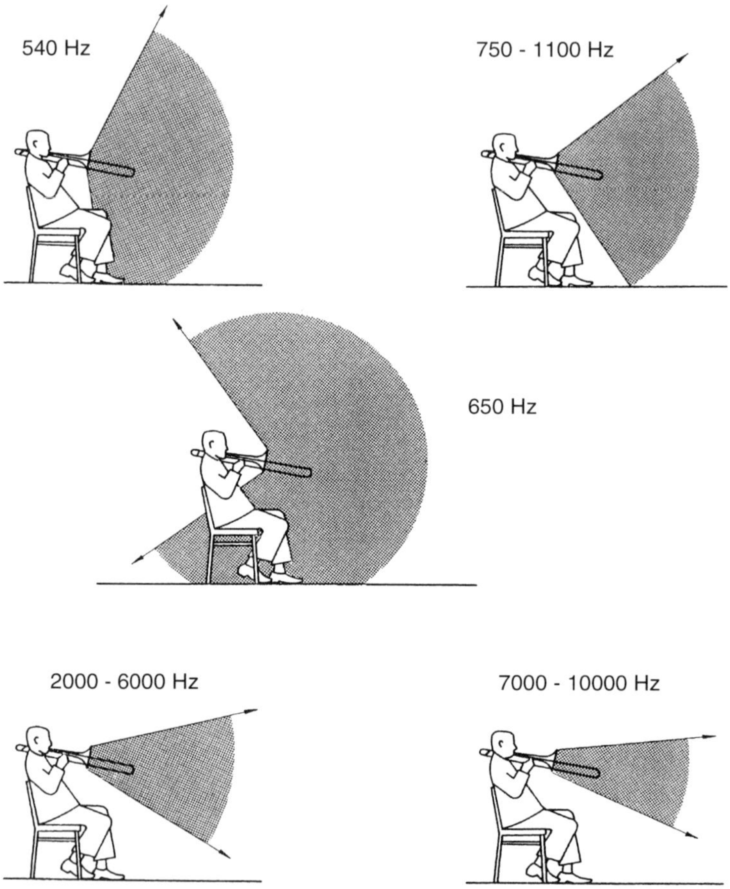

Fig. 7.33 Principal radiation regions (0. . . −3 dB) for trombones in the vertical plane

musicians sitting in front, the tonal effect in the hall, for high frequencies is especially dominated by the direct sound.

This is also shown in the representation for the horizontal plane, where, for frequencies above 2,000 Hz, an angular region of only 45° width of the audience space of the concert hall lies within the principal radiation region; and above 7,000 Hz, i.e., for components, which only in *fortissimo* exhibit greater intensity, a narrowing to 30° occurs. However, for trombones the side wall reflections below about 1,100 Hz and especially below 700 Hz play an important role. In contrast to trumpets, where the strongest energy in a given frequency region is radiated with relatively sharp concentration, for trombones, the principal formant lies in the frequency region for which the intensity toward the side of the player is attenuated only by approximately 3 dB in comparison to the direction of the instrumental axis. These important tone contributions are thus radiated relatively broadly into the hall.

Naturally, in the directions toward the side a somewhat softer tone picture results from the weakening of the higher frequencies in those directions, in comparison to the direction of sight: however, this effect can be advantageous for the sonorous and majestic tonal effect of trombones. In this context Sir Adrian Boult should be remembered, who suggested (1963) the possibility of orienting trombones on the concert stage in such a way that they play at a right angle toward the middle rather than into the hall directly, since then their tone can be more easily merged into the overall sound. This merging of individual instruments within a group, or into the overall sound of the orchestra, is to be understood to mean that trombones should not draw attention by pronounced transients, but rather join a chord unobtrusively by a soft attack without playing with lessened intensity.

The relatively broad radiation of trombones in the frequency region of their principal formant requires special attention in concert halls for which the podium is wider than the area occupied by the orchestra. As already mentioned for other wind instruments, for such cases the danger exists that the intensity for some audience seats is especially high by reason of unimpeded wall reflections, while in other areas of the hall the loudness balance is more equalized. This effect is especially noticeable when the side walls of the podium are not parallel but open in the direction of the hall. Since this effect is more strongly pronounced for trombones than for other wind instruments because of their directional characteristics, it is recommended under such spatial circumstances to locate the trombone section at the highest level not toward the edge, but toward the middle. In rectangular halls or for a rectangular stage such difficulties do not occur.

Since the trombones most frequently are seated in the last row of the orchestra, the question of the influence of a reflecting wall behind them on the tonal impression in the hall naturally arises. Frequently the technical possibility exists to increase the absorptivity by means of a curtain, to the point that sound reflections from the rear wall are practically eliminated.

As expected from the directional characteristics (see Fig. 4.4), rear wall reflections only influence the sound in the hall up to a frequency region of approximately 400 Hz. In analogy to the representation for corresponding conditions in the horn (Fig. 7.26), Fig. 7.34 shows the level increase which can be achieved with a trombone by a hard rear wall; the difference is clearly recognizable. Dampening of the rear wall thus merely affects a decrease of the lower tonal contributions for the trombone, which leads to a somewhat brighter tone color, while the timbre with a reflecting rear wall has more of the nature of the fundamental. However, the principal formant is practically not influenced, so that the loudness is not noticeably changed.

This modification of the tone picture in the hall and also at the position of the conductor need not coincide with the aural impression of the player. Since by virtue of the directional characteristics, the higher frequency contributions already reach the player's ear relatively weakly, an increase of these components by a rear wall reflection is certainly sensed. This in turn has the consequence that the player performs somewhat more softly and the tonal effect in the hall becomes more dull.

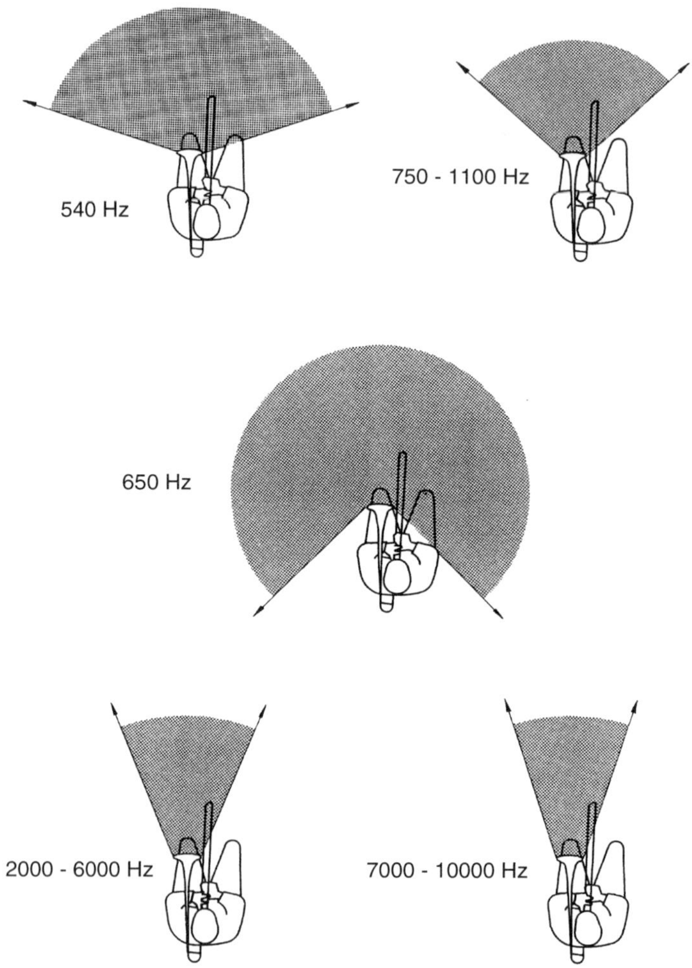

Fig. 7.34 Principal radiation regions (0. . .−3 dB) for trombones in the horizontal plane

7.2.3.4 Tubas

The angular regions of preferred sound radiation for a tuba are represented for three frequency regions and two different vertical planes in Fig. 7.35. In consequence of the slightly angled orientation of the bell axis, slightly different half value widths result for these two planes. Furthermore, it is noted as a reminder, that at frequencies up to approximately 75 Hz the radiation is uniform in all directions. As recognized from the pictures, the direct sound radiated to the audience has special importance in the region of lowest frequencies, where the shadowing by musicians seated before them is certainly minimal. Already in the region of the principal formant the intensity in the horizontal plane is by 3 dB less than toward the top. With increasing frequency the concentrating effect becomes more pronounced;

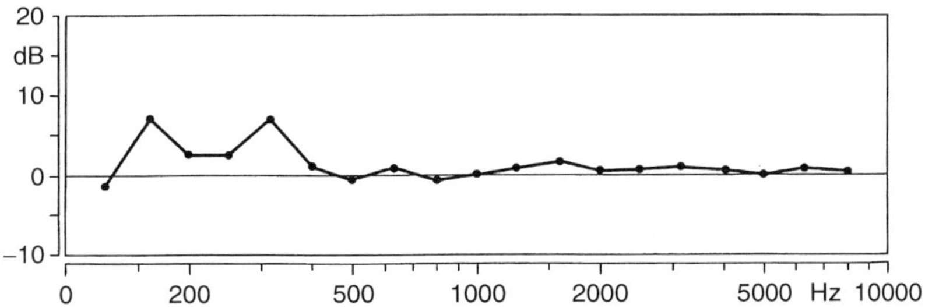

Fig. 7.35 Level increase in the direction of the player by a reflecting rear wall (trombone)

components above 800 Hz are radiated toward the ceiling more strongly by 20 dB than in the horizontal plane.

In view of the low range, as well as by reason of the full and round sound of the tuba, overtones of these frequencies must be considered as very high: this means that they are frequently perceived as undesirable rather than as enriching for the tonal picture. This phenomenon can be noted as annoying when occasionally unfavorably oriented reflection surfaces are responsible for sending the sound contributions of higher frequencies into seats of a limited audience area. Since in this region of higher frequencies, relatively annoying noise mixtures are present, the tone receives a rather rough and strange character. Furthermore, the danger of erroneous localization can arise, i.e., the listener receives an impression localizing the instrument near the ceiling behind the reflecting surface, because the low frequencies with a strong direct sound contribution contribute almost nothing to a sense of direction.

Inasmuch as the reflection surfaces above the orchestra are of great importance for most other instruments, it follows from these considerations that generally considerations of tuba playing technique must include awareness of reflectors, thus no higher dynamic level than *mezzo forte* should be blown to achieve a full, steady tone. Furthermore, consideration should be given to limited absorption possibilities above this portion of the orchestra in light of the narrow concentrations of high frequencies for the tuba, particularly since the trombones, which are frequently seated next to the tubas, would not profit by reflections from the ceiling.

7.2.3.5 Combined Sound of the Brass Sections

There are primarily two considerations for locating the individual brass sections on the concert stage: one is the tonal balance between the French horns and the heavy brass, the other is the spatial effect associated with exchange of motifs. The intensity balance is easily influenced when the trombones are seated behind the French horns. Since, based on directional characteristics, players in both sections hear the instruments of the other section as louder than their own, they are inclined to force the tone – especially at louder dynamic levels. Many horn players consider this seating arrangement, particularly for *fortissimo* passages, only acceptable for doubly occupied horn chairs. The dominance of the other section, as perceived by

Score Example 22 A. Bruckner, 4th Symphony, 3rd movement, measure 9 ff. Excerpt without strings and wood winds

the players, also leads them to regard conductor instructions to reduce volume, as inappropriate in their case. Therefore, placing the trombones behind the woodwinds appears more favorable than behind the French horns, particularly since this also makes it possible to place the first chairs of the trumpets and the trombones close to the wood wind solo players. Thus, for example, for decades, the customary seating for the Vienna philharmonic in the large Musikvereinssaal is as follows: (as seen by the conductor from left to right): Tuba, Trombones (3–2–1), Trumpets (1–2–3), French Horns (1–2–3–4).

The sideways separation of the French horns from the heavy brass also has the advantage that the exchange of motifs occurring between French horns and trumpets, or Trombones, as they jump back and forth, results in greater dimensionality. In fact, even for listeners which are seated farther that 30 m from the orchestra stage, such a spatial effect is clearly sensed. A typical example of this seating technique is given in the Scherzo of the 4th Symphony of A. Bruckner (score example 22), where, however, it should be noted that the third trumpet occasionally coincides with the French horns; it should therefore not be separated too much from that section. In the Sinfonia de Requiem by B. Britten the French Horns are also brought to pronounced opposition with the other brasses in several places, where in part a thematic exchange is carried on between the trumpets and trombones. In that setting, the desired contrast effect can only be achieved by distributing the brass players over the entire width of the stage.

7.2.4 *Timpani*

In spite of the low pitch, timpani are easily localized by the audience because of the time structure of the sound, and this not only for hard blows. For placement within the orchestra, therefore, the question of cooperative sound with other instrument sections becomes as important as the choice to locate the timpani in the middle, behind the orchestra for tonal symmetry. The tonal connection with trumpets in Baroque and Classical music is a natural consequence for this position. The combination with basses is more difficult, be this as opposing voices (see Sect. 7.2.1.4.), or in unison as in the slow introduction of the Freischütz Overture by K. M. von Weber (score example 23). Particularly in this example, it is certainly a question of interpretation, whether the pronounced tone formation accomplished by proximity of the two instrument sections, or the undistinguished tonal effect of spatial separation is preferred, since the latter possibly can express the eerie feeling of this passage more appropriately. Furthermore, the spatial effect of the rolling thunder at the conclusion of the 3rd movement of the Symphonie Fantastique by Berlioz can be increased by distributing the four timpani equally over the entire width of the orchestra.

Certainly the positioning of the timpani in C. Nielsen's 4th Symphony is a problem, since, according to instructions in the score, two pairs of timpani are to be placed on the right and on the left in front of the orchestra, so that "until the end, even when performing at *piano,* a certain threatening character is to be maintained." Without a doubt, this would be an impressive tonal effect, however, in loud passages the timpani would make technical playing situations for strings significantly more difficult, especially for the violins. A possible compromise, though somewhat costly, would be to double the timpani and have the loud passages performed in the rear corners of the orchestra and the soft ones, in contrast, by the timpani in front of the orchestra.

By reason of dynamic balance, timpani, when serving as solo instruments, are better placed behind, rather than in front of the orchestra, particularly since the reflecting rear wall can have an advantageous effect on the harmonic contributions of the timpani sound. For example, this is valid for the concerto for trumpet and timpani by S. Matthus. To visualize this, the principal radiation regions of the most

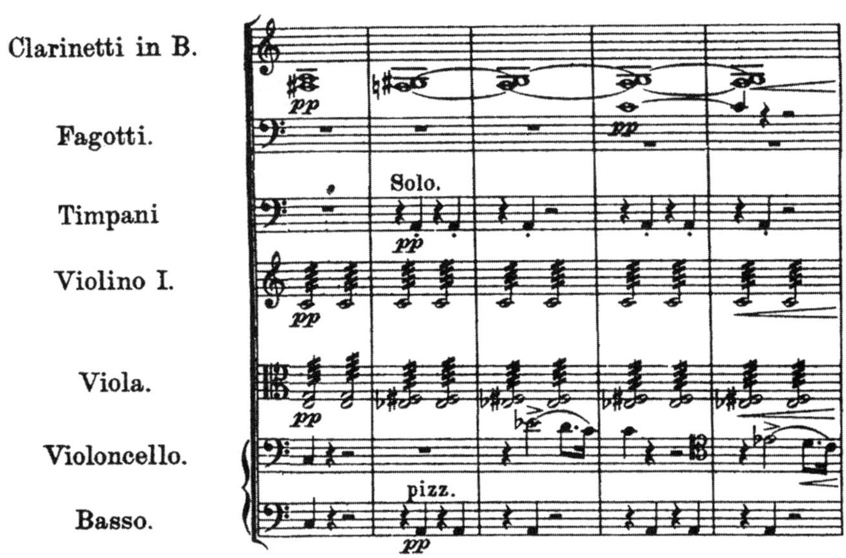

Score example 23 C.M. von Weber, Der Freischütz, overture, measure 25 ff

important components are represented in Fig. 7.36. While the lowest (inharmonic) ring mode is radiated uniformly in all directions, the harmonic partials show clear concentrations with maximum values in the horizontal plane, where the radiation upward for the radial vibrations with increasing index becomes ever more flat. Consequently the rear wall behind the player enhances the concentration of the harmonic partials more than that of the lowest ring mode: thus the tone becomes pure and the pitch more clearly recognized, an effect which is naturally lost when a choir is seated behind the orchestra.

Side wall reflections support the second, however, not the first radial mode, so that the fifth is enhanced in sound in comparison to the sound of the principal mode.

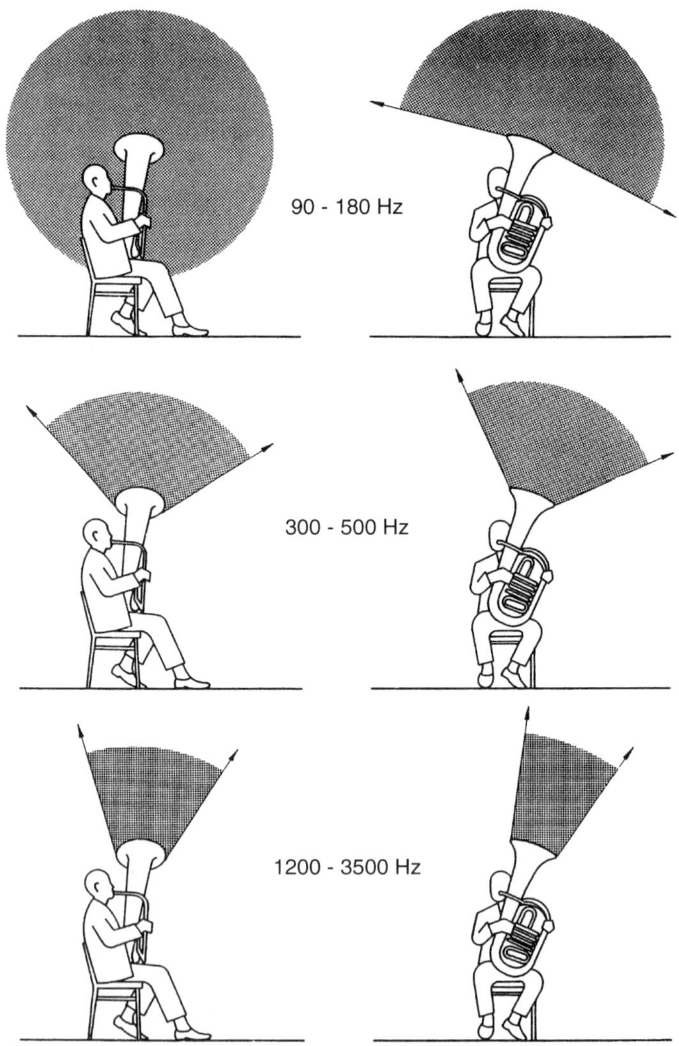

Fig. 7.36 Principal radiation directions (0. . .−3 dB) for a tuba in two vertical planes

Reflections especially include the lowest ring mode and possibly contribute more toward a dull and less pure sound of the timpani. For this, however, it is important that the reflectivity of the ceiling reaches down to 150 Hz, which is often not the case when the ceiling is divided into individual reflectors, or when the ceiling contains large openings for lighting.

When the podium itself, by virtue of its vibration capability, is in a position to enhance the sound of the timpani through structure born sound transmission, this primarily involves the lowest ring mode. For this membrane resonance, the force is transmitted in equal phase to the entire circumference of the kettle rim, and can be transmitted to the floor through the stand. The dull impact noise, though stronger during the attack, will decay more rapidly. This effect, however, is not very pronounced because of the mass difference between the membrane and timpani body.

7.2.5 Grand Pianos

Piano sound receives its characteristic on the one hand by the precision of the attack and on the other hand by the slow decay. Both characteristics are weakened in their aesthetic effects by excessive reverberations in the hall. Consequently, strong direct sound contributions, as well as the stronger partials and the high overtones, which transmit the precision, are essential for convincing sound impression in the audience. Within limits, this even applies to the noise contributions due to articulation. As the representation of the principal radiation angles in a vertical plane show in Fig. 7.37, these conditions, for a grand piano with raised lid are best met for the audience seated toward the right of the performer, whereby the high registers can see additional improvement for an audience seated on a slightly rising floor. In contrast, the higher registers are clearly disadvantaged toward the back of the instrument. Reflections from the hall ceiling enhance the low frequency contributions and thus give the instrument its fullness. Reflections from the ceiling above the stage especially enhance the attack noise as transmitted by the sound board and thus can be annoying both to the audience and to the performer, especially when the delay of the reflection is too long. The unfavorable intensity relation for these directions, between attack noise and harmonic partials, was already mentioned in connection with Fig. 4.27.

The directional effect in the horizontal plane leads to the fact that, for an audience seated outside a preferred region of approximately $\pm 30°$ width, the clarity and transparency of rapid tone sequences decreases, even though the subjective overall loudness is only changed slightly. Since, above all, the low tone contributions are largely preserved, a masking of other instruments by the piano can occur for the seats located further toward the side, where, however, these passages no longer possess the desired precision. The optimal tone picture of the piano with its full brilliance thus can only be heard in the very narrow region of the hall of a width of approximately $\pm 5°$ from both sides of the central axis.

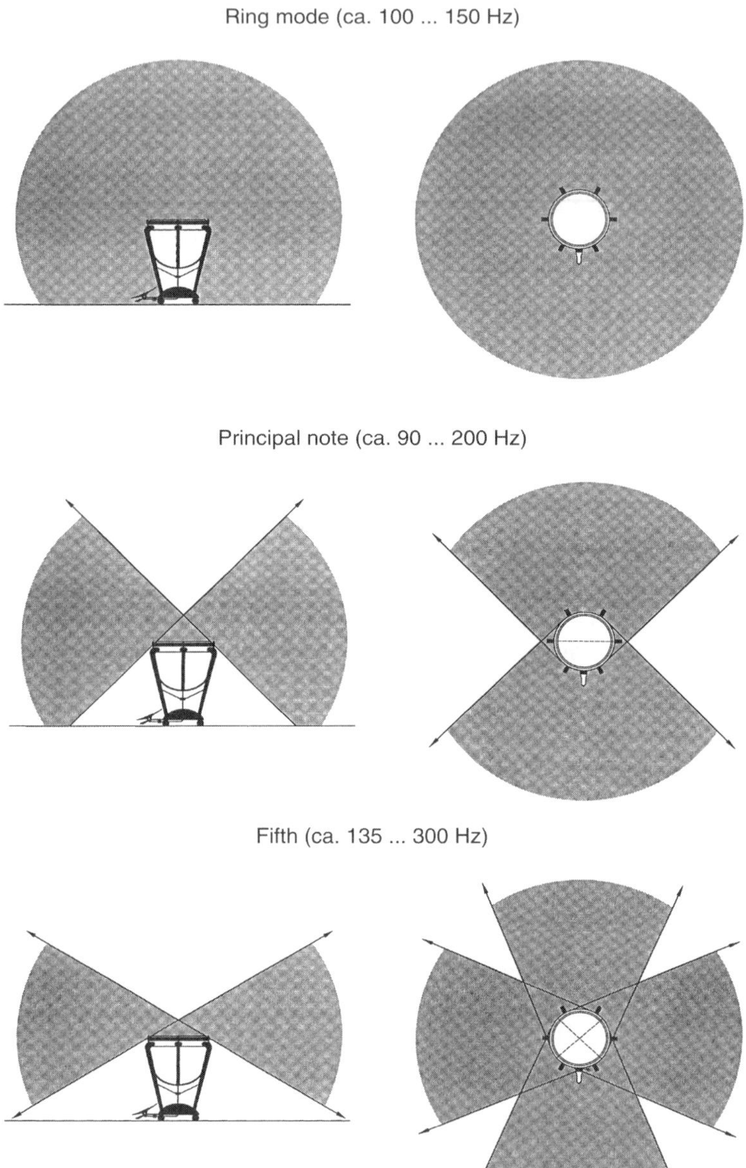

Fig. 7.37 Principal radiation directions (0. . . –3 dB) for timpani in a vertical as well as horizontal plane

Since the directional characteristics result from a cooperation of the direct sound and floor reflections (from the sound board), naturally they are not valid for the near field region in which the pianist, and for piano concertos with orchestras also the conductor, is located (Grützmacher and Lottermoser, 1936). Especially at the location of the conductor, the high tone contributions are largely blocked by the piano lid, so that the conductor hears the piano relatively loudly, but with a dull tone color. However, this effect is slightly diminished when the conductor is not located in the middle behind the piano, but rather approximately toward the extension of the keyboard, as preferred by some conductors, even though this makes eye contact more difficult.

Difficulties concerning positioning occur mostly for works with two pianos. For performance technical reasons pianists value a setting which places keyboards along one line. Of course, this presents the danger that the second piano is blocked in rather strong measure by the lid of the first when this one is open. If, however, the first lid is only half raised, the masking of the second is somewhat reduced, and the high tone contributions of the first are somewhat weakened so that a balance can be reached for the tonal impression in the hall. If, on the other hand, the lid of the first piano is removed entirely, an excessive decrease in high frequency has to be expected. In this case the danger exists that the audience perceives the second piano as more brilliant than the first. However, the tonal impression in the immediate vicinity of the instruments is certainly different from that in the hall.

Under circumstances where sufficiently large and appropriately oriented ceiling surfaces are present, considerations could be given to removing both lids. For this, however, it is a requirement for a good tonal effect, that the ceiling or the reflectors hanging above the instruments are not too high. For ceiling heights above 10 m, precision of the tone entrance is affected in the front portion of the hall by the delayed reflections. Furthermore, the tone can assume an uncomfortable hardness when the attack noise is repeated after a short pause of approximately 10 ms by the delayed reflection. This is especially unpleasant when the relevant surfaces reflect predominantly higher frequencies, because in that case the longer lasting low frequency attack noise (duration around 100 ms) no longer masks the shorter high frequency attack noise (duration around 40 ms) and its reflection, and thus no longer fills the pause between the two (see Fig. 3.21).

For this type of sound radiation by the instruments it is also possible to arrange both pianos in nested fashion such that the pianists face each other. This solution does have the advantage that the sound of the second piano reflected by the floor is not blocked by the other instrument, yet this positioning is not as convenient for the players since it makes precise playing together more difficult.

When the piano is used as an orchestral instrument as is occasionally the case, especially in contemporary compositions, it is generally located further toward the back of the podium. The orientation of the instrument then results from a requirement that the pianist must be able to see the conductor. Choice of position then is closely connected to the nature of the composition of each piece, that is, to the question, to which instrumental section the piano is assigned by the structure of the composition. Frequently it is the percussion section, often also the celeste or harp, which are to merge into a uniform sound with the piano. For the brilliance of the piano sound, it is naturally advantageous to locate the piano with open lid on the left side of the orchestra (as seen by the conductor). On the other hand, removal of the lid is recommended when the piano is located on the right side even though at that location the tonal effect is not as precise or bright and for many piano models the high registers then simply no longer sound right. Even for compositions for which a close contact between piano and harp on the one hand and basses on the other is required, as for example in the Sinfonia da Requiem by B. Britten, it is advantageous from a tonal standpoint, to place the piano on the left side of the stage, which, however, would demand so-called European seating of strings or a positioning of the basses in front of the rear wall of the podium.

7.2.6 *Harps*

Traditionally the harp is placed on the left side of the orchestra behind the violins. This may be partly for tonal reasons which relate above all to playing with violins or – for European seating – also with celli. Certainly the optical impression also plays a role: the instrument is shown in its full form, the head of the player and the music stand are located behind the harp; in contrast, frontal positioning would be less attractive.

As the principal radiation regions show, (represented in Fig. 7.38 for the horizontal plane) the sound radiation is largely symmetrical. Only in the middle frequencies the left side of the player is slightly preferred. It is, however, noteworthy that in this frequency region, which is particularly important for tone color and loudness impression, the frontal sound radiation dominates over the side. The most encompassing spectrum is radiated at an angle toward the front. It would therefore be optimal for a soloist to be seated at an angle of just under 45° with respect to the edge of the podium. In front of the player, by reason of the weaker radiation above 2,000 Hz, the harp sound is less hard in the attack, and thus possibly even somewhat rounder, which at least for recording purposes can be of interest.

For the customary positioning within the orchestra the wall behind the performer, which is most often the left side wall of the hall, presents the most important reflection surface for the harp. Walls which are located toward the side as seen by the player are less effective. Ceiling reflections especially support a middle frequency region from about 400 to 1,000 Hz and thus contribute to the fullness of the sound. However, all these acoustic aspects are not of sufficient importance to prevent primary consideration of the overall sound cooperation with other instruments like the celeste, piano, or basses as already discussed in Sect. 7.2.5.

7.2.7 *Combined Sound of the Orchestra*

7.2.7.1 Balance Between Sections

The multiplicity of instruments, with their different tonal ranges and tone colors, as well as differing preferred radiation directions, combine into the orchestral "body of sound." From an acoustical standpoint, this results in a very complicated sound

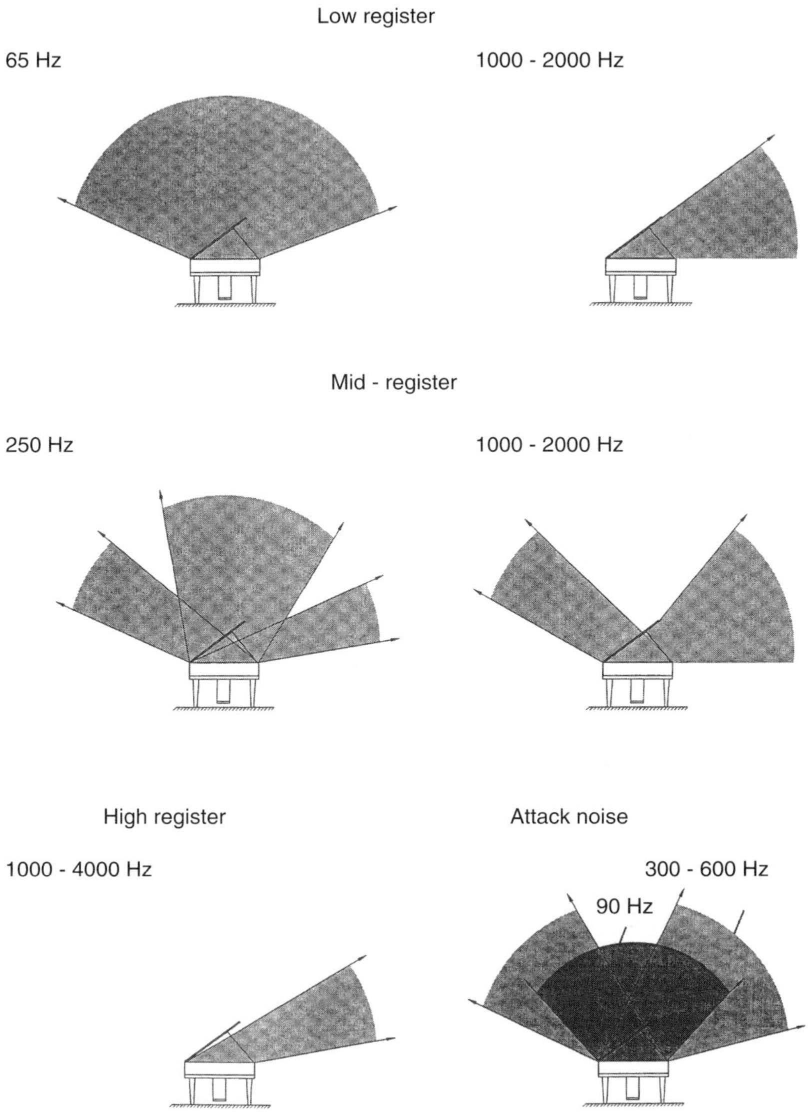

Fig. 7.38 Principal radiation directions (0. . .−3 dB) of a concert grand piano for the most important partials and the attack noise

source. Added to that is the fact that a portion of the instrumental voices are occasionally represented by only two players while others are formed by larger groups. Finally, the spatial spread of the orchestra on the podium also plays a role. In particular, different distances of individual instruments from the conductor result in a tone picture at the conductor's desk which is often sensed as essentially different than that at various locations in the audience.

In spite of that, the listener expects a balanced orchestral sound which conveys the composition in the interpretation of the conductor as convincingly as possible. This balance is essentially related to three factors: the intensity or loudness, the tone color, and the clarity. These three components should possess the "right" balance between the individual instrument groups. This must not change throughout the hall, so that the tone picture is not misrepresented.

In the earlier context of discussing sound radiation relationships for individual musical instruments, preferred directions, and thus directions for which greatest clarity is expected in the hall, were singled out. This should not lead to the immediate conclusion that maximum clarity for all instrument groups of the orchestra is basically to be considered as a tonal optimum; rather, different composition styles place different demands on the transparency of the sound.

Inasmuch as increasing concert hall size diminishes clarity, the task is given to create a balance in large halls by appropriate seating within the orchestra, wherein on the one hand preferred directions of instruments are utilized, and on the other hand the mutual blocking by musicians is reduced by increased tier elevations. Figure. 7.39 shows the podium of the large Musikvereinssaal in Vienna in which even the strings are seated in elevated tiers. As a whole, the elevation difference within the podium is 1.8 m; in the old Berlin Philharmonic – destroyed in the war – the elevation tiers even covered a level difference of 2.8 m (Winkler and Tennhardt, 1993). All such possibilities for increasing the clarity should be utilized for performance in large halls, especially for the performance of classical works (originally written for and performed in smaller halls), which, by their very nature, require a

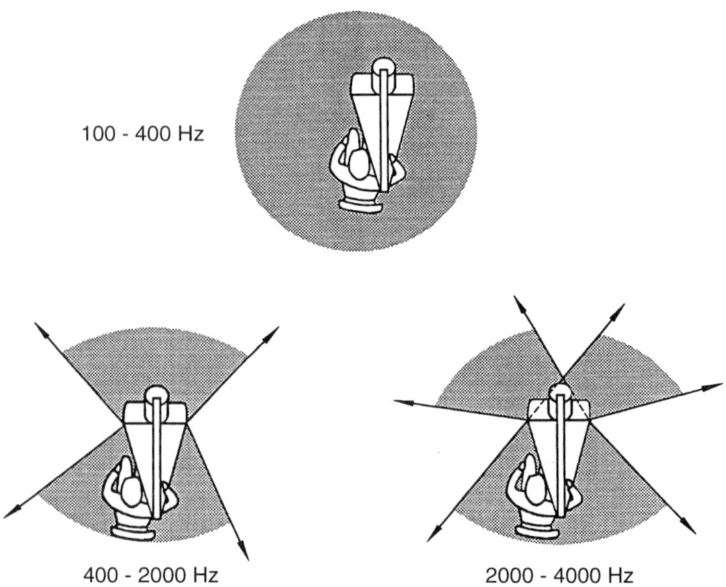

Fig. 7.39 Principal radiation directions (0. . .−3 dB) of a harp in the horizontal plane

transparent tonal picture. In contrast, for romantic music, a more uniform overall sound results, with slightly masked initial transients.

A steeper tier arrangement on the podium naturally also influences the loudness relationships between individual sections. The higher the winds are seated, the freer the path of the direct sound is toward the audience in a flat, main floor, and the stronger they will appear there, in comparison to the strings. This means, that for a small number of strings, the danger of brass dominance can occur and thus a more flat seating on the stage is an advantage for the balance. For a larger number of strings, in contrast, higher tiers for the brass make it easier to create a dynamic balance.

Circumstances are somewhat different when the audience seats rise steeply or galleries are present. The audience in elevated seats always have the advantage of receiving the direct sound of the winds directly (as long as they are not seated to the side of or behind the orchestra). In that setting, the steepness of the tiers on the podium, only plays a minor role. Nevertheless, the tonal picture should not deviate too much from that, on the main floor. Therefore, under such hall conditions, it is absolutely important to provide good direct sound for the flat rows in the audience, thus the wind steps need to be tiered more steeply. Furthermore, blocking by strings can also be reduced by having them seated at different levels. With small numbers of strings the winds naturally must adapt their intensity to that of the strings.

If a large portion of the audience seats are located behind the orchestra then naturally steep tiers of wind rows lead to increased blocking of the strings. With regards to intensity balance, this is nevertheless not perceived as detrimental, since the largest portion of wind instruments radiate sound preferable in the forward direction and are thus heard relatively softly behind the orchestra. Thus, for sufficiently steeply rising audience seats, a relatively even balance can be achieved, even behind the orchestra, provided the horns are placed as close to the middle as possible so that they are shadowed toward the back by other players; furthermore, the horns have to be treated very carefully with respect to dynamics. As a last resort one could also consider placing sound absorbing structures behind this instrument group.

Some investigations which were carried out in the mid 1960s in the Stadthalle in Braunschweig will show in detailed form to what extent the balance between individual instrument sections can be influenced by changing the seating arrangements and the podium step arrangement. The goal of these investigations was to find an optimal arrangement for this hall, which was new at the time. Figure. 7.40 shows the distribution of players on the podium in a floor diagram, as well as sections through the elevation arrangement, as used initially at the time of hall dedication, and as it was changed after the first acoustical measurements. Initially all strings sat at the same level (1.15 m above the main floor) – according to the Furtwängler seating, all woodwinds and horns were located on the first step, percussion and heavy brass on the second level. The tonal effect of the orchestra in the hall was investigated at four different locations in the audience and compared with the relationship at the location of the conductor. These four audience positions are marked by microphone symbols in Fig. 7.41 in the floor plan portion of the figure.

Fig. 7.40 Orchestra stage in the Vienna Musikvereinssaal. The center of the first step is removed for the piano

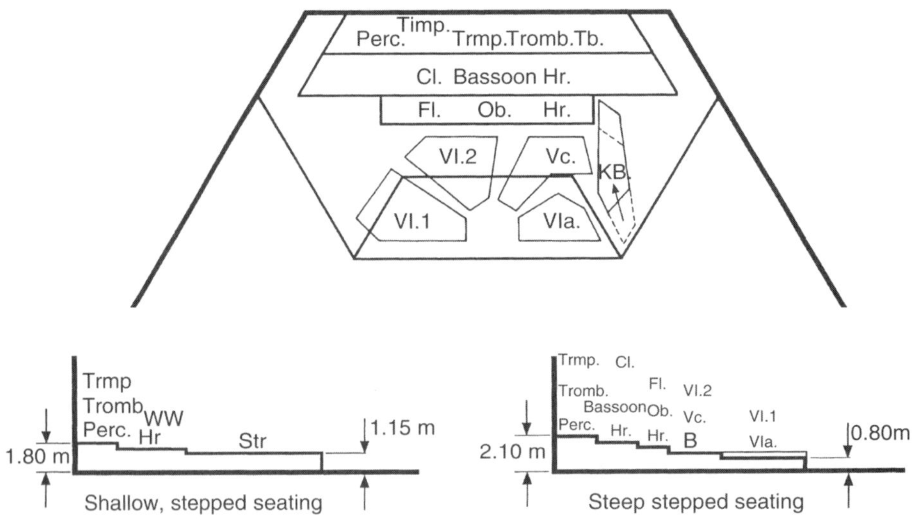

Fig. 7.41 Orchestral arrangement on the podium of the large hall of the Stadthalle Braunschweig

Beginning with the assumption that the tone picture at the location of the conductor is essentially responsible for the dynamics of the performance of the individual groups, the sound level difference for the most important instruments between the conductor and the individual audience locations was determined. The results are shown in the small diagrams for violins (I and II), the cellos and basses, the woodwinds, and the horns. In contrast to the violins and celli in the Beethoven Hall (Figs. 7.7 and 7.42) the frequency dependence is not represented here. The size of the individual blocks merely corresponds to the sound level difference in dB between the level at the conductor's podium and the relevant seat in the audience.

It is noted that in the rear portion of the rising main floor and especially in the gallery, the strings arrive relatively weakly, in the gallery the bass section is by almost 15 dB softer than at the conductor's position. In contrast, the winds are less attenuated in both audience locations. It appears especially loud on the right side of the main floor since the sound is reflected by the angled side walls of the hall for these seats.

In order to obtain an objective measure for the balance between individual instrument groups at the different audience locations, i.e., a value, which does not depend on the loudness of the individual voices, in each case the difference between the results for the most and least attenuated instrument groups is indicated next to the diagrams. At the location of the conductor this value would be 0 dB: in the central front of the main floor a shift of the dynamic balance by 3.8 dB occurs in comparison to the tonal impression of the conductor where the violins retreat most strongly and the horns stand out most loudly. The greatest change in balance of 5.6 dB occurs in the gallery, as indicated.

These tonal disadvantages can be significantly diminished by some changes in the seating arrangement and level adjustments on the podium. The new arrangement is shown in the small drawings on the right of Fig. 7.40. The strings are seated on two levels, with the lower dropped to a level of 0.8 m above the main floor. The woodwinds and horns are positioned on an additional step, and the players of the

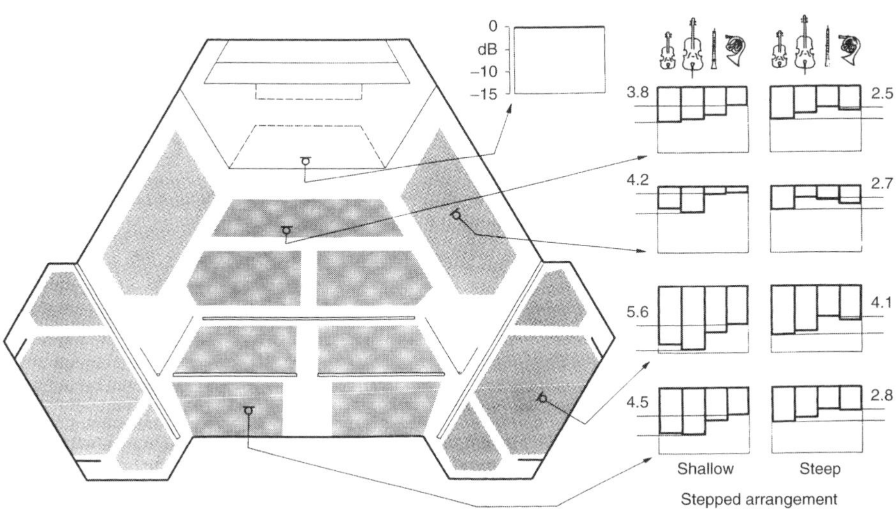

Fig. 7.42 Level differences between different audience locations and the location of the conductor for four instrument sections and two types of orchestral seating (Stadthalle Braunschweig)

heavy brass and percussion are placed on the fifth step. When only two or four horns are used, the trumpets and trombones move forward a row, and are thus placed next to the bassoons.

The floor plan is hardly changed, only the basses are moved as far as possible to the rear which not only aids their own tone but also brings about that the horns are shadowed slightly in relationship to the right side wall. Thus, the undesirable reflections are reduced to a more suitable measure. This steeper step arrangement especially improves the freer sound radiation of the strings, but it also affects the tone of the woodwinds favorably.

The improvement of acoustical conditions achieved by these changes (without construction efforts) is expressed in the diagrams on the right of Fig. 7.41. In the upper ground floor and the gallery, the intensity of the violins and the low strings is raised by several dB, the winds also become somewhat stronger. Thus, the frequently criticized disadvantage of the first arrangement, namely insufficient loudness at these locations, is significantly ameliorated. In contrast, the winds experience significantly stronger damping in the area of the side main floor for the second arrangement and thus no longer dominate as strongly in the overall sound. However, the bass group has become louder which can be considered as undesirable from the tonal standpoint: however, this is the only disadvantage resulting from this change considered for the 16 values for four instrument groups at four locations. In particular, these balanced values, which lie below 3 dB except for the galleries, point to the increased even balance of the orchestral sound.

When the intensity of each instrument group is compared for the four different locations in the hall, a positive influence of the change in orchestral arrangement can also be noted. In particular, the level of the winds differed in the first seating arrangement by 7 dB while the differences for the second seating arrangement were reduced to 3.5 dB. In this regard, even for the strings, an improvement is noted.

In the level considerations so far, the spectral structure of the sound was not considered. However, the freer radiation of violins and woodwinds also raises the overtone contribution of the sounds in addition to the overall level, and thus the brilliance of the instrument, as perceived in the hall. Thus, for violins for example, in the upper main floor in the gallery, a gain in level of 5 dB is noted for components above 1,000 Hz which contrasts with an increase in overall level of only 3 dB. The free radiation of the oboes as observed in the front rows in the main floor also shows an additional rise from 2–3 dB for the higher frequency contributions.

As an example for the frequency level dependent changes caused by the alternative seating arrangement the intensity differences at the individual audience locations for the celli and basses is represented in Fig. 7.43. Again the location of the conductor is the reference point – i.e., the zero line in the diagram. The tonal changes can be explained by the orchestral seating arrangement as follows:

Central main floor: Since the basses are pulled back from the edge, the distance to the audience in these seats increases, consequently the lowest tone contributions become weaker. The steeper seating arrangements on the other hand give freer radiation to the celli, so that the higher frequencies arrive more strongly, and the tone as a whole is somewhat brightened.

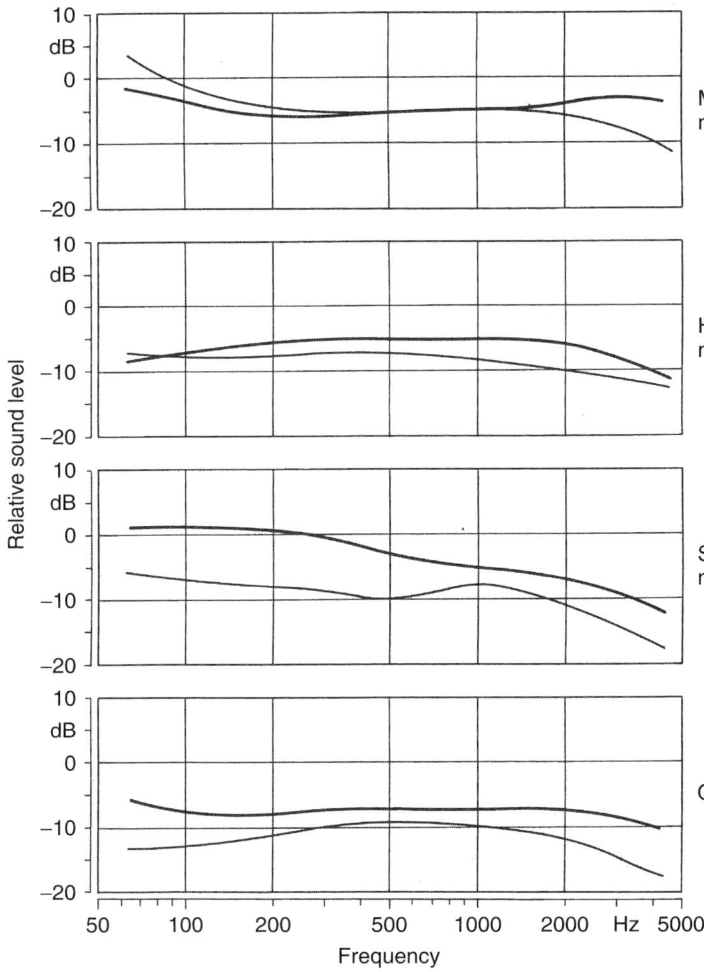

Fig. 7.43 Level difference of celli in different hall locations in relationship to the level at the location of the conductor. (Stadthalle Braunschweig). *Thin line*: seating with shallow steps; *thick line*: seating with steep steps

Upper main floor: Turning the basses, as determined by the new placement, leads to an advantageous radiation of the low components in the direction of this audience location. The steeper arrangement of the celli, furthermore, effects a level gain in the mid and high frequency regions.

Side main floor: Turning the basses leads to an increase of the low tone contributions, which, however, as mentioned, in this location is evaluated as negative. However, the higher frequencies of the celli reach this location more strongly. Since the basses are moved further back, the shadowing of the celli is reduced

Gallery: Turning the basses and the higher seating of the celli affects – as in the upper main floor – a level gain at all frequencies.

Disadvantages of a flat orchestral seating arrangement are not only noticeable in the relationship to the intensity balance between instrument sections, but also in the precision of the tonal onsets. In Fig. 7.44 the structure of the beginning chord of a Mozart symphony (score example 24) is given by means of sonagrams taken at the five relevant locations in thc hall. Here again the frequency of the overtones is recorded in the upper direction while the intensity (in the relatively wide filter regions) is characterized by the degree of blackness. As the time scale shows, these partial pictures, in each case, barely show the first quarter of the first measure.

While at the location of the conductor the entrance of all frequency regions – and thus for all instrument groups – occurs at the same time and sounds correspondingly precise, the front edge of the sonagrams for the other locations in the hall shows different delays at different frequencies. In the center of the main floor the $B_5^\flat$ (with a fundamental around 950 Hz) arrives somewhat later than the lower voices. The reason for this, in the case of strings, lies partially in the *arpeggio* beginning, which has the upper note appearing slightly later, but partially also in the fact that components at around 1,000 Hz are predominantly radiated toward the ceiling. The fact that the *arpeggio* delay – in contrast to the location of the conductor – is so pronounced, also lies in the circumstance, that the 1st oboe, which also plays this $B_{5,}^\flat$ is shadowed for the middle main floor by performers sitting in front and is thus heard predominantly by the delayed ceiling reflection.

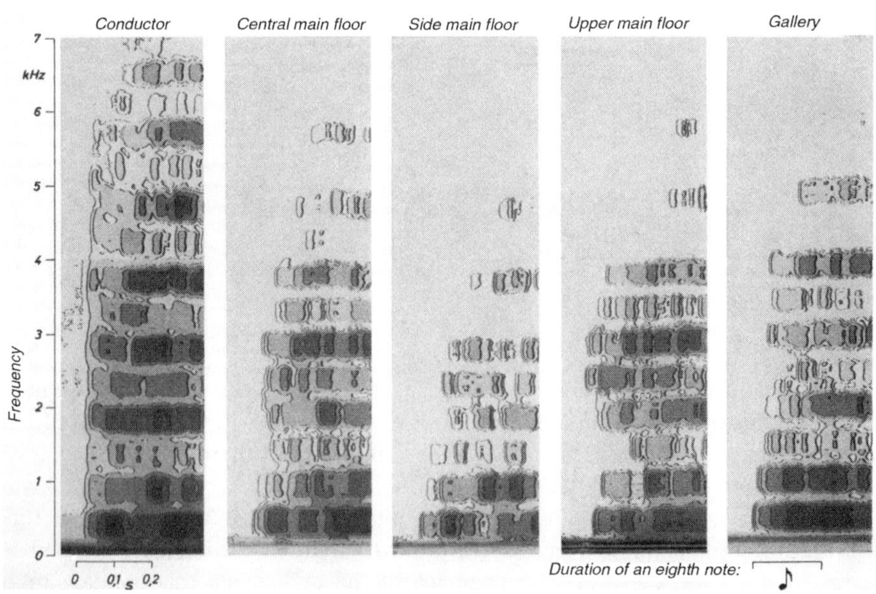

Fig. 7.44 Sonagram of a beginning chord of the 1st movement of Symphony K319 by W.A. Mozart (see score example 24) recorded at five different locations in the hall

Score example 24 W.A. Mozart, Symphony B^b major K 319 beginning of the 1st movement

This effect is even more pronounced in the side of the main floor where the bass group (including the horns) enters especially strongly and early. The delay time of approximately 1/16 s again points toward shadowing and the detour via ceiling reflections.

In the upper main floor the entrance sounds relatively exact: at this location, as expected from the directional characteristics and the location of formants, strong direct sound portions in the region around 3–4 kHz arrive here from the winds, while the principal energy of the violins is directed to this location by a ceiling reflection. The bass group is delayed only slightly so that the chord receives a closed tonal entrance.

In contrast, the sound of the bass group arrives in the gallery first, which can be ascribed to the spatial separation of the instrument groups in the orchestra: the delay of about 1/30 s for the upper voices is quite small and is hardly noticed as disturbing. By reason of the fact that violins as well as woodwinds are directed toward the higher lying gallery with strong direct sound contributions, a relatively precise tone entrance results.

As a whole, these investigations show that a podium elevation of approximately 80 cm is acoustically more advantageous for the front string desks than a height of more than 1 m, and that a multileveled orchestra podium leads to a better tonal balance and greater clarity of the tone picture in the hall. These experiences were also confirmed by more recent studies in the Berlin Schauspielhaus (Winkler and Tennhardt, 1993): experienced test listeners attest to a clear improvement of balance and clarity for increased stepping and lowering of the front podium area to 80 cm, a further, though only slight improvement, was brought by additional, slightly downward angled edge elements protruding above the podium. This kind of steeper steps also brings advantage for mutual listening of orchestra musicians, particularly for the sound impression of woodwinds on the strings. Contact between celli and first violins is also easier for the American seating, while second violins and violas hear each other less (Winkler and Tennhardt 1994).

7.2.7.2 Newer Seating Arrangements

Considerations so far all began with the classical seating arrangement which finds the strings in front and the winds in steps behind. Naturally, new groupings can be considered for which this system is either totally or partially abandoned. The principal effort of L. Stokowski was directed toward enabling the best possible sound radiation for all instruments within the orchestra into the hall. This means the preference of the direct sound in the overall sound and aims at the same time toward an unimpeded transmission of the entire tonal spectrum, especially as high frequency components. This naturally still leaves the problem of dynamic balance between instrument groups of differing strengths.

Building on the intentions of Stokowski, Veneklasen (1986) proposed and carried out experimental tests of several variations for orchestral seating. These experiments included current understanding of sound power and directional characteristics of the individual instruments. In this context it is particularly interesting to consider a comparison between the customary American seating and a seating arrangement for which, in each case the first desk of the string groups is located as usual, close to the conductor, however, additional strings are positioned on steps generally reserved for winds. To the left of the conductor, seated behind the concert master, are the woodwinds (with the first chair always near the edge), to the right behind the first cello chair are the horns followed by the heavy brass). In comparison to the standard seating, this seating arrangement resulted in the following changes in regards to loudness balance in the audience (Seattle Opera House): the horns, as the strongest group among the brasses become weaker by 3 dB, since they have to make due without rear wall reflections. The brass group as a whole was reduced by 5 dB, not least because they had to perform in the direction perpendicular to the hall axis. In contrast, the strings did not lose intensity. The balance therefore was perceived as improved: problems of mutual listening were not considered in this context.

Another experiment conducted during a season opening concert in the same hall extended the Stokowski string sequence, however, grouped the strings to the left only to slightly past the middle, so that the celli faced toward the front. Four woodwind groups sat to the right of the conductor on risers behind each other: horns and heavy brass were seated in steps toward the rear on the right third of the podium. The concert public was impressed by the choir-like total sound of the strings which even for *tutti* performance at the higher dynamic levels was not squelched. No measurements were taken for this experiment. Because of marginal acceptance by musicians, this seating was not continued after the premier performance. No detailed reasons were given.

A not yet realized seating arrangement even departs from the central position of the conductor; the conductor is located close to the right wall of the stage. Seated on the front level (with sight direction toward the right) are the woodwinds in four rows behind each other followed by horns, heavy brass, and the timpani, which form the end at the left edge of the front. On the first higher step – on which the conductor stands toward the right – the eight chairs of the 1st violins are positioned behind each other. On three additional steps, the 2nd violins, violas, and celli in appropriate width. The basses are found toward the left rear.

All these variants reveal an effort toward dynamic balance within the orchestra, emphasizing a choir-like string sound. They evidently depart strongly from a striving toward special balance, which is typical for the orchestral sound of the Classical and Romantic repertoire. How far this brings new perspectives or new development is still an open question.

7.2.7.3 Spatial Effects

While treating several instrument groups, a number of examples were already shown, for which a spatial separation of the voices permits an increased transparency of the tonal picture for thematic exchanges. In similar fashion it is naturally also possible that themes between instruments of different groups are exchanged as in a dialogue, and a corresponding spatial sound effect becomes desirable. However, even for a single run through a motif from high to low instruments (or in the opposite direction), in spite of varying tone colors, a spatial wandering or possibly jumping of the sound source within the orchestra can present a certain charm. Thus, recently even for electronic music reproductions, spatial effects have been deliberately introduced. In this context the separation of performers on the stage does not necessarily have to be large, particularly when single instruments are considered, for the ear can already sense an angular change of the incoming sound of approximately 3° (compare Sect. 1.2.4.).

A typical example for such a passage is represented by the excerpts given in score example 25 from the Concerto for Orchestra Op. 38 by P. Hindemith. To achieve the tonal effect of this passage it is certainly recommended that flutes, clarinets, and trumpets are placed adjacent to each other (even if they are located at different levels on the podium), and not directly behind each other. Otherwise, the listener has difficulties to follow the individual voices, particularly after the triplet run, even if they are clearly distinguishable in tonal character. Likewise, the dialogue between flute and trombone in A. Copland's 3rd Symphony only reaches its full effect when both players are not located behind each other but side by side.

On the other hand, of course, the task is often given to form a new tone color with closed and uniform effect from several individual instruments. Typical examples of different combinations of this nature are found in the Bolero by M. Ravel, such as in the 6th pass through the theme, with trumpet (*con sordino*) and flute in octaves, or the instrumentation with horn, piccolo, and celeste as given in score example 1. In such cases a spatially close contact is demanded for tonal reasons, but naturally also to make a rhythmic and intonation precise combined playing easier to achieve. A combination of low voices is less critical in the context of spatial-tonal con-

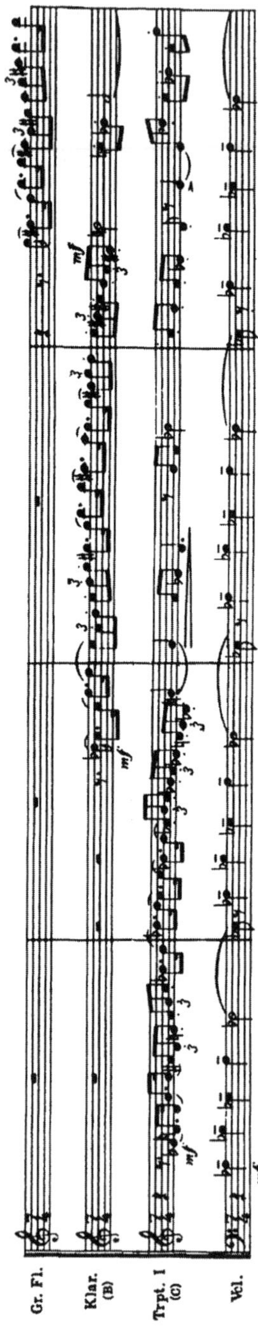

Score example 25 P. Hindemith, Concerto for Orchestra, beginning of the 4th movement, score excerpt without timpani

siderations, especially when instruments with full and round tone color are involved. A combination of a wind voice with strings is also less critical – at least in concert halls where the audience sits only in front of the orchestra – since these groups are not localized as sharply because of their wide distribution on the stage. However, a narrow separation is required by the technical demands in many classical works for the trumpets and the timpani, especially when rhythmic accents occur which are placed into an already existing orchestra chord, as for example the third attack of the large fermata shortly before the end of the 1st movement (measure 332) of the 8th symphony by L. van Beethoven.

Particular spatial effects can be achieved (and usually are intended by the composer), when compositions are written for two orchestras or instrument groups. For example, the symphonies for two orchestras by Joh. Chr. Bach or the Concerto for two string orchestras by M. Tippett belong in this category. Both composers place two instrumental bodies in opposition to each other without giving one of them a dominant position. Consequently the players would be most advantageously located symmetrically on the two halves of the podium. This lateral separation by an imaginary central line already is sufficient for a spatial effect which however can still be somewhat increased by moving the violins on both sides by 1–2 m away from the conductor, which is particularly advantageous in large halls.

As is the case for the two string sections of orchestras with German seating arrangement, for this two sided distribution of both instrument bodies, the high strings of the first orchestra (the left one as seen by the conductor) receives more brilliance than the corresponding voices of the right orchestra. This phenomenon also differentiates the tonal picture and it is therefore important to be aware that the low strings and possibly the winds must be adapted to this tonal picture. Certainly the basses sound more open from the left side if they are located in the two rear corners of the podium, however, for the front facing celli of both groups, a particular dynamic adaptation is required, since they are not differentiated from each other by their timbre. For the Bach symphonies it would also make sense to provide for a freer radiation possibilities for the flutes of the first orchestra (in Op. 18, No. 1 and 3) than for the flutes of the second orchestra, while for the symphony Op. 18, No. 5, the balance between the two groups is taken care of automatically by the tonal difference between the oboes and the flutes.

Similar relationships for strings also exist in the Petite Symphony Concertante by F. Martin, or for the Music for String Instruments, Percussion and Celeste by B. Bartok, however, in this case the separation on two sides of the string bodies are additionally enhanced in their spatial effect by locating the keyboard and percussion instruments as well as the harp, predominantly in the middle. For reason of intensities, the harpsichord should occupy the more favored position in comparison to the piano in the Martin composition. In the Bartok composition, the central group should be positioned to be as small as possible and located on the lowest level since the two voices of the string bodies are not always treated as opposite poles (as for example at the beginning of the 2nd movement), but often performed in unison.

In contrast to this, for the Serenata Notturna K239 by W. A. Mozart, which in the subtitle is designated as a serenade for two orchestras, a horizontal separation is inappropriate. Since the "first orchestra" only involves four strings in solo fashion, rather, a tonal spatial broadening of the central core in the direction of the overall sound is approached, as is characteristic of baroque concerti grossi for strings. However, for a sufficiently large number of strings in the main orchestra it is advantageous in the Mozart Serenade to position all four soloists on the left side of the conductor, which is an advantage for the viola and, above all for the bass (playing without cello support).

In contrast, a larger spatial separation is required for those pieces for which instruments are placed as an echo in opposition to the main orchestra, as for example in the Divertimento for two string trios by J. Haydn for which the passages of the first group are repeated by the second group with different delays. This compositional technique is shown in yet increased form in the serenade for four orchestras K286 by W. A. Mozart for which each section contains strings and two horns.

In an ideal case, a concert hall provides the possibility for such works to locate the echo orchestra in different, distant, musician galleries, as is the case for the Berlin Philharmonic. On the other hand, when arranging the echo orchestra near the main group is unavoidable, one should attempt to weaken the direct sound of the echo by movable walls and by correspondingly unfavorable orientation of musicians so that the echo is clearly differentiated from the original sound by a greater reverberation. In that case, however, if at all possible, the echo group should not consist of fewer players than the main orchestra, since a smaller ensemble with its more transparent sound would contradict the less precise distance impression of an echo.

A more reverberant sound naturally also is created when the respective instruments perform in a staircase or a foyer and the sound (including the reverberation from the side room) reaches the actual concert hall through an open door or studio window. This procedure is also very well suited for individual instruments whose sound in the sense of the composition comes from a distance. Well-known examples for this are given by the trumpet signals in the 2nd and 3rd Leonore Overture by L. Beethoven. The posthorn solo in the 3rd symphony by G. Mahler or the oboe "from the distance" in the Symphonie Fantastique by H. Berlioz, as already mentioned can be performed very expressively in this fashion. Likewise, the excessively direct effect of the second horn pair in the echo symphony in E^b major by Stamitz is removed by such an arrangement, since, based on the directional characteristics, even the first pair of horns include the reverberation of the hall very strongly in the tonal picture.

When direct visual connection with the conductor is not possible, such contact can be established by a TV monitor, however, for reasons of intonation it is absolutely imperative that the performer of the echo hears the orchestra with sufficient loudness and overtone content which in extreme cases can be achieved by a loudspeaker or earphones (compare the comments in the section about stage music, Sect. 9.4.2). On the one hand, the distance impression of the echo can be influenced by having the player orient the more or less strong direct sound portions toward the opening into the hall, which for example in the 3rd Leonore Overture can be used to increase the effect when repeating the trumpet signal. On the other hand, it is often also possible to make the actual location of the echo source more obscure by opening several doors. Furthermore, the reverberation of the ante-room can be adapted to the desired tonal demands with appropriate curtains or carpets. Finally, the distance impression can be enhanced by means of a curtain in front of the opening to the hall which weakens the high frequency components of the echo.

7.2.8 Singing Voices

7.2.8.1 Choirs

For oratorios, and for symphonic works with choir, in addition to the problems of the orchestra, the question of the arrangement of singers also arises, and in many cases this can also become the principal concern. Thus, the clarity of the tonal picture as received by the listener, as well as the loudness, especially in larger halls, i.e., the transmission of sound energy, plays an important role for the vocal soloist. For choirs, on the other hand, above all clarity and tone color – both with sufficient homogeneity – are significant.

A choir consists of a multiplicity of incoherent sound sources; a choir's sound thus possesses a character which is not spatially differentiated and the individual singers are acoustically not separately located. Even localizing the individual voice groups of a choir in space often becomes difficult at great distances. The overall sound primarily gives the impression of filling the room, which is also supported by reflections from side walls. Nevertheless, depending on arrangement of vocal sections, a difference is perceived with regards to spatial tone balance. When female voices are located in the front and male voices behind, a higher degree of symmetry in the tonal picture is achieved than when voices are arranged from left to right in the sequence of soprano-alto-tenor-bass. The latter arrangement, however, corresponds better to the American orchestral arrangement.

Since a choir is always located behind the orchestra, floor reflections are almost never effective for singers. Ceiling reflections and especially side wall reflections, however, can contribute to enhancing high frequency contributions as a look at angular regions of strongest sound radiation shows (see Fig. 4.33). Wall surfaces which lie within an angle of approximately $\pm 60^\circ$ of the direction of sight of the singers are particularly effective. Hanging ceiling reflectors must be located sufficiently high that they are no longer in the field of view of the singers. Disregarding the fact that relatively low reflector positions direct important reflections toward the audience, they are perceived by choral singers as extremely disadvantageous when they convey the feeling that a portion of the sound above the reflectors vanishes and is lost to the sound in the hall (Burd and Haslam, 1994).

In order to insure free sound radiation toward the front, and to avoid mutual shadowing by choir members the steps for the individual rows of singers should rise at an angle of approximately 45° in the optimal case i.e., they should be as high as they are wide. At the same time, this provides an unimpeded sound spreading in the direction of the reflecting side walls.

The more shallow the choir is arranged, the less favorable the sound radiation of the strongest tone contributions, directed downward – by reason of directional characteristics. Clarity and precision of the choral sound and articulation suffer especially since the reflections from the ceiling, which are not energetically influenced, gain in intensity, as perceived in the hall, in relation to the direct sound. Thus a larger portion of the radiated sound energy falls into the reverberation of the hall; in halls with very little reverberation this can be seen as positive while then the excessively direct tonal impression is reduced. A flat choral arrangement is decidedly unfavorable in very high halls because there the delayed ceiling reflection influences the clarity of the articulation when the direct sound is weakened by the rows of singers standing in front of each other. The critical limit for hall height (above the level of the singer risers) lies at around 8–10 m depending on the spatial arrangement of the audience region.

For the listener, the wall behind the choir enhances especially the middle and low frequencies. Portable walls which are designed to shield the singers from the empty room behind the choir should extend at least 0.5 m above the heads of the last row of singers. For mutual listening of singers, rear wall reflections are especially important, particularly when there are no additional close wall or ceiling reflections. As already explained, reflection surfaces with distances to the singers of 2.5–6 m especially support the acoustic contact.

The influence of a more or less tightly spaced choir on mutual listening is rather limited since only the direct sound of the immediate neighbor stands out above the sound of the entire choir (Ternström, 1991b). The "individual diffuse-field distance" of the individual singer, i.e., the distance for which the direct sound of the singer has the same level as the statistical sound field of all partners, lies, depending on choir size at one-third to one-fifth of the normal diffuse-field distance for the corresponding hall, i.e., for a concert hall between 1.7 and 1.0 m. This means that placing the singers more closely, the number of directly audible vocal colleagues is only raised minimally. The influence on the uniformity of the choral sound in the hall lies within very narrow boundaries.

7.2.8.2 Vocal Soloists

For positioning vocal soloists, depending on special circumstances, there are three possibilities: in front of the orchestra next to the conductor, behind the orchestra raised in front of a rear wall, or in the middle of the choir gallery. Advantages and disadvantages of these positions must be evaluated from the point of view of sound radiation (loudness, vocal color, and clarity) as also from the standpoint of contact with the choir or spatial separation from the choir.

Placement next to the conductor, in addition to good contact with the conductor, offers the opportunity of proximity to the audience and utilization of energetically valuable floor reflections: the latter, however, does require a free distance from the stage edge of at least 2 m. For the singer, the proximity to the supporting voices of the orchestra conveys a feeling of security. To what extent the sound in the hall supports the singer's feeling for vocal control depends on the reflection characteristics of each room and thus can not be evaluated in general for this location near the podium. Reflections from surfaces at a distance of up to 6 m are desirable though not absolutely necessary for the soloist; however, later reflections which establish the connection to the reverberation of the hall are important.

For the audience seated at the side of the orchestra, this vocal soloist position is a disadvantage as is the case for many audience seats located on the sides of very wide halls: as determined by the directional characteristics of the voice, an optimal tonal effect can be expected in the angular region of approximately ±45° relative to the direction of sight. This region which opens conically into the hall reaches the side walls in a 20 m wide hall at approximately 11 m in front of the podium. In a 40 m wide hall it does not reach the wall until 22 m. Furthermore, for seats in the side of the hall the spatial separation of soloist and choir is noticeably annoying when a closed tonal impression of all vocal voices is expected by the composition as for example in the Rhapsody for Alto, male chorus and orchestra by J. Brahms (score example 26).

When vocal soloists are positioned in front of a reflecting rear wall, the direct sound is strengthened at least in the low and middle frequency region by 4–5 dB, by the unnoticeably delayed reflections, so that the diffuse-field distance of the singer is apparently nearly doubled. This increase of the reach of the singer, compensates acoustically for the greater distance to the audience – when compared with placement in front of the orchestra. Particularly in wide halls, this positioning can reduce the number of unfavorable audience seats noticeably, and even for locations to the side of the orchestra the tone quality is improved. On the other hand, naturally, the

Score example 26 J. Brahms, Rhapsody for Alto, male chorus and orchestra, measure 166 ff. (without strings)

visual impression of closeness to the singer is very desirable for the audience, consequently more distantly located soloists need to be optically emphasized as for example by special lighting or color control of the surroundings.

For the singer, the position behind the orchestra differs from the position in front of the podium by the fact that behind the orchestra the hall sound of the voice and the direct sound of the instruments are perceived as coming from similar directions. This makes vocal control more difficult, unless, in addition to the rear wall reflection already mentioned, additional strong reflections, especially from the ceiling, support the impression from the voice. Regardless of the acoustical aspects, one can also imagine situations where the organizer does not want to present a prominent singer at a distance greater than absolutely necessary and thus will insist on positioning the singer next to the conductor.

Positioning the vocal soloist in the center of the choir gallery also combines the advantage of reaching the audience at the side well and the possibility of the tonal integration of the choir and soloist. To what degree the latter is essential, or on the other hand not desirable, depends on the compositional structure of each work. For the soloist it is here also an advantage to have a reflecting rear wall, which for example can be accomplished by the structure of an organ or movable walls. For the singer personally, the reduced distance to the hall ceiling provides an advantage over a location directly behind the orchestra in view of the hall sound coming from above so that the tonal impression of the solo voice is not covered as strongly by the orchestra.

Chapter 8
Acoustic Considerations for Instrumentation and Playing Technique

8.1 Strength of Ensembles

8.1.1 Historical Development

The size of the orchestra available to composers at earlier times for the performance of their works naturally had a decisive influence on the style of instrumentation. Since on the one hand the possibility of number of wind instruments was proscribed by the number of available players, on the other hand the tone of the strings was determined by the size of the individual sections. Nevertheless, undoubtedly the tonal perception by great composers of their important works occasionally transcended the realizable possibilities of the time. A look at the historical development of the orchestra (Schreiber, 1938; Becker, 1962) thus can provide reference points for a performance true to the intent of the work, however, historical indications should certainly not be considered as performance instructions to be observed absolutely, particularly since concerts today are frequently performed under very different spatial conditions.

The typical orchestra of the Baroque period was an enlarged chamber music ensemble. An example for this would be the instrumentation which J.S. Bach demanded from the magistrate in the year 1730 as a minimum for his church music in Leipzig: 2–3 first and second violins each, four violas, two celli, one double bass in addition six woodwinds, three trumpets, and timpani. In the year 1746 the Leipzig Concert Association had a larger number of strings with five first and second violins each and two violas, celli, and basses each. Wind chairs were generally singularly occupied, only for the bassoons were two or three players for each part used to support the *basso continuo* line. Also for the Concerti Grossi by Corelli and Vivaldi an orchestra instrumentation of similar order of magnitude can be assumed as normal, even though reports of the performances with very large string sections are found. In contrast, in Italian opera houses orchestras with more than 30 violins were no exception.

DOI: 10.1007/978-0-387-09517-2_8,

The Mannheim Court, where J. Stamitz served as "Kapellmeister", serves as a starting point of the new orchestra technique which became the foundation of the tonal picture of the classical period. Already in the year 1756, he used string sections of 10 + 10 + 4 + 4 + 2, in addition he had two oboes and bassoons, four flutes and horns, 12 trumpets, as well as timpani. The large number of high (i.e., melody carrying) voices in contrast to the smaller number of lower voices, is particularly noticeable, which stands in strong contrast to the Baroque bass emphasis. This orchestra, which at the time was considered the best in all of Europe, was not even from a standpoint of number of instruments approached by most of the other instrumental bodies.

Thus, during the years of 1761–1765 Haydn, at the court of Eisenstadt or Esterháza, only had an orchestra with 11 strings and five winds which could be supplemented with trumpets and timpani as well as two additional horns from the military band. As the arrangement in Table 8.1 shows, in the course of the following years the instrumentation was strengthened twice; however, for the period after 1780 the division of the strings into individual voices is not known. However, the instrumentation with which Haydn performed his London Symphonies was much larger, where especially the doubling of the woodwinds is noted (Robbins Landon, 1976).

Also the orchestra which was normally available to W.A. Mozart and L.van Beethoven for the performance of their symphonies had a size similar to the orchestra of Count Esterházy. For example, the house orchestra of Count Lobkowitz, which at the time gave the premier performance (not public) of the Eroica consisted of 2 × 4 violins and two violas, celli, and basses. Yet for special occasions, frequently concerts were given with far greater numbers, for which string voices were increased by amateurs. Thus, Mozart (1781) enthusiastically reports a concert of a performance of a C Major symphony (presumably K 338) in Vienna with 40 violins, ten violas, eight celli, and ten basses. Still with the exception of bassoons (six musicians) all wind chairs were singly occupied. The concert, which L. van Beethoven gave in connection with the Vienna Congress in 1814 in the Redoutensaal, in which his 7th Symphony was heard for the first time, also became famous. Again, with single

Table 8.1 Instrumentation of J. Haydn's Orchestra

Hall	Time period	Number of musicians									
		Viol.	Vla.	Vc.	Kb.	Fl.	Ob.	Kl.	Fg.	Hr.	Tr.
Eisenstadt	1760–1765	6	1	2	2	–	2	–	1	2	–
Esterháza	1766–1774	7	2	2	2	–	2	–	1	2	–
	1775–1780	11	2	1	2	1	2	–	1	2	–
	nach 1780	...23...				1	2	–	1	2	–
Hanover Square	1791–1792	14	4	3	4	2	2	–	2	2	2
Rooms, London	1793–1794	14	4	3	4	2	2	2	2	2	2
King's Theatre,	1794	24	6	4	5	4	4	–	4	2	2
London	1795	24	6	4	5	4	4	4	4	2	2

occupancy of wind chairs, string section included 2×18 violins, 14 violas, 12 celli, and seven basses. For the premier performance of the 9th Symphony in the Kärtnertor Theater, the strings had a total of 12 violins, ten violas, and six celli and basses each, which for those times was quite remarkable (Becker, 1962).

Strong string sections naturally shifted the balance of the tone with the strengths of the winds remaining equal, away from their favor. The fact that in only extremely rare cases a doubling of winds was used, may be connected with the circumstance that the needed musicians – in contrast to string players – are not so easily found among amateurs; however, the question remains open, whether the string-emphasis tone did not also correspond to the tonal perception of the composers. Certainly in available sources no clear reference is made concerning negative evaluation of large instrumentations, while Beethoven in sketches for his 9th Symphony notes "orchestral violins etc. will be increased tenfold for the last movement" (Kobald, 1964).

The increasing number of winds, which characterizes the development of the orchestra in the nineteenth century, naturally also leads to increasing the number of strings. A typical example for this is the Leipzig Gewandhaus-Orchestra which can be considered as representative of concert life of those days. In Table 8.2 the total number of winds and strings for several years is represented (Becker, 1962).

All these data belong to the era of the "Old Gewandhaus;" the so-called New Gewandhaus was not opened until the year of 1886. The increase of wind sections, included above all the horns (to four players), and trombones as demanded by many symphonies in the Romantic period. It does not yet include the three and fourfold woodwind numbers and the eight horns used occasionally in the second half of that century. The fact that such an increase of the string sections corresponded fully and clearly to the tonal perception to the composer of that generation is proved by the demands stated by them on different occasions in relationship to instrumentation: for example, H. Berlioz requires string sections of at least 15 + 15 + 10 + 11 + 9 players in his score for the Symphonie Fantastique (1830).

Even today, for standard works of the symphonic repertoire, orchestras, if possible, have a strength as desired during the second half of the nineteenth Century. The compositions of the Romantics, and also already the symphonies of L. van Beethoven were performed with 14–16 first violins (when permitted by the number of available musicians), occasionally even with 18–20. The remaining string sections generally decreased numerically toward the low end. Second violin sections are usually by about two players weaker than the first, and the bass section as the smallest section is represented by 6–10 players. Smaller orchestras, however, must remain below this measure when no possibility of strengthening by temporary help is available.

Table 8.2 Instrumentation of the Gewandhaus-orchestera

Jahr	Bläser	1.V.	2.V.	Vla.	Vc.	Kb.
1781	11	6	6	3	2	2
1839	12	9	8	5	5	4
1865	17	16	14	8	9	5

In contrast, for symphonies by J. Haydn and also by W.A. Mozart (with exception of the last three) quite generally smaller numbers of strings are preferred, even when a larger number of players are available. The first violins are thus reduced to 8–10 and the remaining sections correspondingly lowered so that basses finally are represented by only 3–4 players. Reasons for this lie, on the one hand, in the effort to obtain as transparent a sound as possible. On the other hand, it is desirable to balance the strings against the winds, which in early Classical works have an instrumentation of only 4–6 voices.

8.1.2 Adapting to the Hall

When the tonal picture of the classical orchestra is to be carried over to the acoustical conditions of today's concert halls, one needs to take into account the available spatial conditions relative to the instrumentation. An ensemble, as was available to Haydn for the performance of his symphonies in Esterháza, would not be tonally satisfactory in the width of the modern hall in spite of its structural transparency. The necessary loudness, to permit an audience to experience a *forte,* would simply not be available (Vogel, 1968).

As already explained in the chapter about room acoustical foundations, the perceived loudness is related to energy density, and this in turn to the volume and the reverberation time of the hall, as well as the power of the sound source. In this context the hall characteristics can be combined in simplified form as "room damping index." This specifies the numerical difference between the sound power level (of all participating instruments) and the sound pressure level of the statistical sound field in the hall. In Fig. 8.1 this room damping index is plotted versus the volume of the hall. For halls of equal reverberation time these values lie on the relevant straight line. Thus, for every hall whose size and reverberation time is known, the diagram can indicate by how much louder or softer the same sound source appears there in comparison to other halls. Appropriate points for a series of well-known concert halls and churches are entered in the figure; in each case they are given in reference to the reverberation time at middle frequencies. In some cases the difference between an occupied and empty hall is also indicated. For example, for the Vienna Musikvereinssaal it amounts to approximately 2 dB and is less than 1 dB for the Berlin Philharmonie.

When the absolute values for these two halls are compared, then the Vienna hall is higher by approximately 3 dB: it belongs to the class of halls for which a very full orchestral sound is developed, while the Berlin Philharmonie falls into the group of slightly softer halls. In order to equalize this difference, the power of the sound source would have to be approximately doubled (corresponding to Table 1.1, p. 3). Accordingly, by equal intensity of performance by individual musicians, instrumentation would have to be approximately doubled in the Berlin Philharmonie in order to reach the same loudness as in the Musikvereinssaal. When considering that this value lies by about 6 dB higher in the old Gewandhaus than that in the Vienna Musikvereinssaal, one can imagine how loud even small instrumentations must have sounded at that time. For, even a string instrumentation with 2×4 violins and correspondingly other sections, as required by L. van Beethoven as a minimum for his symphonies, would achieve the same energy density in the Gewandhaus as a string section with 16 first violins in the Vienna Musikvereinssaal; in the Palais Lobkowitz it would have even been noticeably louder.

The significance of the energy density as calculated from reverberation time and hall volume, however, should not be overestimated. Because, taken strictly, these values only apply for the steady state of the hall and thus are not valid for very short notes: for short notes, the direct sound contributions and the first reflections become increasingly important for the loudness impression, thus the directional characteristics of the instruments and consequently the orchestral seating arrangement also plays a role. Furthermore, the examples of sound distribution in concert halls (Figs. 7.7, 7.41–7.43) have also shown how much the sound level within the hall can still fluctuate. The diagram of Fig. 8.1, however, only gives mean values for the relevant halls and can thus only give a summary view for the different halls.

Increasing the instrumentation to twice the number of performers, however, effects not only an increase of energy density by 3 dB and thus an equalization of levels between two halls differing in this measure, but in the sense of the chorus

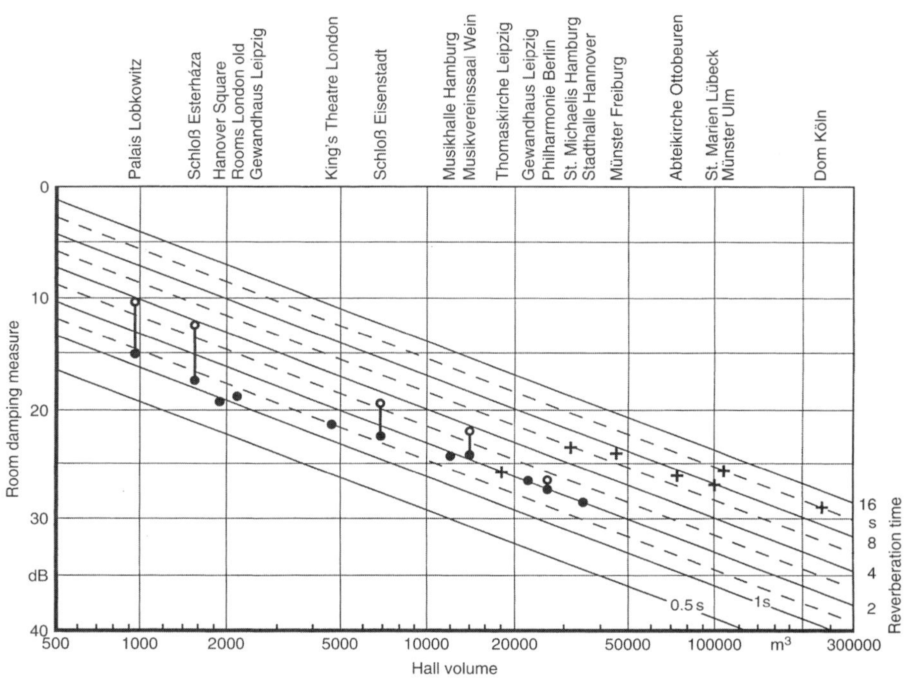

Fig. 8.1 Dependence of room damping index on volume and reverberation time of the hall. *Dots*: fully occupied hall. *Circles*: unoccupied hall. *Crosses*: unoccupied church

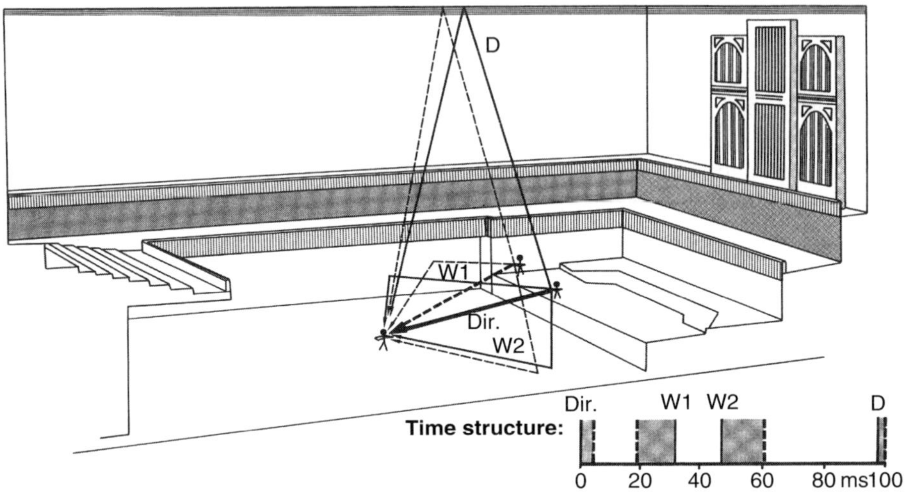

Fig. 8.2 Direct sound and first reflections in a hall originating from the first and last chair of the first violins

effect it leads to an increased bandwidth or a more uniform filling of the existing frequency band which is caused by minor intonation differences as well as vibratos. Thus the sound is perceived as denser and fuller even when the sound level is not changed. This effect is especially pronounced for strings by virtue of their distributed location within each section. For – as shown in Fig. 8.2 for one player on the first and last desk of the first violins – the transit time of the direct sound for the individual players to the audience in the hall is nearly the same (the reference of the time structure picture corresponds to the point in time when the direct sound from the closest player arrives). There are also practically no running time differences for the first ceiling reflections. For the side wall reflections, the picture is quite different: they are fanned out over a time region of more than 10 ms, which results in a very differentiated tone picture, than would be the case, for example, when reproducing an (reverberation free) orchestral recording through a speaker.

As a further consideration for the dynamic impression at an audience location, it is noted that the emotionally perceived intensity of an orchestra *forte* is not determined solely by the loudness level but also by the broadening of the sound which is designated as the "spatial characteristic." As already explained (see Sect. 5.5) the degree of spaciousness does not only depend on the room acoustical properties but also on the loudness level at the location of the listener and thus on the sound power radiated by the orchestra.

This sound power is determined by performance technique and the number of instruments. Starting with the assumption that for the *tutti*-sound, which is important for the spaciousness, all players obey the dynamic instructions for the *forte* in equal manner, the radiated sound energy of the individual instrument types nevertheless differ significantly. Table 8.3 assembles the average sound power levels and

Table 8.3 Mean sound power level L_{wf} and associated power factor k for orchestra insruments in *forte*

	L_{wf} dB	k		L_{wf} dB	k		L_{wf} dB	k
Violin	89	0,8	Flute	91	1,3	French Horn	102	16
Viola	87	0,5	Oboe	93	2	Trumpet	101	13
Cello	90	1	Clarinet	93	2	Trombone	101	13
Double bass	92	1,6	Basson	93	2	Tube	104	25

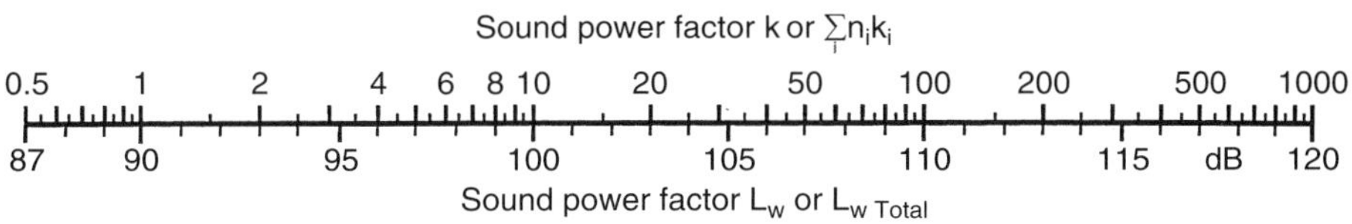

Fig. 8.3 Nomogram for conversion of Sound power factors k and sound power levels

associated power factors (refer to a sound power of 1 mW corresponding to 90 dB) for the orchestra instruments, which in each case apply to a ringing *forte* (see Sect. 2.2.8). These power factors make the determination of the sound power for the whole ensemble easier since an addition of level values is only possible by a recalculation of non-logarithmic values.

When the number of instruments of a section is multiplied by the corresponding power factor for this instrument type, and the values thus obtained are added for all instrument sections, a summarized power factor for the entire orchestra is obtained. Thus the sound power level of an ensemble can be calculated according to the formula

$$L_{\mathrm{wTotal}} = 90\,\mathrm{dB} + 10\log\sum_i n_i k_i\,\mathrm{dB},$$

where n_i is the number of instruments of equal type and k_i the corresponding power factor. This numerical relation between the sound power level and the summarized power factor can be read directly from the nomogram of Fig. 8.3.

Thus for example a summarized power factor of 6.4 results for a group of eight violins, which leads to a sound power level of 98 dB. For a small orchestral instrumentation (number of players per instrument section in the sequence strings-woodwinds – brass: 8, 8, 6, 5, 4 – 2, 2, 2, 2 – 2, 2, 0, 0) in a *forte,* a sound power level of 110 dB results and for a large orchestra (14, 14, 12, 10, 8 – 4, 4, 4, 4 – 4, 3, 3, 1), of 114 dB. From this an average sound pressure level at a *forte* can be calculated at an audience location as (according to the formula explained in Sect. 5.4.1):

$$L_{\mathrm{f}} = L_{\mathrm{wf}} - D_{\mathrm{A}}.$$

In order to give an example of adapting, or in other words transferring original performance condition to other halls, the relevant values for orchestras of J. Haydn

Table 8.4 Level values for the symphonies of J. Haydn

Hall	Time period	D_A dB	L_{wf} dB	L_f dB	Symphonies
Eisenstadt	1760–1765	22,5	106,5	84	2–27, 40, 72 außer 26, 35
Esterháza	1766–1774	17,5	106,5	89	26, 35, 38–59, außer 40
	1775–1780	17,5	107,5	90	60–71
	nach 1780	17,5	108,5	91	73–81
Hanover Square	1791–1992	19,5	109,5	90	93–98
Rooma, London	1793–1794	19,5	110	90,5	99–101
King's Theatre	1794	21	110,5	89,5	102
London	1795	21	111	90	103, 104

are assembled in Table 8.4. The number of winds as required for the majority of symphonies falling into each period form the basis for the instrumentations of Table 8.1. Thus, for example, for the time period of 1761–1775, for the indicated instrumentation, a sound power level of about 106.5 dB results. This corresponds to a *forte* level of 84 dB in Eisenstadt, in Esterháza, however, of 89 dB. The values thus found for the *forte* level can be entered into the spaciousness diagram which contains the degrees of spaciousness for concert halls of J. Haydn (Fig. 8.4). One sees that for a *forte*, two steps of spaciousness are achieved in Eisenstadt in spite of the relatively low level. In spite of similar *forte* levels, the tonal broadening was even more pronounced in the two London halls, in contrast, they were less in Esterháza (Meyer, 1978a).

When symphonic works of earlier times are to be performed in modern concert halls, the possibility exists, with regards to the instrumentation of the orchestra, to transfer considerations concerning the original *forte*-level and the connected spatial conditions to today's concert halls, in order to determine what the sound pressure level of the orchestra must be, in order to give the audience an equivalent *forte*-level impression. Beginning with the assumption that a level increase by 5 dB without raising the spaciousness results quantitatively in the same perceived dynamic change as a rise of spaciousness by one step without level change, then lines of equivalent *forte* impressions can be entered into the spaciousness diagrams of concert halls. In Fig. 8.4 such lines are drawn as broken lines for historic halls. The intersection of these lines with the spaciousness lines of a different hall then indicate at what level an equivalent *forte* level impression is to be expected in this hall.

As examples the spaciousness lines for the Hamburg Musikhalle and the Stadthalle in Hannover are entered in Fig. 8.4. When one examines the broken line passing through the point for the Eisenstadt Hall, one recognizes that the Hamburg Musikhalle requires an increase of the *forte* level by 0.5 dB, and the Hannover Hall, because of its lesser spaciousness requires a rise by 7 dB in comparison to the average level in Eisenstadt. Using the situation in the King's Theater as a reference, the level in Hamburg, because of the more strongly pronounced spaciousness, can be reduced by 2 dB while in Hannover it must be raised by 4.5 dB.

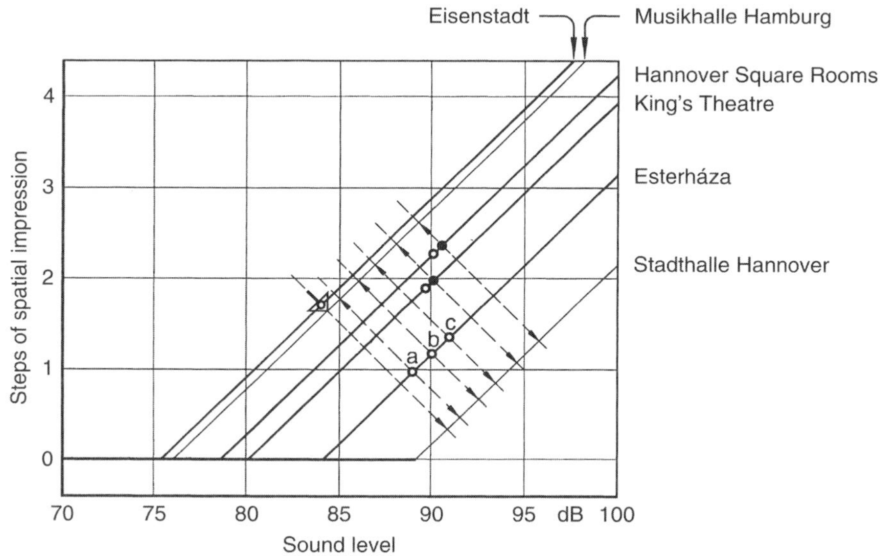

Fig. 8.4 Spaciousness of the orchestral sound in six concert halls in dependence on sound level. The average *forte* level of historical orchestras is given by circles (without clarinets) or dots (with clarinets). a Esterháza 1766–1774, b 1775–1780, c after 1780. *Broken lines*: lines of equivalent *forte* impressions

In order to achieve such changes of the *forte*-level, the orchestra must be adapted either in total numbers or relative to the intensity of the performance, where, however, the room damping index of the relevant halls must be considered. This results in the necessary rise of the sound power given by

$$L_{\mathrm{wk}} - L_{\mathrm{wo}} = (L_{\mathrm{fk}} - L_{\mathrm{fo}}) - (D_{\mathrm{Ak}} - D_{\mathrm{Ao}}).$$

The indices o refer to the original hall and k to the concert hall in which the performance is to be given. The difference between the *forte*-levels in the two halls (L_{wk} and L_{wo}) can be taken from the lines of equivalent *forte*-level impressions in Fig. 8.4. A difference of the hall damping measures ($D_{\mathrm{Ak}} - D_{\mathrm{ao}}$) is to be taken from Fig. 8.1. The relevant values for transferring performance conditions from the four concert halls of Haydn to the newer halls in Hamburg and Hannover are assembled in Table 8.5. As can be seen, for the Hamburg Musikhalle, a rise in sound power level between 1.5 and 3.5 dB is required which corresponds to power ratios from 1.4 to 2.2. For the Stadthalle in Hannover, values between 12 and 14 dB result for the level rise, which is caused by the lesser spaciousness but also by the higher hall damping measure. This corresponds to linear factors from 16 to 25.

While an increase in the number of musicians in the orchestra suggests that the calculated values for the Hamburg Musikhalle can certainly be realized, the high values for the Stadthalle Hannover indicate that there are evidently limits for the

Table 8.5 Calculation example for the instrumentation of the orchestra for performing symphonies of J. Haydn in two new concert halls

Original hall	New concert hall	L_{fk}-L_{fo} dB	D_{Ak}-D_{Ao} dB	L_{wk}-L_{wo} dB	Power factor
Eisenstadt	Musikhalle Hamburg	+0,5	−2	+2,5	1,8
Esterháza		−4	−7	+3	2,0
Hanover					
Square Rooms		−1,5	−5	+3,5	2,2
King's Theatre		−2	−3,5	+1,5	1,4
Eisenstadt	Stadthalle Hannover	+7	−6	+13	20
Esterháza		+2,5	−11	+13,5	22,5
Hanover					
Square Rooms		+5	−9	+14	25
King's Theatre		+4,5	−7,5	+12	16

possible equalization of differing room acoustical conditions. A 20-fold rise of sound energy can neither be achieved by an increased number of players nor by forcing an increase in performance volume, nor possibly by a combination of those two especially when considering the large original instrumentation of the London symphonies. Consequently, the performance of Haydn symphonies will in some sense always be unsatisfying in some halls.

While for strings, based on the relatively large number, a rather fine stepping of instrumentation strength is possible. For woodwinds and horns the only question occurs whether the single voices should be doubly occupied. Wind sections of three or four players for each part are very rare, however, even from Haydn's times authenticated performance instructions for the "Creation" exist, which call for three wind voices in contrast to ten chairs for the two violin sections (Robbins Landon, 1976). Finally, there are conductors who, for example, perform the 5th Symphony by L. van Beethoven with eight horns. Since the winds often have clear solo passages these additional players are only used in tutti-passages. For the standard instrumentation of winds, this creates a chorus effect without causing the undesired tonal impression by slight differences between only two instruments. Thus, in tonal character, the winds come closer to the larger number of strings.

Doubling winds appears particularly sensible in passages where the relevant instruments are already relatively weak because of the pitch of the passages. For example, this is the case for the low register of the flute. Thus, in the 8th Symphony of A. Dvorak in the third movement, measures 35–38, or 119 ff., a single flute (score example 27) will have great difficulties to compete with the overtones of the basses and celli. Frequently, however, for winds, the possibility remains to effect this equalization by appropriate dynamics of performance. However, one also needs to be aware that the tonal spectrum will change in the direction of increased richness in overtones with increased performance intensity. The tone of a single instrument is therefore brighter and more brilliant than the same overall level of several instruments in unison.

Score example 27 A. Dvorak, Symphony No. 8, measure 119 ff

For example, this important tonal difference for the horn can be seen in the partial spectra of Fig. 2.7; in the *pp* spectrum the fundamental is 3 dB weaker than in *mf*; when two players, however, perform this *pp* together, then the partial peaks would be 3 dB higher in the overall spectrum than for the *pp* of the individual instrument. Thus, the fundamental would be as high as for an *mf* for a single horn; however, the other partials would be significantly weaker so that the overall tonal impression would be softer. As much as an increase in brilliance by a raised dynamic level is frequently desirable, in the opposite direction, there are compositions for which a softer tonal characteristic is preferred. Beyond that, the doubling of winds protects the players in large halls from an excessive forcing of the tone for which a reduced tone quality would have to pay.

In order to adapt the number of instruments to the room conditions, in addition to the generally expected loudness levels, and the balance between strings and winds, the frequency dependence of the reverberation is also important: through it the energy density level receives its frequency dependence. The values entered in Fig. 8.1 refer to the middle frequencies. A rise of the reverberation time by a factor of 1.25 for low frequencies, for example, would raise the energy density level by 1 dB. This influence needs to be considered when the sound power of individual instruments is to be calculated to achieve the same balance between high and low registers in different halls. Since, however, the low instruments reach into the frequency of the high instruments with their overtones, such calculations are not totally satisfying; rather, tone color considerations must be taken into account.

In halls for which the upper sound contributions are absorbed in especially strong measure an excessively large number of low instruments would be inappropriate, otherwise the tonal balance would be disturbed. On the other hand, the bass section needs to be strengthened when the hall absorbs the low components strongly: this necessity is naturally perceived in halls for which the reverberation curve at low frequencies is flat and does not rise. Beyond that, the bass section must be strengthened for conditions of increased over all absorption by the concert hall (i.e., the "quieter" it is). For, as indicated by the closeness of equal loudness curves at low frequencies (see Fig. 1.1) a uniform level drop over the entire frequency region effects the oral impression that the low registers are weakened more strongly than the middle and high tonal contributions.

This effect frequently becomes noticeable in open air performances, for which the average loudness level is frequently less that in closed rooms, in which, furthermore, higher components are strengthened by several reflections while low frequencies are refracted around all obstacles and are thus no longer effective for the listener. For serenades in the outdoors for which the orchestra is not placed in a shell which surrounds the performers, the basses consequently must be represented in sufficiently large numbers. From this standpoint it is very understandable why Mozart expressly specifies "cello and bass" for his "Kleine Nachtmusik" which otherwise is not found in any of his other works.

Naturally, the character of the composition is important for the relationship between the number of chairs in the upper voices and the low instruments. The basses frequently go together with the celli or with the low winds. The more they appear as individual voices the more their strength is important to other sections.

Furthermore, the number of bass instruments can be reduced when the violas support the bass line over long passages, as for example in many early classical symphonies, or when the melody of the first violins require only a gentle tonal foundation (without pronounced middle voices). Thus, for example, for an orchestral instrumentation of the Divertimenti for strings and two horns (K247, 287 and 334 by W.A. Mozart) with 8–10 first violins and 2–3 celli and basses each are sufficient for a balanced tonal picture, while for large symphonies the total number of celli and basses should always be somewhat larger than the number of first violins.

Finally, the instrumentation of orchestras also experiences limits when historical instruments are included, which from a standpoint of intensity are not very strong. Starting with the assumption that a clarine is around 10–15 dB softer than a normal trumpet, then the number of strings needed is not much larger than for a chamber orchestra. A similar situation applies to a harpsichord when an electronically amplified model is not available. Also for a Concerto grosso a dynamic stepping between the concertino and the *tutti* requires that the soloists, even when they are developing a full tone, are not strangled by the full sound of the orchestra. Such compositions can thus hardly be performed in a tonally convincing manner in very large halls.

8.2 Dynamics

The dynamic performance indications for scores and voices do not give absolute indication for the requested loudness of the sound level, but are rather to be evaluated as relative. For the performer these directions mean a stepping of the intensity degree within the framework of the dynamic possibility of the instrument, i.e., a step wise division between the limits which can be achieved between the weakest *pp* or the strongest *ff*. These limits, which – as explained in Chap. 2 – depend on the nature of the tone sequence, are once again assembled in Fig. 8.5 for the essential orchestra instruments; the sound power levels entered, are in each case average values for the entire tonal range of the instrument. From this it becomes evident that, especially at higher dynamic levels, the woodwinds, in order of magnitude, lie by 3 dB above the strings and the brasses are by approximately an additional 10 dB stronger. Furthermore, the horn and especially the clarinet are capable of an extreme *pp*. This means, that an indicated *forte,* by nature is played louder by a trumpeter than by a flutist, also a performance indication of *piano* leads to a greater loudness by a string section of an orchestra than for individual players. When playing together in an ensemble, or as a soloist, certainly musicians adapt their performance to the loudness of other instruments, however, in this context the dynamic possibilities of each individual instrument are necessarily diminished.

The listener judges the dynamic level of the tonal impression by the absolute sound level observed at that location in the hall, as well as by the overtone content for which familiarity is obtained, though subconsciously, by experiencing the connection between performance intensity and tone color. In Fig. 8.6 the increase in overtones is assembled, as determined by dynamics, and represented in the form

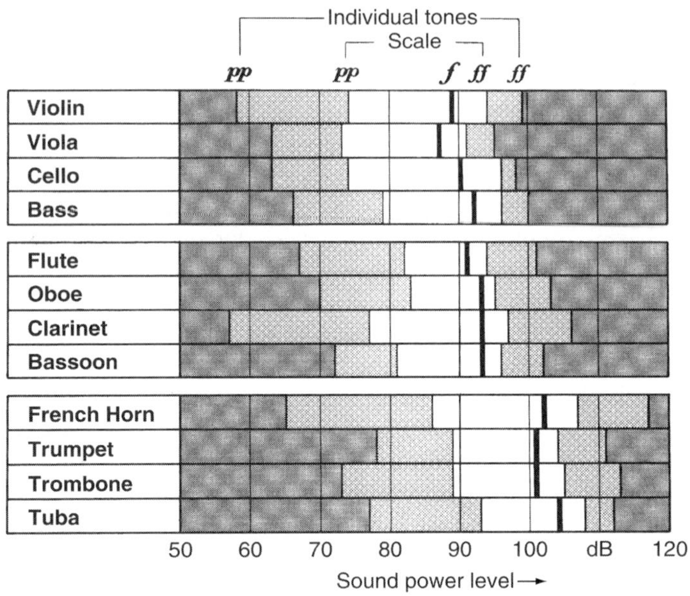

Fig. 8.5 Dynamic range and average sound power levels at a *forte* for orchestral instruments

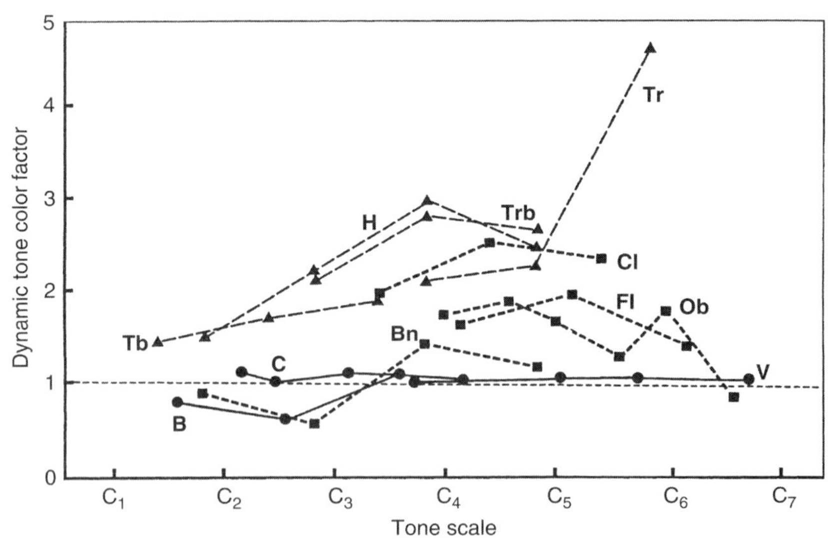

Fig. 8.6 Dynamic tone color factor (relationship of level change of 3,000 Hz components to the level change of its strongest partials) for orchestral instruments in dependence on pitch

of the "dynamic tone color factor" (see Sect. 2.2.7) for orchestral instruments. It is seen from this that the influence of dynamics on the tone color for strings is relatively minor and is strongest for brasses and the clarinet. From this Miśkiewicz and Rakowski (1994) derive an additional broadening of the dynamic region by an apparent value of up to 6 dB. At least, to a lesser degree, therefore, a difference in comparison to the level which the listener imagines as "correct" or "the original loudness" does not lead to an excessively strong tonal sacrifice. However, in the event that the deviations become too large, the intuitive component of the dynamic perception, absolutely essential for the direct personal experience of the production, becomes lost and only a purely intellectual perception remains.

Between the musicians or their instruments and the audience, an important link for sound transmission is added through the room and its acoustical properties. It also affects the dynamic impression, since it influences the sound level as well as the spectrum. The spectral tone change caused by the room is of course noted especially in the form of a more or less strong attenuation of the high frequencies, thus it again weakens the broadening of the dynamic range just mentioned. This means that the dynamic as determined by the spectrum is more apparent to the musician than the audience in the large hall. This applies especially to the strings: when they vary their dynamics primarily by bow pressure they only change the overtone content without strengthening the lower partials. While this type of dynamics is clearly noticeable by the musician – and also by the conductor – for the audience it is mostly lost, while in contrast a dynamic change based on bow speed and location of bow contact is fully noticeable in the hall. From this standpoint the advantage of halls, for which reverberation times up to frequencies around 4,000 Hz hardly drop in comparison to midfrequencies, becomes apparent: they also transmit spectral dynamics to the audience in large measure. A further limitation can be seen for instruments with sharply directed radiation of high frequencies: thus the sound of the trumpet is hardly influenced by the hall for listeners seated within the (diffuse-field distance).

With regards to the dynamic balance within an ensemble, the effect of the high overtones, as just discussed, must certainly be evaluated as critical from a further viewpoint. It is a known fact, that low voices – at least for the clarity of their melodic line – are easily masked by instruments of higher registers: since higher voices generally radiate at a higher level for a certain frequency region than lower voices in the same frequency region, a masking, or at least squelching of these tonal contributions results, so that the perceived loudness of the lower voices again rests only on their strongest (and thus relatively low frequency) components.

The tonal change caused by the room can, within certain limits, be balanced by performance technique adapted to the room. The fact that these possibilities and also their limits are noticed and considered by musicians is shown by an experiment of the Nobel Prize winner G. von Békésy (1968), who had a competent pianist perform the same piece of music first in a very reverberant room, then in a room with nearly optimal reverberation time, and finally, in a relatively dry room. In Fig. 8.7 the measured vibration amplitudes of a grand piano are represented as a measure of the impact strength of the player. The curves show, that as a whole the pianist

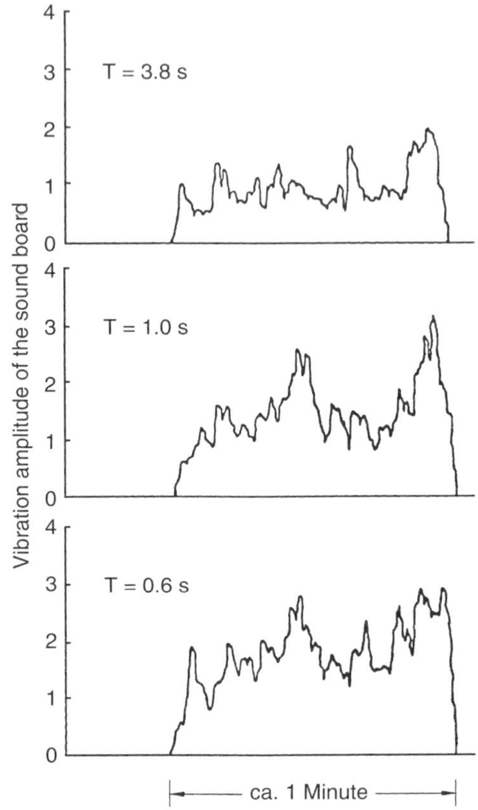

Fig. 8.7 A pianist's adaptation of performance technique to room acoustic conditions in three halls with different reverberation times (after von Békésy, 1968)

plays louder when the reverberation time decreases. However, they also show, that the dynamic is largest for optimal reverberation time; for a lower reverberation time the weakest parts are more strongly enhanced, for long reverberation time, especially the loud parts are damped. Certainly these results required that the player was very familiar with the piece and mastered it technically. Inexperienced players neglected to consider the room acoustics after an average of ten seconds.

The lower limit for the softest possible *pp*, which also for the listener is still sufficiently clearly recognizable, is generally determined by the instrument in smaller and medium-sized halls, since the partials of a sound must exceed the noise contributions, and furthermore the tone cannot drop abruptly. In large halls, in contrast, the noise level of the hall noise plays an important role. It depends on the sound isolation from the surroundings and on the quality of the air conditioning. The audience itself is often surprisingly quiet, particularly for *pp* passages. For most halls, the noise level (when occupied) lies between 35 and 45 dB (Winckel, 1962a; Cremer, 1964); with that, however, differing limits are established for the dynamics which naturally reflect on the performance technique of an orchestra.

Thus, Winckel (1962a) during a concert tour of the Cleveland orchestra under G. Szell found, that in a concert hall with a noise level of 40 dB, a *pp* passage could be

Table 8.6 Sound level in the Stadthalle Braunschweig

Bruckner, 9. Sinfonie, 1. Satz	D	M	H	S	E	
Takt 1, *pp*, Streicher-Tremolo	51	48	46	53	45	dB
Takt 63, *ff*, Tutti	102	99	96	103	96	dB
Takt 247, *ff*, Tutti ohne hohes Holz und Pauken	100	95	92	99	90	dB

reduced to 42 dB, while the same passage in another hall with a noise level of 50 dB was performed at a loudness level of 55 dB. In each case, these values were applicable to an audience seat in the middle region of the hall. The amount by which this level can vary within the room is shown by measurement results from the Braunschweig Stadthalle (see Table 8.6), where the *pp* at the beginning of the 9th symphony by A. Bruckner was by 6 dB stronger at the conductor's desk than at the quietest place, and that near the basses it was heard by an additional 2 dB louder. This example clearly shows that the lower limit of the dynamics is determined by the level relationship between the orchestra sound and the noise at the quietest location within the hall. For very uneven sound distribution, this can lead to the circumstance, that for many seats, a *pp* could appear as too loud.

In halls with all too little reverberation, for which, above all, the high frequencies are absorbed, not only a dull tone color results, but the masking by the room of initial transients and uncertainties is missing. The player perceives this as a poor attack of the instrument and reacts with a more firm beginning. Consequently, it is not possible to perform a *piano* as softly as otherwise allowed by the instrument.

The upper limit of the dynamic range of an orchestra also depends, at least partially, on the room acoustical properties of the concert hall, and is different in different locations. Level values for several characteristic measures of the 9th symphony by A. Bruckner are assembled in Table 8.6 as an example, as measured in the locations indicated in Fig. 7.41.

In the Beethoven Hall in Bonn, for comparable passages in the 8th symphony of A. Bruckner (1st movement, measure 69 and measure 125) – for values of 100–101 dB at the conductor's desk – level values between 90 and 92 dB were registered in the hall at five different seats. In this hall, for such instrumentation, the differences between the individual seat locations are not so large, which makes adapting the dynamics easier. For the upper limit of *ff* there exists also a limit determined by the hearing mechanism which should not be exceeded for tone aesthetic reasons. It lies slightly below 100 dB (Winckel, 1962a). When this cannot be attained at all, because of excessively large room absorptions, the dynamic possibilities are diminished – just as is the case for an excessive noise level.

In addition, the dynamic tone effect is influenced at *f* or *ff* by the attainable measure of spaciousness of the tone picture. In this, not only room acoustical conditions play a role, but also the nature of instrumentation. The stronger the side wall reflections are – based on the directional characteristics of an instrument – the stronger this instrument also contributes to the development of the spatial tone. In this sense the strings, the lower reed instruments, as well as the horn and the trombone show a more pronounced effect (while the latter is highly directional at high frequencies, in regions of its strongest tone contributions it radiates relatively broadly). In contrast, the trumpet radiates its principal formants so strongly directed that its sound does not contribute to its spaciousness so that it almost always stands out of the orchestra sound through its presence and the ability to perceive its location precisely.

Starting with a sound power, the individual instrument in *pp* and the orchestral *tutti* in *ff*, a possible dynamic range of around 60 dB can be derived (ignoring especially loud percussion effects). However, this ideal condition is not always met in practice, so that in most cases only a dynamic range of about 50 dB can be expected. For the various seats in the hall this range generally varies not as much as the absolute level values, thus for the examples of Table 8.6, values of between 50 and 51 dB are observed everywhere.

Especially at the higher dynamic levels, sound power produced can also be influenced – particularly for strings – by the ability of individual players to hear themselves and other members of their section as well as other voices. In Fig. 8.8

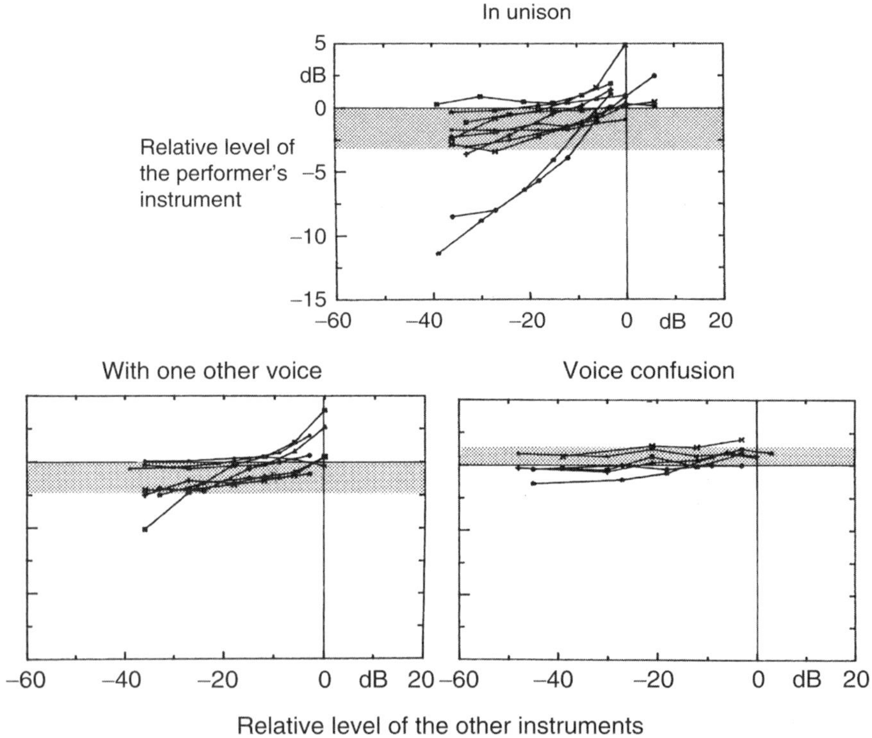

Fig. 8.8 Influence of mutual listening on the radiated sound level: performance strength of individual violinists for varying loudness of other voices. Reference value (0 dB for both axes) is the sound level of the instrument when no other voices are present (after Naylor, 1987)

the level value for sound radiation of the player's instrument in dependence on the level of other instruments as received by the player under consideration for three different musical situations is represented. The reference level for this (for both axes) is that level which the performer generates while playing the relevant musical phrase – thus perceived as appropriate – when no other voices are heard. Measurements performed with several players show an overwhelming uniform tendency, however, there clearly are musicians who respond to unfavorable room acoustical conditions with increased uncertainty.

As the unison diagram shows, the performer's own level drops by up to 3 dB for many players, when the players of the same section are heard too softly. Beyond this laboratory result (Naylor, 1987), in practice it could be expected that this would even lead to a chain reaction, which could cause the drop of the level by clearly more than 3 dB. Lack of reflections in the region of the stage could accordingly influence the fullness of the string sounds noticeably. To what measure this phenomenon becomes effective is – as shown in the extreme curves of these diagrams – likely a question of the quality of the performer or the ensemble. The players are somewhat less sensitive to the loudness of an opposing voice to which they can orient themselves both rhythmically and harmonically. It is interesting to note that they might even react with increased loudness of their own performance when the other voices do not give points of reference for mutual playing based on tone or time structure (see diagram "nonsense").

In order to utilize the dynamic region fully to its lower limit, as determined by hall and noise levels, as well as hearing characteristics, increased demands are placed on the musician and the quality of the instruments. For in *pp* even the slightly too high loudness of an individual player raises the overall level noticeably. Consequently, some Vienna oboe players, for example, play the D_5 at the beginning of the 3rd symphony by Bruckner occasionally without lower joint because then the attack for these instruments is easier.

As the combination of sound pressure for individual instrument sections shows in Fig. 8.9 for *forte* or even *fortissimo*, the brasses are the strongest section, though for the *ff* level it is already reduced by 3 dB when compared to the technically possible *ff* in this representation in consideration of the dynamic balance. With an orchestration of two woodwinds each, this instrument section is clearly weaker than the strings; the flutes are especially disadvantaged. With four woodwinds in each section, the sound power level would be raised by 3 dB, yet it would not reach the strings, however, the difference in relationship to the strings would be noticeably diminished; a doubling of the woodwinds (with only twice the woodwind chairs) certainly seems to be justified from this viewpoint. Nevertheless, the technical dominance of the brasses would naturally be maintained. For the conductor the question therefore remains, should the brass be somewhat reduced in order to achieve a more equal balance or should general increases be built on the naturally heavy weight of the brasses? A certain amount of caution is however advised since excessive forcing leads to a shift in the higher overtones for the brasses. Thus, the sound certainly appears louder on the orchestra stage, however, it loses strength and substance as perceived in the hall: this effect is particularly pronounced in the trombones.

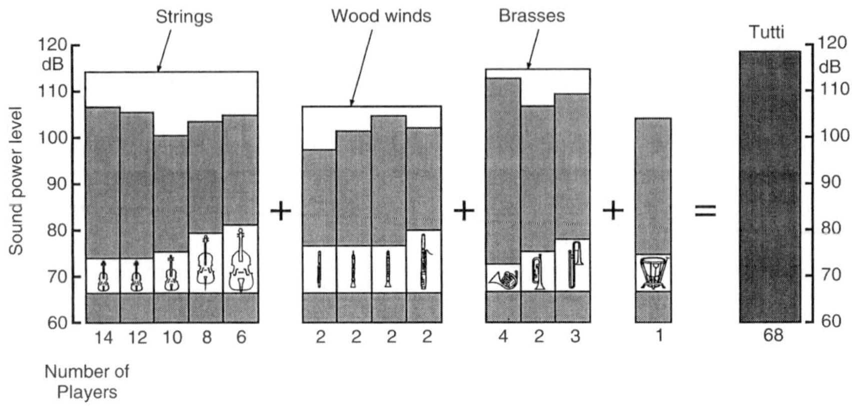

Fig. 8.9 Sound power level of an orchestra in *ff*

The reverberation in large halls supports the strings in *forte*-passages to the extent that bow direction changes during long notes or *legato* runs (above all for uneven execution within the section) are so strongly washed out that they no longer are apparent. However, even for soloists such adapting to the tonal demands is possible, as for example the performance technique of the famous violinist G. Kulenkampff showed during his later years.

For a smaller number of musicians and less full instrumentation, as for example found in many classical symphonies, naturally a different dynamic range is found than for works with very large orchestras. In addition, in many compositions by Haydn and Mozart the extreme dynamic steps *pp* and *ff* are not present and at any rate, further increases were not indicated in those days. Consequently, the *piano* can rise more clearly above the noise level and an *ff* only reaches up to around 90 dB.

Thus, F. Winckel determined a maximum of 90, 90, 88, 85, and 80 dB for the performance of the "Posthorn-Serenade" K 320 by W.A. Mozart in five American concert halls during the tour of the Cleveland Orchestra mentioned earlier. The last value, however, should be considered as insufficient for a convincing tonal experience. One, therefore, has to be satisfied with the fact, that a performance of classical orchestra works which is satisfying from a tonal standpoint, is simply not possible in some extremely large concert halls, when even with relatively large numbers of strings, favorable seating, and increased performance volume, the sound level required for the individual dynamic steps cannot be reached in the hall. The level values for a *crescendo* in symphony K 319 by Mozart are assembled in Table 8.7 as an example for the dynamic range with classical orchestration, score example 28 shows the corresponding measures.

In this result it is notable, that the width of the *crescendo* differs for different locations in the hall. The reason for this lies above all in the fact that the high frequency components of the strings, which are typical for a *forte,* are more strongly attenuated in the hall than near the conductor. This narrows the dynamic range. Furthermore, in the middle of the main floor, because of the shallow step arrange-

Table 8.7 Schallpegel in der Stadthalle Braunschweig

Mozart, Symphony K 319	D	M	H	S	E	
1. Satz, Takt 117	73	72	67	68	65	dB
1. Satz, Takt 121	93	86	84	89	83	dB
Dynamikspanne des *crescendo*	20	14	17	19	18	dB

Score example 28 W.A. Mozart, Symphony B^b major K 319, 1st movement, measure 117 *ff*

ment on the orchestra podium (see Fig. 7.40, left example), the strong shadowing of the woodwinds by the musicians sitting before them is noticed in such a way that the "register *crescendo,*" as determined by the instrumentation, is not fully effective because the oboes and bassoons are dominated by the strings.

The masking effect of the ear plays an important role in connection with balance between the individual instrument sections as well as the transparency of the orchestra sound. Since a sinusoidal tone of lower frequency weakens the loudness of the higher tone in the air, or possibly even renders it inaudible, the middle voices can be masked by the stronger lower partials of the lower instruments within the orchestra sound, if they do not possess sufficiently strong overtones above the masking region of the low instrument. As was seen in Fig. 1.2, the region of this masking effect, however, becomes broader with increasing level, thus, a softer sounding mixture of voices appears more transparent than a loud one.

Yet even high instruments can, as already mentioned, mask lower voices at least partially. To explain this process, the audible thresholds for a high violin tone are represented in Fig. 8.10. With its spectrum, it affects strong masking for all frequen-

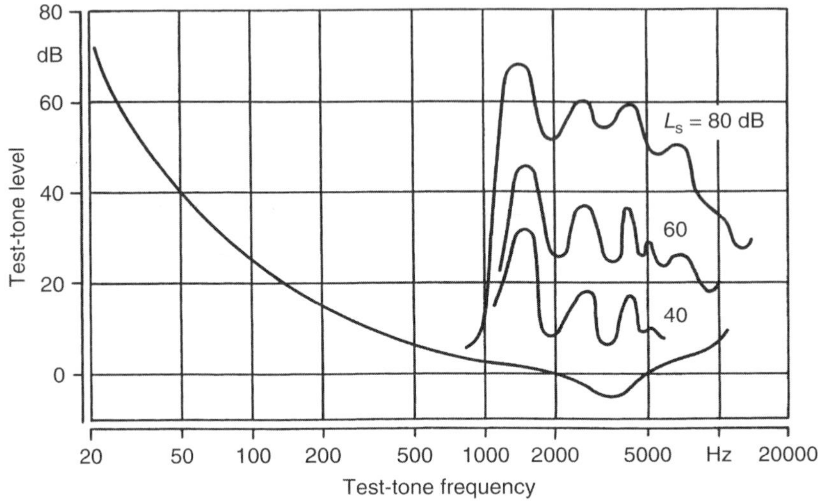

Fig. 8.10 Masked threshold of a sinusoidal tone ("test tone") which is masked by high violin tones performed with variable loudness (After Zwicker, 1982)

cies higher than its fundamental. This affects above all those tonal contributions of low instruments which especially in the attack determine a convincing *forte*. Consequently, the lower instruments appear dull and appear to be almost playing a *piano*. This masking can only be avoided by performance technique, where in the upper voices, attention is paid to a soft tone color, lacking in high overtones, which, in strings, for example, is created by bowing near the finger board.

The mutual masking of several independent voices decreases, the farther they are separated in frequency and the less they overlap in the overtone regions. A typical example for utilizing this effect is found in the Prelude to the Meistersinger, in which at one place three motifs appear simultaneously, the first measures of this passage are given in score example 29.

The "Meistersinger motif" is played at a very low register by the basses, the tuba, and the bassoons while the "Love motif" is presented by the first violins and the first clarinet in a very high soprano register with support by the celli and the first horn. The attention of the ear is drawn, on one hand, by the smooth melody line without breaks (also emphasized by the instructions "*ma molto espressivo*") while on the other hand, the "Meistersinger motif" ("*molto marcato*") is emphasized by accents, in order not to cover the higher register by long sustained notes of high intensity. It is, therefore, recommended, in order to avoid mutual masking, to shift the tonal weight in the lower voices to the lower octave and in contrast to that, to reduce the intensity of the horn and the celli in the upper voices when the spatial conditions are not favorable for a transparent tonal picture.

The shortened "King David motif" separates itself from the two other themes by an extremely short *staccato*, which is predominantly shaped by the initial transients,

Score example 29 Wagner, "Meistersinger von Nürnberg", Prelude to the first act, measure 157 ff

and consequently cannot cover the longer notes of the other voices, since they are audible without interference at least after these initial transient accents. Since the entire passage also moves at a very low loudness level, no spaciousness of the tone picture is developed. Because of that, the localization sharpness for the individual instruments or instrument sections is increased, which contributes significantly to the transparency of the voice mixture.

An opposite effect is occasionally experienced with concertos for trumpet with orchestra. The trumpet sound appears relatively soft for the conductor located at the side of the soloist – particularly in consideration of the closeness of the instrument-, so that the tendency exists to reduce the loudness of the orchestra more than necessary. With this, the orchestra sound loses volume and spaciousness for the audience; it thus no longer forms a contrast to the penetrating direct sound of the trumpet, thus the solo trumpet dominates far stronger above the orchestra than would correspond to a more adjusted balance.

8.3 Performance Technique

8.3.1 Articulation and Tone Presentation

The acoustical characteristics of the room affect not only the average loudness level of the individual dynamic steps, but also influence the fine structure of the dynamic flow and thus naturally, especially the development of the initial transient and the decay of the tone. In order to visualize these processes, the variation range for the initial transient times of the most important instrument groups are compared to the running times of the first reflections in large concert halls in Fig. 8.11. In addition, the representation shows three time regions of musical importance; the "tone separation" is the time between the onset of two tones following each other at very high *tempi* (up to 14 notes per second). Here the individual notes can possibly even overlap slightly: for the aural impression it is therefore important that the higher overtones are not present for the entire tone duration and thus contribute to the clear separation of the individual notes: this "overtone separation" has an order of magnitude of 20–40 ms. The "synchronicity" of ensemble playing gives a region by which simultaneous tone entrances can or may be shifted relative to each other within ensemble performance. Occasionally this is a consciously applied means of formation in order to make chord entrances either more precise or more soft and round.

As a comparison of initial transients and the delay times of room reflections shows in Fig. 8.11 the temporal development of individually notes is stretched by the room and the articulation noise is lengthened. Thus, it can become critical when the articulation noise is so short and the delay of the first room reflection is so long so that direct sound and reflections no longer follow each other seamlessly and a doubling occurs: this makes the attack hard or rough. This is particularly dangerous for the sound of a piano in the presence of concentrated room reflections.

Similarly the decay times of instruments are juxtaposed to usual reverberation times in concert halls and churches in Fig. 8.12. This picture clearly shows that a large portion of the instrument decays faster than the room, so that the listener perceives the decay sound spatially (from the diffused sound field). There are, however, characteristic exceptions for which the decay sound appears to come predominantly directly from the instrument. Aside from the open strings of low

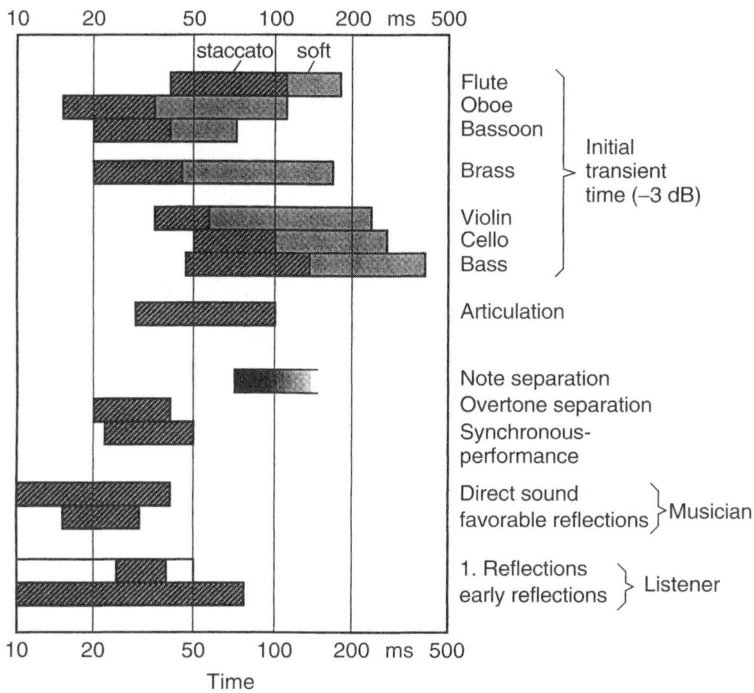

Fig. 8.11 Characteristic regions for time structure of tone onsets

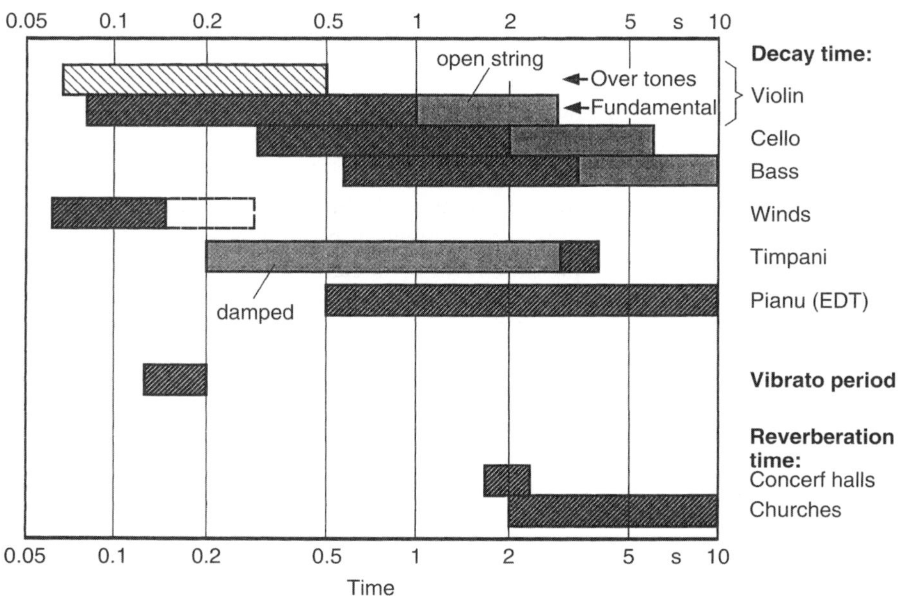

Fig. 8.12 Characteristic regions for time structure for quasistationary states and decays

Score example 30 J. A. Dvorák, "Othello" overture, measure 8 ff

string instruments and timpani, especially the middle register of the piano belongs to these exceptions. An excessively long decay of the hall, therefore, robs this instrument of a portion of its characteristic sound. In addition, this picture also shows that winds, more so than strings, require appropriate hall reverberations so that their tone is not too dry and the tonal transitions become smooth.

When the reverberation time of the hall and the decay time of the instrument have similar orders of magnitude, the listener perceives an integrated decay without differentiating between instrument and hall. Under such circumstances, the reverberation of the hall appears especially short when the sound source abruptly stops "unexpectedly" without essential decay of its own. This, for example, is the case after *forte*-breaks of strings when they are played *con sordino*. The damping of the low tone contributions by the mute is added to the damping of the strings affected by the bow. Such notes, e.g., in the Othello-Overture by A. Dvorak (score example 30) should therefore not be stopped too abruptly in order to avoid a discrepancy with normally performed chord breaks, in view of the tonal effect in the hall.

Since the stationary sound pressure level in the hall is only reached gradually by the arrival of sound reflections with different delay times, short notes never reach the same full level expected by the player or the conductor based on their own aural impressions, as long notes played at the same loudness.

An example for this, is the temporal level development of a passage from a Mozart symphony represented in Fig. 8.10, in which only the first violins are playing. The sound level at the location of the conductor shows four peaks clearly separated by pauses for the *staccato* one-eighth notes, and a longer block for the connected concluding phrase. The average level of the A_6 is as high as the level of the first three *staccato*- tones. The $G_6^{\#}$ is stronger by nearly 10 dB. In the rear of the hall the intensity relations, however, look entirely different: the individual one-eighth notes are, of course, broadened by the reverberation, however, their level lies by about 10 dB below that for the 3rd measure. The fact that the peak of the $G_6^{\#}$ is more flattened is related to the circumstance that the microphone located near the conductor responds in preference to the first desks and an accent by individual players is more pronounced at this location than in the rear of the hall where the sound from the entire section arrives uniformly.

While in the example shown, a dynamic center of gravity is musically meaningful in the 3rd measure, one does recognize from the comparison of the temporal level sequences that the short notes must be performed more forcefully than the long ones in large rooms in order for the audience to perceive them as equally loud. This requirement applies in special measure for very short notes of less than one-eighth of a second duration which approximately corresponds to a 1/16 note (not shortened) at MM. ♩ = 120, because below this limit the ear no longer registers the full loudness which it would perceive for longer durations for the same level.

Furthermore, for dotted rhythms and other similar figures without pauses, the short notes can drown in the reverberation of the long ones if they are not sufficiently strongly accentuated. A typical passage where occasionally, even with renowned conductors, the short notes are no longer heard, is given in score example 31. The difficulty here arises above all through the circumstance that the principal notes also carry accents and the 1/16 notes are only performed by the violins and violas while the winds perform only the accentuated one-fourth notes. This problem also arises shortly before the conclusion of the 1st movement of the Great C-major Symphony by

Score example 31 J. Brahms, 2nd Symphony, 3rd movement, measure 48 ff

Score example 32 L. von Beethoven, Symphony No. 8, measure 357 ff

F. Schubert (measure 673, 675, 677). However, in that case, this problem can also be solved by not only emphasizing the up-beats but separating them clearly against the preceding *sforzato*-tones.

The rise of the sound level for long notes finds a certain parallel in sequences of the same chord. Within such a chord sequence the performance volume which actually remains constant appears to increase in strength from chord to chord, as perceived by the audience in the hall. A typical example for this is given by the three orchestral chords in measures 358 and 360 in the 1st movement of the 8th symphony by L. von Beethoven (score example 32). If the last chord is not performed slightly shorter or somewhat weaker, the danger exists that it receives an emphasis heard by the audience which is not perceived at the location of the conductor. The effect is especially clearly pronounced in this passage because a pause follows the 3rd chord so that the reverberation is heard for a relatively long time and thus an impression of energy-rich tone is created.

The time sequence of the level shown in Fig. 8.13 also clarifies how washed out a *staccato* sounds in the rear of the hall, though it appears sharp to the performer, even though in this representation the effect of the 1st wave front, which ordinarily conveys precision, does not come to full expression. Nevertheless, it can be seen that the rhythmic structure remains recognizable especially through the steepness of the rising flank, while the value of expression is weakened by tone duration and the pause. Consequently, in large halls with relatively long reverberation, an especially sharply attacked *staccato* is required, which at the same time assumes increased rhythmic precision on the part of the string sections. Nevertheless, the individual *staccato*-tones cannot be played too sharply, i.e., too short, since otherwise on the one hand the danger exists that no smooth connection to the hall sound is developed, and that on the other hand the energy is insufficient for the audience seated in far distant locations.

At any rate, an excessively short *staccato* is frequently avoided by performers because then the instruments no longer speak properly (Öhlberger, 1970). In similar fashion, this applies to excessively sharp accents. Many passages with large numbers of brass in the symphonies of Bruckner, in contrast, only receive a convincing effect when the performance instruction "*marcato*" is performed by the wind players as "*tenuto*."

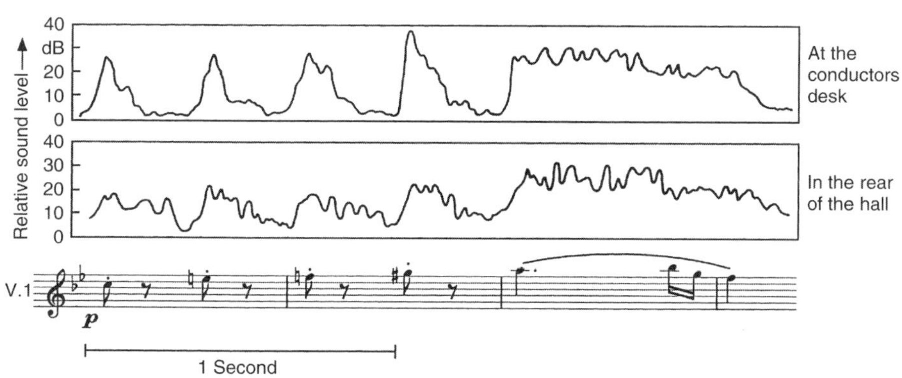

Fig. 8.13 Time sequence of sound level for a violin passage from the 1st movement of symphony K 319 at two locations in the hall (Stadthalle Braumschweig)

In this context, the observation by F. Winckel (1960), that the nature of the beat technique of the conductor has a great influence on the development of the initial transients, is interesting. While a gentle beat leads to a soft entrance, a very precise beat indication results in noisier and harder initial transients whose characteristics are then preserved for the overall sound of the orchestra if a certain variational width (up to approximately 50 ms) is developed within an instrumental section by entrance delays. Since the ear is nearly blocked for short moments by the first noise accents, a hard entrance by the conductor thus creates an apparently exact tone picture when individual instruments enter with a slight delay.

The support of a singer or instrumental soloist by the orchestra has a similar acoustic basis. A very subdued accompaniment certainly avoids that the soloist is lost in the overall sound, however, it results in the feeling of being alone without being carried by the orchestra. A large volume of the accompanying voices certainly does give the soloist increased security but it does become more difficult to penetrate tonally. In order to support the soloist, it is therefore necessary that the chords of the orchestra or the tone sequence of instruments playing *colla parte* are placed under the solo voices with a very soft attack and leave the articulation to the soloist. Thus the soloist stimulates the primary attention of the ear and is nevertheless strengthened energetically by the instruments, which in turn need not play too softly.

Even as the precision of the articulation must be increased with increasing hall size, it is also necessary to emphasize accents in melodic lines or *sforzato*-chords which are located above smooth voices, so that they do not drown in the reverberation. At the same time, care must be given, that not only the dynamic level of the accents must be chosen as higher, but the decay of the emphasis must be adapted to the spatial conditions. The dynamic line of the accented tones cannot fall below the reverberation level which is caused by the intensity peak within the hall, since otherwise the tone appears to be cut short as perceived by the audience. The sharp attack of the accent, therefore, must be followed by a decay of the instrument adjusted to the reverberation, so that the listener can hear the direct sound at all times.

Correspondingly, attention must be paid to the fact that after an abrupt ending of a *forte* passage by the orchestra, only a few instruments are not lost in the reverberation in a *piano* passage following immediately without pause. A typical example for this is found in the 4th symphony by L. von Beethoven: the horns at the conclusion of the 3rd movement as well as the cellos and basses in the 4th movement of measure 120. Instead of an excessively large dynamic jump to the final *piano* level, in such places a transitional *decrescendo* is recommended in the 1st measure after the last *tutti*-chord, which begins with a slightly stronger "*piano*." The ear will not perceive this *decrescendo* (in the hall), since the ear first has to "recover" from the *forte* of the orchestra and it furthermore judges the dynamic level in its distance from the noise level in which it initially also includes the reverberation of the orchestra chord.

8.3.2 Vibrato

Musicians primarily understand the *vibrato* as an artistic means of expression to enliven the sound (Gärtner, 1974). As discussed in the sections about the sound of individual musical instruments, frequency modulations are created by different means. These in turn can lead to instrument specific amplitude variations of individual partials. These descriptions of tones, however, referred only to the sound radiated directly by the instrument without considering the influence of the surrounding room; i.e., they refer to the tone impressions as created in an open space or, within certain limits, to the tonal impression of the performer. For the audience in a hall, however, the impression caused by vibrato is also determined by the characteristics of the hall. This, naturally, is particularly evident in large halls, is however relevant for halls of all sizes, since reflections by walls and ceiling always influence the tonal impression (Meyer, 1994b).

As seen from Fig. 5.12, the sound generated by the player reaches the listener not only by the direct path but also by single reflections from the two side walls, as well as from the ceiling, followed by a multiplicity of reflections which form the reverberation. All these sound paths have different lengths. They thus determine different traveling times for the sound. This means that at a given point in time sound contributions reach the listener which were initially radiated by the source at different times. For a frequency modulated sound, as in the *vibrato*, the sequentially created frequencies therefore arrive simultaneously. For a *vibrato* frequency of 7 Hz, for example, a delay of 30 ms, is not unusual for the first side wall reflection in a concert hall. This corresponds to a scant quarter of a vibrational period, which is the tiny difference between the center frequency and the maximum frequency deviation.

The temporal super position of all these sound contributions is visualized in Fig. 8.14 for the example of a violin tone, by comparing sound spectra measured directly at the instrument to those measured simultaneously at a greater distance (Meyer, 1992). These two examples clearly show how temporally varying frequencies originating near the instrument result in closed frequency bands further back in the hall, which contain the sound radiated by the source into all directions. Here the time structure of the original frequency and amplitude modulation is no longer recognizable. Rather, the resonances of the instrument and possibly hall resonances become evident within these frequency bands. The frequency bands, which are largely stable in time, lend the sound an attribute of a fuller or more voluminous tone – a characteristic which naturally is ever more strongly pronounced with increasing width of the *vibrato*. This tonal impression is perceived only in a limited way at the location of the player, because here the direct sound dominates. On the other hand, thanks to the direct sound, which even at great distances still permits localization, some of the "liveliness" of the tone can still be perceived.

Since the sound field in the room almost exhibits no more level fluctuations in time, the loudness impression for the listener is determined by the overall radiated sound power, which corresponds to the time average of the sound level near the

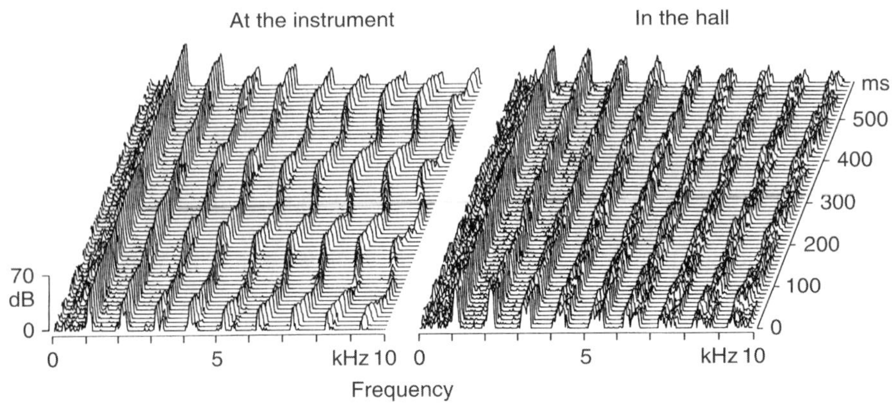

Fig. 8.14 Time structure of a sound spectrum of a violin *vibrato* at the instrument and in the hall (note C_6)

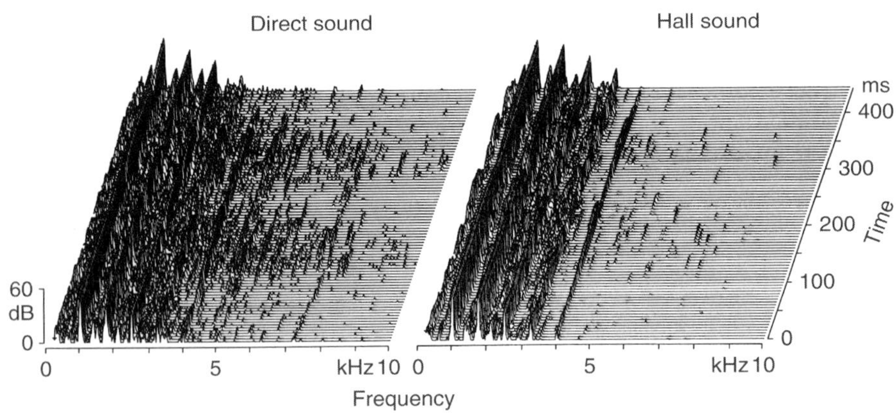

Fig. 8.15 Time structure of a sound spectrum of a flute *vibrato* in the hall, recorded with a directional microphone and a spherical microphone (note C_6)

instrument. In contrast, for tones with pronounced amplitude modulation the player perceives a loudness which corresponds to the maximum value of the sound level during fluctuations in time (see Sect. 1.2.1). For the player, consequently, an increase of the vibrato is connected to a loudness increase. This should not deceive the player to forget that the audience in a larger hall will not sense this increase in loudness – similar to the situation with "bow pressure dynamics."

Since the strength of the direct sound decreases with distance from the sound source, however, the reflected sound energy in its entirety is roughly equal for all seats in the house, the impression of the enlivening effect of the *vibrato* decreases with increasing distance. Therefore, especially the high overtones are of great importance in conveying this to the audience, since at higher frequencies, the energy reflected by walls and ceiling is significantly weaker. This becomes particularly noticeable when the high overtones or also the high frequency noise components all fluctuate in phase.

As an example for this, two sound spectra for a flute *vibrato* are compared in Fig. 8.15: they were recorded simultaneously at the same location in the hall, with two different microphones. The spectrum recorded with the directional microphone practically records only the direct sound. The spectrum recorded with a omnidirectional microphone predominantly records the hall sound. The left partial picture clearly shows how the noise components pulsate in time. Since they reach the listener only in the direct sound with this clarity, they transmit directional information which helps the listener to locate the sound source. For the tonal effect of the flute, this means that this *vibrato*, which at any rate is not very wide, does not raise the volume of the sound, but rather makes it more noticeable within the ensemble or the orchestra.

In order to obtain an overview of differing effects of the *vibrato* on the sound of individual orchestral instruments and singing voices, these must be differentiated with respect to the widths of the *vibrato* and the strength of the amplitude modulations at high frequencies. Since the ear assembles the loudness of higher frequency contributions into "critical bands" (see Sect. 1.2.1.), one option for the measure of audible amplitude modulation is the strength of the fluctuation in that frequency group above 2,000 Hz, for which the largest time differences in level occur.

Thus, the picture of Fig. 8.16 presents the characterization of tonal effects of different kinds of *vibrato*: the farther the instrument is located to the right of the diagram, the stronger the tone volume is enhanced by the *vibrato*: the further it lies to the left, the narrower the sound remains. The farther an instrument is shifted to the top, the more noticeable the sound becomes through the *vibrato*, thus, its presence within the ensemble is enhanced: the further down it lies, the better it melts into the sound of the ensemble. However, it should not be overlooked that in each case the instrument specific strong *vibrato* is entered into the diagram, but naturally weaker forms of the *vibrato* also occur.

In a reed instrument, even a pronounced *vibrato* is so gentle that a narrow ensemble sound is possible. For a brass instrument, a strong *vibrato* – as in the case of a flute – makes the instrument more noticeable, which can also be supported by the directional characteristic of the high frequency components. In contrast, a string and singing *vibrato* above all affects the sound volume. Furthermore, string soloists frequently exceed the represented measure in the width of the *vibrato*. Such a *vibrato* is easily perceived as excessive from close up, whereas an excellent tonal effect is noted in the hall. This can occasionally lead to difficulties with microphone recordings in close proximity. A strongly forced singing *vibrato* becomes extremely noticeable, which in turn influences the tone volume. In exceptional cases, such outstanding presence may be defensible, if this is the only way by which a vocal soloist can penetrate a loud orchestra or choir; within a choir, however, it is extremely disturbing in all situations.

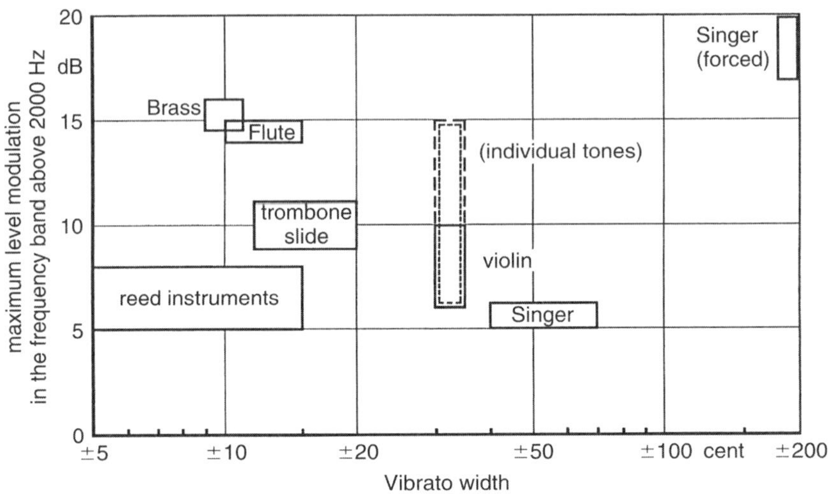

Fig. 8.16 Characteristic values for width and overtone modulation for a pronounced *vibrato*

Individual violin notes can also tend toward standing out excessively when individual overtones incline toward strong amplitude modulation because of a steep resonance edge for the relevant instrument. When an instrument possesses one or more such tones, the player must become aware of them and treat them carefully with respect to the width of the *vibrato*. This also applies to the speed of the *vibrato*, in case the noticeable overtone carries out its *vibrato* at twice the frequency because of the nature of the resonance (see Sect. 3.3.1.3).

8.3.3 Playing Positions of Wind Instruments

In order to achieve special tonal effects, G. Mahler demands at several places in his symphonies that the oboes and clarinets are played with a raised bell because the partial spectrum, based on directional characteristics, exhibits very different amplitude relations in the direction of the instrument axis in comparison to the side, this performance technique changes the spectrum significantly. The envelopes in Fig. 8.17 show a comparison of a low and a somewhat higher clarinet tone for both cases, with the recording microphone located along the raised axis of the clarinet in the one case, and in the other case, along the direction in which the instrument axis would usually be held in relationship to the audience. The splitting of the curve into two branches shows the relationship for the partials of odd (top) and even (bottom) order.

From these diagrams, one recognizes that the high frequency components increase strongly in their intensity by raising the bell, that on the other hand, the amplitude differences between the even and odd partials are diminished. Both phenomena have the effect that the tone color becomes brighter or even more shrill,

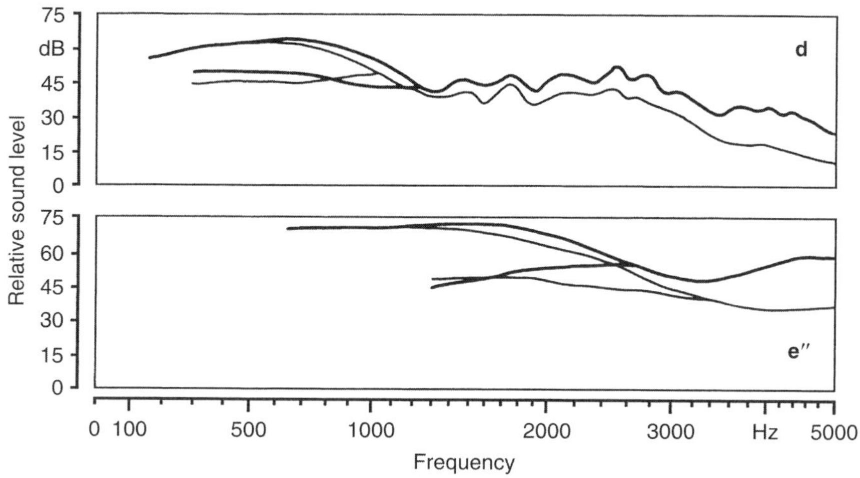

Fig. 8.17 Envelopes of sound spectra of a clarinet for different instrument positions recorded in direction of sight (*horizontal*) of the player. *Thin line*: normal position; *thick line*: bell raised

and their hollow, or covered, character decreases. The example for E_6, which can be considered typical for the entire midregister and the higher tone range of the instrument, especially clarifies, that raising the bell brings the timbre of the B^b clarinet close to that of the high D and E^b clarinets. For example, this effect is utilized in the third symphony by Mahler where in the first movement, three B^b clarinets with raised bells are combined with two E^b clarinets (see score example 33).

When implementing this special performance technique, one must, however, be aware today, that in Mahler's time there were no concert halls which, because of their room acoustical characteristics, favored high frequencies in that manner, like some halls constructed in more recent times. When one hears the stark tone color of raised oboes and clarinets in a hall for which the reverberation time at 4,000 Hz is only slightly less than the value for the midfrequencies, the question arises, if under these circumstances the performance technique demanded by Mahler still corresponds to his tonal intentions, or if it might not be more appropriate to refrain from utilizing the effect.

In addition, in the symphonies by G. Mahler, as well as in some works by I. Stravinsky, frequently the performance instruction "bell up" are found for horns and trumpets (see score example 34). As already mentioned, this technique, however, leads to a relatively rough tonal effect for horns, since the absence of the hand in the bell renders the initial transients as excessively hard. From an intensity standpoint, there is not much to be gained either, since merely the high frequency components are enhanced in their amplitude; however, the sound radiation of these components occurs rather sharply directed toward the ceiling above the stage and the upper part of the rear wall so that the energy is reflected in preference toward the orchestra. The change in tonal effect is thus limited to the character of the instrument and is not suitable to create additional intensity in large *crescendi* as

Score example 33 G. Mahler, 3rd Symphony, 1st movement, excerpt clarinet section

Score example 34 G. Mahler, 4th Symphony, 3rd movement excerpt: without strings and harp beginning 36 measures before the end

occasionally attempted, e.g., in the symphonies of P. Tchaikovski. At best, the effect of open horns can be used to achieve a typical hunting horn color, as for example in the symphonic poem "Le Chasseur maudit" by C. Franck.

In contrast, the raising of the bell for trumpets must be evaluated differently, since for this instrument group no additional changes to the instrument or to the performance techniques are made, but most importantly, the shadowing effect of the music stand is diminished, especially the high and highest frequency contributions are radiated more freely into the hall by this position, as can be explained from the directional characteristics. The rise of sound level, which thus occurs in the direction of sight of the player when blowing equally hard is represented in Fig. 8.18. Above 2,000 Hz, the partials become by more than 10 dB stronger while in the region between 400–1,200 Hz, a weakening occurs by reason of an acoustical exchange between music stand and instrument, which, however, is diminished by distance of stand to player. For the tonal impression, this curve means that raising the bell increases the brilliance and sharpness of the timbre, while some of the strength and substance is lost.

This effect naturally supports the desire for a bright metallic fanfare sound which, for example, toward the end of the Egmont overture corresponds in character to the nature of the work. Also when trumpets are to stand out from the orchestra sound with special brilliance, many conductors require such positioning of the instruments. A typical example for this is found shortly before the end of the first movement of the "Eroica" (measures 654, *ff*) where the trumpets are to rise above the entire orchestra in a large *crescendo*. This performance technique, however, has certain disadvantages when a concert hall is constructed in a manner which leaves a significant fraction of the audience seated outside of the diffuse-field distance for the high trumpet frequencies (see Figs. 6.7 and 7.31). For listeners located to the side or behind the orchestra, this tonally dynamic increase does not become effective in a measure expected by the conductor based on his own hearing impression. In view of the somewhat uniform tonal picture in the entire hall, raising the bell is therefore not advantageous in some halls. A sharper attack with normal positioning would be more appropriate.

Also, in *piano*-passages, in which the trumpets are to sound bright, but also soft, it is not satisfying from a standpoint of tonal quality when the bell is raised to enhance the transparency of the overall impression. In places, as for example in the fine web of voices in the introduction of the last movement of the 1st symphony by J. Brahms (score example 35), preference should rather be given to a downward positioning of the instrument, so that the tone contributions in the region of the a (ah)-formant can be radiated as favorably as possible. The example cited can be considered as especially typical, since based on the pitch of the motif, particularly the octave partials fall in the range in question near 1,000 Hz.

Score example 35 J. Brahms, 1st Symphony, 4th movement, measure 35 ff. Excerpt without strings and percussion

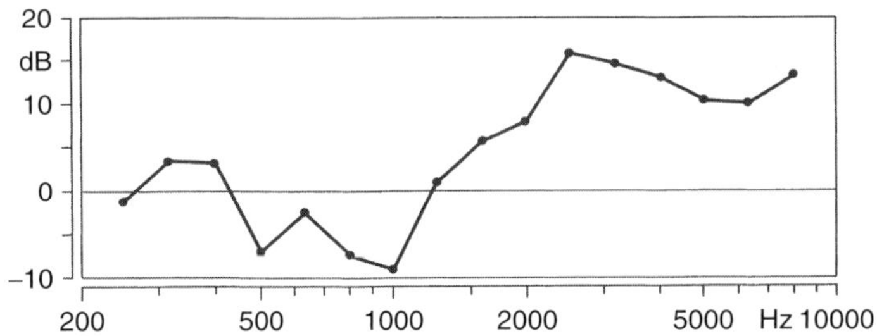

Fig. 8.18 Sound level increase in the direction of sight for the player when raising the bell of the trumpet

In some orchestras, above all in radio studios, music stands with perforated plates are used in order to reduce the shadowing effect, since the low frequencies are refracted around the desk, as also noted in Fig. 8.18, and the high tonal contributions are directed relatively sharply against the stand. The perforated desk only leads to an intensity rise worth mentioning in the middle region around 2,000 Hz (Meyer and Wogram, 1970). Corresponding to its location in the region of the vowel color "e(eh)" this increase could be valued as advantageous from a tonal standpoint because it supports the brightness of the timbre, however, seen as a whole, the acoustic effectiveness of perforated desk, with music on them, is small.

Corresponding to raising the bell of trumpets, the possibility to diminish the shadowing effect of the desk also exists with the trombone; the instrument is positioned so that the bell is located at the side of the desk. As Fig. 8.19 shows, in similarity to trumpets, an intensity rise results for frequencies above approximately 1,200 Hz, while in the region of the principal formants (500–600 Hz) a slight drop is observed. The deviation in comparison to normal positioning, however, is less than for trumpets. In regards to the timbre, the sideways positioning of the trombone leads to a brighter and more clear tone picture, where, however, the sonority decreases. This phenomenon is important for the individual voices in the trombone section.

8.4 Tempo and Room Acoustics

The tempo of a piece of music is an essential element of its character. The usual Italian terms give the tempos desired by the composer only very inaccurately and are to be considered, as in the case of dynamics instructions, as relative. The interpreters are thus left a certain amount of flexibility for the choice of tempos, which, corresponding to the theme and the nature of the composition has to be determined more accurately. Even if the individual movements include metronome specifications these can only be considered as average values around which the tempos are to move during the course of the piece (Winckel, 1960): because small

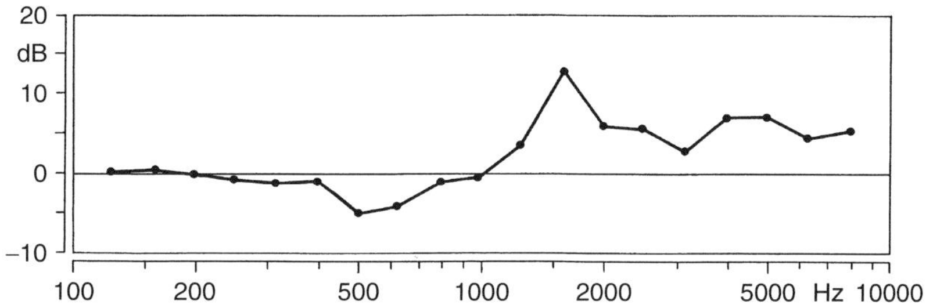

Fig. 8.19 Sound level increase in the direction of sight for the player when positioning the trombone toward the side of the music stand (in contrast to normal positioning)

differences are an important means of expression, which, depending on the style of the work, can be implemented within a narrower or wider framework. Capturing the "correct" tempo, therefore, represents an essential task of interpretation. When Bruno Walter (1959) characterizes this "correct" tempo with the words "that in it, the musical sense, and the importance of the feeling of a phrase can best come to expression and enable technical accuracy," then in this formulation the meaning of personal interpretation finds expression in the same way as the influences of instruments and room acoustics have been considered.

Certainly during the study of the score by a conductor, a firm tempo impression is shaped which best reproduces an individual conception of the character of the piece. In order to accurately realize this tonal representation, optimal room acoustical conditions are required, which do not influence the tonal picture through their own effects, but rather permit the musicians to exhaust the performance technical possibilities of the instruments to their fullest. It was already mentioned earlier that these optimal acoustical conditions for a particular concert hall can be different depending on the style of the music to be performed. Within narrow limits, deviations from optimal conditions can find compensation by appropriate tone formation and articulation, however, finally, even the tempo has to be adapted to the room acoustical conditions, when the tonal perception of the interpreter is to be transmitted to the audience in the hall.

When the reverberation time is somewhat longer than the optimal value for a relevant composition, the tone does increase in fullness, however, it decreases in clarity. Especially in fast passages, therefore, the danger of an impression of blurred immersion in sound exists. Even if one attempts to compensate for these changes in tone picture by a seating arrangement oriented toward the direct sound and by a sharper articulation, nevertheless, as a matter of principle a slower tempo is advantageous for the clarity of the tone picture. Furthermore, it also has the advantage that the articulation can be carried out more precisely, since the pregnancy of the entrance is greater for a firm *staccato* than for a merely "touched" *spiccato*. A series of tests, in which excerpts from two symphony movements were played in a studio with various reverberations and tempo changes of up to 20%

(Tarnóczy et al., 1960), indicate that a somewhat slower tempo in conditions of longer reverberation times is also perceived by the audience as appropriate.

In contrast, raising the tempi of fast movements in conditions of long reverberation times can lead to a tonal sound in which details are no longer perceivable, giving the piece a totally different character. During the performance of virtuoso compositions of the Baroque period, this effect – especially during chamber orchestra concerts in churches – is occasionally utilized and, in spite of sacrificing details, can have a wonderful effect because of the unaccustomed tonal richness.

In slow movements a reduction of clarity caused by the slightly excessive reverberation is generally not noticeably annoying and in this respect is less objectionable. The increased sound fullness rather comes as a support for energetic problems of long notes so that it becomes easier to shape the tone and the movements can be performed more broadly – as also a larger body of strings permits a slower tempo than a smaller ensemble (Walter, 1959). Inasmuch as a longer than optimal reverberation requires a slower tempo in fast movements and also favors a broader playing in slower movements, the relationship between the tempi of the individual movements can be accepted almost without change, maintaining the overall concept of the compositional structure. If, however, the reverberation is significantly longer than normal, then the character of the work also changes somewhat in the broader tempi and in the fuller sound as, for example, can be noticed for Bruckner performances in the cathedral of Chartres or the Abbey Church of Ottobeuren.

In contrast, a short reverberation time presents totally different problems. On the one hand, the high clarity of the tonal picture and the missing room resonances on the other, make it very difficult for the player to get an extremely sharp *staccato* to ring, so that the impression of a poor attack of the instruments is created. This impression often becomes even stronger for the individual players by the fact that it is difficult to hear the fellow players and consequently the clarity of the individual tone entrance is judged very critically by the player himself. Furthermore, because of this – especially for strings – a somewhat firmer, more tone-rich *staccato* is required, rather than a sharp, *spiccato*. This, in turn requires a somewhat slower tempo. Consequently, fast movements must be performed more slowly under conditions of short reverberation times than under acoustically optimal hall conditions, as was already shown by investigations of the tour of the Cleveland orchestra by F. Winckel (1962a).

In addition, one should not fail to recognize that in special cases, halls, poor in reverberation mislead to an attempt to begin a movement too rapidly. When a piece begins with individual chords separated by pauses, these pauses appear longer for shorter reverberation times and thus raise the impression of a slower tempo. The best known example for this is the beginning of the Eroica. However, such movement beginnings occur not only in symphonic music but for example also in several string quartets by J. Haydn (Op. 76, No. 1 and No. 3).

In slow movements a short reverberation time leads to difficulties in developing a full, expressive tone. Consequently, a *piano* already must be played relatively strongly so that at this dynamic level an appropriate sound level is created in the hall, while at a *forte* the required intensity can hardly be achieved. This narrow range of dynamic expression potential, on the one hand, and the tonally energetic difficulties for long held notes, on the other, support a tendency to tighten up the tempo somewhat in order to maintain the melodic flow. This, however, also changes the relationship between the tempi of the individual movements in relation to the conception, with reference to an optimal reverberation, when one attempts to execute each movement as convincingly as possible in the framework of the existing room acoustical qualities. Thus, in halls with overly short reverberation times, not only the dynamics, but also the range of playable tempos is narrowed.

Chapter 9
Acoustical Problems in the Opera House

9.1 Strength of the Orchestra

9.1.1 Historical Development

As was the case with concert orchestras so has the strength of an opera orchestra been subject to the changes of time since the development of the standard works of today's repertoire. This suggests that a look at historical developments could be rewarding. In this context the numerical strength of the strings is most interesting since the winds are always only represented by one instrument and always follow the instrumentation of the individual pieces. Consequently, Table 9.1 represents the number of strings for four historic opera houses from the Mozart era, along with the demands given by R. Wagner for the "Ring" and Verdi around 1870 as "normal for opera performances" (Holl, 1947). In addition, this table contains two examples for currently standard smaller and larger orchestras which will be considered for further considerations.

In the Prague Ständetheater the possibility of increasing the numbers given as normal certainly must have existed at that time, as can already be seen by the instrumentation demands for the "banquet music" in "Don Giovanni." However, this table likely presents the limits of the framework within which Mozart heard or conducted most of his operas – at least north of the Alps. In Italy, however, far larger numbers of strings were not unusual; thus the premiere performance of "Mitridate" in Milano 1771 should be mentioned with 24 violins, six violas, two cellos, and six basses, where especially the relationship between the cellos and basses is noteworthy (Becker, 1962). Also notable is the unusually large number of basses in Verdi's requirements which together with the celli form an extraordinarily large bass group, while Wagner adds only eight basses to the 12 celli and furthermore demands more violins than Verdi. As an extreme for instrumentation the orchestra for R. Strauss' "Elektra" needs to be considered, which, for the premiere performance in 1909 included approximately 115 musicians, among them 3×8 violins, 3×6 violas, 2×6 celli, and eight basses.

J. Meyer, *Acoustics and the Performance of Music*.
DOI: 10.1007/978-0-387-09517-2_9, © Springer Science + Business Media, LLC 2009

Table 9.1 Strength of string sections

Opera Orchestras	Year	1. V.	2. V.	Vla.	C.	B.
Burgtheater Wien	1781	6	6	4	3	3
Drottningholm	1783	8	7	4	4	8
Ständetheater Prag	1787	3	3	2	2	2
Wiedener Freihaustheater	1791	5	4	4	3	3
Wagner (»Ring«)	1876	16	16	12	12	8
Verdi	1870	14	14	14	12	12
Instrumentation »A«	–	8	8	6	5	4
Instrumentation »B«	–	12	12	10	8	6

Quelle: 1. bis 4. Becker, 1962.

9.1.2 Sound Level in the Hall

The diffuse field is essentially responsible for the loudness impression of the orchestra, especially at audience locations for which the direct sound from the orchestra pit is blocked. This can be characterized by an average sound pressure level which is determined by the sound power of the instruments as well as by the room damping index of the opera house. In this context it is first of interest to determine the sound power level of the orchestra (see Sect. 8.1.2). Furthermore, it is important to use that dynamic level as a base, which is most critical for the singer, namely the *forte*.

In order to include winds in addition to the instrumentation comparisons given in Table 9.1, two typical instrumentations will be considered. First, an orchestra with two flutes, oboes, clarinets, and bassoons each, as well as two horns ("winds I") and the same instrumentation for woodwinds, but with four horns, two trumpets, three trombones, and one tuba ("winds II"). For this woodwind section a sound power level at a *forte* of 101.5 dB results, for the two horns in the small orchestra 105 dB, and for the large number of brasses 111.5 dB. These values already indicate that an instrument specific *forte* of brasses cannot be used at full strength in view of the tonal balance of the orchestra.

This is also shown by a look at the sound power level of the strings in Table 9.2. In contrast, the woodwind section is even somewhat weaker than the strings, where the fact that in the operas of Mozart often not all wind players were used at the same time, should not be overlooked. This was done to achieve a large differentiation of tone colors. On the other hand, many operas of the nineteenth century used three or four players in each woodwind section whereby the sound power level can rise by up to 3 dB. The *tutti*-values in Table 9.2 are especially of interest in view of the balance between singers and orchestra, which will be considered later.

The room damping measure of opera houses is determined primarily by the volume of the audience space and only in a smaller measure by its reverberation time, since the latter varies only within relatively narrow limits. This is also shown

Table 9.2 Sound power level at a *forte*

Opera Orchestra	Strings	Str. + Winds I	Str. + Winds II
Burgtheater Wien	103	108	–
Ständetheater Prag	102,5	108	–
Wiedener Freihaustheater	100,5	107,5	–
Wagner	107,5	110	113,5
Verdi	108	110,5	114
Besetzung »A«	104	108,5	113
Besetzung »B«	106	109,5	113

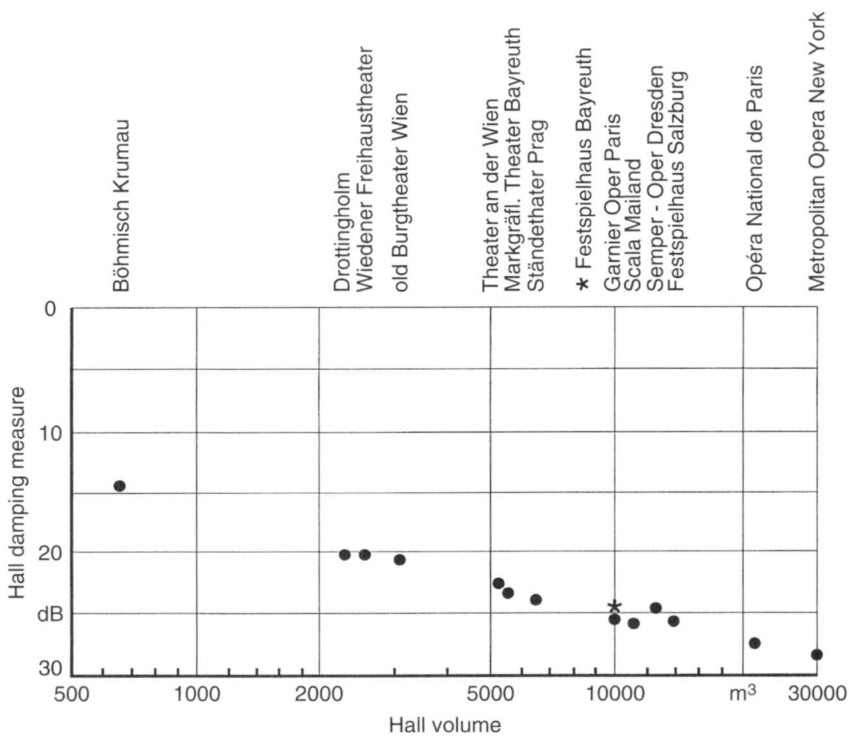

Fig. 9.1 Room damping measure in opera houses (occupied, midfrequencies)

in the fact that in Fig. 9.1 nearly all points lie on a straight line. Aside from the extremely small castle theater in Český Krumlov, the room damping measure in old theaters is approximately 20 dB. In contrast, for typically newer ones it is about 25 dB. This means, that an orchestra of equal size is heard in the old theaters by approximately 5 dB louder – a difference, which corresponds to more than half a dynamic step. The very large theaters in Paris and New York are then again by several dB softer.

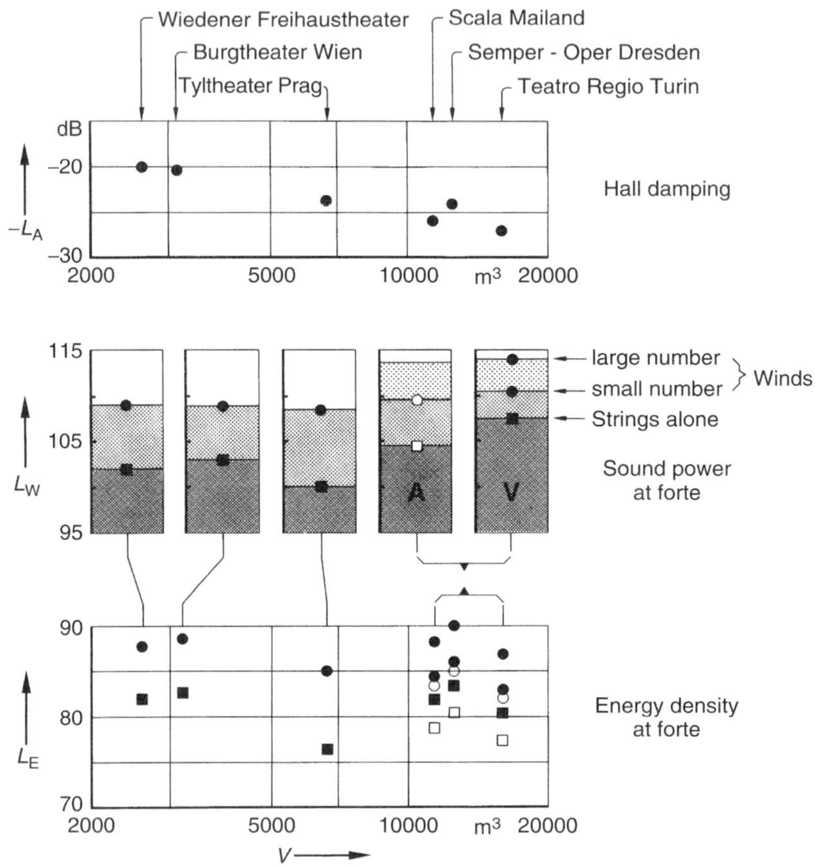

Fig. 9.2 Influence of room acoustics in orchestral strength in the average sound pressure level for a *forte*, in several opera houses

This tendency is visualized by the three diagrams of Fig. 9.2. The upper partial picture contains the room damping measure for the three old and newer theaters, the middle picture the sound power levels of the relevant three historic orchestras, as well as for two modern instrumentations, where in the latter case two different numbers of winds are considered in each case. Furthermore, the sound power level for strings is entered separately. The lower partial picture, finally, contains the average sound pressure level which is achieved in the corresponding rooms by the appropriate orchestras for an average *forte*.

When one considers the levels only for the strings, it is noticed that the relatively small ensembles reach values in the historic Vienna theaters which can only be realized by large Verdi-numbers in the newer opera houses. When one listens today to performances of works from the Mozart era in these theaters for which the number of strings corresponds to version A, then this means that the string sound – even without consideration of the blocking effect by the deep orchestra pit – is by approximately 3 dB softer. The size of the newer opera houses becomes clearly noticeable also for the smaller numbers of winds, especially since a numerical adaptation to the room acoustical conditions is not possible for the winds. One therefore must be aware that the typical number of winds (winds I) used for many Mozart passages in the old Vienna theaters generate approximately the same level, as is reached by the larger number of winds (winds II) in the newer opera houses. This should be considered in the dynamic shaping – at least for pure orchestral passages. On the other hand, one must count on nearly unsolvable dynamic problems, when a performance is attempted with a large number of brasses in a theater of the order of magnitude of only 3,000 m^3, since they cannot be adequately damped.

9.1.3 Sound Level in the Orchestra Pit

While in a concert hall the orchestra is located in the same space as the audience, one must assume that in an opera house the audience hall and the orchestra pit will not function acoustically as a single room, but rather must be considered as two coupled rooms depending on the depth of the pit. A large portion of the sound radiated by the instruments is reflected by the walls of the orchestra pit in such a way that it returns to the musicians, where it finally is absorbed, after multiple reflections. Only the sound radiated into a more or less wide angular region reaches the audience.

When considering that, at least for most musicians, one wall of the orchestra pit is located within a single diffuse-field distance, and a second wall within twice the diffuse-field distance of the instrument, it is not surprising that, because of the reflections connected with the orchestra pit walls, the sound level within the orchestra rises significantly above the levels found at the podium which is located in the free space. Measurements in three opera houses with orchestra pits open to different degrees, or covered, show that the sound level at the center of the orchestra (for equally radiated sound power) is higher by approximately 1.5–3.5 dB when the orchestra is located in a pit, than when it is located on the stage of a corresponding theater, with the depth of the orchestra pits 2.5 m below the front stage edge (Westphal, 1994). The contribution of only slightly delayed reflections is particularly large: sound energy arriving within the first 50 ms – including the performer's own instrument – is approximately 8 dB above what would be observed for the instruments on the stage (O' Keefe, 1994).

This makes mutual listening within the orchestra easier, as long as the dynamics is limited to the lower and middle regions. In contrast, for greater loudness, masking in the ear increases noticeably, this makes mutual hearing of weaker voices more difficult. For *tutti*, the hearing impression is predominantly shaped by the brasses and the remaining voices are mixed into a tonal mass which is difficult to differentiate. Naylor (1987) compares the situation with that in orchestra rehearsal rooms and recommends absorbing wall coverings in the area of loud instruments. This would not change the direct sound radiation in the audience hall, and thus would reduce the loudness in the hall only slightly, and this at most for low and middle frequencies – above all of the winds.

In the locations just in front of the timpani, individual peak levels of up to 115 dB can be observed (Katschke et al., 1981). Low movable walls made of approximately 10 mm thick glass have proven to reduce this objectionable sound for performers located immediately in front of the timpani. Though no direct measurement results are available, one can assume a level reduction of several dB by this in the positions in front of the timpani; in particular, the hardness is removed from the individual blows.

Because of the high loudness level, musicians occasionally use cotton as hearing protection, at least for loud passages (Frei, 1979). For normal cotton a damping of about 5 dB can be expected for frequencies below 500 Hz; above 500 Hz, the damping rises by 5 dB per octave. Special hearing protection cotton enables a damping of about 10 dB below 500 Hz, and above this frequency a rise of again 5 dB per octave, so that by 2,000 Hz 30 dB are achieved (Brinkmann, 1978). Since the effectiveness of the masking effect – as already mentioned – increases for higher loudness, it can be expected that cotton in the ear does indeed reduce the measure of acoustical masking; however, this advantage is diminished by the fact that the self evaluation of the performance becomes significantly more difficult because of the pronounced frequency dependence of the damping. Naturally, this does not change the primarily intended effect, namely that the cotton is to protect the ear from the impact of the excessively high sound level.

9.2 Seating Arrangement in the Orchestra Pit

9.2.1 Customary Arrangements of Instrument Groups

In contrast to concert halls, most earlier opera houses did not offer enough space to locate musicians from a standpoint of tonal considerations. Thus in the eighteenth century it was generally customary to have players seated in two long rows at a table music stand in front of the stage whereby one row had its back toward the stage and the other row its' back toward the audience. At one end, the conductor was located at the harpsichord and around the conductor the musicians of the *basso continuo* sections were grouped. For larger orchestras a second harpsichord was located at the opposite end of the orchestra space along with several cellos, basses, and bassoons.

Figure 9.3 shows this type of orchestral arrangement in the Teatro Regio Turin; the picture shows the condition of the opera house after its reopening in the year 1740. The orchestra space is so narrow that it just barely suffices for the double-sided long music desk; the low instruments at the sides find themselves in a rather crowded space. Note also the openly blown horns on the left edge of the orchestra.

Fig. 9.3 Teatro Regio Turin (about 1740) (Bärenreiter - Verlag Kassel (MGG))

This seating arrangement, as determined by the long score tables persisted, at least in part, into the nineteenth century, as for example in the Covent Garden Opera in London. Construction of somewhat wider orchestra spaces thereafter, however, led to a looser seating arrangement with individual desks, or possibly desks for two players. After the disappearance of the *continuo*-harpsichord, it initially became customary for the conductor to stand close to the stage, in order to have better contact with the singers. As a relic from the general bass era, a portion of the cellos and basses remained in the middle of the orchestra in immediate proximity to the conductor. The remaining voices were strictly divided into the principal sections of strings and winds including percussion in most theaters, each of these occupied about half of the orchestra space. Thus in the Dresden Opera House at the time of

C. M. von Weber, the strings sat on the left side of the conductor and the winds on the right; in contrast, in Munich, the winds were located to the left and the strings on the right (Hoffmann, 1949; Becker, 1962).

While in the Berlin opera, the attempt to achieve a better symmetry by spreading the violins on both sides was not yet permanently established in 1840, by 1810 the orchestra of the Great Opera in Paris was already arranged in that fashion. Thus the first violins sat to the left of the conductor, the second to the right, and most of the other players directly in front. On the other hand, in Germany, the nearly general discontinuation of strict wind-string-separation only occurred when R. Wagner forcefully insisted on the uniform distribution of strings over the entire orchestra and a division of the brasses onto both flanks.

Still today, this system is the basis for seating arrangement of most orchestras. Thus without exception the first violins are seated to the left of the conductor, followed by the other strings in a half circle. Currently a concert arrangement is again more frequent, where the second violins are located toward the right; this location is most frequently occupied by the violas, the basses are then arranged to the right or to the left in front of the stage. The woodwinds and horns are seated on the left side in three rows behind each other, while the heavy brass and percussion are placed on the right.

For operas also utilizing Wagner-tubas, frequently the horns are added to the remaining brass on the right side, while equally large numbers of woodwinds fill out the orchestra on the left flank. A strict wind-string separation, for which even the woodwinds are seated toward the right of the conductor, is found only rarely in larger operas today. However, this seating arrangement was still used in the twentieth century by the Metropolitan Opera in New York, for works of Wagner and Strauss, for example, as well as for Italian operas.

Circumstances for operas with only small orchestras, however, are different. Especially for chamber music-like instrumentation, as for example in "Ariadne on Naxos" by R. Strauss or "Albert Herring" and "The Rape of Lucretia" by B. Britten. It appears natural to place the few strings and winds together – also in view of rhythmic exactness. In the Vienna state opera, occasionally even for Mozart operas the strings are located toward the left, and the winds as well as the timpani on the right, which undoubtedly aids the ensemble playing of horns and woodwinds on the one hand and the trumpets on the other. Some conductors, however, prefer to place the winds toward the front and to stretch the strings correspondingly far apart toward the side. Toscanini, for example, demanded such an arrangement in "Falstaff."

In the course of historical developments, not only has the seating arrangement of opera orchestras been changed several times, but the orchestra halls themselves were changed in their shape. In most baroque theaters the musicians sat in front of the stage at the same level as the front rows of the main floor, and were separated by shoulder high partitions from the audience, as can be seen in Fig. 9.3. In some theaters of the Baroque era the players were already placed by one or two levels lower than the audience.

A sunken orchestra pit, as customary today, only appeared in the nineteenth century; the musicians thus were removed from the field of vision of the audience

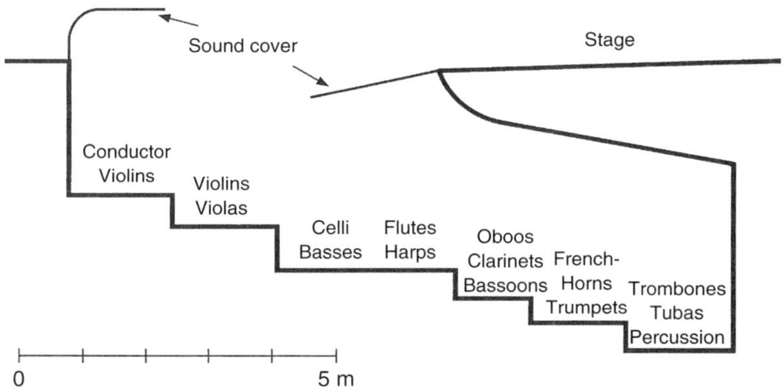

Fig. 9.4 Distribution of instrument sections in the covered orchestra pit in the Bayreuth Festspielhaus

on the main floor and no longer formed a barrier for the visual contact to the stage. Furthermore, the candle illumination for the score desks, which were difficult to cover, were less of a hindrance for the illumination effects on the stage. A far reaching covering of the orchestra space was realized in line with the plans of R. Wagner in the Festspielhaus in Bayreuth, after already in the year 1817 the well-known Berlin architect K. F. Schinkel had made similar suggestions in order to assist in the merging of the individual instruments (Hoffmann, 1949). A cross section of the Bayreuth orchestra pit is shown in Fig. 9.4. A special advantage of this arrangement lies in the fact that the instruments can be located in several levels of significant depth without making the distance between stage and the front audience rows too great. This permits a frontal arrangement of wind rows, and a location of the tonal centers toward the edges is avoided. At the same time the location of the opening in the orchestra cover permits locating the first violins to the right and the second violins to the left of the conductor. Since the height of the orchestra space is nearly 3 m everywhere, the acoustical conditions are significantly more comfortable for the musicians than if they have to sit closely spaced in an opera house of usual construction under the stage which is extended toward the front.

9.2.2 *The Tonal Effect in the Hall*

Sinking the orchestra space, naturally also has a significant effect on the sound of the instruments, especially for the audience on the main floor. Since the low frequency contributions are refracted better around the wall of the orchestra pit (see Fig. 6.18), the sound of the orchestra on the main floor appears darker and less brilliant than in seats from which the players can be seen directly, as would be the case in a concert hall. At the same time the tonal picture is less transparent so that many details in the instrumental voices are lost. This phenomenon was already criticized by Berlioz (1867) for the low-lying orchestra pits, which were new at that time. Also Mozart (1791), after attending a performance of his Magic Flute, the two acts of which he heard at different locations of the "Wiedener Freihaustheater," judged that "the music sounded much better in a box close to the orchestra than in the balcony," which pointed to his preference for a more transparent tonal picture.

Increased transparency can be achieved by lowering the orchestra pit as little as possible, or by a seating arrangement advantageous for direct sound radiation by the instruments into the audience hall. The increased loudness caused by such an "open" instrumental arrangement can find compensation by a reduction in numbers of strings, or by appropriate performance technique. For woodwinds, especially for bassoons, the possibility exists – as mentioned – to absorb a portion of the radiated sound energy by hanging a curtain toward the side, as was earlier customary in the Dresden Staatsoper. The new Festspielhaus in Salzburg does in fact have interchangeable wall elements of varying absorptivity for just such situations (Residenzverlag Salzburg, 1960). Since these absorbers are not effective for the direct sound toward the audience, they also enhance the clarity of the tone picture at the same time.

When deciding on the arrangement of strings, it is important to be aware that the tone of the second violins should not be more intensive, and above all not more brilliant than that of the first violins. For larger numbers of first violins, however, this is generally not a danger, since they usually sit next to each other on two desks, so that only a portion of the players is blocked by the partition separating the orchestra pit. Only for solo passages is it a disadvantage for the concert master to sit directly near the wall, i.e., at the location customary for a concert arrangement; acoustically the location at the first desk of the inner row is better, this seating arrangement is customary in the Vienna Staatsoper, for example.

With small numbers of strings, and locating both string sections next to each other, it is recommended, at least in relatively deep orchestra pits, to place the second violins toward the front at the wall, and to seat the first violins behind them. While this arrangement does not change the tonal impression for the conductor significantly, since he receives the direct sound equally from both sections, it does lead to a significantly better balance of the timbres for the audience. The mutual listening is practically not changed by this arrangement, the slight decrease in sensitivity of the second violin for sound which arrives from the left (see Fig. 1.10) is compensated by reflections from the wall in the orchestra pit located on the right. The second violins can even be heard better by the first violins, only the acoustical contact within that section becomes somewhat more difficult, however, not any more so than it was previously for the second violins.

The seating arrangement in the opera house is not as important for the lower strings. It is merely a disadvantage when unaccompanied passages, as for example the bass passage in the last act of "Othello," comes too far from the edge and thus loses its relation to the stage. The basses, however, should not sit too far below the stage extension, because then the sound radiation – at least at higher frequencies – is influenced, and the desired precision is missed. Thus the cello solo in "Don

Giovanni" should be aligned with the stage position of the singer performing the role of "Zerline," in order to make it easier to perform together. However, in this aria one should also be careful to avoid influencing the radiation of the cello by the harpsichord, so that the sextuplets do not lose their precision through lack of high frequency components.

When the oboes and clarinets are arranged to face toward the front, this means a gain of 3–5 dB in comparison to a sideways orientation for the direct sound reaching the audience. Furthermore, it is often the case that seats toward the side experience more blocking by the partition as far as the direct sound is concerned. Frontal seating is also results in higher intensity for the flutes, while for bassoons there is practically no observable difference. Especially in operas, for which woodwinds in solo passages contribute significantly to the characteristic of the tone picture and for which a very transparent orchestral sound is desirable, the frontal arrangement of this instrument section is recommended.

Since the directions of preferred sound radiation are oriented toward the front and slightly upward for clarinets (and correspondingly also for bassett horns), the tonal effect is not essentially influenced when these instruments are moved up to about 1 m behind the front of the stage. Only for clear solo voices in large arias, as for example in "La Clemenza di Tito" by W. A. Mozart this arrangement is a disadvantage, since then the wind passages are not heard too well on the stage. However, such a placement means a loss in brilliance for oboes and flutes, since the higher frequency tone contributions are reflected too much back to the musicians. It is furthermore important for tone color balance of oboes and clarinets, for both sections to have approximately the same orientation. If for example the oboes are seated with the frontal arrangement and the clarinets to the side (or opposite), then the frontal section dominates the other instruments, which in turn appear dull.

When the bassoons are seated toward the right of the conductor, then for a portion of the main floor seats the danger exists that the tonal impression becomes hard and crisp, because a balance between the high and low frequency contributions is disturbed by reflection from the ceiling or from the surface below the balconies. Since, furthermore, the high frequency components influence a spatial orientation of the sound source most strongly, the impression can arise that the sounds come from above, which naturally influences the closeness of the orchestra sound. This effect is increasingly pronounced, the more the direct sound of the bassoon is blocked from the audience on the main floor, i.e., the deeper the orchestra pit is, and the closer the player sits to the wall close to the audience. Angled separations designed to block the audience view of the orchestra, or sound reflectors on the partition turned toward the stage, can further increase this effect. Especially with small instrumentations, therefore, frontal seating on the right side of the orchestra in front of the stage is to be preferred to a sideways orientation. However, even for a large orchestra, the effects discussed can be noticeable. Furthermore, for bassoon location on the right side, an unpleasant timbre can arise for the audience in those balcony seats which lie in the direction of the instrument axis.

For the standard arrangement of horns on the left side of the orchestra, a large portion of the sound radiated toward the front is already absorbed within the orchestra pit, since the players are mostly seated only slightly raised, or not at all, behind the rows of the woodwinds. Also, the preferred radiation directions of the sound toward the rear can only reach the audience when a closed reflecting surface is present. This condition is not always met by the openings of the orchestra pit. Often curtains close off that wall. In such cases, naturally, strong absorption occurs, which encompasses the entire frequency range when the distance of the curtain from the wall is sufficiently large. When, by reason of space, the horns are set back below the box close to the stage or the stage extension, a further reduction of the higher tone contributions occurs and the sound becomes dull, even though not necessarily too soft. For such spatial situations, a curtain behind the players is certainly inappropriate, particular for operas for which originally the brighter sounding invention horns were intended. If, however, rear wall reflections are undesirable because of the overall intensity, installing a low frequency absorber, which nevertheless reflects high frequencies, could be appropriate. In view of the blocking of the direct sound from the audience, it is absolutely essential in deeper orchestra pits that the first horn sits closest to the stage, this applies both for

Score example 36 W.A. Mozart,The Abduction from the Seraglio, aria of Pedrillo "Frisch zum Kampfe," measure 61 *ff*

placement from the left as well as on the right flank of the orchestra. Furthermore, the other horn players should not have better reflection surfaces behind them than the first chair.

Since the trumpets in an opera orchestra are mostly oriented toward the side, their sound is not as brilliant as for a concert arrangement. However, the difference in tonal effect sensed by the audience is not audible at the location of the conductor, since the conductor is located in the direction of the principal sound radiation, both in the concert setting as well as in an opera. This circumstance must be taken into consideration especially in such passages where the trumpets have to play rhythmically important figures alone, or with very dark sounding instruments, as for example with timpani, which should stand out clearly from the total sound. A typical example for this is an excerpt from the aria of Pedrillo shown in score example 36, where the trumpets and the timpani have to penetrate the *crescendo* of the other voices. Especially in this passage it can easily occur that the singer covers the dotted trumpet motif for the audience, without having this tonal effect audible at the conductor's desk, because the trumpets sound not only louder but also appear to be richer in overtones and correspondingly more clear at this location. A frontal arrangement of the winds would have a very positive effect in such cases.

While for trombones a sideways orientation certainly is advantageous in view of a sonorous tone presentation, for tubas, and also for Wagner tubas, similar phe-

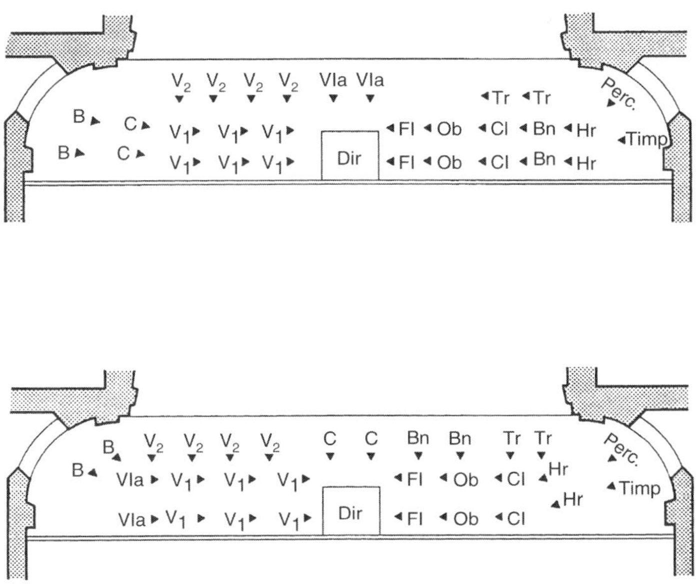

Fig. 9.5 Seating arrangement of the opera orchestra in the "Schloßtheater Drottningholm" for a Mozart opera. *Top*: customary seating in 1969. *Bottom*: seating arrangement improved, based on acoustic principles

nomena can occur as for the bassoon, when their bells are not oriented toward the audience. Especially for Wagner tubas, the soft and round tone color suffers when too many upper partials reach the audience. In situations where the orchestra pit is not coverrchestra pit is not covered, one should therefore attempt to locate the tubas in a frontal direction in the front of the stage or below an audience box at the side. In the latter case, however, the ceiling above the player must be highly absorbent sincc otherwise the reflections become uncomfortable for the musician. For a bass tuba the danger of excessive brightening certainly does not exist – especially when it plays together with the trombones as a bass voice – yet, the choice of location should avoid undesirable reflections of high tone and noise contributions.

As an example of an opera orchestra arrangement, two settings for W.A. Mozart's "Abduction from the Seraglio," are pictured in Fig. 9.5. The upper scheme illustrates the seating arrangement used in the summer of 1969 in the Castle Theatre of Drottningholm, the lower picture contains some change recommendations which would effect tonal improvements. The direction of sight of individual players is in each case indicated by the orientation of the arrow. Complementary to the upper scheme, it should be noted that the solo cellist was moved to the viola desk for the execution of the quartet in the "Marter" Aria, in order to sit more closely to the other soloists. All winds were oriented toward the side, as were the first violins, cellos, and basses, while the second violins and violas sat with their back toward the stage. For this orchestral size and the size of the hall, there are hardly any energy problems, only the bass section was somewhat too weak in the first seating arrangement.

The following tonal advantages are offered by the changed arrangement: both violas and cellos receive a brighter tone color, which is thus better suited for the style of that work. Also the tone of the double basses is radiated into the audience with more overtones; furthermore, these instruments are not located as closely to the front row of the audience. The frontal orientation of the bassoon avoids ceiling reflections of the highest frequency contributions into the audience. As is the case with the basses, locating the low instruments not too closely to the listeners also is a favorable effect. Also placing the trumpets in the frontal orientation fills the demands for a brilliant sound, as discussed earlier, and also the horns receive a somewhat brighter timbre when the wall reflections behind them are not impeded.

The only disadvantage of the revised seating arrangement could be seen in the fact that the cellos and the basses are separated. This could be remedied by moving the second violins clear up to the front of the conductor and continuing with the cellos to the left. The tonal effect of both groups would not suffer by this measure, the only problem would arise from the distance of the solo cello in the Konstanze Aria. It should also be emphasized that the sideways orientation of the first violins and the frontal orientation of the seconds provides a meaningful tonal separation since the orchestra is practically at the same level as the front audience rows and the first violins are not blocked by a partition.

9.3 Balance between Singers and Orchestra

The balance between stage and orchestra, which so strongly determines the musical and dramatic overall impression, is influenced by several factors. For one, the dynamic relationship between a singer and the orchestra naturally depends on the radiated sound power of both sources. For another, the spectral differences between voice and instrumental sound play an important role in the ability of the singer to project above the orchestra. Furthermore, the differences in temporal fine structure of song and orchestra sound are of great importance in order to make the articulation of the singer more clear for the audience, and also to make it easier to localize the singer on the stage. It is also helpful when the listener can perceive a difference in orientation for the singing voice and the orchestral sound.

As shown in Fig. 3.31 for several singing voices, for all pitch ranges the sound power level at a *forte* at the low end of each tonal range is of the order of 85 dB; toward the upper voices it rises up to a value of 100 dB. Peak values in *ff* – especially in the upper register of soprano voices – can even exceed 110 dB. Naturally, there are pronounced individual differences between singers.

When this is compared to the sound power levels of opera orchestras assembled in Table 9.2, one sees that the sound power of a singer at a *forte* does reach that of an opera orchestra with historically small numbers of strings, that, however, on the other hand, the power of the winds at a *forte* corresponds approximately to the peak values of the singers or – depending on vocal register – even exceed it. Certainly the balance between singer and orchestra favors the more advantageous sound radiation from the stage, thus, Westphal (1994) found that the level, created by a broad surfaced orchestral reproduction with speakers in the audience region, rises by 2 dB when this "orchestra" is located on the stage instead of the pit: in the Bayreuth Festspielhaus, with its largely covered orchestra pit, this difference amounted to as much as 3.5 dB.

Measurements, using an impulse – response procedure in two Canadian opera houses did not result in a significant difference between the sound radiation from the stage and from the open orchestra pit: however, for a partially covered orchestra pit an advantage of about 5 dB resulted for sound sources on the stage (O'Keefe 1994). All these data are with reference to the entire sound energy impinging on the audience without consideration for a time structure; they thus only give a total overview over expected loudness relationships between singer and orchestra: however, they already show the possibility that a full orchestra *forte* would cover the singer.

Since acoustical masking in the ear primarily relates to the fact that low frequency contributions render higher ones inaudible or at the very least reduce their effective loudness, particular attention must be given to those opera houses in which the reverberation time of low frequencies increases significantly. The massive sound of the orchestral instruments in the low and middle registers, created by this, which also creates a high degree of spaciousness, can thus decrease the transparency of the tonal picture significantly on the one hand and can, on the other, influence the vowel color

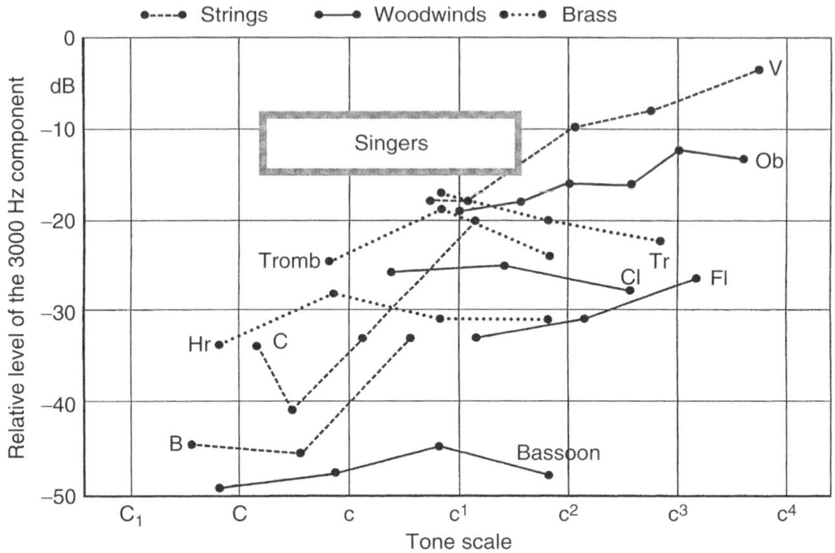

Fig. 9.6 Relative strength of sound power level of 3,000 Hz components (refer to the strongest partials) for a *forte* sound of solo singing voices and orchestra instruments

Score example 37 W.A. Mozart, "Le nozze di Figaro," aria of Marcellina, measure 86 *ff*

of the singing voices in large choral scenes (for example in Verdi operas). Under such room acoustical conditions, a certain dynamic restraint for the full orchestra is recommended as long as the choir or soloists are singing.

From this viewpoint it is all the more surprising that Verdi demands such a large number of basses (see Table 9.1) for his opera performances, since, at least in the past (likely also in Verdi's time) the Milano Scala exhibited a clear rise of the reverberation time from 1.2 s for midfrequencies to 1.5 s for low frequencies

(Beranek, 1996). It is possible that this also reflects a change in the taste of the times, which today leads to an expectation of higher transparency.

The higher frequency spectral contributions are of special importance for the singer's ability to project above the orchestra. As already shown in Fig. 3.30, the singer's formant in the region around 3,000 Hz clearly can stand out above the orchestra sound and can thus draw the attention of the audience to the singing voice. Figure 9.6 contains an overview of the relative strength of tone contributions in the frequency region around 3,000 Hz, with reference to the strongest partials of the corresponding instrument sounds. In the power spectrum of nearly all orchestra instruments 3,000 Hz components lie by 20 dB or more, for the lowest instruments by even 30–50 dB below the strongest partials. Only for the oboe do the 3,000 Hz components in the high register reach to within 12 dB to the strongest partials, and for the violin, this distance amounts to less than 10 dB in the mid and high register. All these values apply to *forte* sounds: with decreasing dynamic level the 3,000 Hz components – at least for winds – drop more rapidly in their level than the lower frequency contributions (see Fig. 8.6).

This means, that violins in *forte* influence the perception of the singer's formant and thus can shift the balance when their sound power is of the same order of magnitude as that of the singer. This danger exists particularly in the lower region of the tonal range of singing voices and is especially noticeable in voices of dark timbre. A typical example of this is the conclusion of the aria of Marcellina in the fourth act of "Figaro" (score example 37) where the violins are an octave above the descending singing voice and rise in a *crescendo* to the *forte* of their theme entrance; the last two notes of the alto voice can thus easily drown.

The audibility of the singer's formant – which for top singers certainly can even be stronger than shown in Fig. 9.6 – is not only assisted by the free sound radiation from the stage but also by the directional effect of the voice. The effect is, that the three fold diffuse-field distance for high frequency contributions approximately reaches the dimensions of the hall (see Fig. 6.19) and thus enables an acoustical localization of the singer even for far distance seats. This effect can further be supported through reflections by the stage floor in front of the singer.

However, this is mostly lost when the singer turns the direction of sight by more than 60° relative to the long axis of the hall. A clear exception to this is only heard when the stage scenery reflects the sound well and these reflections are directed toward the audience without further reflections. This is particularly easily achieved when two wall elements meet in an angle of 90° so that the sound is reflected back to the singer and thus reaches the audiences in the same direction as the direct sound.

Even minor turns of position changes of the singer can be very dangerous when the scenery contains large curved or angled surfaces. These often lead to clearly noticeable focusing effects which in certain audience regions lead to changes in the vocal effect of an order of magnitude of 10 dB louder or softer and thus, depending on the reflection characteristics of the relevant surfaces, also change the timbre of the voice in unnatural ways. Since these tone color changes, and above all dynamic jumps, are neither based on dramatic nor musical reasons they not only have a disturbing effect but raise the impression for the listeners, for whom the connection with the acoustic boundary conditions is not at all obvious, of limited vocal control by the singer.

The free sound radiation from the stage and the effectiveness of reflection surfaces in the neighborhood of the singer on one hand, and the blocking of the direct sound from the orchestra pit on the other hand, lead to the fact that the condition of good balance between singer and orchestra is not as unfavorable as would appear from a comparison between the total sound energy radiated by singers and orchestras. Aside from the spectral differences discussed earlier, the temporal fine structure of the tonal development also favors the singer, thus when one considers the energy comparison of the direct sound and the reflections of the sound reaching the audience within the first 50 ms from the stage and orchestra pit, then the stage has an advantage of 2 dB even from an open orchestra pit, for a partially covered orchestra pit this even rises from 7 to 12 dB. These values are significantly higher than those mentioned above, which also include all reflections arriving later (O'Keefe, 1994).

This means that the clarity of articulation of singing on the stage rises above the tonal entrances of the instruments in the orchestra pit through the room acoustical conditions. This effect can be further supported by performance techniques of the orchestra, by avoiding sharp attacks, thus leaving the dominance to the articulation of the singer. In this context one should also mention the conscious delay of orchestral chords in relation to the entrance of the singer even though use of this means of shaping tones is certainly not uncontroversial.

Finally, it will also be easier for the listener to concentrate on the stage when the direction of the direct sound from the stage can be differentiated from that of the orchestra, where naturally a clear view of the singer is of help. In the sense of the cocktail party – effect, the directional selectivity relates to a level rise for the articulation frequencies. The meaning of this effect is especially notable (in a negative sense) when opera scenes with large instrumentation – as for example in the case of Salome's final song – are performed in a concert hall where the soloist stands directly in front of the orchestra so that directional separation is no longer possible. Singers are also disadvantaged on stage when sight and direct sound are blocked for a portion of the audience and the cocktail effect thus does not apply.

For the singer, certainly the dynamic balance between voice and orchestra is important. A level range of the orchestral sound between 5 dB above and 15 dB below the voice, is required at the ear for intonation reference purposes (see Sect. 6.4.1). The sound pressure level of the vocalist's voice, measured at the ears, is by 5–10 dB below its sound power level. In contrast, the orchestra sound pressure level at the singer's ear is about 20 dB below the sound power level of the orchestra in historic theaters, but about 25 dB in newer theaters (see Fig. 9.1). Starting with the assumption that the sound power of singer and orchestra are approximately equal – not only in loud passages but also for *piano* – then the singer hears just barely enough of the orchestra in a modern opera house. Early reflections of the solo voice (also from within the stage) would thus change the balance in such a way that the intonation of the singer becomes more difficult – unless also the orchestra sound is enhanced by early reflections. As Figs. 6.22 and 6.23 have shown, the likelihood for this circumstance decreases as the singer moves toward the back of the stage. This observation also shows that the ears of the singer cannot be covered through costumes (e.g., closed nun head coverings made of nontransparent cloth). In old theaters, where the orchestra was not lowered, the problem of sound balance at the ear of the singer was not as significant; in the very large new theaters the necessity certainly can arise to amplify the orchestral sound for the stage in order to provide increased security for the singers.

In this context naturally the question also arises whether in large opera houses the solo voice on stage should receive some electronic enhancement for the audience. When the personal ego of the singer is discounted, which would assume no need for amplification, it can certainly be demonstrated that today's pick-up techniques with lapel microphones and wireless transmissions can amply satisfy artistic demands; the director merely needs to be sure that head turns are avoided if possible and replaced by turns of the entire upper body when the microphone is not directly attached to the head.

Reproduction, on the other hand, is extremely problematic. Sound radiation by fixed speakers with different directional characteristics (than the singing voice) creates a sound in the hall which is independent of the position of the singer even when its direct sound, as the first wave front, still permits localizing the singer. Through this and possibly also through the frequency flow of the reproduction, the singing voice receives a tone in the hall which is somewhat unnatural and the connection of the sound to the singer is somewhat disturbed instead of improved. While this certainly applies to the individual singer it becomes even more noticeable for ensembles for which the transparency, and thus the relationship to the individual voices is influenced or possibly even lost entirely.

Loudspeaker support or transmission of singing voices cannot only be defended from an artistic standpoint but can also be extremely effective when a deliberate tonal change or a particular spatial effect is to be achieved. As examples of the first kind, a gentle reverberating of the voice of the Komturstatue in "Don Giovanni" or of the Fafnerdragon in "Siegfried" are mentioned, in order to obtain a certain transcendental effect and a contrast to the human voices on stage. However, here also caution is recommended when manipulating frequency sequences, so that the tonal aesthetic limits of the singing voice are not exceeded. An example of the second kind is the stepping of the distance of the sound from the riding Valkyries, arriving from great distances, by additional electronically created reflections of various delays and amplitudes.

9.4 Arrangement of Choirs and Music on Stage

9.4.1 Musicians in the Scene

Placing of choruses on the opera stage naturally cannot be determined exclusively from tonal viewpoints, as is the case in the concert hall, since their positions frequently are determined by the action on stage. Consequently, acoustic boundary conditions for choir locations should be included for consideration from the very beginning of stage design – at least for operas with significant choir tasks.

For large choirs, energy problems do not arise, thus only questions of a tonal nature as well as problems of hearing the other voices and the orchestra are of concern. For small numbers of choral singers, it could certainly make sense to avoid having the radiated choir sound influenced by distributing the choir over too large an area or having the choir located too low on the stage. These considerations are not important for a persistent choir sound as for example in the temple scene in the first act of "Aida."

As a matter of principle, it is advantageous, both for the tonal effect in the hall, as for the security of the singer to have the individual voice groups separated or mixed any more than the action absolutely demands. Naturally, the direction of sight of the choral singers has an influence on the tonal effect for the audience, however it is not as important as for the soloist. Since the homogeneity of the choral sound rests among other things on the circumstance that no singer stands out through the singer's formant and thus can be identified as a single voice, the higher frequency tone contributions with their pronounced directional effects play a less important role. On the other hand, one must be aware that the choir for sideways orientation can excite the frequently relatively long reverberation of the stage enclosure in such a way that it leads to a veiled effect, and thus frontal arrangement should be preferred, especially when scenery is sparse. It is also always advantageous when different elevation levels can be utilized within large choirs.

Placement of instrumentalists on stage is of course primarily determined by the scenery and the action, thus by the director, however even here acoustic viewpoints can be considered, if they are included sufficiently early in the overall conception. The possibility of establishing contact with the conductor by the use of a TV monitor – as occasionally is done for choirs, the requirement for a direct sight connection and orientation of the performer and the conductor is relaxed.

Trumpets and trombones in the scene generally do not pose particular acoustic difficulties since in comparison to the instruments in the orchestra pit they have freer radiation conditions and can thus achieve desired brilliance. Only for sparse scenery it can happen that they excite reverberation of the stage enclosure when oriented toward the side. For example, this can occur at the beginning of the festival scene in "Die Meistersinger von Nürnberg," when the (sound absorbing) masses are not yet on stage. The "open air" location, however, demands that no significant reverberation is audible. Furthermore, it must be avoided that the sound of the trumpets does not become more dull as the size of the choir increases which actually can only be accomplished by a change of arranging of the instrumental musicians.

The arrangement of several trumpet groups is also interesting, as required in the 3rd act of the opera "Lohengrin," for the effective entrance of the four counts and the kings. At the end, the king's theme of the C-trumpets must dominate above the different trumpet groups (in E^b, F, D, and E). This tonal effect is best achieved when the trumpets of the counts are turned slightly sideways and those of the kings, however, play directly into the audience. In similar manner, the two groups of the "Aida" trumpets (A^b and B) should be arranged to overpower the accompanying instruments in the theme at the conclusion of the triumphal march., because the dynamic instructions *p* or *pp* are frequently not sufficiently observed by auxiliary stage musician.

Trumpets absolutely must be placed in a frontal orientation when performing on stage with closed curtains as for example during the change in the 3rd act of "Die Meistersingers von Nürnberg," since the curtain predominately absorbs the high tonal contributions, otherwise the trumpet sound would be overly deprived of its fanfare character. It is not advantageous, however, to place the players directly behind the curtain, which would cause a point-like localization of the sound source for the audience which would contradict the musical preparation for the spaciousness of the subsequent scene.

Large wind ensembles on the stage could lead to acoustical difficulties, insofar as differing directional characteristics of individual instruments could excite the reverberation of the stage enclosure to different degrees, or also when absorbing decorations make it difficult to establish a balance between the instruments. The Great Opera in Paris, therefore, has available tubas and tenor horns with angled bells, the openings of which can furthermore be turned.

Naturally, great caution is advised when a band is to move on stage as is the case for example in Gounod's "Margarethe" or A. Berg's "Wozzeck." Since the horns mostly play secondary voices they must above all be careful that, during their exit, they do not turn the curve of the horn toward the audience while the melody voices are already leaving the stage. As a result, exit toward the front left is always better than toward the front right, which should already be considered during the design of the scenery.

A single violin on stage has significantly greater difficulties to fill the room. It is, therefore, absolutely essential to strive for a position which optimizes the radiation direction. In this context, one also needs to be careful that the preferred regions around 1,000 Hz as well as the highest frequencies are not prevented from spreading into the audience by the upper edge of the stage frame. This means that the player, if possible should not stand further back on stage than the height of the stage frame; furthermore, it is also an advantage to have several meters of stage floor between the player and the orchestra pit to be used as a reflection surface. It is also generally advantageous for tonal reasons when the singer (as in "Orpheus in the Underworld") can play the instrument himself, rather than have the instrument played behind stage.

In the finale of the first act of "Don Giovanni," three separate orchestras are to be located on stage in such a manner that an intensity balance between the groups is achieved, as well as a sufficiently full overall sound, since during this scene only the cellos and basses are still playing in the orchestra pit. In order not to cover the other two dances with the Minuet, the first orchestra should be situated on the right side of the stage, if possible, since on the left side the danger exists that the two horns stand out too strongly. The frontal arrangement in the background could frequently be problematic since the first orchestra must provide the rhythmic reference for the scene as the only continuously playing ensemble, it thus requires good contact with the conductor. The strings can also be supported from an intensity standpoint by making a portion of the decorations out of wood, thus creating favorable reflection surfaces. If the horns have to stand at the left side because of demands by their director or the scene designers, then they must turn the bell as far to the rear as possible, in this case cloth scenery would be preferred.

When the entire orchestra is to be located behind the actual performance space as for example in Z. Kodaly's "Hary Janos" as a farmer's orchestra, it is almost unavoidable to have a shallow tonal effect. Even with acoustic means, this can hardly be remedied; in such cases one simply must be aware that the scenic effect is bought at the price of tonal influence.

9.4.2 Musicians behind the Scene

The reverberation of the stage enclosure has an even larger effect for singers or musicians behind the decorations, than for musicians on stage. Beyond that, clearly the nature of the scenery is of great importance for the tonal effect in the audience. Painted cloth scenery reflects high frequency contributions to a relatively strong degree while the low components are transmitted practically undiminished. Consequently, the singers or instrumentalists perceive a far more brilliant and brighter tone picture than the conductor or the audience. Since, on the other hand, the tone of the orchestras behind the decorations has a reduced overtone content and is perceived as dull and lacking in overtones, by reason of the damping of the high frequency components during transmission through the scenery, and is thus perceived from the standpoint of pitch as slightly lower, choirs as well as instrumental soloists behind the scenes have a tendency toward lower intonation (Leipp, 1969).

For musicians behind the scenery it is therefore advantageous to transmit the sound of the orchestra by means of speakers with an emphasized brightness in tone color in order to cause them to play with slightly higher intonation; because, based on overtone damping, choirs or instrumental soloists are easily perceived as too low by the audience – even if the intonation is correct behind the stage as far as frequency is concerned. Another possibility for subjective pitch control is to have a colleague in the orchestra pit provide signals for the musician playing behind stage which evaluate the intonation and indicate possible corrections. Such a practice has proven very successful in the last decades in the Vienna State Opera, when the performer behind the stage is no longer able to hear the orchestra properly.

This effect of wrong intonation is naturally especially relevant for string instruments, since they do not possess a predetermined tone scale and can be freer in their intonation than wind instruments with their limited region of adjustment (Meyer, 1966c). However, the opinion is frequently voiced that a violin must be tuned slightly higher for a solo behind the stage in comparison to the orchestra (e.g., as in E. Humperdinck's "Königskinder"). In order to compensate for the damping of overtone contributions by the scenery it is naturally recommended to have the scenery material stretched relatively loosely. Furthermore, when choosing a musician for such a solo, one should consider using a bright or possibly even sharp-sounding instrument, the sound of which is not unduly influenced by curtailing the overtone content, or may possibly even gain by it. This of course applies in a corresponding way to small string ensembles behind stage as for example in R. Strauss' "Capriccio" even though the problem of intonation does not exist in the same degree since a closed harmonic setting appears more firm in relationship to the orchestra than a solo voice.

For the same reason the Ball-music in several Verdi operas, or the old Vienna dance orchestra in the "Rosenkavalier," do not pose an acoustic problem, since at the same time the damped tonal effect corresponds to the perception that the music comes from the hall located behind the scenery. In large opera houses these stage music pieces are often performed in a distant practice room and then transmitted by loudspeaker groups (Martin, 1962). It is however important that the reverberation time of the recording hall is not too long so that the speakers in cooperation with the reverberation behind the stage can achieve a believable spatial effect; if the reverberation is expected to come exclusively from the speakers, a significantly greater technical effort is required.

Thus in the Vienna States Opera many stage music pieces are recorded in the organ and practice hall which, at a volume of about 3,000 m^3 has an average reverberation time of 1.4 s when empty; this value drops slightly when the hall is occupied, and for low frequencies the reverberation time is shorter than 1 s at any rate.

The apparent distance of a player behind the scene depends on the precision of the entrance, since also in free air, for a long sound path the attacks are somewhat washed out through a series of delayed reflections. Consequently the reverberation of the stage enclosure can cause a distance impression for scenes which are performed outdoors, even though in nature there is no actual reverberation. The more directly the sound reaches the audience through the frame of the stage, i.e., the larger the intensity relation between the direct sound and the reverberation of the stage enclosure, the closer the musician appears to be. This effect can be utilized for example in the English horn solo in "Tristan and Isolde." Brass instruments can also be brought to have a stronger effect because of their sharper sound directions, and this not only for individual instruments, as for example in the trumpet signal in "Fidelio," but also for groups of winds, as for example in the Hunting-Fanfares of the 12 horns in the first act of "Tannhäuser."

On the other hand the tonal effect for most listeners can be unsatisfying, when stage music is played in a sideways direction directly behind the stage frame because in a large section of the audience the tonal difference between players in the orchestra pit is too small to be noticed, especially for winds. In this case the more effective perception at the conductor's desk is certainly misleading.

Appendix
Table for Angular Dependence of the Statistical Directivity Factor

Direction	Trumpets				Trombones			
	2000 Hz	6000 Hz	10 000 Hz	15 000 Hz	500 Hz	1000 Hz	3000 Hz	10 000 Hz
0°(Bell axis)	2,30	4,40	4,70	6,60	1,60	2,10	4,50	6,10
10°	2,21	3,85	4,40	4,40	1,59	2,05	3,90	5,15
20°	1,92	3,18	3,35	3,05	1,55	1,85	3,0	3,20
30°	1,85	2,35	1,85	1,60	1,51	1,60	2,0	1,67
40°	1,78	1,30	1,10	0,87	1,47	1,36	1,30	1,21
50°	1,30	0,86	0,75	0,65	1,32	1,22	0,95	0,50
60°	1,10	0,60	0,50	0,56	1,18	1,00	0,53	0,27
70°	0,94	0,39	0,47	0,51	1,05	0,90	0,53	0,23
80°	0,85	0,24	0,32	0,46	0,94	0,84	0,54	0,29
90°(side)	0,75	0,15	0,22	0,28	0,84	0,73	0,44	0,30

Direction	Singing voice (horizontal)						
	125 Hz	250 Hz	500 Hz	1000 Hz	2000 Hz	4000 Hz	8000 Hz
0° (front)	1,15	1,29	1,33	1,02	1,23	1,66	1,97
20°	1,25	1,41	1,40	1,08	1,38	1,62	2,0
40°	1,11	1,24	1,33	1,15	1,62	1,32	1,97
60°	0,91	1,06	1,14	1,06	1,41	1,17	1,14
80°	0,99	1,02	1,15	1,23	1,08	1,40	1,12
90° (side)	0,86	0,91	1,0	1,20	0,88	1,04	0,67
100°	0,94	0,96	0,94	1,13	0,85	0,97	0,61
120°	0,95	0,86	0,76	0,90	0,71	0,72	0,45
140°	0,84	0,69	0,55	0,62	0,48	0,43	0,25
160°	0,71	0,65	0,48	0,46	0,25	0,28	0,12
180° (rear)	0,78	0,68	0,47	0,50	0,27	0,25	0,09

Direction	Singing voice (vertical)						
	125 Hz	250 Hz	500 Hz	1000 Hz	2000 Hz	4000 Hz	8000 Hz
−40°	1,07	1,18	1,31	1,48	0,90	1,10	1,42
−20°	1,33	1,35	1,35	1,45	2,0	1,94	2,30
0° (front)	1,15	1,29	1,33	1,02	1,23	1,44	1,97
20°	1,18	1,29	1,18	1,08	1,28	1,65	1,80
40°	1,48	1,42	1,10	1,28	1,33	1,35	1,40
60°	1,42	1,25	1,0	1,05	1,23	1,13	1,13
90° (up)	1,08	1,08	0,98	0,70	0,78	0,84	0,54
120°	1,05	0,98	0,90	0,91	0,77	0,60	0,34
140°	1,0	0,88	0,76	0,91	0,43	0,29	0,17
160°	0,74	0,66	0,58	0,58	0,31	0,24	0,11
180° (rear)	0,75	0,65	0,45	0,48	0,31	0,28	0,10
200°	0,84	0,67	0,52	0,44	0,31	0,21	0,08
220°	0,78	0,78	0,53	0,63	0,21	0,13	0,06

References

Ahnert, W., Steinke, G., Hoeg, W. und Steffen, F. (1986): Moderne Methoden der Beschallung von Sälen und Freilichtspielstätten. Kulturbauten H. 1/86, S. 2

Allen, W. A. (1980): Music stage design. J. Sound Vib. 69, S. 143

Alrutz, H. und Gottlob, D. (1978): Der Einfluß der frühen seitlichen Reflexionen auf die Räumlichkeit. Fortschritte der Akustik – DAGA '78, Bad Honnef DPG-GmbH, S. 579

Ando, Y. (1985): Concert Hall Acoustics. Springer, Berlin

Askenfelt, A. (1986): Stage floors and risers – supporting resonant bodies or sound traps? In: S. Ternström, Acoustics for Choir and Orchestra. Royal Academy of Music, Stockholm

Askenfelt, A. and Jansson, E. (1990): From touch to string vibrations. In: A Askenfelt, Five lectures on the Acoustics of the Piano. Royal Swedish Academy of Music No 64, Stockholm

Askenfelt, A. (1993): Observations on the transient components of the piano tone. Proc. SMAC '93, Stockholm, S. 297

Atal, B. S., Schroeder, M. R. and Sessler, G. M. (1965): Subjektive reverberation time and its relation to sound decay. Kongreßber. 5. ICA, Liege

Backus, J. (1961): Vibrations of the reed and the air column in the clarinet. J. Acoust. Soc. Am. 33, S. 806

Backus, J. (1963): Acoustical investigations of the clarinet. Sound 2(3), S. 22

Bagenal, H. and Bursar, G. (1930): Bach's music and church acoustics. J. R. Inst. Brit. Architects, S. 154

Bagenal, H. und Wood, A. (1931): Planning for good Acoustics. Methuen, London

Barron, M. (1971): The objective effects of first reflections in concert halls – the need for lateral reflections. J. Sound Vib. 14(4), S. 475

Barron, M. (1978): The Gulbenkian great hall, lisbon, ii – an acoustic study of a concert hall with variable stage. J. Sound Vib. 59, S. 481

Barron, M. (1993): Auditorium Acoustics and Architectural Design. Taylor & Francis, London

Barron, M. and Lee, L.-J. (1988): Energy relations in concert auditoriums. J. Acoust. Soc. Am. 84, S. 618

Becker, H. (1962): Art. »Orchester« in MGG, Kassel

Bell, A. J. and Firth, I. M. (1989): The directivity of the concert harp. Acustica 69, S. 26

Benade, A. H. (1976): Fundamentals of Musical Acoustics. Dover Publications, New York

Beranek, L. L. (1962/1995): Music, Acoustics and Architecture. Krieger Publishing Company, New York (2. Auflage in Vorbe-reitung für 1995)

Beranek, L. L., Johnson, F. R., Schultz, T. J. and Watters, G. B. (1964): Acoustics of Philharmonic Hall, New York during its first season. J. Acoust. Soc. Am. 36, S. 1247

Berlioz, H. (1864): Der Orchesterdirigent. Deutsche Ausgabe Leipzig

Blauert, J. (1970): Zur Trägheit des Richtungshörens bei Laufzeit- und Intensitätsstereophonie. Acustica 23, S. 287

Blauert, J. (1974): Räumliches Hören, Stuttgart

Blauert, J. and Lindemann, W. (1986): Explorative studies on auditory spaciousness. Proc. Vancouver Symp. Acoust. Theatre Plann., S. 39

Blaukopf, K. (1957): Hexenküche der Musik, Teufen/St. Gallen, Wien

Blaukopf, K. (1960): Raumakustische Probleme der Musiksoziologie. Grav. Blätter 5, H. 19/20, S. 163

Bork, I. und Meyer, J. (1988): Zum Einfluß der Spieltechnik auf den Klang der Querflöte. Tibia 13, S. 179

Bork, I. (1989): Longitudinalschwingungen von Klaviersaiten. Jahresber. Phys.-Techn. Bundesanstalt, S. 129

Bork, I. (1991a): Modalanalyse von Schallfeldern. Acustica 75, S. 154

Bork, I. (1991b): Klang und Schallabstrahlung der Querflöte. Ber. 16. Tonmeistertagung Karlsruhe 1990, München, London, S. 351

Bork, I. (1993): Akustische Untersuchungen an Klavieren und Flüge ln. Ber. 17 Tonmeistertagung Karlsruhe 1992, München, London, S. 751

Bork, I., Marshall, A. H. und Meyer, J. (1994): Abstrahlung des Anschlaggeräusches beim Flügel. Ber. 18. Tonmeistertagung Karlsruhe 1994, München, London

Borris, S. (1969): Die großen Orchester. Hamburg, Düsseldorf

Boult, Sir A. (1963): Thoughts on Conducting. Phoenix House, London

Bradley, J. S. (1976): Effects of bow force and speed on violin response. J. Acoust. Soc. Am. 60, S. 24

Briner-Aimo, E. (1966): Ungelöste Probleme der Klangübertragung. Grav. Blätter, H. 27/28, S. 162

Brinkmann, K. (1978): Gehörschützer – Sicherheitstechnische Anforderungen, Prüfungen, Meßergebnisse. Mod. Unfallverh. H. 22, Vulkan-Verlag, Essen

Brückmann, M. (1984): Akustische Daten der Alten Oper Frankfurt/Main. Fortschritte der Akustik – DAGA '84, Bad Honnef DPG-GmbH, S. 371

Burd, A. and Haslam, L. (1994): The relationship of choir and orchestra in concert halls. Proc. Inst. Acoust. 16, S. 479

Burghauser, J. und Spelda, A. (1971): Die akustischen Grundlagen der Instrumentation, Regensburg

Canac, F (1967): L'Acoustique des théatres antiques, Paris

Clark, M. and Luce, D. (1965): Intensities of orchestral instrument scales played at prescribed dynamic markings. J. Audio Eng. Soc. 13, S. 151

Clements, P. (1999): Reflections on an ideal: tradition and change of the grosser Musikvereinssaal, Vienna. Acoust. Bull.-Heft, S. 5

Cohen, E. (1992): Acoustics of practice rooms, paper presentation, at 92nd AES Convention Wien, Preprint 3347

Coltman, J. W. (1971): Effect of material on flute tone quality. J. Acoust. Soc. Am. 49, S. 520

Conklin, H. A. (1990): Piano design factors – their influence on tone and acoustical performance. In: A. Askenfelt, Five Lectures on the Acoustics of the Piano. Royal Swedish Academy of Music No 64, Stockholm

Creighton, H. (1978): Music colleges, Design Experience. Proc. Inst. Acoust. London, S. 17.4

Cremer, L. (1953): Die Plexiglas-Reflektoren im neuen Herkulessaal der Münchener Residenz. Die Schalltechnik 13, Nr. 5, S. 1

Cremer, L. (1961): Statistische Raumakustik, Stuttgart

Cremer, L. (1964): Die raum- und bauakustischen Maßnahmen beim Wiederaufbau der Berliner Philharmonie. Die Schalltechnik 24, Nr. 57, S. 1

Cremer, L. (1981): Physik der Geige, Stuttgart

Cremer, L. und Müller, H. A. (1964): »Bemerkungen zur Akustik« in dem Repräsentationsband der Meistersingerhalle in Nürnberg, Nürnberg

Cremer, L. und Müller, H. A. (1978): Die wissenschaftlichen Grundlagen der Raumakustik. Band 1, 2. Aufl., Stuttgart

Cremer, L., Keidel, L. und Müller, H. A. (1956): Die akustischen Eigenschaften des großen und des mittleren Saales der neuen Liederhalle in Stuttgart. Acustica 6, S. 466

Cremer, L., Kürer, R. und Plenge, G. (1968): Impulsmessungen in Freilufttheatern. Phys. Verh. 19, S. 129

Creuzberg, H. (1953): Die neue Sitzordnung der Sinfonie-Orchester. Das Musikleben II, S. 81

Dahlstedt, S. (1974): Electronic reverberation equipment in the Stockholm Concert Hall. J. Audio Eng. Soc. 22, S. 627

Davies, W. J. and Lam, Y. W. (1994): New attributes of seat dip attenuation. Appl. Acoust. 41, S. 1

Ditters von Dittersdorf, C.: Lebensbeschreibung, seinem Sohne in die Feder diktiert. Neuausgabe, Regensburg 1940

Dünnwald, H. (1988): Ableitung objektiver Qualitätsmerkmale aus Messungen an alten und neuen Geigen. In: J. Meyer, Qualitätsaspekte bei Musikinstrumenten, Celle

Dunn, H. K. and Farnsworth, D. W. (1939): Exploration of pressure field around the human head during speech. J. Acoust. Soc. Am. 10, S. 184

Elfrath, Th. (1992): Bestimmung der akustischen und schwingungstechnischen Eigenschaften von Cembali. Diss. Techn. Univ. Braunschweig

Elkin, E. (1955): The Old Concert Rooms of London. Edward Arnold, London

Estill, J., Fujimura, O., Erickson, D., Zhang, T. and Beechler, K. (1993): Vocal tract contributions to voice qualities. Proc. Stockholm Music Acoustics Conf., S. 161

Fasold, W., Tennhardt, H.-P. und Winkler, H. (1982): Die Raumakustik im Neuen Gewandhaus Leipzig. Bauten der Kultur H. 1/82, S. 14

Fasold, W., Sonntag, E. und Winkler, H. (1987): Bau- und Raumakustik, Berlin

Fasold, W., Tennhardt, H.-P und Winkler, H. (1991): Ergänzende raumakustische Maßnahmen im Großen Konzertsaal des Schauspielhauses Berlin. Bauforschung – Baupraxis H. 287, S. 23

Fasold, W., Küstner, E., Tennhardt, H.-P. und Winkler, H. (1981): Akustische Maßnahmen im Neuen Gewandhaus Leipzig. Bauforschung – Baupraxis H. 117, S. 9

Fasold, W., Lehmann, U., Tennhardt, H.-P und Winkler, H. (1986): Akustische Maßnahmen im Schauspielhaus Berlin. Bauforschung – Baupraxis H. 181, S. 5

Firth, I. M. (1977): On the acoustics of the harp. Acustica 37, S. 148

Fleischer, H. (1988): Die Pauke – Mechanischer Schwinger und akustischer Strahler. For-schungsber. 01/88 Univ. BW München

Fleischer, H. (1991): Akustische Untersuchungen an Orchesterpauken. Forschungsber. 02/91 Univ. BW München

Fleischer, H. (1992): Zur Rolle des Kessels bei Pauken. Forschungsber. 01/92 Univ. BM München

Fletcher, H. and Bassett, I. G. (1978): Some experiments with the bass drum. J. Acoust. Soc. Am. 64, S. 1570

Fletcher, H., Blackham, E. D. and Gertsen, O. N. (1965): Quality of violin, viola, cello and bass-viol tones. J. Acoust. Soc. Am. 37, S. 851

Fletcher, N. H. (1977): Analysis of the design and performance of harpsichords. Acustica 37, S. 139

Fletcher, N. H. and Rosssing, Th. D. (1991): The Physics of Musical Instruments. Springer, Berlin

Fransson, F. (1966/67): The source spectrum of double-reed woodwind instruments. Quartly Progress and States Report 4/66 and 1/67 des Speech Transmission Laboratory der KTH, Stockholm

Forsyth, M. (1987): Buildings for Music. MIT, Cambridge

Franz, G., Ising, H. und Meinusch, P. (1969/70): Schallabstrahlung von Orgelpfeifen. Acustica 22, S. 226

Frei, J. (1979): Die Gehörbelastung des Orchestermusikers in der Konzert- und Opernformation der Tonhalle Zürich. Diss. Univ. Zürich

Friesenhagen, A. (1993): Ein englisches Oratorium als Wegbereiter. Das Orchester 41, S. 927

Fry, D. W. (1978): The auditorium and the singer. Proc. Inst. Acoust., London, S. 17.5

Furrer, W. (1972): Raum- und Bauakustik, Lärmabwehr 3. Aufl., Basel, Stuttgart

Furtwängler, W. (1965): Podiumsgespräche in der Berliner Hochschule für Musik. Sendung des NDR mit Originalaufnahmen der Gespräche

Fütterer, Th. (1988): The acoustics of the new chamber music hall in Berlin. Proc. Inst. Acoust. 10, Teil 2, S. 339

Gabler, W. (1955): Akustische Gesichtspunkte beim Entwurf und Bau von Bühnendekorationen. Bühnentechn. Rundschau, H. 1, S. 8

Gabler, W. (1962): Zur Akustik der Kirchen. Die Schalltechnik 22, S. 4

Gade, A. C. (1989a): Investigations of musician's room acoustic conditions in concert halls. Acustica 69, S. 193

Gade, A. C. (1989b): Acoustical survey of eleven European concert halls. Techn. Univ. Denmark Report No. 44

Gärtner, J. (1974): Das Vibrato unter besonderer Berücksichtigung der Verhältnisse beim Flötisten, Regensburg

Gilford, Ch. (1972): Acoustics for Radio and Television Studios. Peter Peregrinus, London

Gottlob, D. (1973): Vergleich objektiver akustischer Parameter mit Ergebnissen subjektiver Untersuchungen an Konzertsälen. Diss. Univ., Göttingen

Graham, R. N. (1992): Symphony Hall Birmingham: a fusion of architecture and acoustics. Proc. Inst. of Acoust. 14, Teil 2, S. 73

Graner, H. (1988): Persönliche Mitteilung an den Verfasser (Messung des Ingenieurbüros Graner und Partner, Bergisch Gladbach)

Grützmacher, M. und Lottermoser, W. (1936): Neuere Untersuchungen an Flügeln. Akust. Z. 1, S. 49

Haas, H. (1951): Einfluß eines Einfach-Echos auf die Hörsamkeit von Sprache. Acustica 1, S. 49

Hadamowsky, H. (1958): Wiener Bläserstil, Wien (Eigenverlag)

Hoeg, W. und Steinke, G. (1972): Stereofonie-Grundlagen, Berlin

Hoffmann, H. (1949): Art. »Aufführungspraxis« in MGG, Kassel

Holl, K. (1947): Giuseppe Verdi, Lindau i.B.

Husson, R. (1952): L'Acoustique des Salles. Ann. Télécomm, 7, S. 16

Irion, H. (1979): Gehörschäden durch Musik. Moderne Unfallverhütung 23, S. 89

Jahn, G. (1963): Zum Unterschied zwischen einohrigem und beidohrigem Hören. HF und Ela 72, S. 15

Januschka, J. (1969): Persönliche Mitteilung an den Verfasser. (Messung des VÜZORT, Prag)

Jordan, V. L. (1968): Einige Bemerkungen über Anhall und Anfangsnachhall in Musikräumen. Appl. Acoust. 1, S. 29

Jordan, V. L. (1979): Acoustical criterion – effencieny of lateral reflections. Proc. 3. Symp. FASE Dubrovnik, S. 101

Junius, W. (1959): Raumakustische Untersuchungen mit neueren Meßverfahren in der Liederhalle Stuttgart. Acustica 9, S. 289

Karlsson, K., Lundquist, P. G. and Olaussen, T. (1983): The Hearing of symphony orchestra musicians. Scand. Audiol. 12, S. 257

Karsai, M. (1974): The acoustical reconstruction of teaching studios at the Hungarian academy of music. Kongr.-Ber. 8 ICA, London

Katschke, N., Neubauer, Ch. und Behrmann, C.-A. (1981): Untersuchungen des Schallpegels in Orchestern. Das Orchester 29, S. 437

Keet, W. d. V. (1968): The influence of early lateral reflections on the spatial impression. Kongr.-Ber. 6. ICA, Tokio

Keibs, L. und Kuhl, W. (1959): Zur Akustik der Thomaskirche in Leipzig. Acustica 9, S. 365

Keller, W. und Widmann, M. (1979): Kommunikationssysteme im ICC Berlin. Funkschau 51, S. 858

Kern, E. (1972): Rückkopplungsphänomene zwischen Musiker und Musikinstrument. Nova acta Leopoldina 37, S. 574

Kihlman, T. and Kleiner, M. (1980): Scale models in room acoustics – what accuracy is needed? paper presented at 100th meeting of ASA

Kleis, D. (1979): Nachhallbeeinflussungsanlagen. Ext. publ. Audio. Eng. Exp. Group, Philips Breda

Kobald, K. (1964): Beethoven, Wien

Koornhof, G. W. and van der Walt, A. J. (1993): The influence of touch on piano sound. Proc. SMAC '93, Stockholm, S. 318

Kürer, R. und Kurze, U. (1967): Integrationsverfahren zur Nachhallauswertung. Acustica 19, S. 313

Kuhl, W. (1954a): Über Versuche zur Ermittlung der günstigsten Nachhallzeit großer Musikstudios. Acustica 4, S. 618

Kuhl, W. (1954b): Durchführung und Ergebnisse eines Ringversuchs zur Ermittlung der günstigsten Nachhallzeit großer Musikstudios. Ber. 3. Tonmeister-Tagung Detmold, S. 49

Kuhl, W. (1959): zitiert nach Beranek (1962)

Kuhl, W. (1965): Das Zusammenwirken von direktem Schall, ersten Reflexionen und Nachhall bei der Hörsamkeit von Räumen und bei Schallaufnahmen. Rundfunktechn. Mitt. 9, S. 170

Kuhl, W. (1978): Räumlichkeit als Komponente des Raumeindrucks. Acustica 40, S. 167

Kuhl, W. und Kath, V. (1963): Akustische Anforderungen an ein Konzertstudio und ihre Realisierung beim Großen Sendesaal des NDR in Hannover. Rundfunktechn. Mitt. 7, S. 270

Kunitz, H. (1957): Die Instrumentation, Teil 4: Klarinette, Leipzig

Kunitz, H. (1959): Die Instrumentation, Teil 9: Tuba, Leipzig

Kunitz, H. (1961): Die Instrumentation, Teil 6: Horn, Leipzig

Kurtovic, H. and Gurganov, M. (1979): Computer calculated initial reverberation time of some open air musical theaters. Kongr. Ber. FASE '79, Dubrovnik, S. 149

Kuttruff, H. (1978): Gelöste und ungelöste Fragen der Konzertsaalakustik. Vortr. Rhein.-Westf. Akad. Wiss., N 278, Köln, S. 7

Kuttruff, H. (1991): Room acoustics, 3. Aufl., London

Kuwano, S., Namba, S. and Yamasaki, T. (1991): Effect of temporal pattern of non-steady state sounds on loudness. J. Acoust. Soc. Jpn. (E) 12, S. 229

Lamberty, D. C. (1978): Music practice rooms. Proc. Inst. Acoust. London, S. 17.7

Lamparter, H. und Brückmann, M. (1989): Akustik des neugestalteten Hörfunkstudios I des Hessischen Rundfunks in Frankfurt a. M. Fortschritte der Akustik – DAGA '89, Bad Honnef DPG-GmbH, S. 451

Lehmann, P. R. (1962): The harmonic structure of the tone of the basson. Diss., Michigan

Lehmann, P. (1976): Über die Ermittlung raumakustischer Kriterien und deren Zusammenhang mit subjektiven Beurteilungen der Hörsamkeit. Diss., Berlin

Lehmann, U. (1975): Untersuchung zur Bestimmung des Raumeindrucks bei Musikdarbietungen und Grundlagen der Optimierung. Diss. TU Dresden

Leipp, E. (1965): Le Violon, Paris

Leipp, E. (1969): Un diapason electronique nouveau a l'Opera de Paris. GAM-Bul. Nr. 40

Lessig, E. (1965): Richtungsfaktoren additiver elektroakustischer Strahlersysteme. HF und Ela 74, S. 211

Lifschitz, S. (1925): Optimum reverberation for an auditorium. Phys. Rev. 25, S. 391

Lottermoser, W. (1952): Nachhallzeiten in Barockkirchen. Acustica 2, S. 109

Lottermoser, W. (1958): Das Ausgleichsverhalten von Geigen und seine Beziehung zu der Resonanzkurve. Acustica 8, S. 91

Lottermoser, W. (1960): Die Akustik des Raumes und der Orgel in der Frauenkirche zu Dresden. Arch. f. Musikwiss. 17, S. 71

Lottermoser, W. (1983): Orgeln, Kirchen und Akustik. Frankfurt a. M.

Lottermoser, W. und Meyer, J. (1958): Verdeckungseffekt bei Orgelspektren. Acustica 8, S. 398

Lottermoser, W. und Meyer, J. (1960): Frequenzmessungen an gesungenen Akkorden. Acustica 10, S. 181

Lottermoser, W. und Meyer, J: (1961): Über das Anstrichgeräusch bei Geigen. Instrumentenbau-Zeitschrift 15, S. 382

Lottermoser, W. und Meyer, J. (1962a): Akustische Messungen an elektronischen Kirchenorgeln. Bulletin des SEV 53, S. 657

Lottermoser, W. und Meyer, J. (1962b): Die d'Egville-Geige von Guarneri del Gesù – Resonanzmessung und Dendrochronologie. Instrumentenbau-Zeitschrift 16, S. 270

Lottermoser, W. und Meyer, J. (1965): Raumakustische Grundlagenmessungen zur Planung von Orgeln. Das Musikinstrument 14, S. 723

Lottermoser, W. und Meyer, J. (1966): Orgelakustik in Einzeldarstellungen. Das Musikinstrument, Frankfurt/Main

Lottermoser, W. und Meyer, J. (1968): Über den Klang der Stradivari-Geige »Prince Khevenhüller«. Instrumentenbau-Zeitschrift 22, S. 140
Luce, D. and Clark, M. (1965): Durations of attack transients of nonpercussive orchestral instruments. J. Audio Eng. Soc. 13, S. 194
Marshall, A. H (1967): A note on the importance of room cross-section in concert halls. J. Sound Vib. 5, S. 100
Marshall, A. H. and Meyer, J. (1985): The directivity and auditory impressions of singers. Acustica 58, S. 130
Marshall, A. H., Gottlob, D. and Alrutz, H. (1978): Acoustical conditions preferred for ensemble. J. Acoust. Soc. Am. 64, S. 1437
Martin, D. W. (1942): Directivity and the acoustic spectra of brass wind instruments. J. Acoust. Soc. Am. XIII S. 309
Martin, D. W. (1962): Supplementary sound for opera. Sound 1H(1), S. 25
Marx, B. und Tennhardt, H.-P. (1991): Raum- und bauakustische Aspekte bei der Rekonstruktion der Deutschen Staatsoper Berlin. Bauforschung – Baupraxis H. 287, S. 15
Melichar, A. (1981): Der vollkommene Dirigent. München, Wien 1981
Melka, A. (1970): Messungen der Klangeinsatzdauer bei Musikinstrumenten. Acustica 23, S. 108
Mertens, P. H. (1975): Die Schumannschen Klangfarbengesetze und ihre Bedeutung für die Übertragung von Sprache und Musik, Frankfurt
Meyer, E. (1965): Zur Akustik von Theater- und Konzerträumen. Phys. Blätter, Hbchst, S. 368
Meyer, E. und Cremer, L. (1933): Über die Hörsamkeit holzausgekleideter Räume. Zeitschr. techn. Physik 14, S. 500
Meyer, E. und Kuttruff, H. (1959): Zur akustischen Gestaltung der neuerbauten Beethovenhalle in Bonn. Acustica 9, S. 465
Meyer, E. und Kuttruff, H. (1963): Reflexionseigenschaften durchbrochener Decken (Modelluntersuchungen an der Reflektoranordnung der neuen Philh.-Hall N.Y.). Acustica 13, S. 183
Meyer, J. (1964a): Geräuschanteile im Klangspektrum der Musikinstrumente. Das Musikinstrument 13, S. 685
Meyer, J. (1964b): Die Richtcharakteristiken von Geigen. Instrumentenbau-Zeitschrift 18, S. 275
Meyer, J. (1965a): Die Richtcharakteristik des Flügels. Das Musikinstrument 14, S. 1085
Meyer, J. (1965b): Die Richtcharakteristiken von Klarinetten. Das Musikinstrument 14, S. 21
Meyer, J. (1965c): Die Richtcharakteristiken von Violoncelli. Instr.-Bau-Zeitschr. 19, S. 281
Meyer, J. (1966a): Der Klang des Heckelphons. Instrumentenbau-Zeitschrift 20, S. 197
Meyer, J. (1966b): Die Richtcharakteristiken von Oboen und Fagotten. Das Musikinstrument 15, S. 958
Meyer, J. (1966c): Akustik der Holzblasinstrumente in Einzeldarstellungen, Frankfurt
Meyer, J. (1967a): Die Richtcharakteristiken von Bratschen und Kontrabässen. Instrumenten-bau-Zeitschrift 21, S. 3 und S. 116
Meyer, J. (1967b): Akustische Untersuchungen über den Klang des Horns. Das Musikinstrument 16, S. 32 und S. 199
Meyer, J. (1968): Akustische Untersuchungen über den Klang alter und neuer Fagotte. Das Musikinstrument 17, S. 1259
Meyer, J. (1975): Die Wirksamkeit von Reflexionsflächen in der Nähe eines Orchesters. Tag.-Ber. 10. Tonmeistertagung, Köln, S. 71
Meyer, J. (1976/77): Der Einfluß der richtungsabhängigen Schallabstrahlung der Musikinstrumente auf die Wirksamkeit von Refexions- und Absorptionsflächen in der Nähe des Orchesters. Acustica 36, S. 147
Meyer, J. (1977): Der akustische Raum. In: Umgang mit Raum, Gütersloh, S. 41
Meyer, J. (1978a): Raumakustik und Orchesterklang in den Konzertsälen Joseph Haydns. Acustica 41, S. 145
Meyer, J. (1978b): Physikalische Aspekte des Geigenspiels, Siegburg
Meyer, J. (1979): Die Tonhöhenempfindung bei musikalischen Klängen in Abhängigkeit vom Grad der Gehörschulung. Acustica 42, S. 189

Meyer, J. (1982a): Zum Klangphänomen der altitalienischen Geigen. Acustica 51, S. 1
Meyer, J. (1982b): Zum Hör-Erlebnis des Musikers im Konzertsaal. In: M. Krause: Tiefenstruktur der Musik. Techn. Univ. Berlin
Meyer, J. (1985): Akustik der Gitarre in Einzeldarstellungen, Frankfurt a. M.
Meyer, J. (1986): Some problems of opera house acoustics. Proc. Vancouver Symp. Acoust. Theatre Plann., S. 13
Meyer, J. (1987): Gedanken zur Sitzordnung der Streicher im Orchester. Das Orchester 35, S. 249
Meyer, J. (1988a): Kammermusik in drei Räumen. In: J. Meyer, Qualitätsaspekte bei Musikinstrumenten, Celle
Meyer, J. (1988b): Some aspects of opera house acoustics. Proc. Inst. of Acoust. 10, Teil 2, S. 237
Meyer, J. (1990): Zur Dynamik und Schalleistung von Orchesterinstrumenten. Acustica 71, S. 277
Meyer, J. (1991): Die spektrale Feinstruktur von Vibratoklängen. Proc. 9th FASE Symp. Balatonfüred, S. 285
Meyer, J. (1992): Zur klanglichen Wirkung des Streicher-Vibratos. Acustica 76, S. 283
Meyer, J. (1994a): Understanding the orchestral stage environment from the musician's, singers's and conductor's point of view. Proc. W. C. Sabine Centennial Symp., Cambridge MA, S. 93
Meyer, J. (1994b): Vibrato sounds in large halls. Proc. SMAC '93, Stockholm, S. 117
Meyer, J. (2000): Zur Raumakustik in Johann Sebastian Bachs Kirchen. Ber. 21. Tonmeistertagung Hannover, S. 1064
Meyer, J. (2002): Acoustics of Gothic Churches. Proc. Forum Acoust. Sevilla, paper RBA-05-002-IP
Meyer, J. (2003): Kirchenakustik. Frankfurt am Main
Meyer, J. und Lottermoser, W. (1961): Über die Möglichkeiten einer klanglichen Beurteilung von Flügeln. Acustica (Ak. Beihefte), S. 291
Meyer, J. und Wogram, K. (1969): Die Richtcharakteristiken des Hornes. Das Musikinstrument 18, H. 6, S. 1
Meyer, J. und Wogram, K. (1970): Die Richtcharakteristiken von Trompete, Posaune und Tuba. Das Musikinstrument 19, S. 171
Meyer, J. und Biassoni de Serra, E. C. (1980): Zum Verdeckungseffekt bei Instrumentalmusikern. Acustica 46, S. 130
Meyer, J. und Angster, J. (1981): Zur Schalleistungsmessung bei Violinen. Acustica 49, S. 192
Meyer, J. und Melka, A. (1983): Messung und Darstellung des Ausklingverhaltens von Klavieren. Das Musikinstrument 32, S. 1049
Miśkiewicz, A. and Rakowski, A. (1994): Sound level versus sound pressure level: A comparison of musical instruments. J. Acoust. Soc. Am. 96, S. 3375
Mommertz, E. (1993): Einige Messungen zur streifenden Schallausbreitung über Publikum und Gestühl. Acustica 79, S. 42
Mozart, W. A. (1781): Brief vom 1 April an seinen Vater
Mozart, W. A. (1791): Brief vom 7/8 Oktober an seine Frau
Mühle, Ch. (1965): Akustische Untersuchungen an einer D-Clarine, einer D- und einer B-Trompete. Das Orchester 9, S. 296
Müller, H. A. (1969): Persönliche Mitteilung an den Verfasser (Messung von Müller BBM, München)
Müller, H. A. und Opitz, U. (1986): Raumakustische Gestaltung. arcus H. 1/86, S.
Müller, H. A. und Vian, J. P. (1989): The acoustics of the new "Opera de la Bastille" in Paris. Proc. 13th ICA Belgrad, Vol. 2, S. 191
Müller, U. (1971): Untersuchungen zu den Strukturen von Klängen der Clarin- und Ventiltrompete, Regensburg
Müller, U. (1982): Die Klangerzeugung bei Becken. Das Musikinstrument 31, S. 1424
Nagata, M. (1989): Nankohall. In: E. McCue and R. H. Talaske, Acoustical Design of Music Education Facilities. Syracuse, New York
Nakamura, S. (1992): A preliminary study on the directional characteristics of room resonance perceived by solo performers. Proc. Int. Symp. Mus. Acoust. Tokyo, S. 255

Naylor, G. M. (1987): Musical and acoustical influences upon achievement of ensemble. Diss. Univ. Edinburgh

Neupert, W.-D. (1971): Physikalische Aspekte des Combaloklanges. Das Musikinstrument 20, S. 857

Niese, H. (1956): Untersuchung über die Knallform bei raumakustischen Impulsmessungen. HF und Ela 65, S. 98

Öhlberger, K. (1970): Artikulationsprobleme des Bläsers bei der Wiedergabe der Werke Mozarts. Wiener Figaro 38, H. 5, S. 18

O'Keefe, J. (1994): Modern stage acoustics measurements in orchestra pits. Proc. W. C. Sabine Centenial Symp., Cambridge, MA, S. 219

Olson, H. F. (1967): Music, Physics and Engineering. 2. Aufl., New York

Opitz, U. (1993): Architektur und Akustik. Bauwelt 44, S. 2378

Parkin, P. H. and Morgan, K. (1965): »Assisted Resonance« in the Royal Festval Hall London. J. Sound Vib. 2, S. 74

Parkin, P. H. and Morgan, K. (1970): »Assisted Resonance« in the Royal Festval Hall, London. J. Acoust. Soc. Am. 48, S. 1025

Parkin, P. H., Scholes, W. E. and Derbyshire, A. G. (1952): The reverberation times of ten British concert halls. Acustica 2, S. 97.

Paumgartner, B. (1966): Das instrumentale Ensemble, Zürich

Planyavsky, A. (1984): Die Geschichte des Kontrabasses, Tutzing

Plenge, G. und Schwarz, N. (1967): Über die Ausklingzeit von Musikinstrumenten. Kongreßber. 4. Akust. Konferenz, Bundapest

Prame, E. (1993): Measurements of the vibrato rate of 10 singers. Proc. SMAC '93, Stockholm, S. 122

Prante, H., Remmers, H. und Mellert. V. (1990): Experimente zur binauralen Lautstärkeempfindung im diffusen Schallfeld. Fortschritte der Akustik – DAGA '90, Bad Honnef DPG-GmbH, S. 711

Pratt, R. L. and Bowsher, J. M. (1978): The subjective assessment of trombone quality. J. Sound Vib. 57, S. 425

Pravica, P. (1979): Sound pressure levels in open air theatres. Ungedruckter Vortrag FASE '79, Dubrovnik

Prince, D. and Talaske, R. (1994): Variation of room acoustic measurements as a function of source location and directivity. Proc. W. C. Sabine Centennial Symp., Cambridge, MA, S. 211

Quantz, J. J. (1752): Versuch einer Anweisung, die Flöte traversière zu spielen, Berlin

Rakowski, A. (1966): Opening Transients in tones of the flute. Bul. Soc. Amis Sc. et Lt. Poznan, Ser. B, 19, S. 157

Rakowski, A. (1967): Musical Instruments as natural sources of sound for the reverberation time measurements. Arch. Akustyki, 3, S. 179

Rasch, R. A. (1979): Synchronization in performed ensemble music. Acustica 43, S. 121

Redfearn, S. W. (1940): Some acoustical source observer problems. Phil. Mag. J. Soc. 7, S. 223

Reichardt, W. (1968): Grundlagen der Technischen Akustik, Leipzig

Reichardt, W. und Schmidt, W. (1967): Die Wahrnehmung der Veränderung von Schallfeldparametern bei der Darbietung von Musik. Acustica 18, S. 274

Reichardt, W., Kohlsdorf, E. und Mutscher, H. (1955): Die optimale Nachhallzeit für Studioräume. HF u. ELA 64, H. 1, S. 18

Reichardt, W., Schmidt, W. und Lehmann, U. (1972): Harter und weicher Klangeinsatz bei Musik. Acustica 26, S. 253

Reichardt, W., Abdel Alim, O. und Schmidt, W. (1975): Zusammenhang zwischen Klarheitsmaß C und anderen objektiven raumakustischen Kriterien. Z. el. Inf. Energ. Techn. 5, S. 144

Reinecke, H.-P. (1953): Über den doppelten Sinn des Lautheitsbegriffes beim musikalischen Hören. Diss. Hamburg

Rindel, J. H. (1992): Acoustic design of reflectors in auditoria. Proc. Inst. Acoust. 14, Teil 2, S. 119

Robbins Landon, H. C. (1955): The Symphonies of Joseph Haydn, London

Robbins Landon, H. C. (1976): Haydn: Chronicle and Works, Vol. 3, Thames & Hudson, London
Roederer, J. G. (1977): Physikalische und psychoakustische Grundlagen der Musik. Springer, Berlin
Rossing, Th. D. (1982a): Acoustics of bar percussion instruments. Percussive Notes 19, Nr. 3, S. 6
Rossing, Th. D. (1982b): The Sience of Sound. Reading MA, Menlo Park CA
Rossing, Th. D., Bork, I., Zhao, H. and Fystrom, D. O. (1992): Acoustics of snare drums. J. Acoust. Soc. Am. 92, S. 84
Saunders, F. A. (1946): Analyses of tones of a few wind instruments. J. Acoust. Soc. Am. 18, S. 395
Schelleng, J. C. (1973): The bowed string and the player. J. Acoust. Soc. Am. 53, S. 26
Schirmer, W. (1963): Die Richtcharakteristik des Ohres. HF und Ela 72, S. 39
Schlosser, H. and Krieger, A. (1993): Mozarts »Zauberflöte« in der Waldbühne zu Berlin – ein zau-berhaftes Erlebnis. Tonmeister-Inform. Sept./Okt. 1993, S. 10
Schmidt, W. (1985): Die neue Semperoper in Dresden – die Raumakustik im Zuschauerraum. Kulturbauten H. 2/85, S. 20
Schreiber, H. (1958): Der große Sendesaal des Hessischen Rundfunks. Rundfunktechn. Mitt. 2, S. 29
Schreiber, O. (1938): Orchester und Orchesterpraxis in Deutschland zwischen 1780 und 1850. Diss. Friedrich-Wilhelms-Univ. Berlin
Schroeder, M. R. (1979): Binaural dissimilarity and optimum ceilings for concert halls. J. Acoust. Soc. Am. 65, S. 958
Schubert, P. (1969): Die Wahrnehmbarkeit von Rückwürfen bei Musik. HF und Ela 78, S. 230
Schultz, Th. J, and Watters, G. B. (1964): Propagation of sound across audience seating. J. Acoust. Soc. Am. 36, S. 885
Schultz, Th. J. (1981): Persönliche Mitteilung an den Verfasser
Seraphim, H. P. (1958): Untersuchungen über die Unterschiedsschwelle exponentiellen Abklingens von Rauschbandimpulsen. Acustica 8, S. 280
Shankland, R. S. and Shankland, H. K. (1971): Acoustics of St. Peter's and Patriarchal Basilicas in Rome, J. Acoust. Soc. Am. 50, S. 389
Singer, H. (1958): Die Akustik des alten Burgtheaters. Maske und Kothurn 4, S. 220
Singer, H. (1959): Das ideale Mozarttheater. Phono 5, H. 4
Skudrzyk, E. (1954): Die Grundlagen der Akustik, Wien
Smith, R. A. and Mercer, D. M. A. (1973): The effect of lip pressure and air pressure on the into-nation and tone quality of the bassoon. J. Sound Vib. 30, S. 261
Smith, R. A. and Mercer, D. M. A. (1974): Possible causes of woodwind tone colour. J. Sound Vib. 32, S. 347
Spelda, A. (1968): Pizzicato smyčcových nástroju. Hud. Veda, S. 49
Stensson, K. (1968): Persönliche Mitteilung an den Verfasser (Messung des Schwedischen Rund-funks Stockholm)
Strauss, R. (1905): rev. Neuausgabe der Instrumentationslehre von Borlioz, Leipzig
Strong, W. and Clark, M. (1967): Synthesis of wind instrument tones. J. Acoust. Soc. Am. 41, S. 39
Stumpf, K. (1970): Die Viola d'amore in der neuen Musik. Das Orchester 18, S. 405
Sundberg, J. (1977): The acoustics of the singing voice. Sci. Am. 236, S. 82
Sundberg, J. (1979): Chest wall vibrations in singers. Tagungsber. 18. Akust. Konferenz, Cesky Krumlov, S. 155
Sundberg, J. (1990): What's so special about singers? J. Voice 4, S. 107
Sundberg, J. (1991): The Science of Musical Sounds. Academic press, San Diego
Tarnóczy, T. (1943): Resonanzdaten der Vokalresonatoren. Akust. Z. 8, S. 22
Tarnóczy, T. (1991): Einführung in die musikalische Akustik, Budapest
Tarnóczy, T., Járfas, T. und Lukács, M. (1960): Neuere subjektiv-akustische Untersuchungen über die Nachhallzeit. Elektron. Rundschau 14, S. 223
Tennhardt, H.-P. und Winkler, H. (1994): Raumakustische Probleme bei der Planung von Orchesterproberäumen. Fortschritte der Akustik – DAGA '94, Bad Honnef DPG-GmbH, S. 245

Terhardt, E. (1973): Tonhöhenwahrnehmung und harmonisches Empfinden. Fortschritte der Akustik – DAGA '72, Berlin, S. 59

Terhardt, E. (1974): On the perception of periodic sound fluctuations. Acustica 30, S. 201

Terhardt, E. and Stoll, G. (1978): Bewertung des Wohlklangs verschiedener Schalle. Fortschritte der Akustik, Tagungsbericht – DAGA '78, Berlin, S. 583

Ternström, S. (1989): Long-time average spectrum characteristics of different choirs in different rooms. Quarterly Progress and States Report 3/89 des Speech Transmission Laboratory der KTH Stockholm, S. 15

Ternström, S. (1991a): Perceptual evaluations of voice scatter in unison choir sounds. Quarterly Progress and States Report 2/1991 des Speech Transmission Laboratory der KTH Stockholm, S. 41

Ternström, S. (1991b): Physical and acoustic factors that intoract with the singer to produce the choral sound. J. voice 5, S. 128

Ternström, S. (1993): Perceptual evaluations of voice scatter in unison choir sounds. J. Voice 7, S. 129

Ternström, S. and Sundberg, J. (1983): How loudly should you hear your collegues and yourself. Quarterly Progress and States Report 4/83 des. Speech Transmission Laboratory der KTH Stockholm, S. 16

Theile, G. (1980): Über die Lokalisation im überlagerten Schallfeld. Diss. TU Berlin

Thienhaus, E. (1934): Neuere Versuche zur Klangfarbe und Lautstärke von Vokalen. Zeitschr. Techn. Phys. 15, S. 637

Thienhaus, E. (1954): Stereophonische Übertragung klangschwacher Instrumente im Konzertsaal. Acustica 4, S. 253

Thienhaus, E. (1962): Art. »Raumakustik« im MGG, Kassel

Titze, I. and Story, B. (1993): The lowa singing synthesis. Proc. SMAC '93, Stockholm, S. 294

Trendelenburg, F. (1961): Einführung in die Akustik. 3. Auflage, Berlin, Göttingen, Heidelberg

Veneklasen, P. S. (1986): Has symphony orchestra performance reached its Zenith? P. S. Veneklasen Research Foundation Santa Monica California

Venzke, G. (1959): Die Raumakustik der Kirchen verschiedener Baustilepochen. Acustica 9, S. 151

Völker, E. J. (1988): Zur Akustik von Orchester-Proberäumen. Fortschritte der Akustik – DAGA '88. Bad Honnef, DPG-GmbH, S. 733

Vogel, M. (1968): Die Zukunft der Musik, Düsseldorf

von. Békézy, G. (1968): Feedback phenomena between the stringed instrument and the musician, Rockefeller Univ. Rev.

von. Bismarck G. (1974): Timbre of steady sounds – a factorial investigation of its verbal attributes. Acustica 30, S. 146

Vos, J. and Rasch, R. (1981): The perceptual onset of musical tones. Percept. Psychophys. 29, S. 323

Wagner, R. (1911): Mein Leben, 2. Teil, München

Walter, B. (1959): Von der Musik und vom Musizieren, Frankfurt

Weinreich, G. (1977): Coupled piano strings. J. Acoust. Soc. Am. 62, S. 1474

Weisse, K. und Gelies, O. (1979): Akustik weltberühmter Musikräume. Technik am Bau H. 8/79

Westphal, W. (1994): Zur Schallverteilung in Musiktheatern im Nah- und Fernfeld einer ausgedehnten Schallquelle. Acustica 80, S. 226

Weyer, R.-D. (1976): Time-Frequency-Structures in the Attack-Transients of Piano and Harpsichord Sounds. Acustica 35, S. 232

Widholm, G. und Sonneck, G. (1988): Wiener Horn versus Doppelhorn, WWV Wien

Wilkens, R. (1975): Mehrdimensionale Beschreibung subjektiver Beurteilungen der Akustik von Konzertsälen. Diss. TU Berlin

Winckel, F. (1955): Die besten Konzertsäle der Welt. Baukunst und Werkform 8, S. 751

Winckel, F. (1960): Phänomene des musikalischen Hörens, Berlin und Wunsiedel

Winckel, F. (1962a): Optimum acoustic criteria of concert halls for the performence of classical Music. J. Acoust. Soc. Am. 34, S. 81

Winckel, F. (1962b): Über den Einfluß der Deckenhöhe auf die Klangqualität in Konzertsälen. 4, ICA-Kongreß Kopenhagen, M 37

Winckel, F. (1963): Von der Akustik im Kirchenraum. Kunst und Kirche 26, S. 18

Winckel, F. (1969): Nachrichtenverarbeitung unter kybernetischen Aspekten. In: Handbuch für HF- und E-Techniker, Bd. 8, Berlin

Winckel, F. (1971): How to measure the effectiveness of stage singers voices. Folia phoniat. 23, S. 228

Winckel, F. (1974): Neuere Erkenntnisse in der Raumakustik, ORF-Seminar, Wien

Winckel, H. (1979): Probleme der Orchesteraufstellung in großen Sälen. Tagungsber. 18. Akust. Konf., Cesky Krumlov, S. 195

Winkler, H. (1992): Das Sehen beim Hören. Fortschritte der Akustik – DAGA '92, Bad Honnef DPG-GmbH, S. 181

Winkler, H. und Tennhardt, H.-P. (1993): Balance der Instrumentengruppen im Orchester, Bestimmung und Beeinflussung. Fortschritte der Akustik – DAGA '93, Bad Honnef DPG-GmbH, S. 207

Winkler, H. und Tennhardt, H.-P. (1994): Gegenseitiges Hören der Musiker im Orchester. Fortschritte der Akustik – DAGA '94, Bad Honnef DPG-GmbH, S. 237

Winkler, K. und Kaetel, K. (1990): Natürlicher Klang von innen heraus. Das Orchester 38, S. 507

Wirth, H. (1958): »Kammermusik« in MGG, Kassel

Wogram, K. (1979): Diskrepanz zwischen Tragfähigkeit und Hörbarkeit von Blechblasinstrumenten. Tagungsber. 18 Akust. Konf., Cesky Krumlov, S. 55

Wogram, K. (1984): Akustische Untersuchungen an Klavieren. In: H. Junghanns, Der Piano- und Flügelbau. Frankfurt a. M

Woolford, D. H. and Carterette, E. C. (1989): Hearing impairment among orchestral musicians and music performance. Proc. 1st Int. Conf. Music Perception and Cognition, S. 287

Zwicker, E. (1982): Psychoakustik. Berlin, Heidelberg, New York O. V. (1960): Das neue Salzburger Festspielhaus. Residenzverlag Salzburg, ohne Herausgeber

Subject Index

Author, Composer, and Composition Index

* Page numer of score examples.

Printed in the United States of America